Drew Wade
'75

5 99
Physics
As-is

Relativistic Quantum Fields

INTERNATIONAL SERIES IN PURE AND APPLIED PHYSICS
Leonard I. Schiff, *Consulting Editor*

Adler, Bazin, and Schiffer Introduction to General Relativity
Allis and Herlin Thermodynamics and Statistical Mechanics
Becker Introduction to Theoretical Mechanics
Bjorken and Drell Relativistic Quantum Mechanics
Bjorken and Drell Relativistic Quantum Fields
Chodorow and Susskind Fundamentals of Microwave Electronics
Clark Applied X-rays
Collin Field Theory of Guided Waves
Evans The Atomic Nucleus
Feynman and Hibbs Quantum Mechanics and Path Integrals
Ginzton Microwave Measurements
Green Nuclear Physics
Gurney Introduction to Statistical Mechanics
Hall Introduction to Electron Microscopy
Hardy and Perrin The Principles of Optics
Harnwell Electricity and Electromagnetism
Harnwell and Livingood Experimental Atomic Physics
Henley and Thirring Elementary Quantum Field Theory
Houston Principles of Mathematical Physics
Kennard Kinetic Theory of Gases
Lane Superfield Physics
Leighton Principles of Modern Physics
Lindsay Mechanical Radiation
Livingston and Blewett Particle Accelerators
Middleton An Introduction to Statistical Communication Theory
Morse Vibration and Sound

Morse and Feshbach Methods of Theoretical Physics
Muskat Physical Principles of Oil Production
Present Kinetic Theory of Gases
Read Dislocations in Crystals
Richtmyer, Kennard, and Lauritsen Introduction to Modern Physics
Schiff Quantum Mechanics
Seitz The Modern Theory of Solids
Slater Introduction to Chemical Physics
Slater Quantum Theory of Matter
Slater Quantum Theory of Atomic Structure, Vol. I
Slater Quantum Theory of Atomic Structure, Vol. II
Slater Quantum Theory of Molecules and Solids, Vol. 1
Slater and Frank Electromagnetism
Slater and Frank Introduction to Theoretical Physics
Slater and Frank Mechanics
Smythe Static and Dynamic Electricity
Stratton Electromagnetic Theory
Thorndike Mesons: A Summary of Experimental Facts
Tinkham Group Theory and Quantum Mechanics
Townes and Schawlow Microwave Spectroscopy
White Introduction to Atomic Spectra

The late F. K. Richtmyer was Consulting Editor of the series from its inception in 1929 to his death in 1939. Lee A. DuBridge was Consulting Editor from 1939 to 1946; and G. P. Harnwell from 1947 to 1954.

James D. Bjorken
Professor

Sidney D. Drell
Professor

Stanford Linear Accelerator Center
Stanford University

McGraw-Hill Book Company

New York St. Louis San Francisco Toronto London Sydney

Relativistic Quantum Fields

Relativistic Quantum Fields

Library of Congress Catalog Card Number 64-7726

1112131415 FGBP 7765

ISBN 07-005494-0

Preface

The propagator approach to a relativistic quantum theory pioneered in 1949 by Feynman has provided a practical, as well as intuitively appealing, formulation of quantum electrodynamics and a fertile approach to a broad class of problems in the theory of elementary particles. The entire renormalization program, basic to the present confidence of theorists in the predictions of quantum electrodynamics, is in fact dependent on a Feynman graph analysis, as is also considerable progress in the proofs of analytic properties required to write dispersion relations. Indeed, one may go so far as to adopt the extreme view that the set of all Feynman graphs *is* the theory.

We do not advocate this view in this book nor in its companion volume, "Relativistic Quantum Mechanics," nor indeed do we advocate any single view to the exclusion of others. The unsatisfactory status of present-day elementary particle theory does not allow one such a luxury. In particular, we do not wish to minimize the importance of the progress achieved in formal quantum field theory nor the considerable understanding of low-energy meson-nucleon processes given by dispersion theory. However, we give first emphasis to the development of the Feynman rules, proceeding directly from a particle wave equation for the Dirac electron, integrated with hole-theory boundary conditions.

Three main convictions guiding us in this approach were the primary motivation for undertaking these books:

1. The Feynman graphs and rules of calculation summarize quantum field theory in a form in close contact with the experimental numbers one wants to understand. Although the statement of the theory in terms of graphs may imply perturbation theory, use of graphical methods in the many-body problem shows that this formalism is flexible enough to deal with phenomena of nonperturbative character (for example, superconductivity and the hard-sphere Bose gas).

2. Some modification of the Feynman rules of calculation may well outlive the elaborate mathematical structure of local canonical quantum field theory, based as it is on such idealizations as fields defined at points in space-time. Therefore, let us develop these rules first, independently of the field theory formalism which in time may come to be viewed more as a superstructure than as a foundation.

3. Such a development, more direct and less formal—if less compelling—than a deductive field theoretic approach, should bring quantitative calculation, analysis, and understanding of Feynman graphs into the bag of tricks of a much larger community of physicists than the specialized narrow one of second quantized theorists. In particular, we have in mind our experimental colleagues and students interested in particle physics. We believe this would be a healthy development.

Our original idea of one book has grown in time to two volumes. In the first book, "Relativistic Quantum Mechanics," we develop a propagator theory of Dirac particles, photons, and Klein-Gordon mesons and perform a series of calculations designed to illustrate various useful techniques and concepts in electromagnetic, weak, and strong interactions. These include defining and implementing the renormalization program and evaluating effects of radiative corrections, such as the Lamb shift, in low-order calculations. The necessary background for this book is provided by a course in nonrelativistic quantum mechanics at the general level of Schiff's text "Quantum Mechanics."

In the second book, "Relativistic Quantum Fields," we develop canonical field theory, and after constructing closed expressions for propagators and for scattering amplitudes with the *LSZ* reduction technique, return to the Feynman graph expansion. The perturbation expansion of the scattering amplitude constructed by canonical field theory is shown to be identical with the Feynman rules in the first book. With further graph analysis we study analyticity properties of Feynman amplitudes to arbitrary orders in the coupling parameter and illustrate dispersion relation methods. Finally, we prove the finiteness of renormalized quantum electrodynamics to each order of the interaction.

Without dwelling further on what we do, we may list the major topics we omit from discussion in these books. The development of action principles and a formulation of quantum field theory from a variational approach, spearheaded largely by Schwinger, are on the whole ignored. We refer to action variations only in search of symmetries. There is no detailed discussion of the powerful developments in axiomatic field theory on the one hand and the purely *S*-matrix approach, divorced from field theory, on the other. Aside from a discussion of the Lamb shift and the hydrogen atom spectrum in the first book, the bound-state problem is ignored. Dynamical applications of the dispersion relations are explored only minimally. A formulation of a quantum field theory for massive vector mesons is not given—nor is a formulation of any quantum field theory with derivative couplings. Finally, we have not prepared a bibliography of all the significant original papers underlying many of the developments recorded in these books. Among the following recent excellent books or monographs is to be found the remedy for one or more of these deficiencies:

Schweber, S.: "An Introduction to Relativistic Quantum Field Theory," New York, Harper & Row, Publishers, Inc., 1961.

Jauch, J. M., and F. Rohrlich: "The Theory of Photons and Electrons," Cambridge, Mass., Addison-Wesley Publishing Company, Inc., 1955.

Bogoliubov, N. N., and D. V. Shirkov: "Introduction to the Theory of Quantized Fields," New York, Interscience Publishers, Inc., 1959.

Akhiezer, A., and V. B. Bereztetski: "Quantum Electrodynamics," 2d ed., New York, John Wiley & Sons, Inc., 1963.

Umezawa, H.: "Quantum Field Theory," Amsterdam, North Holland Publishing Company, 1956.

Hamilton, J.: "Theory of Elementary Particles," London, Oxford University Press, 1959.

Mandl, F.: "Introduction to Quantum Field Theory," New York, Interscience Publishers, Inc., 1960.

Roman, P.: "Theory of Elementary Particles," Amsterdam, North Holland Publishing Company, 1960.

Wentzel, G.: "Quantum Theory of Field," New York, Interscience Publishers, Inc., 1949.

Schwinger, J.: "Quantum Electrodynamics," New York, Dover Publications, Inc., 1958.

Feynman, R. P.: "Quantum Electrodynamics," New York, W. A. Benjamin, Inc., 1962.

Klein, L. (ed.): "Dispersion Relations and the Abstract Approach to Field Theory," New York, Gordon and Breach, Science Publishers, Inc., 1961

Screaton, G. R. (ed.): "Dispersion Relations; Scottish Universities Summer School," New York, Interscience Publishers, Inc., 1961.

Chew, G. F.: "S-Matrix Theory of Strong Interactions," New York, W. A. Benjamin, Inc., 1962.

In conclusion, we owe thanks to the many students and colleagues who have been invaluable critics and sounding boards as our books evolved from lectures into chapters, to Prof. Leonard I. Schiff for important initial encouragement and support to undertake the writing of these books, and to Ellen Mann and Rosemarie Stampfel for marvelously cooperative secretarial help.

James D. Bjorken
Sidney D. Drell

Contents

11

General Formalism

Intuitive and correspondence arguments were used in "Relativistic Quantum Mechanics" in developing the propagator approach and giving practical rules for calculating, in perturbation theory, interactions of relativistic particles. We now turn to a systematic derivation of these rules from the formalism of quantized fields. Our motivation is first to "patch up the holes" in our arguments in the propagator approach and then to develop a formalism which might be applied to problems for which perturbation theory is not adequate, such as processes involving strongly coupled mesons and nucleons.

Our approach is best illustrated by the electromagnetic field. The potentials $A^\mu(x)$ satisfy the Maxwell wave equations and may be considered as describing a dynamical system with an infinite number of degrees of freedom. By this we mean that $A^\mu(x)$ at each point of space may be considered an independent generalized coordinate. To make the transition from classical to quantum theory, we must, according to the general principles proclaimed in Chap. 1,[1] elevate coordinates and their conjugate momenta to operators in the Hilbert space of possible physical states and impose quantum conditions upon them. This is the canonical quantization procedure. It is a straightforward extension to field functions, which obey differential wave equations derivable from a lagrangian, of the quantization procedure of nonrelativistic mechanics. When it is done, there emerges a particle interpretation of the electromagnetic field—in the sense of Bohr's principle of complementarity.

If photons emerge in such a natural way from the quantization of the Maxwell field, one is led to ask whether other particles whose existence is observed in nature are also related to force fields by the same quantization procedure. On this basis Yukawa predicted the existence of the π meson from knowledge of the existence of nuclear forces. Conversely, it is natural from this point of view to associate with each kind of observed particle in nature a field $\varphi(x)$ which satisfies an assumed wave equation. A particle interpretation of the field φ is then obtained when we carry through the canonical quantization program.

In such a program we must first define the momenta $\pi(x)$ conjugate to the field coordinates $\varphi(x)$. We do this in terms of a lagrangian, from which the wave equation for each field $\varphi(x)$ as well as the conjugate momenta are derivable. Applying the canonical quantization procedure with the commutator condition of Chap. 1, we obtain field quanta, such as photons, which obey Bose statistics. In order to

[1] References to Chaps. 1 to 10 or parts thereof are references to the companion volume, "Relativistic Quantum Mechanics."

describe Fermi particles which obey an exclusion principle with a similar quantum field formalism, it turns out to be necessary only to replace the quantum commutator conditions by anticommutator relations.

In this way a unified formalism which provides a basis for the description of both kinds of particles can be constructed. An additional attractive feature of the lagrangian approach which will be seen shortly is that it leads directly to the conservation laws.

11.1 Implications of a Description in Terms of Local Fields

Before continuing and exploring the consequences of applying the quantization procedure to classical fields which satisfy wave equations, it is perhaps worthwhile to discuss the implications of such a program. The first is that we are led to a theory with differential wave propagation. The field functions are continuous functions of continuous parameters $\mathbf{x}$ and t, and the changes in the fields at a point $\mathbf{x}$ are determined by properties of the fields infinitesimally close to the point $\mathbf{x}$.

For most wave fields (for example, sound waves and the vibrations of strings and membranes) such a description is an idealization which is valid for distances larger than the characteristic length which measures the granularity of the medium. For smaller distances these theories are modified in a profound way.

The electromagnetic field is a notable exception. Indeed, until the special theory of relativity obviated the necessity of a mechanistic interpretation, physicists made great efforts to discover evidence for such a mechanical description of the radiation field. After the requirement of an "ether" which propagates light waves had been abandoned, there was considerably less difficulty in accepting this same idea when the observed wave properties of the electron suggested the introduction of a new field $\psi(x)$. Indeed there is no evidence of an ether which underlies the electron wave $\psi(\mathbf{x},t)$. However, it is a gross and profound extrapolation of present experimental knowledge to assume that a wave description successful at "large" distances (that is, atomic lengths $\approx 10^{-8}$ cm) may be extended to distances an indefinite number of orders of magnitude smaller (for example, to less than nuclear lengths $\approx 10^{-13}$ cm).

In the relativistic theory, we have seen that the assumption that the field description is correct in arbitrarily small space-time intervals has led—in perturbation theory—to divergent expressions for the electron self-energy and the "bare charge." Renormalization theory has sidestepped these divergence difficulties, which may be indicative of the failure of the perturbation expansion. However, it is widely felt that the divergences are symptomatic of a chronic disorder in the small-distance behavior of the theory.

We might then ask why local field theories, that is, theories of fields which can be described by differential laws of wave propagation, have been so extensively used and accepted. There are several reasons, including the important one that with their aid a significant region of agreement with observations has been found, examples of which have already appeared in the discussions of the companion volume. But the foremost reason is brutally simple: there exists no convincing form of a theory which avoids differential field equations.

A theory of the interaction of relativistic particles is necessarily of great mathematical complexity. Because of the existence of creation and annihilation processes it is at once a theory of the many-body problem. At the present time one knows how to develop only approximate solutions to this problem, and therefore the predictions of any such theory are incomplete and at best somewhat ambiguous.

Faced with this situation, the most reasonable course to steer in constructing theories is to retain the general principles which have worked before in a more restricted domain. In this case, this includes the prescription for quantization which strongly involves the existence of a hamiltonian H. However, since H generates infinitesimal time displacements according to the Schrödinger equation, we are led to a description with differential development in time. Lorentz invariance then requires a differential development in space as well. A hamiltonian may well not exist for a nonlocal "granular" theory; if it does not, the link connecting us with the quantization methods of nonrelativistic theories is broken.

If we simply retain the notion of a Lorentz-invariant microscopic description in terms of continuous coordinates $\mathbf{x}$ and t, we expect that the influence of interactions should not propagate through space-time with velocity faster than c. This notion of "microscopic causality" strongly forces us into the field concept. Even if there is a granularity at small distances, if we are to retain microcausality the influence of one "granule" upon the next must be retarded; the most natural way to describe this is with additional fields. The problem thus becomes more complicated, without corresponding gain in understanding.

There is no concrete experimental evidence of a granularity at small distances.[1] There is likewise nothing but positive evidence that

[1] In quantum electrodynamics there exists an agreement between theory and experiment to very great precision in both low- and high-energy processes. See, for example, R. P. Feynman, *Rept. Solvay Congr., Brussels*, Interscience Publishers, Inc., New York, 1961.

special relativity is correct in the high-energy domain, and furthermore, there is, if anything, positive evidence[1] that the notion of microscopic causality is a correct hypothesis. Since there exists no alternative theory which is any more convincing, we shall hereafter restrict ourselves to the formalism of local, causal fields. It is undoubtedly true that a modified theory must have local field theory as an appropriate large-distance approximation or correspondence. However, we again emphasize that the formalism we develop may well describe only the large-distance limit (that is, distances $> 10^{-13}$ cm) of a physical world of considerably different submicroscopic properties.

11.2 The Canonical Formalism and Quantization Procedure for Particles

To preface our development, we recall the familiar path to the quantization of a classical dynamical system in particle mechanics. For purposes of illustration consider the one-dimensional motion of a particle in a conservative force field. We let q be the (generalized) coordinate of the particle, $\dot{q} = dq/dt$ the velocity, and $L(q,\dot{q})$ the lagrangian. According to Hamilton's principle, the dynamics of the particle is determined by the condition

$$\delta J = \delta \int_{t_1}^{t_2} L(q,\dot{q})\, dt = 0 \qquad (11.1)$$

Equation (11.1) states that the actual physical path $q(t)$ which the particle follows in traversing the interval from (q_1,t_1) to (q_2,t_2) is that along which the action J is stationary. Thus small variations from this path, $q(t) \rightarrow q(t) + \delta q(t)$, as shown in Fig. 11.1, leave the action unchanged to first order in the variation.

Hamilton's principle leads directly to the Euler-Lagrange equations of motion[2]

$$\frac{d}{dt}\frac{\partial L}{\partial \dot{q}} - \frac{\partial L}{\partial q} = 0 \qquad (11.2)$$

In order to carry out the formal quantization of this equation, we rewrite it in hamiltonian form. We do so by defining the momentum p conjugate to q,

$$p = \frac{\partial L}{\partial \dot{q}} \qquad (11.3)$$

[1] We mean by this the experimental verification of the dispersion relations for forward pion-nucleon scattering, to be discussed in Chap. 18.

[2] Cf. H. Goldstein, "Classical Mechanics," Addison-Wesley Publishing Company, Inc., Reading, Mass., 1950. The form of (11.2) applies for no higher than the first derivative of the coordinates appearing in L.

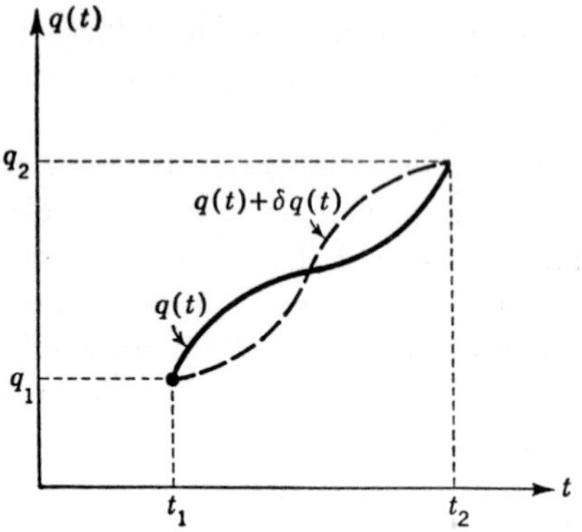

Fig. 11.1 Variation in path with fixed end points for action principle.

and introducing the hamiltonian by the Legendre transformation

$$H(p,q) = p\dot{q} - L(q,\dot{q}) \tag{11.4}$$

In terms of H, the equation of motion (11.2) becomes

$$\{H,q\}_{\text{PB}} = \frac{\partial H}{\partial p} = \dot{q} \qquad \text{and} \qquad \{H,p\}_{\text{PB}} = -\frac{\partial H}{\partial q} = \dot{p} \tag{11.5}$$

where $\{\quad\}_{\text{PB}}$ means a Poisson bracket.

To quantize (11.5), we let q become a hermitian operator in a Hilbert space and replace p by $-i\,\partial/\partial q$ so that the conjugate momentum and coordinate satisfy a commutator relation

$$[p,q] = -i \tag{11.6}$$

corresponding to the classical Poisson bracket $\{p,q\}_{\text{PB}} = 1$. With this definition, p is also hermitian. The dynamics of the particle is contained in the Schrödinger equation

$$H(p,q)\Psi(t) = i\frac{\partial \Psi(t)}{\partial t} \tag{11.7}$$

where Ψ is the wave function, or state vector, in the Hilbert space. If we specify the initial state Ψ at an arbitrary time, say $t = 0$, the Schrödinger equation determines the state and hence physical expectation values at all future times.

This formulation of the time development of the motion of the particle, with the time dependence carried in Ψ while the operators p and q are not time-dependent, is known as the Schrödinger picture. Alternatively, we may express the time development of the motion in a different language in which the operators $p(t)$ and $q(t)$ carry the time dependence instead of the state vectors Ψ. This is known as the

Heisenberg picture and is equivalent to the Schrödinger picture, as can be shown formally by constructing the unitary transformation which relates the two pictures to each other. Formally integrating the Schrödinger equation (11.7), we have

$$\Psi_S(t) = e^{-iHt}\Psi_S(0) \equiv e^{-iHt}\Psi_H \tag{11.8}$$

where the operator e^{-iHt} is unitary for hermitian H and gives the time development of $\Psi_S(t)$. The value of Ψ_S at $t = 0$ is the Heisenberg state vector $\Psi_H \equiv \Psi_S(0)$. The time-independent operators O_S are transformed into the time-dependent Heisenberg operators according to

$$O_H(t) = e^{iHt}O_S e^{-iHt} \tag{11.9}$$

The unitary operator transformation (11.9) is so constructed as to keep matrix elements and thereby physical observables invariant.

The solution of a dynamical problem in quantum theory consists in finding, at a later time t, matrix elements of operators which represent physical observables if given their matrix elements at some initial time, say $t = 0$. In the Schrödinger picture we do this by solving (11.7) for the time development of the wave function. In the Heisenberg picture, on the other hand, we solve the equation of motion for the time development of the Heisenberg operator which, by (11.9), is[1]

$$\frac{dO_H(t)}{dt} = i[H,O_H(t)] \tag{11.10}$$

As long as we deal with energy eigenfunctions in the nonrelativistic theory there is little practical difference between the two schemes. According to (11.9), $H_H(t) = H_S = H$, and in the absence of external time-varying forces (11.10) shows that $dH/dt = 0$. For energy eigenfunctions the Schrödinger wave function is $\psi_n(q,t) = e^{-i\omega_n t}u_n(q)$, and the corresponding Heisenberg function is $u_n(q)$.

In the relativistic field theory, however, we shall find the Heisenberg picture to be more convenient, since the explicit representation of the state vector Ψ is considerably more awkward than in the nonrelativistic case and the dynamics of the operators is easier to describe than the dynamics of Ψ. In addition, in developing the field theory we shall see the Lorentz invariance more readily in the Heisenberg picture, which puts the time together with the space coordinates in the field operators.

[1] For operators which are explicit functions of the time coordinate, (11.10) becomes

$$\frac{dO_H(t)}{dt} = i[H,O_H(t)] + \frac{\partial O_H}{\partial t}$$

In the Heisenberg picture it follows from (11.6) and (11.9) that the fundamental commutator conditions retain the form

$$[p(t),q(t)] = -i \tag{11.11}$$

for an arbitrary time t, and we may write

$$p(t) = -i\frac{\partial}{\partial q(t)} \qquad \text{and} \qquad q(t) = i\frac{\partial}{\partial p(t)}$$

Thus the dynamics expressed by (11.10) for the canonical coordinate and momentum

$$\frac{dp(t)}{dt} = i[H,p(t)] \qquad \frac{dq(t)}{dt} = i[H,q(t)] \tag{11.12}$$

can now be recast in the form of the *classical* equations of motion for the operators

$$\frac{dp(t)}{dt} = -\frac{\partial H}{\partial q(t)} \qquad \frac{dq(t)}{dt} = \frac{\partial H}{\partial p(t)}$$

This formulation of the quantum-dynamical problem in terms of Heisenberg operators is quite similar to the original classical formulation, which we shall often turn to and imitate. The dynamical variables in the quantum theory are hermitian operators $p(t)$ and $q(t)$ which satisfy equations of motion identical to the classical equations (11.5). To completely determine the dynamical problem, we must specify the matrix elements of the operators p and q at the initial time. The initial conditions on $p(0)$ and $q(0)$ posed in the classical theory must be supplemented here by the requirement that the commutator conditions (11.11) are fulfilled at $t = 0$ for any physical state. Since by postulate the physical states of a system—for example, the energy eigenstates—form a complete set, the initial commutator condition may be stated as an operator requirement as in Eq. (11.11).

To illustrate this procedure, we quantize the simple harmonic oscillator in one dimension in the Heisenberg picture. The hamiltonian is

$$H = \tfrac{1}{2}(p^2 + \omega_0^2 q^2) \tag{11.13}$$

with equations of motion

$$p(t) = \frac{dq(t)}{dt} \equiv \dot{q}(t) \qquad \ddot{q} + \omega_0^2 q = 0$$

To solve for the coordinates, we introduce the convenient linear combinations

$$a = \sqrt{\frac{1}{2\omega_0}}\,(\omega_0 q + ip) \qquad a^\dagger = \sqrt{\frac{1}{2\omega_0}}\,(\omega_0 q - ip) \qquad (11.14)$$

Since $q^\dagger = q$ Hermitian
$p^\dagger = p$

in terms of which the equations of motion are

$$\dot{a}(t) = -i\omega_0 a(t) \qquad \dot{a}^\dagger(t) = +i\omega_0 a^\dagger(t)$$

The solutions are

$$a(t) = a_0 e^{-i\omega_0 t} \qquad a^\dagger(t) = a_0^\dagger e^{+i\omega_0 t} \qquad (11.15)$$

where a_0 and $a_0^\dagger$ are time-independent operators which satisfy commutator relations following from (11.11):

$$[a(t),a^\dagger(t)] = [a_0,a_0^\dagger] = 1 \qquad [a(t),a(t)] = [a_0,a_0] = 0 \qquad (11.16)$$

$$[a^\dagger(t),a^\dagger(t)] = [a_0^\dagger,a_0^\dagger] = 0$$

In terms of a and $a^\dagger$ the hamiltonian is $\longrightarrow$ since commutes w/ $a_0, a_0^\dagger$ is a c-number

so commutes

$$H = \tfrac{1}{2}\omega_0(a^\dagger a + a a^\dagger) = \tfrac{1}{2}\omega_0(a_0^\dagger a_0 + a_0 a_0^\dagger) \qquad (11.17)$$

$\omega_0 a_0 a_0^\dagger - \tfrac{1}{2}\omega_0$

Since any state Ψ may be expanded in stationary eigenfunctions Ψ_n of H, we need only find the properties of Ψ_n. From the commutation relations we have   cf. above (11.5)

$$[H,a_0] = -\omega_0 a_0 \qquad \text{and} \qquad [H,a_0^\dagger] = +\omega_0 a_0^\dagger \qquad (11.18)$$

Thus if $H\Psi_n = \omega_n \Psi_n$, $H a_0^\dagger \Psi_n = (\omega_n + \omega_0) a_0^\dagger \Psi_n$ and we generate an infinite series of states of higher energy—starting with a given Ψ_n corresponding to energy eigenvalue ω_n—by successive applications of the operator $a_0^\dagger$: $a_0^\dagger \Psi_n = \Psi_{n+1}$. Likewise, if $\omega_{n+1} = \omega_n + \omega_0$

$$H\Psi_n = \omega_n \Psi_n, \quad H a_0 \Psi_n = (\omega_n - \omega_0) a_0 \Psi_n$$

and by successive applications of the operator a_0 we generate states of lower energy, $a_0\Psi_n = \Psi_{n-1}$. This series must terminate, however, since the hamiltonian (11.13) is a sum of squares of hermitian operators and therefore can have no negative eigenvalues. The ground state of lowest energy is found by the requirement that $a_0\Psi_0 = 0$. The energy of the ground state is then

$$H\Psi_0 = \tfrac{1}{2}\omega_0 a_0 a_0^\dagger \Psi_0 = \tfrac{1}{2}\omega_0 [a_0,a_0^\dagger]\Psi_0 = \tfrac{1}{2}\omega_0 \Psi_0$$

and the energy of the nth state

$$\Psi_n \equiv (a_0^\dagger)^n \Psi_0 \qquad (11.19)$$

is

$$\omega_n = (n + \tfrac{1}{2})\omega_0$$

The energy spectrum is nondegenerate for the one-dimensional oscillator, and the different states (11.19) are mutually orthogonal:[1]

obtained
from (1.21

$$(\Psi_n, \Psi_m) = \delta_{nm}(n!)(\Psi_0, \Psi_0) \tag{11.20}$$

The matrix elements of $a_0^\dagger$ in this representation are found by considering

$~n+1$

$$\left\langle \Psi_n \left| \frac{H}{\omega_0} + \frac{1}{2} \right| \Psi_n \right\rangle = \langle \Psi_n | a_0 a_0^\dagger | \Psi_n \rangle = |\langle \Psi_{n+1} | a_0^\dagger | \Psi_n \rangle|^2 = n + 1$$

We may then take

$$\langle \Psi_{n+1} | a_0^\dagger | \Psi_n \rangle = \sqrt{n + 1} = \langle \Psi_n | a_0 | \Psi_{n+1} \rangle \tag{11.21}$$

At an arbitrary time t, the matrix element of the Heisenberg operator is

$$\langle \Psi_{n+1} | a^\dagger(t) | \Psi_n \rangle = e^{i\omega_0 t} \langle \Psi_{n+1} | a_0^\dagger | \Psi_n \rangle \quad = e^{i\omega_0 t} \sqrt{n+1}$$

$$= \langle e^{-i\omega_{n+1} t} \Psi_{n+1} | a_0^\dagger | e^{-i\omega_n t} \Psi_n \rangle \tag{11.22}$$

In this special example only those matrix elements which change n by $\Delta n = \pm 1$ are nonzero. Equations (11.13) to (11.22) constitute the complete quantum-mechanical solution. They describe the time development of the operators, the possible physical states, and the matrix elements of the operators in the energy representation.

since $x, p,$
H, etc., com-
be expressed
or terms of
in a and $a^\dagger$

It is straightforward to generalize the discussion thus far to a system with n degrees of freedom. We introduce n hermitian operators $q_i(t)$, $i = 1, \ldots, n$ in the Heisenberg picture and n conjugate momenta $p_i(t)$. The dynamics is again given by the $2n$ classical equations of motion.

$$-\frac{\partial H}{\partial q_i} = \frac{dp_i}{dt} \qquad \frac{\partial H}{\partial p_i} = \frac{dq_i}{dt} \qquad i = 1, \ldots, n \tag{11.23}$$

Specifying the matrix elements of p_i and q_i at an initial time, say $t = 0$, with the restriction that they fulfill the commutator conditions

$$[p_i(0), q_j(0)] = -i\delta_{ij}$$

$$[p_i(0), p_j(0)] = 0 \tag{11.24}$$

$$[q_i(0), q_j(0)] = 0$$

completely defines the quantum-dynamical problem. The quantum form of (11.23) is, as in (11.12),

$$\dot{p}_i(t) = i[H, p_i(t)] \qquad \dot{q}_i(t) = i[H, q_i(t)] \tag{11.25}$$

[1] Notice that the Ψ_n are not normalized.

for each of the n independent coordinates and momenta describing a system with n degrees of freedom.

11.3 Canonical Formalism and Quantization for Fields[1]

Taking the limit $n \to \infty$ brings us to a field theory with the field at each point of space being considered an independent generalized coordinate. A simple example of this limit in classical physics is the weighted vibrating string. For a finite number N of beads along the string we solve N coupled oscillator equations for the motion; in the limit $N \to \infty$ we come to the limit of a continuous string described by a displacement field $\varphi(\mathbf{x},t)$ which varies continuously as a function of position $\mathbf{x}$ along the string and of time t. By its value, φ measures the amplitude of displacement of the string from rest at $(\mathbf{x},t)$; by its time derivative $\partial\varphi(\mathbf{x},t)/\partial t$, its velocity at $(\mathbf{x},t)$.

Following this analogy, in terms of a canonical formalism we expect $\varphi(\mathbf{x},t)$ to play the role of the coordinate $q_i(t)$ and $\partial\varphi(\mathbf{x},t)/\partial t$ to correspond to $\dot{q}_i(t)$. The discrete label i is replaced by the continuous coordinate variable $\mathbf{x}$, and in the Heisenberg picture the field is a function of both the space and time coordinates $x = (\mathbf{x},t)$. It is in this treatment of both space and time coordinates on the same footing that we see the advantage of the Heisenberg representation for explicitly maintaining the Lorentz covariance of the formalism. The only vestige of a preferred role for the time coordinate lies in the statement of initial conditions and of commutation relations at, say, $t = 0$. The surface $t = 0$ is a noncovariant element in the theory. Even this, however, may be removed by the covariant notion of a space-like surface on which to specify initial conditions and commutators (Fig. 11.2). A space-like surface is a three-dimensional surface σ whose normal η_μ is everywhere time-like, that is, $\eta_\mu\eta^\mu = 1 > 0$. By convention we shall always choose η^μ in the future light cone, that is, $\eta^0 > 0$. In our future development, the words "at a given time t" and "on a space-like surface σ" are interchangeable; in this way we may give our initial conditions a covariant statement.

The road to the quantization of a classical field theory starts with the field equations. Knowing these, we seek a lagrangian which, via Hamilton's principle, reproduces them. Having the lagrangian, it is then possible to identify canonical momenta and carry out the quantization procedure in accord with (11.24). With this step the

[1] W. Heisenberg and W. Pauli, *Z. Physik*, **56,** 1 (1929); G. Wentzel "Quantum Theory of Fields," Interscience Publishers, Inc., New York, 1949.

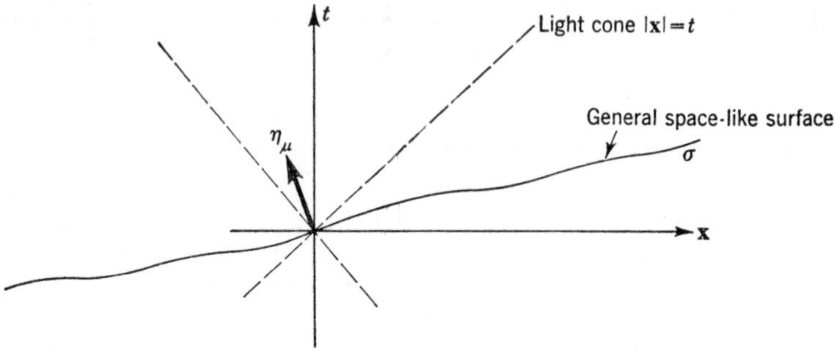

Fig. 11.2 General space-like surface σ with normal η_μ.

fields $\varphi_i(\mathbf{x},t)$, and their canonical momenta $\pi_i(\mathbf{x},t)$, become operators in a Hilbert space, operating upon state vectors Φ. We postulate, as in the one-particle quantum mechanics mentioned at the very beginning of Chap. 1, that the physical states Φ form a complete set in the Hilbert space. Most frequently we shall encounter Φ in the Heisenberg picture as eigenstates of the hamiltonian constructed from the fields φ_i and momenta π_i in analogy with the particle procedure:

$$H(\varphi_i,\pi_i)\Phi_n = E_n\Phi_n$$

We first recall the construction of a lagrangian L from equations of motion in classical point mechanics. We can then imitate these same steps to construct a lagrangian in field theory. Starting from Newton's law

$$m_i\ddot{q}_i = -\frac{\partial}{\partial q_i} V(q_1, \ldots, q_n)$$

we multiply by δq_i and sum over $i = 1, \ldots, n$:

$$\sum_{i=1}^{n} m_i\ddot{q}_i\,\delta q_i = -\sum_{i=1}^{n} \frac{\partial V}{\partial q_i}\,\delta q_i = -\delta V$$

Integrating next over the time interval t_1 to t_2 with fixed end points of the particle trajectory, that is, $\delta q_i(t_1) = \delta q_i(t_2) = 0$, we find after one partial integration

$$\int_{t_1}^{t_2} dt \left(\sum_{i=1}^{n} m_i\dot{q}_i\delta\dot{q}_i - \delta V\right) = 0$$

This is just Hamilton's principle with a lagrangian

$$L = \tfrac{1}{2} \sum_{i=1}^{n} m_i\dot{q}_i^2 - V$$

We follow this exact same procedure in the case of a classical field $\varphi(x)$ which we take for illustrative purposes to satisfy a free Klein-Gordon wave equation

$$\left(\frac{\partial^2}{\partial t^2} - \nabla^2 + m^2\right)\varphi = 0 \tag{11.26}$$

The summation over q_i above is here replaced by a three-dimensional integral over space coordinates $\mathbf{x}$. First multiply the field equation (11.26) by an infinitesimal variation in the amplitude of the field at x,

$$\delta\varphi(x) = \varphi'(x) - \varphi(x)$$

and integrate over all coordinates $\mathbf{x}$ and over the time interval from t_1 to t_2:

$$\int_{t_1}^{t_2} dt \int_{-\infty}^{\infty} d^3x \left(\frac{\partial^2\varphi}{\partial t^2} - \nabla^2\varphi + m^2\varphi\right)\delta\varphi = 0$$

Again the variations are restricted to vanish at the end points t_1 and t_2: $\delta\varphi(t_1) = \delta\varphi(t_2) = 0$. We assume also that the system is localized in space[1] so that there are no contributions from distant surfaces at $\mathbf{x} \to \pm\infty$. This gives

$$\int_{t_1}^{t_2} dt \int_{-\infty}^{\infty} d^3x\, \delta\left[+\frac{1}{2}\left(\frac{\partial\varphi}{\partial t}\right)^2 - \frac{1}{2}|\nabla\varphi|^2 - \frac{1}{2}m^2\varphi^2\right] = 0$$

or

$$\delta \int_{t_1}^{t_2} d^4x\, \mathfrak{L}\left(\varphi, \frac{\partial\varphi}{\partial x^\mu}\right) = 0$$

with

$$\mathfrak{L}\left(\varphi, \frac{\partial\varphi}{\partial x^\mu}\right) \equiv \frac{1}{2}\left(\frac{\partial\varphi}{\partial x^\mu}\frac{\partial\varphi}{\partial x_\mu} - m^2\varphi^2\right) \tag{11.27}$$

$\mathfrak{L}$ is a Lorentz-invariant functional of the fields and first derivatives φ and $\partial\varphi/\partial x^\mu$; it is called the lagrangian density. The lagrangian L which plays the same role as in the particle mechanics is seen to be the volume integral of the density:

$$L \equiv \int_{-\infty}^{\infty} d^3x\, \mathfrak{L}\left(\varphi, \frac{\partial\varphi}{\partial x^\mu}\right)$$

In a general theory, we shall always assume the field equations may be derived from some lagrangian density $\mathfrak{L}$. In analogy with Hamilton's principle for particle mechanics (11.1), we ask that the action be stationary for fields which are solutions of the actual equations

[1] We may also accomplish this by enclosing the system in a box with periodic boundary conditions.

of motion, that is,

$$\delta \int_{t_1}^{t_2} dt \int d^3x\, \mathfrak{L} = 0 \tag{11.28}$$

Upon carrying out the variation in (11.28) by varying the fields in the integration interval but keeping the boundary values at t_1 and t_2 fixed as in the particle mechanics case, we obtain the Euler-Lagrange equations of the field development. Specifically, for a simple system described by a single field φ and by a lagrangian density $\mathfrak{L}$ which is a functional of φ and $\partial\varphi/\partial x^\mu$ only, (11.28) gives

$$\int_{t_1}^{t_2} dt \int d^3x \left[\mathfrak{L}\left(\varphi + \delta\varphi, \frac{\partial\varphi}{\partial x^\mu} + \delta\,\frac{\partial\varphi}{\partial x^\mu}\right) - \mathfrak{L}\left(\varphi, \frac{\partial\varphi}{\partial x^\mu}\right)\right]$$
$$= \int_{t_1}^{t_2} dt \int d^3x \left[\frac{\partial\mathfrak{L}}{\partial\varphi}\,\delta\varphi + \frac{\partial\mathfrak{L}}{\partial(\partial\varphi/\partial x^\mu)}\,\delta\left(\frac{\partial\varphi}{\partial x^\mu}\right)\right] = 0$$

Integrating by parts, and using the relation

$$\delta\,\frac{\partial\varphi}{\partial x^\mu} = \frac{\partial}{\partial x^\mu}\,(\varphi + \delta\varphi) - \frac{\partial\varphi}{\partial x^\mu} = \frac{\partial}{\partial x^\mu}\,(\delta\varphi)$$

gives

$$\int_{t_1}^{t_2} d^4x\, \delta\varphi \left[\frac{\partial\mathfrak{L}}{\partial\varphi} - \frac{\partial}{\partial x^\mu}\,\frac{\partial\mathfrak{L}}{\partial(\partial\varphi/\partial x^\mu)}\right] = 0$$

For arbitrary variations $\delta\varphi$, this leads to the field equation

$$\frac{\partial}{\partial x^\mu}\,\frac{\partial\mathfrak{L}}{\partial(\partial\varphi/\partial x^\mu)} - \frac{\partial\mathfrak{L}}{\partial\varphi} = 0 \tag{11.29}$$

For the special choice of $\mathfrak{L}$ in (11.27), (11.29) is just the Klein-Gordon equation (11.26).

Field equations derived in this way are generally local differential equations. If $\mathfrak{L}$ contains higher than first-order derivatives of the fields, the field equations will be of higher than second order.[1] As long as $\mathfrak{L}$ contains only a finite order of derivatives of $\varphi(x)$, the fields satisfy differential equations and the theory is "local." In accepting here this hypothesis of local action we recall the reservations discussed in the introductory paragraphs of this chapter. It is likely that the canonical development presented here applies only in the sense of a correspondence principle for large distances if nature is nonlocal or granular in the small.

[1] We shall adhere in what follows to the requirement that $\mathfrak{L}$ shall be a functional only of the fields and first derivatives. In addition, $\mathfrak{L}$ shall have no explicit dependence on the space-time coordinates. This means that we consider "closed" systems only, that is, systems which do not exchange energy and momentum with applied external sources.

In all cases the choice of lagrangians is dictated by the specific field equations that are desired. Equation (11.27) was designed to give the Klein-Gordon equation; other examples of interest to us are the Dirac and Maxwell lagrangians. For systems described by more than one independent field, say, $\varphi_r(x)$, $r = 1, \ldots, n$, we obtain n field equations

$$\frac{\partial \mathcal{L}}{\partial \varphi_r} - \frac{\partial}{\partial x^\mu} \frac{\partial \mathcal{L}}{\partial(\partial \varphi_r / \partial x^\mu)} = 0 \qquad r = 1, \ldots, n \tag{11.30}$$

from Hamilton's principle by independently varying each field, $\delta\varphi_r(x)$.

Turning now to the canonical formalism and quantization procedure, we use the lagrangian $\mathcal{L}$ to define a canonical momentum as in the particle mechanics. In order to exhibit explicitly the complete parallel between the field and particle mechanics, we revert to a system with a finite number of degrees of freedom by dividing the three-dimensional space into cells of volume ΔV_i and defining the ith coordinate $\varphi_i(t)$ by the average value of $\varphi(x)$ over the ith cell

$$\varphi_i(t) \equiv \frac{1}{\Delta V_i} \int_{(\Delta V_i)} d^3x \, \varphi(\mathbf{x},t)$$

Denoting by $\dot{\varphi}_i(t)$ the average of $\partial\varphi(\mathbf{x},t)/\partial t$ over the ith cell, we rewrite the lagrangian as

$$L = \int d^3x \, \mathcal{L} \to \sum_i \Delta V_i \, \bar{\mathcal{L}}_i(\dot{\varphi}_i(t), \varphi_i(t), \varphi_{i\pm s}(t), \ldots) \tag{11.31}$$

The different φ_i are all independent degrees of freedom; the values of $\varphi_{i\pm s}(t)$ in neighboring cells appear in $\bar{\mathcal{L}}_i$, so that appropriate differences can be taken in forming the $\nabla\varphi$. Since only the one time derivative $\dot{\varphi}_i(t)$ appears in each term $\bar{\mathcal{L}}_i$, the canonical momenta are simply

$$p_i(t) = \frac{\partial L}{\partial \dot{\varphi}_i(t)} = \Delta V_i \frac{\partial \bar{\mathcal{L}}_i}{\partial \dot{\varphi}_i(t)} \equiv \Delta V_i \, \pi_i(t) \tag{11.32}$$

The hamiltonian defined as in (11.4) is then

$$H = \sum_i p_i \dot{\varphi}_i - L = \sum_i \Delta V_i \, (\pi_i \dot{\varphi}_i - \bar{\mathcal{L}}_i) \tag{11.33}$$

Returning now to continuum notation, we define the momentum conjugate to $\varphi(\mathbf{x},t)$ by

$$\pi(\mathbf{x},t) = \frac{\partial \mathcal{L}(\varphi,\dot{\varphi})}{\partial \dot{\varphi}(\mathbf{x},t)} \tag{11.34}$$

Its cell average gives $\pi_i(t)$ in (11.32). The hamiltonian is written as a volume integral over a hamiltonian density $\mathcal{H}(\pi,\varphi)$ defined as suggested by (11.33)

$$H = \int d^3x\, \mathcal{H}(\pi(\mathbf{x},t),\varphi(\mathbf{x},t)) \qquad \mathcal{H} = \pi\dot{\varphi} - \mathcal{L} \tag{11.35}$$

Having identified the canonical momenta, we now follow the quantization procedure of replacing dynamical variables $\varphi_i(t)$, $p_i(t)$ by hermitian operators which satisfy commutation relations analogous to (11.24):

$$\begin{aligned}[\varphi_i(t),\varphi_j(t)] &= [p_i(t),p_j(t)] = 0\\ [p_i(t),\varphi_j(t)] &= -i\delta_{ij}\end{aligned}$$

or in terms of $\pi_i(t)$

$$[\pi_i(t),\varphi_j(t)] = -\frac{i\delta_{ij}}{\Delta V_i}$$

These become in continuum language

$$\begin{aligned}[\varphi(\mathbf{x},t),\varphi(\mathbf{x}',t)] &= 0\\ [\pi(\mathbf{x},t),\pi(\mathbf{x}',t)] &= 0\\ [\pi(\mathbf{x},t),\varphi(\mathbf{x}',t)] &= -i\delta^3(\mathbf{x}-\mathbf{x}')\end{aligned} \tag{11.36}$$

where the Dirac delta function emerges as the limit of $\delta_{ij}/\Delta V_i$, as $\Delta V_i \to 0$, according to its definition

$$\int d^3x'\, \delta^3(\mathbf{x}-\mathbf{x}')f(\mathbf{x}') = f(\mathbf{x})$$

Equations (11.34) to (11.36), in addition to the equations of motion, provide the basis of canonical quantum field theory. In order to generalize this development to physical systems described with several independent fields $\varphi_r(\mathbf{x},t)$, we introduce momenta conjugate to each field by

for $\mathcal{L}(\pi, \varphi, \frac{\partial\varphi \ldots}{\partial\pi \ldots})$ see end of Goldstein

$$\pi_r(\mathbf{x},t) = \frac{\partial\mathcal{L}}{\partial\dot{\varphi}_r(\mathbf{x},t)} \tag{11.37}$$

and the hamiltonian density by

$$\mathcal{H}(\pi_r \cdots, \varphi_r \cdots) = \sum_{r=1}^{n} \pi_r\dot{\varphi}_r - \mathcal{L} \tag{11.38}$$

For the quantization conditions we introduce the commutators

$$\begin{aligned}[\varphi_r(\mathbf{x},t),\varphi_s(\mathbf{x}',t)] &= 0\\ [\pi_r(\mathbf{x},t),\pi_s(\mathbf{x}',t)] &= 0\\ [\pi_r(\mathbf{x},t),\varphi_s(\mathbf{x}',t)] &= -i\delta_{rs}\delta^3(\mathbf{x}-\mathbf{x}')\end{aligned} \tag{11.39}$$

Finally, to complete this purely imitative transcription of the particle mechanics to establish a quantum field formalism, we find from (11.37) to (11.39) that (11.25) transcribes to

$$\dot{\pi}_r(\mathbf{x},t) = i[H,\pi_r(\mathbf{x},t)] \qquad \dot{\varphi}_r(\mathbf{x},t) = i[H,\varphi_r(\mathbf{x},t)] \tag{11.40}$$

11.4 Symmetries and Conservation Laws

The lagrangian formulation provides a convenient and systematic way of identifying and extracting constants of the motion in classical field theory. One can show that, starting from a scalar lagrangian, there is a conservation theorem and a constant of the motion corresponding to each continuous symmetry transformation which leaves the lagrangian density $\mathcal{L}$ and the equations of motion invariant in form. Such a theorem[1] (known as Noether's theorem), which permits observed selection rules in nature to be described directly in terms of symmetry requirements in $\mathcal{L}$, is useful as a guide for the introduction of interaction terms in developing new theories. We are therefore interested in its application to quantized field theories.

We first discuss the conservation laws which result from translational invariance in a classical field theory. Under an infinitesimal displacement

$$x'_\mu = x_\mu + \epsilon_\mu \tag{11.41}$$

the lagrangian $\mathcal{L}$ changes by the amount

$$\delta\mathcal{L} = \mathcal{L}' - \mathcal{L} = \epsilon_\mu \frac{\partial \mathcal{L}}{\partial x_\mu} \tag{11.42}$$

On the other hand, if $\mathcal{L}$ is translationally invariant, it has no explicit coordinate dependence and we write $\mathcal{L} = \mathcal{L}(\varphi_r, \partial\varphi_r/\partial x_\mu)$ so that

$$\delta\mathcal{L} = \sum_r \left[\frac{\partial \mathcal{L}}{\partial \varphi_r(x)} \delta\varphi_r + \frac{\partial \mathcal{L}}{\partial(\partial\varphi_r/\partial x_\mu)} \delta\left(\frac{\partial \varphi_r}{\partial x_\mu}\right) \right] \tag{11.43}$$

where

$$\delta\varphi_r = \varphi_r(x+\epsilon) - \varphi_r(x) = \epsilon_\nu \frac{\partial \varphi_r(x)}{\partial x_\nu} \tag{11.44}$$

Equating these two expressions and using the Euler-Lagrange equations

$$\frac{\partial \mathcal{L}}{\partial \varphi_r} - \frac{\partial}{\partial x^\mu} \frac{\partial \mathcal{L}}{\partial(\partial\varphi_r/\partial x^\mu)} = 0 \tag{11.45}$$

gives

$$\epsilon_\mu \frac{\partial \mathcal{L}}{\partial x_\mu} = \frac{\partial}{\partial x_\mu} \left[\sum_r \frac{\partial \mathcal{L}}{\partial(\partial\varphi_r/\partial x_\mu)} \epsilon_\nu \frac{\partial \varphi_r}{\partial x_\nu} \right] \tag{11.46}$$

Since this holds for arbitrary displacements ϵ_μ, we can write

$$\frac{\partial}{\partial x_\mu} \mathfrak{I}_{\mu\nu} = 0 \tag{11.47}$$

[1] See E. L. Hill, *Rev. Mod. Phys.*, **23**, 253 (1957), for a detailed discussion of this theorem.

with the energy-momentum stress tensor $\mathfrak{J}_{\mu\nu}$ defined by

$$\mathfrak{J}_{\mu\nu} = -g_{\mu\nu}\mathfrak{L} + \sum_r \frac{\partial \mathfrak{L}}{\partial(\partial\varphi_r/\partial x_\mu)} \frac{\partial \varphi_r}{\partial x^\nu} \tag{11.48}$$

From this differential conservation law one finds the conserved quantities

$$P_\nu = \int d^3x\, \mathfrak{J}_{0\nu} = \int d^3x \left[\sum_r \pi_r \frac{\partial \varphi_r}{\partial x^\nu} - g_{0\nu}\mathfrak{L}\right] \tag{11.49}$$

$$\frac{\partial P_\nu}{\partial t} = 0$$

We have already seen in (11.37) and (11.38) that $\mathfrak{J}_{00}$ is the hamiltonian density

$$\mathfrak{J}_{00} = \sum_r \pi_r \dot{\varphi}_r - \mathfrak{L} = \mathfrak{H} \tag{11.50}$$

and $$\int d^3x\, \mathfrak{J}_{00} = H$$

so that we may identify P_ν as the conserved energy-momentum four-vector.

In the same manner we may construct the angular-momentum constants of the motion by considering an infinitesimal Lorentz transformation.

$$x'_\nu = x_\nu + \epsilon_{\nu\mu}x^\mu \qquad \epsilon_{\mu\nu} = -\epsilon_{\nu\mu} \tag{11.51}$$

The practical test of Lorentz invariance is to make the replacement[1]

$$\varphi_r(x) \to S^{-1}_{rs}(\epsilon)\varphi_s(x') \tag{11.52}$$

in the equations of motion and to determine whether they then take the same form in the primed coordinate system as they did in the unprimed system. Here $S_{rs}(\epsilon)$ is a transformation matrix for the fields φ_r under the infinitesimal Lorentz transformation (11.51) and differs from the unit matrix if the field is not a scalar field. We have already seen an example of this for the Dirac equation, where we recall from (2.17) that[2]

$$S_{rs}(\epsilon) = \delta_{rs} + \tfrac{1}{8}[\gamma^\mu,\gamma^\nu]_{rs}\epsilon_{\mu\nu}$$

We now take over the test (11.52) into the lagrangian theory and demand that the lagrangian density be a Lorentz scalar and therefore

[1] Henceforth repeated indices r and s for field components are understood to be summed.

[2] For the Dirac theory, see (2.11).

remain form invariant under the replacement (11.52), that is

$$\mathfrak{L}\left(S_{rs}^{-1}\varphi_s(x'),\ \frac{\partial}{\partial x_\mu} S_{rs}^{-1}\varphi_s(x')\right) = \mathfrak{L}\left(\varphi_r(x'),\ \frac{\partial \varphi_r(x')}{\partial x'_\mu}\right) \tag{11.53}$$

This will guarantee the form invariance of the equations of motion, which are derived from $\mathfrak{L}$ by an invariant action principle. We write for an infinitesimal transformation

$$\delta\varphi_r(x) = S_{rs}^{-1}(\epsilon)\varphi_s(x') - \varphi_r(x) = \varphi_r(x') - \varphi_r(x) - \tfrac{1}{2}\epsilon_{\mu\nu}\Sigma_{rs}^{\mu\nu}\varphi_s(x)$$

with the definition

$$S_{rs}(\epsilon) = \delta_{rs} + \tfrac{1}{2}\Sigma_{rs}^{\mu\nu}\epsilon_{\mu\nu} \tag{11.54}$$

Expanding (11.53) about x we find, using the Euler-Lagrange equations (11.45)

$$\mathfrak{L}(x') - \mathfrak{L}(x) = \epsilon^{\mu\nu}x_\nu \frac{\partial \mathfrak{L}}{\partial x^\mu} = \frac{\partial}{\partial x^\mu}\left[\frac{\partial \mathfrak{L}}{\partial(\partial\varphi_r/\partial x^\mu)}\,\delta\varphi_r\right] \tag{11.55}$$

Equations (11.54) and (11.55) lead to the conservation law

$$\begin{aligned}\frac{\partial}{\partial x^\mu}\,\mathfrak{M}^{\mu\nu\lambda} &= \frac{\partial}{\partial x^\mu}\left[(x^\lambda g^{\mu\nu} - x^\nu g^{\mu\lambda})\mathfrak{L} + \frac{\partial \mathfrak{L}}{\partial(\partial\varphi_r/\partial x^\mu)}\left\{\left(x^\nu \frac{\partial}{\partial x_\lambda} - x^\lambda \frac{\partial}{\partial x_\nu}\right)\varphi_r \right.\right.\\ &\qquad\qquad \left.\left. + \Sigma_{rs}^{\nu\lambda}\varphi_s\right\}\right]\\ &= \frac{\partial}{\partial x^\mu}\left[(x^\nu \mathfrak{J}^{\mu\lambda} - x^\lambda \mathfrak{J}^{\mu\nu}) + \frac{\partial \mathfrak{L}}{\partial(\partial\varphi_r/\partial x^\mu)}\,\Sigma_{rs}^{\nu\lambda}\varphi_s\right] = 0\end{aligned} \tag{11.56}$$

The conserved angular momentum is

$$M^{\nu\lambda} = \int d^3x\, \mathfrak{M}^{0\nu\lambda} = \int d^3x[(x^\nu \mathfrak{J}^{0\lambda} - x^\lambda \mathfrak{J}^{0\nu}) + \pi_r\Sigma_{rs}^{\nu\lambda}\varphi_s] \tag{11.57}$$

$$\frac{\partial M^{\nu\lambda}}{\partial t} = 0$$

Proceeding in the same spirit, one obtains additional conservation laws if the lagrangian possesses "internal symmetries," that is, if under *local* transformations

$$\varphi_r(x) \to \varphi_r(x) - i\epsilon\lambda_{rs}\varphi_s(x) \tag{11.58}$$

the lagrangian density remains invariant. In (11.58) the λ_{rs} are constant coefficients independent of x^μ and ϵ is an infinitesimal parameter. The diagonal components of the matrix λ correspond to simple phase changes of the fields, while the others mix the different field amplitudes which appear symmetrically in $\mathfrak{L}$. If $\mathfrak{L}$ is invariant under the substitution (11.58), we find by repeating the steps (11.42) to (11.46)

$$0 = \delta\mathcal{L} = \frac{\partial\mathcal{L}}{\partial\varphi_r}\,\delta\varphi_r + \frac{\partial\mathcal{L}}{\partial(\partial\varphi_r/\partial x^\mu)}\,\delta\,\frac{\partial\varphi_r}{\partial x^\mu}$$
$$= -i\epsilon\,\frac{\partial}{\partial x^\mu}\left[\frac{\partial\mathcal{L}}{\partial(\partial\varphi_r/\partial x^\mu)}\,\lambda_{rs}\varphi_s\right] \qquad (11.59)$$

Thus for each "internal symmetry" operation (11.58) which leaves $\mathcal{L}$ invariant there is a differentially conserved current

$$\frac{\partial J^\mu(x,\lambda)}{\partial x^\mu} = 0 \qquad (11.60)$$

with

$$J^\mu(x,\lambda) = -i\,\frac{\partial\mathcal{L}}{\partial(\partial\varphi_r/\partial x^\mu)}\,\lambda_{rs}\varphi_s \qquad (11.61)$$

along with a conserved "charge"

$$Q(\lambda) = -i\int d^3x\,\pi_r\lambda_{rs}\varphi_s \qquad \frac{\partial Q(\lambda)}{\partial t} = 0 \qquad (11.62)$$

Going over now to quantum field theory, we must ask whether we may still apply the classical result that a scalar $\mathcal{L}$ guarantees Lorentz invariance of the theory and provides, via the Noether theorem, the energy-momentum and angular-momentum constants of the motion. In the quantum theory the field amplitudes $\varphi_r(x)$ become operators upon state functions, or vectors, in a Hilbert space. If we impose the requirements of Lorentz covariance on the *matrix elements* of these operators, which represent physical observables as viewed in two different Lorentz frames, we come to certain operator restrictions on the $\varphi_r(x)$. For a quantum field theory a scalar $\mathcal{L}$ is not sufficient to guarantee relativistic invariance, but we must also verify that the fields obey these operator requirements.

To show how these requirements arise, we take as a physical observable the matrix elements of the field operator $\varphi_r(x)$ between two state functions

$$(\Phi_\alpha,\varphi_r(x)\Phi_\beta) \qquad (11.63)$$

For arbitrary states, labeled by α and β, this complete set of amplitudes in quantum field theory replaces the classical field amplitudes $\varphi_r(x)$. The analogous role in the Schrödinger quantum mechanics is played by the matrix elements of the coordinate $q(t)$. To an observer in another Lorentz frame, related by a coordinate transformation

$$x^{\mu\prime} = a^\mu{}_\nu x^\nu + b^\mu$$

the amplitudes (11.63) are

$$(\Phi'_\alpha, \varphi_r(x')\Phi'_\beta) \tag{11.64}$$

as expressed in terms of the state vectors Φ'_α and Φ'_β representing the same physical states α and β to the observer in the primed system and in terms of the field operators $\varphi_r(x')$ at the transformed point x'. The amplitudes (11.64) are the quantum theory correspondence of the classical fields $\varphi'_r(x') \equiv S_{rs}\varphi_s(x)$ in the primed system. This classical transformation law is now expressed by

$$(\Phi'_\alpha, \varphi_r(x')\Phi'_\beta) = S_{rs}(a)(\Phi_\alpha, \varphi_s(x)\Phi_\beta) \tag{11.65}$$

and provides the mathematical law of communication between the two Lorentz observers. We require then that there exist a unitary operator $U(a,b)$ which accomplishes the desired transformation of the state vectors between the two Lorentz frames so that by the equation

$$\Phi'_\alpha = U(a,b)\Phi_\alpha \tag{11.66}$$

we can relate the corresponding states. The field operators then transform according to

$$U(a,b)\varphi_r(x)U(a,b)^{-1} = S_{rs}^{-1}(a)\varphi_s(ax + b) \tag{11.67}$$

as follows from (11.65).

Considering first displacements, in particular, we have

$$U(b)\varphi_r(x)U(b)^{-1} = \varphi_r(x + b) \tag{11.68}$$

where $U(b)$ is the unitary operator generating the coordinate displacements. For infinitesimal displacements $x^{\mu\prime} = x^\mu + \epsilon^\mu$, we may write

$$U(\epsilon) = \exp(i\epsilon_\mu \mathsf{P}^\mu) \approx 1 + i\epsilon_\mu \mathsf{P}^\mu \tag{11.69}$$

where P_μ is a hermitian operator. Equation (11.68) now reduces to

$$i[\mathsf{P}^\mu, \varphi_r(x)] = \frac{\partial \varphi_r(x)}{\partial x_\mu} \tag{11.70}$$

Correspondence with classical canonical mechanics and nonrelativistic Schrödinger theory [Eq. (11.10)] suggests the identification of P^μ in (11.69) and (11.70) with the energy-momentum four-vector $\mathsf{P}^\mu \equiv P^\mu$. Since we have derived an explicit form (11.49) for P^μ, we can explicitly check in any theory whether this identification follows from the commutation relations imposed in the quantization procedure. Thus we can compute directly from the commutators whether (11.69) remains an operator identity and whether the components of **P** commute with H, that is

$$[H,\mathbf{P}] = 0 \tag{11.71}$$

so that **P** remains a constant of the motion.

If (11.70) and (11.71) are consistent with the commutation relations, the quantum theory is displacement invariant; if not, either a $\mathbf{P}_\mu$ satisfying (11.70) and (11.71) must be found by some other means, the commutator condition modified, or the theory abandoned. For the theories we consider, the P_μ and $M_{\mu\nu}$ found by Noether's procedure will be found to be satisfactory.

In a similar way we construct the analogous statement for invariance of the quantum field theory under Lorentz transformations. The unitary operator which generates the infinitesimal Lorentz transformation $x^{\mu\prime} = x^\mu + \epsilon^\mu{}_\nu x^\nu$ is written

$$U(\epsilon_{\mu\nu}) = 1 - \frac{i}{2}\,\epsilon_{\mu\nu}\mathbf{M}^{\mu\nu} \tag{11.72}$$

where $\mathbf{M}^{\mu\nu}$ is a hermitian operator which satisfies an operator equation according to (11.67):

$$\varphi_r(x) - \frac{i}{2}\,\epsilon_{\mu\nu}[\mathbf{M}^{\mu\nu},\varphi_r(x)] = S_{rs}^{-1}(\epsilon^{\mu\nu})\varphi_s(x + \epsilon x)$$

With the help of (11.54) this reduces to

$$i[\mathbf{M}^{\mu\nu},\varphi_r(x)] = x^\mu \frac{\partial \varphi_r}{\partial x_\nu} - x^\nu \frac{\partial \varphi_r}{\partial x_\mu} + \Sigma_{rs}^{\mu\nu}\varphi_s(x) \tag{11.73}$$

Again we rely on correspondence with classical and with nonrelativistic theory to identify $\mathbf{M}^{\mu\nu}$, which generates the Lorentz transformation, with the angular-momentum tensor (11.57): $\mathbf{M}^{\mu\nu} = M^{\mu\nu}$. The space components of (11.73) are nothing else than the familiar commutation relations in nonrelativistic quantum mechanics of the angular-momentum operator $\mathbf{L} = (M^{12},M^{23},M^{31})$ which generates the three-dimensional spatial rotations. The consistency with the commutation relations of this identification of the angular-momentum tensor $\mathbf{M}^{\mu\nu}$ with $M^{\mu\nu}$ is the additional requirement here for invariance under Lorentz transformations. It can be explicitly checked as in the discussion of $\mathbf{P}^\mu$.

For most field theories now generally discussed in physics the lagrangian approach and Noether's theorem can be carried over directly to the quantum domain without difficulty. It is from this that they derive their practical usefulness, as will be illustrated in succeeding chapters.

11.5 Other Formulations

The approach given above uses the classical lagrangian as a "crutch" in order to derive consistent field equations and commutation relations. We emphasize that the physics lies in the field equations and their solutions, the commutation relations, and the properties of the states of the system.

It is possible also to formulate the theory *ab initio* in terms of a quantum action principle; in such theories the lagrangian plays a more central role. This powerful, but more abstract, approach to local field theory has been discussed extensively in the literature, especially by Schwinger.[1]

Conversely, it is possible to formulate the theory without the mention of lagrangians. A general approach from an axiomatic viewpoint has been given by Lehmann, Symanzik, and Zimmermann.[2]

Problems

The wave equation for a massive spin-1 particle is

$$\left[g_{\mu\nu}(\Box + \mu^2) - \frac{\partial}{\partial x^\mu}\frac{\partial}{\partial x^\nu}\right]\varphi^\nu(x) = 0$$

from which follows:

$$\frac{\partial \varphi^\nu}{\partial x^\nu} = 0$$

1. From this equation construct the lagrangian density

$$\mathcal{L} = -\frac{1}{2}\left(\frac{\partial \varphi^\nu}{\partial x^\mu}\right)\left(\frac{\partial \varphi_\nu}{\partial x_\mu}\right) + \frac{\mu^2}{2}\varphi_\nu\varphi^\nu + \frac{1}{2}\left(\frac{\partial \varphi^\nu}{\partial x^\nu}\right)^2$$

2. Construct the hamiltonian density

$$\begin{aligned}\mathcal{H} &= \frac{1}{2}\pi_i{}^2 - \frac{1}{2}(\nabla\varphi_0)^2 + \frac{1}{2}\left(\frac{\partial \varphi^j}{\partial x^i}\right)^2 - \frac{\mu^2}{2}\varphi_0{}^2 + \frac{\mu^2}{2}\varphi_j{}^2 - \frac{1}{2}\pi_0{}^2\\ &= -\frac{1}{2}\pi_\mu\pi^\mu - \frac{1}{2}(\nabla\varphi_\mu)\cdot\nabla\varphi^\mu - \frac{\mu^2}{2}\varphi_\mu\varphi^\mu\end{aligned}$$

where

$$\pi_\mu = \frac{\partial \mathcal{L}}{\partial \dot{\varphi}^\mu}$$

[1] J. Schwinger, *Phys. Rev.*, **91**, 713 (1953), and R. E. Peierls, *Proc. Roy. Soc.* (*London*), **A214**, 143 (1952).

[2] H. Lehmann, K. Symanzik, and W. Zimmermann, *Nuovo Cimento*, **1**, 1425 (1955), **6**, 319 (1957).

and

$$\pi_0 = \nabla \cdot \dot{\boldsymbol{\varphi}} = \nabla_i \dot{\varphi}^i$$
$$\pi_i = \dot{\varphi}^i$$

3. Check that the Hamilton equations of motion

$$\frac{\partial \mathcal{H}}{\partial \pi_\mu} = \dot{\varphi}^\mu \qquad \frac{\partial \mathcal{H}}{\partial \varphi^\mu} = -\dot{\pi}_\mu$$

with the subsidiary condition

$$\pi_0 = \nabla \cdot \boldsymbol{\varphi} = -\dot{\varphi}_0$$

reproduce the original wave equation.

4. Supposing that $\mathcal{L}$ is *not* invariant under some internal symmetry operation such as (11.58), relate the change in $\mathcal{L}$ to the divergence of the associated current.

$$\pi^\mu = -\dot{\phi}^\mu$$

$$-\dot{\pi}_\mu = \frac{\partial \mathcal{H}}{\partial \varphi^\mu} - \partial_i \frac{\partial \mathcal{H}}{\partial (\varphi^\mu{}_{,i})}$$

see Goldstein pg. 362

eqn. (12.4) is time indep't:

Gauss-Stokes: $\int d^4x \, A^\mu{}_{,\mu} = \int d^3S_\mu A^\mu$ (flat space)

let $A^\mu = \varphi \psi^{,\mu}$

$\int d^4x \, (\varphi_{,\mu} \psi^{,\mu} + \varphi \psi^{,\mu}{}_{\mu}) = \int d^3S_\mu \varphi \psi^{,\mu}$

switch $\varphi \leftrightarrow \psi$ and substract

$\int_V d^4x \, (\varphi \Box \psi - \psi \Box \varphi) = \int_{\partial V} d^3S_\mu \, (\varphi \psi^{,\mu} - \psi \varphi^{,\mu})$

Now choose ∂V = two t = constant surfaces, V = 4-space between them.

Use $\Box \psi = -m^2 \psi$, $\Box \varphi = -m^2 \varphi \Rightarrow$ left side $\to 0$

$0 = \int_{t=b} d^3x \, (\varphi \overleftrightarrow{\partial^\mu} \psi) - \int_{t=a} d^3x \, (\varphi \overleftrightarrow{\partial^\mu} \psi)$

↑ from outward normal

$\Rightarrow \int d^3x \, \varphi \overleftrightarrow{\partial^\mu} \psi$ = indep't of time

In fact, it's independent of any space-like hypersurface Σ over which we evalute $\int d^3\Sigma_\mu (\varphi \overleftrightarrow{\partial^\mu} \psi)$

12

The Klein-Gordon Field

12.1 Quantization and Particle Interpretation

A real scalar field $\varphi(x)$ which satisfies a free Klein-Gordon equation

$$(\Box + m^2)\varphi(x) = 0 \tag{12.1}$$

is the simplest of all fields and has already been used for purposes of illustration. The lagrangian density leading to (12.1) is

$$\mathcal{L} = \frac{1}{2}\left(\frac{\partial \varphi}{\partial x^\mu}\frac{\partial \varphi}{\partial x_\mu} - m^2\varphi^2\right) \tag{12.2}$$

and the conjugate momentum is

$$\pi = \frac{\partial \mathcal{L}}{\partial \dot{\varphi}} = \dot{\varphi} \tag{12.3}$$

With the canonical quantization procedure, π and φ become hermitian operators satisfying equal-time commutators

$$\begin{aligned} [\varphi(\mathbf{x},t),\varphi(\mathbf{x}',t)] &= [\pi(\mathbf{x},t),\pi(\mathbf{x}',t)] = 0 \\ [\pi(\mathbf{x},t),\varphi(\mathbf{x}',t)] &= -i\delta^3(\mathbf{x} - \mathbf{x}') \end{aligned} \tag{12.4}$$

The resulting quantum field theory is invariant under displacements and Lorentz transformations of the coordinates, as we verify by direct calculation of the commutators (11.70) and (11.73). The hamiltonian following from (12.2) and (12.3) is

$$\begin{aligned} P^0 &= H = \int d^3x\, \mathcal{H}(\pi,\varphi) \\ \mathcal{H}(\pi,\varphi) &= \pi\dot{\varphi} - \mathcal{L} = \frac{1}{2}\left[\pi(\mathbf{x},t)^2 + |\nabla\varphi(\mathbf{x},t)|^2 + m^2\varphi(\mathbf{x},t)^2\right] \end{aligned} \tag{12.5}$$

and the momentum operator is

$$\mathbf{P} = -\int \pi\nabla\varphi\, d^3x \tag{12.6}$$

Using the commutation relations (12.4), we find

$$i[P^\mu,\varphi(\mathbf{x},t)] = \frac{\partial\varphi(\mathbf{x},t)}{\partial x_\mu}$$

as required. In an analogous way we compute $M^{\mu\nu}$ by (11.57) and confirm the relation (11.73), with $\Sigma_{rs}^{\mu\nu} = 0$ for a scalar field.

In order to discuss further the properties of the quantized Klein-Gordon field, we want to construct a complete set of state vectors Φ from the algebra of the operators assigned by the commutators. It is convenient to do this by forming the eigenvectors of momentum and energy.

To this end we observe that an arbitrary solution of (12.1) may be expanded as a Fourier integral over elementary plane-wave solutions:

$$\varphi(\mathbf{x},t) = \int \frac{d^3k}{\sqrt{(2\pi)^3 2\omega_k}} [a(k)e^{i\mathbf{k}\cdot\mathbf{x}-i\omega_k t} + a^\dagger(k)e^{-i\mathbf{k}\cdot\mathbf{x}+i\omega_k t}]$$
$$\equiv \int d^3k\, [a(k)f_k(x) + a^\dagger(k)f_k^*(x)] \tag{12.7}$$

with

$$\omega_k = +\sqrt{k^2 + m^2} \qquad \text{and} \qquad f_k(x) = \frac{1}{\sqrt{(2\pi)^3 2\omega_k}} e^{-ik\cdot x}$$

Classically for a real field $\varphi(x)$, $a^\dagger(k)$ is the amplitude complex conjugate to $a(k)$. In quantum field theory the amplitudes become operators with $a^\dagger(k)$ hermitian conjugate to $a(k)$. The algebra of $a(k)$ and $a^\dagger(k)$ is determined by rewriting the commutator requirements (12.4) on the φ field in terms of the $a(k)$ coefficients.

Inverting the expansion (12.7) and solving for the coefficients, we find[1]

$$\int f_k^*(\mathbf{x},t)\varphi(\mathbf{x},t)\, d^3x = \frac{1}{2\omega_k} [a(k) + a^\dagger(-k)e^{2i\omega_k t}]$$
$$\int f_k^*(\mathbf{x},t)\dot{\varphi}(\mathbf{x},t)\, d^3x = \frac{-i}{2} [a(k) - a^\dagger(-k)e^{2i\omega_k t}] \tag{12.8}$$

This gives

$$a(k) = \int f_k^*(\mathbf{x},t)[\omega_k\varphi(\mathbf{x},t) + i\dot{\varphi}(\mathbf{x},t)]\, d^3x$$
$$= i\int d^3x \left[f_k^* \frac{\partial\varphi(\mathbf{x},t)}{\partial t} - \left(\frac{\partial f_k^*}{\partial t}\right)\varphi(\mathbf{x},t)\right]$$
$$= i\int d^3x\, f_k^*(\mathbf{x},t)\overleftrightarrow{\partial_0}\varphi(\mathbf{x},t) \tag{12.9}$$

where the notation $\overleftrightarrow{\partial_0}$ is defined by the last line

$$a(t)\overleftrightarrow{\partial_0}b(t) = a(t)\frac{\partial b}{\partial t} - \left(\frac{\partial a}{\partial t}\right)b(t)$$

The right-hand side of (12.9) is independent of time, by Green's theorem, since $f_k(\mathbf{x},t)$ and $\varphi(\mathbf{x},t)$ are both solutions[2] of (12.1). This form is

[1] f_k and f_k^* satisfy the orthogonality conditions of (9.6):

$$\int f_k^*(\mathbf{x},t)\, i\overleftrightarrow{\partial_0} f_{k'}(\mathbf{x},t)\, d^3x = \delta^3(\mathbf{k} - \mathbf{k}')$$
$$\int f_k(\mathbf{x},t)\, i\overleftrightarrow{\partial_0} f_{k'}(\mathbf{x},t)\, d^3x = 0$$

[2] This is true more generally if we superpose plane waves (12.7) to form wave packets in making the expansion of $\varphi(\mathbf{x},t)$ in terms of orthogonal functions. See H. Lehmann, K. Symanzik, and W. Zimmermann, *Nuovo Cimento*, **1**, 1425 (1955).

reminiscent of the inner product encountered in the propagator discussion of the Klein-Gordon equation in Chap. 9.

The commutation relations now follow from (12.4) and (12.9). Since $a(k)$ is time-independent, we may choose a common time to evaluate, with the aid of (12.3) and (12.4), the commutator

$$\begin{aligned}[a(k),a^\dagger(k')] &= \int d^3x\, d^3y\, [f_k^*(\mathbf{x},t)\overleftrightarrow{\partial}_0\varphi(\mathbf{x},t), f_{k'}(\mathbf{y},t)\overleftrightarrow{\partial}_0\varphi(\mathbf{y},t)] \\ &= +i\int d^3x\, f_k^*(\mathbf{x},t)\overleftrightarrow{\partial}_0 f_{k'}(\mathbf{x},t) = \delta^3(\mathbf{k}-\mathbf{k}')\end{aligned}$$

Similarly,

$$\begin{aligned}[a(k),a(k')] &= (i)^2\int d^3x\, d^3y\, [f_k^*(\mathbf{x},t)\overleftrightarrow{\partial}_0\varphi(\mathbf{x},t), f_{k'}^*(\mathbf{y},t)\overleftrightarrow{\partial}_0\varphi(\mathbf{y},t)] \\ &= -i\int d^3x\, f_k^*(\mathbf{x},t)\overleftrightarrow{\partial}_0 f_{k'}^*(\mathbf{x},t) = 0\end{aligned}$$

and

$$[a^\dagger(k),a^\dagger(k)] = 0 \tag{12.10}$$

The total energy and momentum for the free Klein-Gordon field take simple forms in terms of these expansion coefficients. A straightforward calculation using (12.5) to (12.7) gives[1]

$$\begin{aligned}H &= \tfrac{1}{2}\int d^3k\, \omega_k[a^\dagger(k)a(k) + a(k)a^\dagger(k)] \\ \mathbf{P} &= \tfrac{1}{2}\int d^3k\, \mathbf{k}[a^\dagger(k)a(k) + a(k)a^\dagger(k)]\end{aligned} \tag{12.11}$$

At this stage we see that the hamiltonian is a continuous sum of terms

$$H_k = \tfrac{1}{2}\omega_k[a^\dagger(k)a(k) + a(k)a^\dagger(k)]$$

[1] In the interest of formal covariance, we may rewrite the plane-wave expansions in an invariant notation, using the identity

$$\int \frac{d^3k}{2\omega_k} = \int d^4k\, \delta(k^2 - m^2)\theta(k_0)$$

With the notation $A(k) = \sqrt{2\omega_k}\, a(k)$, we find

$$\varphi(\mathbf{x},t) = \frac{1}{(2\pi)^{3/2}}\int d^4k\, \delta(k^2 - m^2)\theta(k_0)[A(k)e^{-ik\cdot x} + A^\dagger(k)e^{ik\cdot x}]$$

and

$$A(k) = \frac{i}{(2\pi)^{3/2}}\int d^3\sigma\, \eta_\mu e^{ik\cdot x}\frac{\overleftrightarrow{\partial}}{\partial x_\mu}\varphi(x)$$

with σ a flat space-like surface and η_μ its normal. Likewise,

$$P^\mu = \tfrac{1}{2}\int d^4k\, \delta(k^2 - m^2)\theta(k_0)k^\mu[A^\dagger(k)A(k) + A(k)A^\dagger(k)]$$

The $A(k)$, like φ, are Lorentz scalars.

This is what Schweber uses: $[A_k, A_{k'}^\dagger] = 2\omega\,\delta^3(\vec{k}-\vec{k}')$

each of which has the form of a hamiltonian for a simple harmonic oscillator of frequency ω_k. Indeed, the $a^\dagger(k)$ and $a(k)$ are simply the raising and lowering operators discussed in the example of a particle oscillator in Chap. 11 and satisfy the same commutation relations, aside from the normalization convention. In order to clarify the procedure with regard to normalization and to show the complete equivalence with our earlier quantum description of a simple one-dimensional oscillating particle, we again revert to a discrete notation. Dividing up k space into cells of volume ΔV_k, we write

$$\int d^3k \rightarrow \sum_k \Delta V_k \qquad \text{and} \qquad \delta^3(\mathbf{k} - \mathbf{k}') \rightarrow \frac{\delta_{kk'}}{\Delta V_k} \tag{12.12}$$

H then becomes a sum of oscillator hamiltonians H_k for each cell in momentum space:

$$H = \sum_k H_k = \sum_k \tfrac{1}{2}\omega_k(a_k^\dagger a_k + a_k a_k^\dagger) \qquad a_k = \sqrt{\Delta V_k}\, a(k) \tag{12.13}$$

with

$$[a_k, a_{k'}^\dagger] = \delta_{kk'} \qquad [a_k, a_{k'}] = [a_k^\dagger, a_{k'}^\dagger] = 0 \tag{12.14}$$

This analogy with the harmonic oscillator should not be surprising, since the classical Klein-Gordon wave field can be described by its normal-mode expansion. The normal modes and coordinates are just simple harmonic oscillators. What we have done here is simply to quantize each of these oscillators, $a(k)$.

Upon quantization we expect that the classical field energy will become a sum over discrete oscillator energies. To determine the energy eigenvalues and construct the energy eigenfunctions, consider each oscillator hamiltonian H_k individually. Since H is a sum of mutually commuting terms H_k for each wave number $\mathbf{k}$ and frequency $\omega_k = \sqrt{\mathbf{k}^2 + m^2}$, the energy eigenfunctions will be products of eigenfunctions Φ_k of each H_k. General state vectors Φ can be built of superpositions of such products over all $\mathbf{k}$ values, according to the completeness postulate of Chap. 1 transcribed to the field hamiltonian.

The solution to the oscillator eigenvalue problem for each $\mathbf{k}$ may be characterized by an integer $n_k = 0, 1, 2, \ldots$ in terms of which the energy eigenfunctions and eigenvalues are (in the discrete notation)

$$H_k\Phi_k(n_k) = \omega_k(n_k + \tfrac{1}{2})\Phi_k(n_k) \tag{12.15}$$

$$\Phi_k(n_k) = \frac{1}{\sqrt{n_k!}}\,(a_k^\dagger)^{n_k}\Phi_k(0) \tag{12.16}$$

$\Phi_k(0)$ is the ground state, defined here by

$$a_k\Phi_k(0) = 0 \tag{12.17}$$

and the states are normalized to

$$(\Phi_k(n_k), \Phi_k(n_k')) = \delta_{n_k, n_k'}$$

The momentum operator may be similarly decomposed

$$\mathbf{P} = \sum_k \mathbf{P}_k = \sum_k \tfrac{1}{2}\mathbf{k}(a_k^\dagger a_k + a_k a_k^\dagger) \tag{12.18}$$

with $\quad \mathbf{P}_k\Phi_k(n_k) = \mathbf{k}(n_k + \tfrac{1}{2})\Phi_k(n_k) \qquad n_k = 0, 1, 2, \ldots$

The energy-momentum eigenfunctions Φ are products of the Φ_k for each momentum cell, and they are characterized by integers n_k for each $\mathbf{k}$:

$$\begin{aligned} \Phi(n_{k_1} \cdots n_{k_\alpha} \cdots) &= \prod_k \Phi_k(n_k) \\ P^\mu\Phi(\cdots n_{k_\alpha} \cdots) &= \sum_k k^\mu(n_k + \tfrac{1}{2})\Phi(\cdots n_{k_\alpha} \cdots) \end{aligned} \tag{12.19}$$

The ground state, that is, the state of lowest energy, is that for which all $n_{k_\alpha} = 0$:

$$\Phi_0 = \prod_k \Phi_k(0) \tag{12.20}$$

None of the normal modes of the field are excited in this state, which physically represents the vacuum.

The energy of the vacuum is

$$E = \sum_k \tfrac{1}{2}\omega_k \tag{12.21}$$

and is badly divergent, being the sum of zero-point energies for an infinite number of oscillators, one for each normal mode, or degree of freedom, of the field. This is the first of a number of divergences we shall encounter in field theory. It is the easiest one to remove, simply by subtracting an infinite constant from H to cancel $\sum_k \tfrac{1}{2}\omega_k$. This can be done because absolute energies are not measured observables; only energy differences have physical meaning. According to (12.17) and (12.20) this infinite constant is just the vacuum expectation value of the energy and is automatically removed by rewriting the energy-momentum operator as

$$P'_\mu = P_\mu - (\Phi_0, P_\mu \Phi_0) = \sum_k k_\mu a_k^\dagger a_k \tag{12.22}$$

or in continuum language

$$P'_\mu = \int d^3k\, k_\mu a^\dagger(k) a(k) \tag{12.23}$$

At any time before the commutator conditions are imposed on the fields, P_μ and P'_μ are identical, since classically the normal-mode amplitudes commute and there is no zero-point energy.

Replacing P_μ by P'_μ in the quantum theory is identical to rewriting the operator factors in L and P_μ so that the positive-frequency parts of φ

$$\varphi^{(+)}(x) = \int d^3k\, a(k) f_k(x) \tag{12.24a}$$

always stand to the right of the negative-frequency parts

$$\varphi^{(-)}(x) = \int d^3k\, a^\dagger(k) f_k^*(x) \tag{12.24b}$$

This ordering of factors, known as normal ordering, is denoted by

$$:\varphi\varphi: = \varphi^{(-)}\varphi^{(-)} + 2\varphi^{(-)}\varphi^{(+)} + \varphi^{(+)}\varphi^{(+)} \tag{12.25}$$

It is clear from (12.17) and its hermitian conjugate that the vacuum expectation value of an operator vanishes when its factors are in normal order. The only effect of normal ordering here is to remove the infinite zero-point energy from the theory and to define the zero of energy as the energy of the vacuum state Φ_0.

From (12.19) and (12.22) we find for the eigenvalues of P'_μ:

$$P'_\mu \Phi(\cdots n_{k\alpha} \cdots) = \sum_k n_k k_\mu \Phi(\cdots n_{k\alpha} \cdots) \qquad n_k = 0, 1, 2, \ldots \tag{12.26}$$

The different eigenstates for each normal mode k carry four-momenta corresponding to n_k quanta, each with four-momentum k^μ and mass m according to the Einstein relation $k_\mu k^\mu = m^2$. Here we see the emergence of a particle picture from the canonical quantization procedure. The integer n_k is called the occupation number of the kth momentum state, and specification of the numbers of quanta $n_{k\alpha}$ gives a complete description of the eigenstate $\Phi(\cdots n_{k\alpha} \cdots)$.

It is convenient to introduce a number operator

$$N_k = a_k^\dagger a_k \tag{12.27}$$

with integer eigenvalues

$$N_k\Phi(\cdots n_k \cdots) = n_k\Phi(\cdots n_k \cdots) \qquad n_k = 0, 1, 2, \ldots \tag{12.28}$$

and in terms of which the energy-momentum operator is[1]

$$P^\mu = \sum_k k^\mu N_k \tag{12.29}$$

From the commutators (12.14), we verify that

$$[N_k, a_{k'}^\dagger] = \delta_{kk'}a_{k'}^\dagger \qquad \text{and} \qquad [N_k, a_{k'}] = -\delta_{kk'}a_{k'} \tag{12.30}$$

Combined with (12.29), this shows that $a_k^\dagger$ is a creation operator for a quantum of momentum k^μ and that it produces a state with $n_k + 1$ quanta of this momentum from a state with n_k such quanta:

$$\begin{aligned} P_\mu a_k^\dagger \Phi(\cdots n_k \cdots) &= a_k^\dagger[P_\mu + k_\mu]\Phi(\cdots n_k \cdots) \\ &= \Big(\sum_{k'} n_{k'}k'_\mu + k_\mu\Big)a_k^\dagger\,\Phi(\cdots n_k \cdots) \end{aligned}$$

Similarly, a_k destroys a quantum with k^μ and, in particular, operating on a state containing zero such quanta destroys the state, $a_k\Phi_k(0) = 0$, according to (12.17).

The only nonvanishing matrix elements of the operators a_k and $a_k^\dagger$ connect states with occupation numbers $n_k' = n_k \pm 1$, as we recall from the oscillator example in Chap. 11:

$$\begin{aligned} (\Phi_k(n_k'), a_k\Phi_k(n_k)) \equiv \langle n_k'|a_k|n_k\rangle &= \sqrt{n_k}\;\delta_{n_{k}',n_k-1} \\ \langle n_k'|a_k^\dagger|n_k\rangle &= \sqrt{n_k+1}\;\delta_{n_{k}',n_k+1} \end{aligned} \tag{12.31}$$

12.2 Symmetry of the States

The canonical quantization procedure applied to the classical free Klein-Gordon field has yielded a many-particle description in terms of numbers of quanta. For the free field, N_k and H commute and the number of quanta is a constant of the motion. The interesting physical problems are encountered when we add interaction terms changing the occupation numbers n_k. In the present free-field discussion there remains the problem of showing that the quanta obey symmetric, or Bose-Einstein, statistics.

[1] Henceforth, we drop the primes when normal-ordering operators, for example, P_μ.

An arbitrary state is constructed by superposing the

$$\Phi(\cdots n_{k_\alpha} \cdots) = \prod_k \frac{1}{\sqrt{n_k!}} (a_k^\dagger)^{n_k}\Phi_k(0) \tag{12.32}$$

for different occupation numbers n_k. The state (12.32) is described completely by the numbers of quanta n_k for each k. The individual quanta are indistinguishable, since all $a_k^\dagger$ mutually commute, according to (12.14), and the ordering of operators $a_k^\dagger$ is immaterial. This is reflected in the symmetry of the expansion coefficients for the different states. For an arbitrary state we write, reverting back to continuum normalization,[1]

$$\Phi = \left[c_0 + \sum_{n=1}^{\infty} \frac{1}{\sqrt{n!}} \int d^3k_1 \cdots d^3k_n\, c_n(k_1, \ldots, k_n) \times a^\dagger(k_1)a^\dagger(k_2) \cdots a^\dagger(k_n)\right] \Phi_0 \tag{12.33}$$

The factors $1/\sqrt{n!}$ are inserted for convenience to give a simple form to the normalization condition on the c_n:

$$1 = (\Phi,\Phi) = |c_0|^2 + \sum_{n=1}^{\infty} \int d^3k_1 \cdots d^3k_n\, |c_n(k_1,k_2, \ldots ,k_n)|^2 \tag{12.34}$$

The c_n's describe the momentum distribution of that component of the state containing n quanta. They are the momentum-space wave functions for an assembly of n identical particles with a given set of k_α. Owing to the commutativity of the $a^\dagger(k)$ among each other in (12.33), these wave functions are symmetric functions of their arguments

$$c(\cdots k_i \cdots k_j \cdots) = +c(\cdots k_j \cdots k_i \cdots) \tag{12.35}$$

As noted above, it is only the *numbers* of quanta with the various k values that characterize a state. The quanta are indistinguishable and the probability of quantum a having momentum k_i and quantum b having k_j is the same as that for a and b interchanged:

$$|c(\cdots k_i \cdots k_j \cdots)|^2 = |c(\cdots k_j \cdots k_i \cdots)|^2 \tag{12.36}$$

[1] We assume here that the probability of two particles being in precisely the same state k is infinitesimally small, so that in the continuum limit $n_k \to 1$ or 0. In highly degenerate systems, such as the ground state of a free Bose gas, where all quanta are in the state $k = 0$, it is easiest to remain in the discrete normalization. For relativistic fields, the states of interest are generally scattering states, where this question does not arise.

The symmetry condition (12.35), which is a consequence of the commutation algebra of the $a^\dagger(k)$, shows that the quanta emerging from the canonical quantization prescription obey symmetric, or Bose-Einstein, statistics.

12.3 Measurability of the Field and Microscopic Causality

Classically, the field $\varphi(x)$ is an observable and its strength at a point x can be measured. With the reservations discussed in the introduction to field quantization in Chap. 11, we have introduced the concept of a local field operator $\varphi(x)$ defined at a point x in the quantum domain.

In the quantum theory, in contrast with the classical theory, there are limitations on our ability to make precise measurements of the field strengths, owing to the commutation relations. For example, a precise measurement of field strengths at two different space-time points x and y is possible only if the commutator $[\varphi(x),\varphi(y)]$ vanishes.

Having constructed the explicit solutions of the free Klein-Gordon field in (12.7), we can evaluate the field commutator with the aid of (12.10):

write as two integrals, let $\vec{k} \to -\vec{k}$ in second integral, then can factor out ω.

$$[\varphi(x),\varphi(y)] = \int \frac{d^3k\, d^3k'}{(2\pi)^3 \sqrt{2\omega_k \cdot 2\omega'_k}} \left([a(k),a^\dagger(k')]e^{-ik\cdot x + ik'\cdot y}\right.$$
$$\left. + [a^\dagger(k),a(k')]e^{ik\cdot x - ik'\cdot y}\right)$$
$$= \int \frac{d^3k}{(2\pi)^3 2\omega_k} \left(e^{-ik\cdot(x-y)} - e^{ik\cdot(x-y)}\right)$$
$$= -\frac{i}{(2\pi)^3} \int \frac{d^3k}{\omega_k} e^{i\mathbf{k}\cdot(\mathbf{x}-\mathbf{y})} \sin \omega_k(x_0 - y_0)$$
$$\equiv i\Delta(x-y) \tag{12.37}$$

$\Delta(x-y)$ is one of a menagerie of invariant singular functions which are collected and discussed in Appendix C. Its Lorentz invariance is apparent in (12.37), where an invariant exponential is integrated over the invariant volume element

$$\int \frac{d^3k}{2\omega_k} = \int d^4k\ \delta(k^2 - m^2)\theta(k_0)$$

Introducing the odd function

$$\epsilon(k_0) = \begin{cases} +1 & k_0 > 0 \\ -1 & k_0 < 0 \end{cases} \tag{12.38}$$

which is invariant for time-like vectors $k^2 > 0$, we can put Δ in a more compact form

$$\Delta(x - y) = -i \int \frac{d^4k}{(2\pi)^3}\, \delta(k^2 - m^2)\epsilon(k_0)e^{-ik\cdot(x-y)} \qquad (12.39)$$

As required by its definition in terms of the commutator on the left-hand side of (12.37), Δ is a solution of the free Klein-Gordon equation and is an odd function of its argument

$$(\square_x + m^2)\Delta(x - y) = 0 \qquad \Delta(x - y) = -\Delta(y - x) \quad (12.40)$$

It follows from (12.37), as well as (12.4), that the equal-time commutator of two field amplitudes vanishes: $\Delta(\mathbf{x} - \mathbf{y},\ 0) = 0$. From Lorentz invariance we know then that

$$\Delta(x - y) = 0 \qquad \text{for all } (x - y)^2 < 0 \qquad (12.41)$$

and two fields separated by a space-like interval commute with each other. Therefore, at two points which cannot be connected by a light signal or by any physical disturbance, that is, $(x - y)^2 < 0$, the field strengths φ, if interpreted as physical observables, can be measured precisely and independently of each other. The time derivative of Δ is singular at the origin

$$\left.\frac{\partial\Delta(x - y)}{\partial x_0}\right|_{x_0 = y_0} = -\delta^3(\mathbf{x} - \mathbf{y}) \qquad (12.42)$$

and (12.42) combined with (12.37) reproduces the canonical commutators, (12.4).

The condition of vanishing of the commutators for all space-like intervals, (12.41), no matter how small, is referred to as the condition of *microscopic causality*. In order to associate any physical content with this mathematical result, we must assume that it makes sense to attach physical meaning to the measurement of a field strength *at a point*, a concept already criticized in earlier paragraphs.[1]

12.4 Vacuum Fluctuations

We have already noted that the field quantization is essentially the quantization of an infinite assemblage of harmonic oscillators; the

[1] For quantum electrodynamics N. Bohr and L. Rosenfeld, *Kgl. Danske Videnskab. Selskab. Mat.-Fys. Medd.*, **12**, 8 (1933), *Phys. Rev.*, **78**, 794 (1950), have made a detailed analysis of the physical meaning of the commutation relations in terms of physical measurement processes.

vacuum energy was seen to be analogous to the zero-point energy of these oscillators. In an energy eigenstate of an oscillator the coordinate q is not sharp, that is,

$$(\Psi_n, q^2 \Psi_n) > (\Psi_n, q\Psi_n)^2 = 0 \tag{12.43}$$

This is also true in field theory; the coordinates $\varphi(x)$ fluctuate. For example, in the ground state

$$\Delta_+(x,y) = \langle 0|\varphi(x)\varphi(y)|0\rangle \neq 0 \tag{12.44}$$

although

$$\langle 0|\varphi(x)|0\rangle = 0 \tag{12.45}$$

We may evaluate $\Delta_+(x,y)$ by using (12.7), (12.10), and (12.17); we find

$$\begin{aligned}\Delta_+(x,y) &= \int \frac{d^3k\, d^3k'}{(2\pi)^3 \sqrt{2\omega_k 2\omega_{k'}}}\, e^{-ik\cdot x} e^{ik'\cdot y} \langle 0|a(k)a^\dagger(k')|0\rangle \\ &= \int_{(k_0 = +\omega_k)} \frac{d^3k}{(2\pi)^3 2\omega_k}\, e^{-ik\cdot(x-y)} = \Delta_+(x - y)\end{aligned}$$

As $y \to x$, this approaches a quadratically diverging expression for the vacuum fluctuations

$$\langle 0|\varphi^2(x)|0\rangle = \Delta_+(0) = \int \frac{d^3k}{(2\pi)^3 2\omega_k} \tag{12.46}$$

Unlike the zero-point energy encountered earlier, this divergence cannot be completely eliminated by a simple subtraction. In fact, we have already seen that the vacuum fluctuations lead to observable, finite physical effects in the Lamb shift, which was discussed from this point of view in Chap. 4.

We may make this troublesome result—the divergence of (12.46)—less unpleasant with the observation that one cannot in fact measure the square of a field amplitude *at a point.* In order to probe a single isolated point of space-time, one needs infinitely large frequencies and infinitesimally short wavelengths—and these are not to be achieved at less than infinite energies. Also, in practical calculations the fact that (12.46) diverges will cause no serious difficulties. However, it is disturbing to find our formalism full of expressions such as $\mathcal{L}$ and P^μ which, like (12.46), involve products of field operators evaluated at the same space-time point. It should be kept in mind that only products of fields averaged over finite regions of space-time may exist mathematically and have physically observable meaning. We interpret results like (12.46) as indicative of the limitations of a continuous-field description—it is an idealization which provides an adequate description of the physical world only in the sense of the correspondence principle for large space-time intervals. It remains for experiment to show how small are the space-time lengths before quantitative revisions in the theories are required.

12.5 The Charged Scalar Field[1]

Having discussed the quantum theory of one real free Klein-Gordon field, we may adapt the results to describe the charged particle discussed in Chap. 9. Such a particle was described in terms of a complex wave function

$$\varphi(x) = \frac{1}{\sqrt{2}}[\varphi_1(x) + i\varphi_2(x)]$$

with φ_1 and φ_2 real. We first consider then two identical noninteracting real fields of this type. The field equations

$$(\Box_x + m^2)\varphi_1(x) = 0 \qquad (\Box_x + m^2)\varphi_2(x) = 0 \tag{12.47}$$

follow from the lagrangian density

$$\mathfrak{L} = \tfrac{1}{2}:\left(\frac{\partial\varphi_1}{\partial x_\mu}\frac{\partial\varphi_1}{\partial x^\mu} - m^2\varphi_1^2 + \frac{\partial\varphi_2}{\partial x_\mu}\frac{\partial\varphi_2}{\partial x^\mu} - m^2\varphi_2^2\right): \tag{12.48}$$

where $:\cdots:$ denotes the normal product as defined by (12.24) and (12.25). The canonical momenta are found as before

$$\pi_1 = \dot\varphi_1 \qquad \pi_2 = \dot\varphi_2 \tag{12.49}$$

and the canonical commutation relations are

$$[\varphi_k(x),\varphi_j(y)] = i\delta_{kj}\Delta(x - y) \tag{12.50}$$

Since the hamiltonian is the sum of two terms of the form in (12.23), the energy eigenstates are a direct product of the independent eigenstates of the hamiltonians for quanta of types 1 and 2. The numbers of particles of types 1 and 2 are separately conserved in the absence of interaction terms, and it is again convenient to label the states by the eigenvalues of the number operators:

$$N_1(k) = a_1^\dagger(k)a_1(k) \qquad N_2(k) = a_2^\dagger(k)a_2(k) \tag{12.51}$$

As in (12.30) and (12.31), $a_i^\dagger(k)$ and $a_i(k)$ create and destroy quanta of type i and momentum $\mathbf{k}$, respectively, and therefore connect states differing by ± 1 for these occupation numbers.

[1] W. Pauli and V. F. Weisskopf, *Helv. Phys. Acta*, **7**, 709 (1934).

All of the remarks of the first paragraph apply for arbitrary masses m_1 and m_2 appearing in the field equations (12.47). As a special consequence of the identity of the two masses $m_1 = m_2 = m$, we may replace these two equations by one wave equation for a complex field

$$\varphi = \frac{1}{\sqrt{2}}(\varphi_1 + i\varphi_2) \qquad \varphi^* = \frac{1}{\sqrt{2}}(\varphi_1 - i\varphi_2) \tag{12.52}$$

φ and φ^* satisfy the Klein-Gordon equation

$$(\square + m^2)\varphi = 0 \qquad (\square + m^2)\varphi^* = 0 \tag{12.53}$$

and in terms of the complex coordinates φ and φ^*, $\mathfrak{L}$ becomes

$$\mathfrak{L} = :\frac{\partial\varphi^*}{\partial x_\mu}\frac{\partial\varphi}{\partial x^\mu} - m^2\varphi^*\varphi: \tag{12.54}$$

The canonical momenta for these coordinates are

$$\pi = \frac{\partial\mathfrak{L}}{\partial\dot\varphi} = \dot\varphi^* = \frac{\dot\varphi_1 - i\dot\varphi_2}{\sqrt{2}} \qquad \text{and} \qquad \pi^* = \frac{\partial\mathfrak{L}}{\partial\dot\varphi^*} = \dot\varphi = \frac{\dot\varphi_1 + i\dot\varphi_2}{\sqrt{2}}$$

The hamiltonian density is then

$$\mathfrak{H} = \pi\dot\varphi + \pi^*\dot\varphi^* - \mathfrak{L} = \pi^*\pi + (\nabla\varphi^*)\cdot(\nabla\varphi) + m^2\varphi^*\varphi \tag{12.55}$$

and the commutation relations are

$$[\varphi(x),\varphi(y)] = 0 = [\varphi^*(x),\varphi^*(y)] \qquad [\varphi(x),\varphi^*(y)] = i\Delta(x-y) \tag{12.56}$$

They reduce at equal times to the nonvanishing canonical commutators

$$[\pi(\mathbf{x},t),\varphi(\mathbf{x}',t)] = [\pi^*(\mathbf{x},t),\varphi^*(\mathbf{x}',t)] = -i\delta^3(\mathbf{x}-\mathbf{x}')$$

Fourier-transforming the solutions to k space, we write, following (12.7),

$$\begin{aligned}\varphi(x) &= \int \frac{d^3k}{\sqrt{(2\pi)^3 2\omega_k}}[a_+(k)e^{-ik\cdot x} + a_-^\dagger(k)e^{ik\cdot x}] \\ \varphi^*(x) &= \int \frac{d^3k}{\sqrt{(2\pi)^3 2\omega_k}}[a_+^\dagger(k)e^{ik\cdot x} + a_-(k)e^{-ik\cdot x}]\end{aligned} \tag{12.57}$$

with

$$\begin{aligned}a_+(k) &= \frac{1}{\sqrt{2}}[a_1(k) + ia_2(k)] \qquad a_+^\dagger(k) = \frac{1}{\sqrt{2}}[a_1^\dagger(k) - ia_2^\dagger(k)] \\ a_-(k) &= \frac{1}{\sqrt{2}}[a_1(k) - ia_2(k)] \qquad a_-^\dagger(k) = \frac{1}{\sqrt{2}}[a_1^\dagger(k) + ia_2^\dagger(k)]\end{aligned} \tag{12.58}$$

Equations (12.7) and (12.57) differ in that $\varphi(x)$ is a complex field and hence upon quantization becomes a non-hermitian operator; by (12.58), $a_-^\dagger(k) \neq a_+^\dagger(k)$.

The commutator relations for the $a_\pm(k)$ are readily constructed in parallel with (12.10):

$$\begin{aligned}
[a_+(k),a_+^\dagger(k')] &= [a_-(k),a_-^\dagger(k')] = \delta^3(\mathbf{k} - \mathbf{k}') \\
[a_+(k),a_-^\dagger(k')] &= [a_-(k),a_+^\dagger(k')] = 0 \\
[a_\pm(k),a_\pm(k')] &= [a_\pm^\dagger(k),a_\pm^\dagger(k')] = 0
\end{aligned} \tag{12.59}$$

Evidently the $a_\pm(k)$ and the $a_1(k)$, $a_2(k)$ satisfy the identical algebra, and the number operators formed from them have the same form and the same integer eigenvalues. Returning to the convention of a discrete normalization to define the number operators for the + and − quanta, we write

$$N_k^+ = a_{+,k}^\dagger a_{+,k} \qquad N_k^- = a_{-,k}^\dagger a_{-,k} \tag{12.60}$$

in terms of which

$$P_\mu = \sum_k k_\mu(N_k^+ + N_k^-) \tag{12.61}$$

There is a complete parallel here with the earlier discussion of one field; for example,

$$\begin{aligned}
N_k^+[a_{+,k}\Phi(\cdots n_k^+ \cdots, \cdots n_k^- \cdots)] \\
&= a_{+,k}(N_k^+ - 1)\Phi(\cdots n_k^+ \cdots, \cdots n_k^- \cdots) \\
&= (n_k^+ - 1)[a_{+,k}\Phi(\cdots n_k^+ \cdots, \cdots n_k^- \cdots)]
\end{aligned}$$

and the state of lowest energy, the vacuum, contains no quanta of either type, so that

$$a_{\pm,k}\Phi_0 = 0 \tag{12.62}$$

The operators $a_{\pm,k}$ are destruction operators for the + and − quanta of momentum k, respectively, and the $a_{\pm,k}^\dagger$ are the corresponding creation operators. In the normal-ordered form the destruction operators stand to the right of the creation operators as in (12.61).

At this point it clearly does not matter at all whether we describe the fields in terms of their hermitian amplitudes φ_1 and φ_2 or in terms of their complex ones, φ, φ^*. The states may be labeled equally well by numbers of quanta of types 1 and 2, or + and −.

The wave equation (12.53), for a complex field φ, reminds us that we were able to identify a conserved current

$$j^\mu = i(\varphi^*\nabla^\mu\varphi - \varphi\nabla^\mu\varphi^*)$$

with
$$\frac{\partial j^\mu}{\partial x^\mu} = 0$$

and
$$Q = i\int d^3x\,(\varphi^*\dot{\varphi} - \varphi\dot{\varphi}^*) = \text{const} \tag{12.63}$$

when discussing the Klein-Gordon equation in Chap. 9. We may readily confirm that Q remains a constant of the motion in the present quantum theory following from (12.54) by expanding (12.63) in k space and showing that it commutes with H. We find

$$Q = \int d^3k\,[a_+^\dagger(k)a_+(k) - a_-^\dagger(k)a_-(k)]$$

or in the discrete notation

$$Q = \sum_k (N_k^+ - N_k^-) \tag{12.64}$$

and by (12.59), $[Q,P_\mu] = 0$.

According to (12.64) the + and − quanta each carry +1 and −1 unit of charge Q, respectively. Thus $[P_\mu,a_+^\dagger(k)] = +k_\mu a_+^\dagger(k)$ and $[Q,a_+^\dagger(k)] = +a_+^\dagger(k)$, and $a_+^\dagger(k)$ is an operator which increases the energy by k^μ and the charge by +1; that is, it is a creation operator for a quantum of four-momentum k^μ and charge +1. Similarly, $a_+(k)$ is an annihilation operator for such quanta and $a_-^\dagger(k)$ and $a_-(k)$ are the creation and annihilation operators, respectively, for quanta with momentum k^μ and charge −1.

The quanta of charge +1 and −1 appear symmetrically in the theory according to (12.59), (12.61), and (12.64). In order to attach a physical significance to the charge Q, we must introduce couplings which distinguish between different signs and magnitudes of charge. In discussing the Klein-Gordon equation in Chap. 9, the current j^μ was coupled to the electromagnetic field and Q was identified as the electric charge. More generally, we identify the quanta with positive eigenvalues of Q as the *particles* and those with negative eigenvalues as the *antiparticles*. The charge symmetry of the quantum field theory is then equivalent to a statement of the symmetry of the theory between particle and antiparticle interchange. The complex field amplitudes provide a convenient basis for constructing the charge eigenstates. These may be the π^+ and the π^- which are created from the vacuum by the $a_+^\dagger(k)$ and $a_-^\dagger(k)$, respectively. Also, we may use this theory to describe electrically neutral particles of zero spin which differ by "strangeness" charge, such as the K^0 and $\bar{K}^0$.

12.6 The Feynman Propagator

In the propagator approach to a theory of the charged Klein-Gordon particle in the companion volume, we were led to the Feynman Green's function by the physical boundary condition that only positive-frequency solutions propagate forward in time out of an interaction. In order to see what plays the role of the Feynman propagator in the quantized field theory version of the charged Klein-Gordon particle, we consider in this formalism the space-time development of a state containing one quantum. To form a one-particle state (unnormalized) of charge $+1$, we operate on the vacuum with $\varphi^*(\mathbf{x},t)$:

$$\Psi_+(\mathbf{x},t) = \varphi^*(\mathbf{x},t)\Phi_0 \equiv \varphi^*(\mathbf{x},t)|0\rangle \tag{12.65}$$

Only the creation, or negative-frequency, part of $\varphi^*(x)$ survives in (12.65) according to (12.62), and therefore we may write

$$\Psi_+(\mathbf{x},t) = \varphi^{*(-)}(\mathbf{x},t)|0\rangle \tag{12.66}$$

where

$$\begin{aligned} \varphi^{*(-)}(\mathbf{x},t) &= \int \frac{d^3k}{\sqrt{(2\pi)^3 2\omega_k}}\, a_+^\dagger(k) e^{ik\cdot x} \\ \varphi^{(-)}(\mathbf{x},t) &= \int \frac{d^3k}{\sqrt{(2\pi)^3 2\omega_k}}\, a_-^\dagger(k) e^{ik\cdot x} \end{aligned} \tag{12.67}$$

denote the creation, or negative-frequency, parts and $\varphi^{*(+)}$ and $\varphi^{(+)}$ the corresponding positive-frequency parts of the fields (12.57).

The amplitude for the state (12.65) to propagate forward in time to $(\mathbf{x}',t')$ for $t' > t$ is given by the projection

$$\begin{aligned} \theta(t'-t)\langle\Psi_+(\mathbf{x}',t')|\Psi_+(\mathbf{x},t)\rangle &= \langle 0|\varphi(\mathbf{x}',t')\varphi^*(\mathbf{x},t)|0\rangle\theta(t'-t) \\ &= \langle 0|\varphi^{(+)}(\mathbf{x}',t')\varphi^{*(-)}(\mathbf{x},t)|0\rangle\theta(t'-t) \end{aligned} \tag{12.68}$$

Equation (12.68) is the matrix element for the creation of a quantum of charge $+1$ at $(\mathbf{x},t)$ and its reabsorption into the vacuum at $\mathbf{x}'$ and at a later time $t' > t$. Another way of increasing the charge by $+1$ unit at $(\mathbf{x},t)$ and of lowering it by -1 at $(\mathbf{x}',t')$ is to create a quantum of charge -1 at $(\mathbf{x}',t')$ and to propagate it to $\mathbf{x}$, where it is reabsorbed into the vacuum at a time $t > t'$. The amplitude for this is given by

$$\begin{aligned} \theta(t-t')\langle\Psi_-(\mathbf{x},t)|\Psi_-(\mathbf{x}',t')\rangle &= \langle 0|\varphi^*(\mathbf{x},t)\varphi(\mathbf{x}',t')|0\rangle\theta(t-t') \\ &= \langle 0|\varphi^{*(+)}(\mathbf{x},t)\varphi^{(-)}(\mathbf{x}',t')|0\rangle\theta(t-t') \end{aligned} \tag{12.69}$$

The Feynman propagator is formed by adding together the amplitudes (12.68) and (12.69):

$$i\Delta_F(x'-x) = \langle 0|\varphi(x')\varphi^*(x)|0\rangle\theta(t'-t) + \langle 0|\varphi^*(x)\varphi(x')|0\rangle\theta(t-t') \tag{12.70}$$

states $\varphi|0\rangle$ and $\varphi^*|0\rangle$ are not localized at $\vec{x},t$. see Schweber. [If they were, causality would be violated, contrary to 12.37, 41.]

Inserting the expansions (12.57), we verify the identity of (12.70) with the expression for the Feynman propagator (9.10) and (9.11) used earlier in the propagator approach in Chap. 9:

$$i\Delta_F(x'-x) = \int \frac{d^3k}{2\omega_k(2\pi)^3}\,[\theta(t'-t)e^{-ik\cdot(x'-x)} + \theta(t-t')e^{ik\cdot(x'-x)}]$$
$$= i\int \frac{d^4k}{(2\pi)^4}\frac{1}{k^2-m^2+i\epsilon}\,e^{-ik\cdot(x'-x)} \tag{12.71}$$
$$(\Box_{x'}+m^2)\Delta_F(x'-x) = -\delta^4(x'-x)$$

In this form the Lorentz invariance of the Feynman propagator is displayed explicitly. Because the field operators commute for space-like intervals according to (12.37) and (12.41), their products can be time-ordered in a Lorentz-invariant way as in (12.70). As a convenient shorthand for this time-ordering operation, we introduce a T operator with the definition

$$T(a(x)b(x')) = a(x)b(x')\theta(t-t') + b(x')a(x)\theta(t'-t) \tag{12.72}$$

The T operator carries the instruction that the field operators at the earliest times stand to the right and may be generalized to apply to a product of any numbers of operators. The Feynman propagator is then written

$$i\Delta_F(x'-x) = \langle 0|T(\varphi(x')\varphi^*(x))|0\rangle \tag{12.73}$$

or equivalently in terms of hermitian fields

$$i\delta_{ij}\Delta_F(x'-x) = \langle 0|T(\varphi_i(x')\varphi_j(x))|0\rangle \tag{12.74}$$

Just as in the one-particle theories, the Feynman propagator plays a central role in calculations of transition amplitudes in quantum field theory. The propagation of a particle from $\mathbf{x}$ to $\mathbf{x}'$ when $t' > t$, and the antiparticle from $\mathbf{x}'$ to $\mathbf{x}$ when $t > t'$, is described by $\Delta_F(x',x)$. This is the same physical interpretation discussed in detail for the Feynman propagator and boundary conditions in Chaps. 6 and 9 for the particle and antiparticle solutions.

Problems

1. Confirm that $-(i/2)\epsilon^{\mu\nu}[M_{\mu\nu},\varphi] = \delta\varphi$ for a scalar field.
2. Compute

$$\langle 0|\bar{\varphi}^2|0\rangle$$

with

$$\bar{\varphi} = \frac{1}{V}\int_V d^3x\,\varphi(x)$$

and V a spherical region of radius R.

13

Second Quantization of the Dirac Field

13.1 Quantum Mechanics of n Identical Particles

We have explored the consequences of quantizing a classical field according to the canonical procedure. In this way we obtained a consistent description of particles which obey Bose-Einstein statistics. The formalism, flexible enough to allow creation and destruction of particles, successfully circumvented the difficulties in the one-particle theory of the negative-energy solutions and of the negative probabilities.

It would be natural at this point to apply this formalism to develop similar many-particle theories, starting with different lagrangians which lead to the nonrelativistic Schrödinger equation, perhaps, or to the Dirac equation for particles with spin ½. Such a program is bound to lead us astray, however, because we have seen that from the canonical quantization procedure there emerge particles obeying Bose-Einstein statistics, whereas the spin-½ particles such as electrons and nucleons are observed to obey Fermi-Dirac statistics and an exclusion principle. We shall therefore have to change some of the steps. The same changes which lead to the Fermi-Dirac statistics are also required on other grounds, as we shall see, and we thereby arrive at a connection between spin and statistics which is one of the significant achievements of quantum field theory.

In order to come directly to the required changes in the quantization procedure, we shall turn it around and begin with a many-particle theory for fermions based on the n-body Schrödinger equation, which we attempt to reformulate as a theory of quantized fields. Instead of quantizing a classical field theory as in the preceding chapter in order to arrive at a many-particle theory, we start with the latter and seek the form of a quantum field theory in accord with the exclusion principle.[1]

Our starting point is the Schrödinger equation for n identical noninteracting particles:

$$i \frac{\partial \Psi}{\partial t}(\mathbf{x}_1, \ldots, \mathbf{x}_n; t) = \boldsymbol{H}\Psi \tag{13.1}$$

with $\boldsymbol{H} = \sum_{i=1}^{n} H(\mathbf{x}_i, p_i)$ given by a sum of one-body terms, all of the same form. Variables may be separated in such a problem and a particular solution is a product

$$\Psi(\mathbf{x}_1, \ldots, \mathbf{x}_n; t) = \prod_{i=1}^{n} u_{\alpha_i}(\mathbf{x}_i, t) \tag{13.2}$$

[1] P. Jordan and O. Klein, *Z. Physik*, **45**, 751 (1927); P. Jordan and E. P. Wigner, *Z. Physik*, **47**, 631 (1928); and V. Fock, *Z. Physik*, **75**, 622 (1932).

of solutions $u_\alpha(\mathbf{x},t)$ of the single-particle Schrödinger equation

$$H u_\alpha(\mathbf{x},t) = i \frac{\partial u_\alpha(\mathbf{x},t)}{\partial t} \tag{13.3}$$

The general solution of (13.1) is built up by superposing product solutions (13.2), and it may be written *the number of eigentn. solu's of (13.3)*

$$\Psi(\mathbf{x}_1, \ldots ,\mathbf{x}_n; t) = \frac{1}{\sqrt{n!}} \sum_{\alpha_1, \ldots, \alpha_n = 1}^{N} c(\alpha_1, \ldots ,\alpha_n)$$

$$\times u_{\alpha_1}(\mathbf{x}_1,t) \cdots u_{\alpha_n}(\mathbf{x}_n,t) \tag{13.4}$$

with the $1/\sqrt{n!}$ inserted for later convenience. **N** is the number of single particle levels. If the $u_{\alpha_i}(\mathbf{x}_i,t)$ are assumed to be members of an orthonormal set, the normalization condition on the expansion coefficients is

$$\frac{1}{n!} \sum_{\alpha_1, \ldots, \alpha_n = 1}^{N} |c(\alpha_1, \ldots ,\alpha_n)|^2 = 1 \tag{13.5}$$

The spectrum of coefficients c determines the n-particle state in (13.4) and must satisfy an indistinguishability principle for identical particles. This requires the density $|\Psi(\mathbf{x}_1, \ldots ,\mathbf{x}_n; t)|^2$ to be invariant under all interchanges of its arguments and, therefore, that Ψ itself must be either symmetric or antisymmetic upon such interchanges.[1] Correspondingly, $c(\alpha_1, \ldots , \alpha_n)$ must be symmetric or antisymmetric upon interchange of α_i

$$c(\cdots \alpha_i \cdots \alpha_j \cdots) = \pm c(\cdots \alpha_j \cdots \alpha_i \cdots) \tag{13.6}$$

These two sign alternatives lead to the Bose-Einstein or the Fermi-Dirac statistics, respectively.

An enormous amount of information is contained in (13.6), since if one knows one of the coefficients $c(\alpha_1, \ldots ,\alpha_n)$ in (13.4), (13.6) immediately provides knowledge of $n! - 1$ additional coefficients. One may then construct a more compact expansion than (13.4) by assigning a natural ordering to the states α and defining

$$\bar{c}(\alpha_1, \ldots ,\alpha_n) = \begin{cases} c(\alpha_1, \ldots ,\alpha_n) & \alpha_1 < \alpha_2 < \cdots < \alpha_n \\ 0 & \text{otherwise} \end{cases}$$

---

[1] If Ψ is *itself* allowed to be a column vector, more general kinds of statistics are possible. See in this connection H. S. Green, *Phys. Rev.*, **90**, 270 (1953), and O. W. Greenberg and A. Messiah, to be published.

For the case of the Fermi-Dirac statistics, (13.4) may be rewritten as

$$\Psi(\mathbf{x}_1, \ldots, \mathbf{x}_n; t)$$

$$= \frac{1}{\sqrt{n!}} \sum_{\alpha_1, \ldots, \alpha_n = 1}^{N} \tilde{c}(\alpha_1, \ldots, \alpha_n) \sum_P \delta_P u_{\alpha_1}(\mathbf{x}_1,t) \cdots u_{\alpha_n}(\mathbf{x}_n,t)$$

$$= \frac{1}{\sqrt{n!}} \sum_{\alpha_1, \ldots, \alpha_n = 1}^{N} \tilde{c}(\alpha_1, \ldots, \alpha_n) \begin{Vmatrix} u_{\alpha_1}(\mathbf{x}_1,t) & \cdots & u_{\alpha_n}(\mathbf{x}_1,t) \\ \cdots & \cdots & \cdots \\ u_{\alpha_1}(\mathbf{x}_n,t) & \cdots & u_{\alpha_n}(\mathbf{x}_n,t) \end{Vmatrix} \tag{13.7}$$

with $\sum_P$ = sum over all permutations P of the α_i

δ_P = sign of the permutation P

13.2 The Number Representation for Fermions

The information contained in the wave functions (13.4) or (13.7) is not *which* particles have *which* quantum numbers, but *how many* of the n indistinguishable particles are in the various quantum levels. In this we see an analogy with the quantum theory description of the Klein-Gordon field. The state of the field, or n-particle system, is described in terms of the *numbers* of quanta, or particles, in each single-particle state. The difference here lies only in the fact that for antisymmetric solutions these numbers are 0 or 1 for each state. Recognizing this parallel, we aim to express the dynamics of the n-partclie fermion system (13.1) in the language of quantum field theory.

To begin, we change notation in (13.7) so that the sum over levels $\alpha_1, \ldots, \alpha_n = 1, \ldots, N$ is replaced by a sum over the occupation numbers n_α, which tell whether the individual levels α are occupied ($n_\alpha = 1$) or vacant ($n_\alpha = 0$). We introduce the notation $\Psi(\mathbf{x}_1, \ldots, \mathbf{x}_n; n_1, \ldots, n_N, t)$ for the Slater determinant [as in (13.7)] formed from single-particle wave functions u_{α_i} of n particles in levels α_i; the n columns of the determinant are so arranged that the u_{α_i} appear in ascending or natural order in $\alpha_1 < \alpha_2 < \cdots < \alpha_n$ and

$$n_\alpha = \begin{cases} 1 & \text{if } \alpha = \alpha_i \text{ for some } i \\ 0 & \text{otherwise} \end{cases}$$

For example, if there are seven levels and particles in levels 2, 4, and 5 ($\alpha_1 = 2$, $\alpha_2 = 4$, $\alpha_3 = 5$),

$$\Psi(\mathbf{x}_1,\mathbf{x}_2,\mathbf{x}_3; 0\,1\,0\,1\,1\,0\,0; t) = \begin{Vmatrix} u_2(\mathbf{x}_1,t) & u_4(\mathbf{x}_1,t) & u_5(\mathbf{x}_1,t) \\ u_2(\mathbf{x}_2,t) & u_4(\mathbf{x}_2,t) & u_5(\mathbf{x}_2,t) \\ u_2(\mathbf{x}_3,t) & u_4(\mathbf{x}_3,t) & u_5(\mathbf{x}_3,t) \end{Vmatrix} \tag{13.8}$$

The wave function (13.7) is now simply

$$\Psi(\mathbf{x}_1,\mathbf{x}_2, \ldots ,\mathbf{x}_n; t) = \frac{1}{\sqrt{n!}} \sum_{n_1, \ldots ,n_N=0}^{1} c'(n_1, \ldots ,n_N) \\ \times \Psi(\mathbf{x}_1,\mathbf{x}_2, \ldots ,\mathbf{x}_n; n_1, \ldots ,n_N; t) \quad (13.9)$$

with

$$c'(n_1, \ldots ,n_N) = \tilde{c}(\alpha_1, \alpha_2, \ldots ,\alpha_n) \quad (13.10)$$

with the n_α defined as before.

The normalization condition now reads

$$\sum_{n_1, \ldots ,n_N=0}^{1} |c'(n_1, \ldots ,n_N)|^2 = 1 \quad (13.11)$$

indicating that $c'(n_1, \ldots ,n_N)$ is to be interpreted as the probability amplitude for a given population distribution $\{n_1, \ldots ,n_N\}$.

In order to continue this transcription of the n-particle fermion system into the quantum field theory language, we seek a convenient way of building up n-particle wave functions from a vacuum state. It is already clear from the propagator development of the Dirac hole theory that the creation and destruction of particles plays a central role in the dynamics. Inclusion of interparticle interactions in (13.1) will lead to transitions between states of different quantum numbers, and therefore we are interested in amplitudes to destroy a particle in one state α and create one in another state α'.

To this end we follow the lead of the Klein-Gordon theory and introduce creation and destruction operators designed to build and connect such states. First we define a vacuum state Φ_0. The vacuum contains no particles, and hence no energy or momentum, corresponding to solutions of (13.1) or (13.3). Operating on Φ_0, the creation operator is so defined as to generate a one-particle state Φ_α with quantum numbers α

$$a_\alpha^\dagger \Phi_0 = \Phi_\alpha \equiv |0\,0 \cdots 1 \cdots\rangle \quad (13.12)$$

Before relating these states and operators to the wave function language of (13.1), we construct a simple and convenient representation. Because of the exclusion principle the state α is either empty or full—and we represent these two possibilities by $\begin{bmatrix} 0 \\ 1 \end{bmatrix}_\alpha$ and $\begin{bmatrix} 1 \\ 0 \end{bmatrix}_\alpha$, respectively. The vacuum state is then a product of the column matrices for each state empty

$$\Phi_0 = \prod_{\alpha=1}^{N} \begin{bmatrix} 0 \\ 1 \end{bmatrix}_\alpha \quad (13.13)$$

and the one-particle state is

$$\Phi_{\alpha'} = \begin{bmatrix}1\\0\end{bmatrix}_{\alpha'} \prod_{\alpha\neq\alpha'} \begin{bmatrix}0\\1\end{bmatrix}_{\alpha} \tag{13.14}$$

The creation operator $a_{\alpha'}^{\dagger}$ can then be represented by a 2×2 matrix in the space of the α'th state which produces $\begin{bmatrix}1\\0\end{bmatrix}_{\alpha'}$ from $\begin{bmatrix}0\\1\end{bmatrix}_{\alpha'}$. Any matrix of the form $\begin{bmatrix}x & 1\\ y & 0\end{bmatrix}_{\alpha'}$, with arbitrary x and y, accomplishes this. Because of the exclusion principle we want $a_{\alpha'}^{\dagger}$ operating on an occupied state $\begin{bmatrix}1\\0\end{bmatrix}_{\alpha'}$ to destroy the state, yielding a null vector; consequently, we set $x = y = 0$:

$$a_{\alpha'}^{\dagger} = \begin{bmatrix}0 & 1\\ 0 & 0\end{bmatrix}_{\alpha'} \tag{13.15}$$

In a similar way we construct the annihilation operator a_{α} from the conditions $a_{\alpha'}\begin{bmatrix}1\\0\end{bmatrix}_{\alpha'} = \begin{bmatrix}0\\1\end{bmatrix}_{\alpha'}$ and $a_{\alpha'}\begin{bmatrix}0\\1\end{bmatrix}_{\alpha'} = 0$. We find

$$a_{\alpha'} = \begin{bmatrix}0 & 0\\ 1 & 0\end{bmatrix}_{\alpha'} \tag{13.16}$$

which is the hermitian conjugate of $a_{\alpha'}^{\dagger}$. A simple set of anticommutation relations[1] for the $a_{\alpha'}$ and $a_{\alpha'}^{\dagger}$ follows from (13.15) and (13.16)

$$\{a_{\alpha'},a_{\alpha'}\} = 0 \tag{13.17a}$$

$$\{a_{\alpha'}^{\dagger},a_{\alpha'}^{\dagger}\} = 0 \tag{13.17b}$$

$$\{a_{\alpha'},a_{\alpha'}^{\dagger}\} = \begin{bmatrix}1 & 0\\ 0 & 1\end{bmatrix} \equiv 1 \tag{13.17c}$$

The exclusion principle has led to anticommutation relations between the creation and annihilation operators in place of the analogous commutators (12.10) of the canonical quantization procedure for bosons. Equations (13.17*a*) and (13.17*b*) express the impossibility of removing or of introducing two fermions into the same state. The two eigenvalues of the product $a_{\alpha'}^{\dagger}a_{\alpha'} = \begin{bmatrix}1 & 0\\ 0 & 0\end{bmatrix}$ are 1 for

[1] The representation in (13.15) to (13.17) shows the complete analogy between $a_{\alpha'}$, $a_{\alpha'}^{\dagger}$ and the two-component Pauli spin matrices $\sigma_x - i\sigma_y$, $\sigma_x + i\sigma_y$. The vacuum state (13.13) corresponds to a spin state with all spins down; $a_{\alpha'}^{\dagger}$ flips the α'th state from spin down to spin up, an occupied state here corresponding to the state of spin up.

an occupied state α' and 0 for an empty state. This is the number operator, denoted by

$$N_{\alpha'} = a^\dagger_{\alpha'} a_{\alpha'}$$

It differs from the corresponding number operator for bosons, since it has just the two eigenvalues 0 and 1.

We can now link the operator language with the one-particle wave functions $u_\alpha(x)$ in a simple way. To do this, we introduce a field operator with the definition

$$\chi(\mathbf{x},t) = \sum_{\alpha=1}^{N} u_\alpha(\mathbf{x},t) a_\alpha \qquad \chi^*(\mathbf{x},t) = \sum_{\alpha=1}^{N} u^*_\alpha(\mathbf{x},t) a^\dagger_\alpha \tag{13.18}$$

The wave function $u_{\alpha_i}(\mathbf{x},t)$ is just the matrix element of the field operator $\chi(\mathbf{x},t)$ between the vacuum Φ_0 and the one-particle state Φ_{α_i}:

$$(\Phi_0, \chi(\mathbf{x},t), \Phi_{\alpha_i}) = u_{\alpha_i}(\mathbf{x},t) \tag{13.19}$$

We can continue in the same spirit and construct from the fields the n-particle wave function (13.8) and the hamiltonian operator with the same eigenvalue spectrum as H in (13.1). For this we must consider states containing several particles. As constructed in (13.15) and (13.16), the operators a_{α_i} and $a^\dagger_{\alpha_i}$ commute with the a_{α_j} and $a^\dagger_{\alpha_j}$ for $j \neq i$, since they operate on different states. For example, for $i \neq j$,

$$a^\dagger_{\alpha_i} a^\dagger_{\alpha_j} \Phi_0 = \Phi_{\alpha_i \alpha_j} = \begin{bmatrix} 1 \\ 0 \end{bmatrix}_{\alpha_i} \begin{bmatrix} 1 \\ 0 \end{bmatrix}_{\alpha_j} \prod_{\alpha \neq \alpha_i, \alpha_j} \begin{bmatrix} 0 \\ 1 \end{bmatrix}_\alpha = \Phi_{\alpha_j \alpha_i} \tag{13.20}$$

If we continue to work directly with the operators $a^\dagger_{\alpha_i}$, we shall obtain a representation which is chained to a particular ordering of the states α. For whereas $\Phi_{\alpha_i \alpha_j}$ in (13.20) is symmetric under interchange of α_i and α_j, the state Ψ defined above in (13.8) is antisymmetric. It is therefore, from a mathematical point of view, convenient to modify the operators[1] a_α, $a^\dagger_\alpha$ so that they satisfy anticommutation relations for different states as well as for the same state (13.17). This means, in particular, for different states α and α' the modified operators $b^\dagger_\alpha$ should satisfy

$$b^\dagger_\alpha b^\dagger_{\alpha'} |0\rangle = -b^\dagger_{\alpha'} b^\dagger_\alpha |0\rangle \tag{13.21}$$

in contrast to $a^\dagger_\alpha a^\dagger_{\alpha'} |0\rangle = +a^\dagger_{\alpha'} a^\dagger_\alpha |0\rangle$. To accomplish this desired sign change and at the same time preserve the interpretation of $b^\dagger_\alpha$ as a creation operator, we write

$$b^\dagger_\alpha = a^\dagger_\alpha \eta_\alpha \tag{13.22}$$

[1] Jordan and Wigner, *op. cit.*

where η_α is an operator diagonal in the number representation in which we work. Inserting (13.22) into (13.21), we find that (13.21) is satisfied provided

$$a_{\alpha_i}^\dagger \eta_{\alpha_i} = -\eta_{\alpha_i} a_{\alpha_i}^\dagger \qquad \text{for } \alpha_i < \alpha_j \tag{13.23}$$

$$a_{\alpha_j}^\dagger \eta_{\alpha_i} = \eta_{\alpha_i} a_{\alpha_j}^\dagger$$

[handwritten margin notes:]
$a_\alpha \eta_{\alpha'} = -\eta_{\alpha'} a_\alpha,$ $\alpha' > \alpha$
$a_\alpha \eta_{\alpha'} = +\eta_{\alpha'} a_\alpha$
for $\alpha > \alpha'$
$\alpha_j \geq \kappa_i \longrightarrow$
and $[\eta_\alpha, \eta_{\alpha'}] = 0$

The diagonal operator $(1 - 2N_\alpha)$ anticommutes[1] with $a_\alpha^\dagger$; we are then led to write

$$\eta_{\alpha_i} = \prod_{\alpha=1}^{\alpha_i-1} (1 - 2N_\alpha) = \prod_{\alpha=1}^{\alpha_i-1} \begin{bmatrix} -1 & 0 \\ 0 & 1 \end{bmatrix}_\alpha \tag{13.24}$$

therefore

$$\eta_{\alpha_i} \Psi(n_1 \cdots n_N) = \prod_{\alpha=1}^{\alpha_i-1} (-1)^{n_\alpha} \Psi(n_1 \cdots n_N)$$

The operator

$$b_\alpha = \eta_\alpha a_\alpha = a_\alpha \eta_\alpha$$

is the hermitian conjugate of $b_\alpha^\dagger$ and is interpreted as a destruction operator. As a creation operator, $b_{\alpha_i}^\dagger$ generates an occupied state $\begin{bmatrix} 1 \\ 0 \end{bmatrix}_{\alpha_i}$ from an unoccupied one $\begin{bmatrix} 0 \\ 1 \end{bmatrix}_{\alpha_i}$ with an amplitude $+1$ or -1 depending upon whether an even or odd number of particles occupy the states α for which $\alpha < \alpha_i$; thus a state in which levels α_i and α_j are occupied is antisymmetric under interchange of the labels α_i and α_j. These operators b_α and $b_\alpha^\dagger$ satisfy the same anticommutation relations as the a_α and $a_\alpha^\dagger$

$$\{b_\alpha^\dagger, b_\alpha^\dagger\} = \{b_\alpha, b_\alpha\} = 0 \qquad \{b_\alpha, b_\alpha^\dagger\} = 1 \tag{13.25}$$

Furthermore, the general commutation relations

$$\{b_\alpha, b_{\alpha'}^\dagger\} = \delta_{\alpha\alpha'} \qquad \{b_\alpha, b_{\alpha'}\} = \{b_\alpha^\dagger, b_{\alpha'}^\dagger\} = 0 \tag{13.26}$$

follow by our construction; they can be verified explicitly. For example, for $\alpha > \alpha'$

$$b_\alpha b_{\alpha'} + b_{\alpha'} b_\alpha = \eta_\alpha a_\alpha a_{\alpha'} \eta_{\alpha'} + \eta_{\alpha'} a_{\alpha'} a_\alpha \eta_\alpha$$
$$= -a_\alpha a_{\alpha'} \eta_\alpha \eta_{\alpha'} + a_{\alpha'} a_\alpha \eta_\alpha \eta_{\alpha'} = 0 \qquad \alpha > \alpha'$$

The number operator in terms of the b's is simply

$$N_\alpha = a_\alpha^\dagger a_\alpha = b_\alpha^\dagger b_\alpha \tag{13.27}$$

---

[1] The operator $(1 - 2N_\alpha)$ is here the analogue of the Pauli spin operator $-\sigma_z$.

The states

$$(b_N^\dagger)^{n_N} \cdots (b_\alpha^\dagger)^{n_\alpha} \cdots (b_1^\dagger)^{n_1}\Phi_0 \equiv \Phi(n_1, \ldots ,n_N) \qquad (13.28)$$

are eigenfunctions of the number operator and form a complete orthonormal set:

$$(\Phi(n_1', \ldots ,n_N'),\Phi(n_1, \ldots ,n_N)) = \prod_{\alpha=1}^{N} \delta_{n_\alpha n_\alpha'}$$

Superposing to form a general state, we write

$$\Phi = \sum_{n_1, \ldots ,n_N=0}^{1} c'(n_1, \ldots ,n_N)(b_N^\dagger)^{n_N} \cdots (b_1^\dagger)^{n_1}\Phi_0 \qquad (13.29)$$

where the expansion coefficients $c'(n_1, \ldots ,n_N)$ are the probability amplitudes for a given population distribution

$$(\Phi,\Phi) = \sum_{n_1, \ldots ,n_N=0}^{1} |c'(n_1, \ldots ,n_N)|^2 \qquad (13.30)$$

and may therefore be identified with the coefficients in (13.8) and (13.9).

Nonvanishing matrix elements of the operators b_α and $b_\alpha^\dagger$ are

$$\begin{aligned}\langle\Phi(n_1', \ldots ,n_N'),b_{\alpha_k}^\dagger\Phi(n_1, \ldots ,n_N)\rangle &= \langle\Phi_0, b_{\alpha_1'} \cdots b_{\alpha_{n'}'}b_{\alpha_k}^\dagger b_{\alpha_n}^\dagger \cdots b_{\alpha_1}^\dagger\Phi_0\rangle \\ &= \begin{cases}(-)^{n'-k'} & \text{if } n_\alpha = n_\alpha' \text{ for } \alpha \neq \alpha_k \text{ and } n_{\alpha_k} = 0;\ n_{\alpha_k}' = 1 \\ 0 & \text{otherwise}\end{cases}\end{aligned}$$

$$\begin{aligned}\langle\Phi(n_1', \ldots ,n_N'),b_{\alpha_k}\Phi(n_1, \ldots ,n_N)\rangle \\ = \begin{cases}(-)^{n-k} & \text{if } n_\alpha = n_\alpha' \text{ for } \alpha \neq \alpha_k \text{ and } n_{\alpha_k}' = 0;\ n_{\alpha_k} = 1 \\ 0 & \text{otherwise}\end{cases}\end{aligned} \qquad (13.31)$$

where $n = \sum_{\alpha=1}^{N} n_\alpha$, $n' = \sum_{\alpha=1}^{N} n_\alpha'$, and α_k is the kth member of the ordered set $\{\alpha_1,\alpha_2, \ldots ,\alpha_n\}$ and $\alpha_{k'}$ the k'th member of the set $\{\alpha_1', \ldots ,\alpha_{n'}'\}$.

In the same way as we formed one-particle wave functions (13.19), we are now in a position to construct antisymmetric n-particle wave functions. For this we introduce the field operator $\varphi(\mathbf{x},t)$ as in (13.18) but with the a_α replaced now by the mutually anticommuting b_α:

$$\varphi(\mathbf{x},t) = \sum_{\alpha=1}^{N} u_\alpha(\mathbf{x},t)b_\alpha \qquad \varphi^*(\mathbf{x},t) = \sum_{\alpha=1}^{N} u_\alpha^*(\mathbf{x},t)b_\alpha^\dagger \qquad (13.32)$$

Forming the matrix element of a product of n field amplitudes $\varphi(\mathbf{x}_i,t)$ at a common time t between the vacuum and a general state Φ

given by (13.29), we find after repeated application of (13.31)

$$\frac{1}{\sqrt{n!}}\,(\Phi_0,\varphi(\mathbf{x}_1,t)\,\cdots\,\varphi(\mathbf{x}_n,t)\Phi) = \Psi(\mathbf{x}_1,\,\ldots\,,\mathbf{x}_n;t) \tag{13.33}$$

where Ψ is the antisymmetric n-particle wave function in (13.9). The anticommutation relations of the creation and annihilation operators may be reexpressed as anticommutation relations of the operators[1] φ and φ^*

$$\begin{aligned}
\{\varphi(\mathbf{x},t),\varphi(\mathbf{x}',t)\} &= 0\\
\{\varphi^*(\mathbf{x},t),\varphi^*(\mathbf{x}',t)\} &= 0\\
\{\varphi(\mathbf{x},t),\varphi^*(\mathbf{x}',t)\} &= \sum_{\alpha,\alpha'=1}^{N} \{b_\alpha,b_{\alpha'}^\dagger\}u_\alpha(\mathbf{x},t)u_{\alpha'}^*(\mathbf{x}',t)\\
&= \sum_{\alpha=1}^{N} u_\alpha(\mathbf{x},t)u_\alpha^*(\mathbf{x}',t) = \delta^3(\mathbf{x}-\mathbf{x}')
\end{aligned} \tag{13.34}$$

With the states and operators in hand, we are now equipped to cast off from our moorings to the wave function language and to develop the dynamics, that is, (13.1), in terms of the quantized field formalism. Equation (13.3) can be rewritten as a linear differential equation

$$H(\mathbf{x})\varphi(\mathbf{x},t) = i\,\frac{\partial\varphi(\mathbf{x},t)}{\partial t} \tag{13.35}$$

for the field operator introduced in (13.32). In analogy with the procedure applied to the Klein-Gordon theory, (13.35) may be considered as a field equation for a classical field φ, derived perhaps from an appropriate lagrangian. Then upon imposing the relations (13.34), we reinterpret (13.35) as an operator equation, the main difference from the Klein-Gordon development being our use here of anticommutators instead of commutators, resulting in Fermi-Dirac instead of Bose-Einstein statistics. This procedure is known as the *second quantization.* In the first quantization, classical particle coordinates are replaced by quantum-mechanical operators acting on wave functions; now we have taken the one-particle Schrödinger equation, interpreted it as a field equation, and then imposed quantum conditions on the field amplitudes, which become operators satisfying (13.34). However, we have shown above that the content of this formalism is the same as that of the many-body Schrödinger equation. The expansion coefficients $c'(n_1, \ldots, n_N)$ in both (13.9) and (13.29) describe the state of the n-particle

[1] It is here that we see the utility of the $b_\alpha^\dagger$ operators instead of the $a_\alpha^\dagger$. The simple form of Eqs. (13.33) and (13.34) would not exist had $a_\alpha^\dagger$ been used instead of $b_\alpha^\dagger$ in (13.28).

system in either language. For the total energy we have in the wave function language, by (13.9),

$$\int d^3x_1 \cdots d^3x_n \, \Psi^*(\mathbf{x}_1, \ldots, \mathbf{x}_n; t) \sum_{i=1}^{n} H(\mathbf{x}_i, \mathbf{p}_i) \Psi(\mathbf{x}_1, \ldots, \mathbf{x}_n; t)$$
$$= \sum_{n_1, \ldots, n_N = 0}^{1} |c'(n_1, \ldots, n_N)|^2 \left(\sum_{\alpha=1}^{N} n_\alpha E_\alpha \right) \quad (13.36)$$

Reverting to the operator language, we recognize (13.36) as the expectation value in the state (13.29) of the hamiltonian operator

$$\begin{aligned} \mathbf{H} &= \int d^3x \, \varphi^*(\mathbf{x},t) H(\mathbf{x},\mathbf{p}) \varphi(\mathbf{x},t) \\ &= \sum_{\alpha=1}^{N} N_\alpha \int d^3x \, u_\alpha^* H u_\alpha \\ &= \sum_{\alpha=1}^{N} N_\alpha E_\alpha \end{aligned} \quad (13.37)$$

$\mathbf{H}$ is just the hamiltonian operator for the field, as we might have expected.

To round out our discussion of this formalism, we construct the Green's function for a particle to propagate from $(\mathbf{x},t)$ to $(\mathbf{x}',t')$ with $t' > t$, for comparison with Chap. 6. For this we need the amplitude for a particle to be created from the vacuum at $(\mathbf{x},t)$ and to be destroyed later at $(\mathbf{x}',t')$.

A single particle localized at the point $\mathbf{x}$ has a wave function proportional to

$$\psi_{\mathbf{x}}(\mathbf{x}',t) = \delta^3(\mathbf{x} - \mathbf{x}') = \sum_{\alpha=1}^{N} u_\alpha(\mathbf{x}',t) u_\alpha^*(\mathbf{x},t) \quad (13.38)$$

The corresponding state in the second quantized version is, by the comparison of (13.38) with (13.9) and (13.29),

$$\Psi_1(\mathbf{x},t) = \sum_{\alpha=1}^{N} u_\alpha^*(\mathbf{x},t) b_\alpha^\dagger \Phi_0 = \varphi^*(\mathbf{x},t)\Phi_0 \quad (13.39)$$

The Green's function is given by projecting $\Psi_1(\mathbf{x},t)$ for the particle produced at $(\mathbf{x},t)$ on the one-particle state $\Psi_1(\mathbf{x}',t')$ at a later time $t' > t$:

$$\begin{aligned} G(\mathbf{x}',t';\mathbf{x},t) &= -i(\Psi_1(\mathbf{x}',t'),\Psi_1(\mathbf{x},t))\theta(t' - t) \\ &= -i\theta(t' - t)\langle 0|\varphi(\mathbf{x}',t')\varphi^*(\mathbf{x},t)|0\rangle \end{aligned}$$

Using (13.31) and (13.32), this reduces to

$$G(\mathbf{x}',t';\mathbf{x},t) = -i\theta(t'-t)\sum_{\alpha} u_\alpha(\mathbf{x}',t')u_\alpha^*(\mathbf{x},t)$$

and coincides with (6.28) as introduced in the discussion of retarded Green's functions in the propagator theory in Chap. 6.

In concluding this discussion of the correspondence between the many-body and the second quantized field theory versions of the Schrödinger theory, we point out two of the advantages of the field-theoretic formulation. These merits are measured by ease of calculating physically interesting matrix elements and have led to extensive applications of the field theory methods to nonrelativistic many-body problems in recent years. First of all, the operators b_α and $b_\alpha^\dagger$, by their algebraic rules, automatically do the bookkeeping necessary to preserve the antisymmetry of the wave functions. Second, they provide the desired flexibility for a natural and simple description of physical systems with varying numbers of particles.

13.3 The Dirac Theory

We return now to the case specifically at hand, namely, the Dirac equation. In order to maintain a close parallel with the discussion of the Klein-Gordon theory, we derive it from a lagrangian by an action principle. The four components of the field ψ_α plus those of the adjoint $\bar{\psi}_\alpha$ are treated as eight independent variables.

Starting from the free Dirac equation

$$(i\not\nabla - m)\psi = 0 \tag{13.40}$$

we construct the lagrangian by left-multiplying by $\delta\bar{\psi}$ and integrating over all space-time between t_1 and t_2

$$0 = \int_{t_1}^{t_2} d^4x\, \delta\bar{\psi}(x)(i\not\nabla - m)\psi(x) = \delta\int_{t_1}^{t_2} d^4x\, \bar{\psi}(x)(i\not\nabla - m)\psi(x) \tag{13.41}$$

from which we identify the lagrangian density

$$\mathcal{L}(x) = \bar{\psi}(x)(i\not\nabla - m)\psi(x) \tag{13.42}$$

Variation of the action (13.41) with respect to ψ gives the adjoint equation

$$\bar{\psi}(-i\overleftarrow{\not\nabla} - m) = 0$$

The canonical procedure gives the momentum conjugate to ψ

$$\pi_\alpha = \frac{\partial \mathcal{L}}{\partial \dot{\psi}_\alpha} = i\psi_\alpha^\dagger \tag{13.43}$$

Since (13.41) contains no derivatives of $\bar{\psi}$, we find no conjugate momentum to $\psi_\alpha^\dagger$; Eq. (13.43) shows that $i\psi_\alpha^\dagger$ is itself the conjugate momentum to ψ_α. The resulting form of the hamiltonian density is

$$\mathcal{H} = \pi\dot{\psi} - \mathcal{L} = \psi^\dagger(-i\boldsymbol{\alpha}\cdot\boldsymbol{\nabla} + \beta m)\psi = \psi^\dagger i\frac{\partial}{\partial t}\psi \tag{13.44}$$

where the last form follows from the Dirac equation. The form of $\mathcal{H}$ as the one-particle hamiltonian operator bracketed by the field and its hermitian conjugate agrees with that to which the nonrelativistic discussion led us in (13.37).

The full array of conservation laws for energy, momentum, and angular momentum follow automatically from the displacement and Lorentz invariance of $\mathcal{L}$ and can be calculated from the definitions (11.48) and (11.57):

$$\mathfrak{T}^{\nu\mu} = i\bar{\psi}\gamma^\nu\frac{\partial}{\partial x_\mu}\psi \tag{13.45}$$

which leads by (11.49) to the energy and momentum constants of the motion

$$H = \int \mathfrak{T}^{00}\,d^3x = \int \psi^\dagger(-i\boldsymbol{\alpha}\cdot\boldsymbol{\nabla} + \beta m)\psi\,d^3x$$

by (13.44) and

$$\mathbf{P} = \int \psi^\dagger(-i\boldsymbol{\nabla})\psi\,d^3x \tag{13.46}$$

The angular momentum density $\mathfrak{M}^{\mu\nu\lambda}$ and conserved angular momentum $M^{\nu\lambda}$ are

$$\mathfrak{M}^{\mu\nu\lambda} = i\bar{\psi}\gamma^\mu\left(x^\nu\frac{\partial}{\partial x_\lambda} - x^\lambda\frac{\partial}{\partial x_\nu} + \Sigma^{\nu\lambda}\right)\psi \tag{13.47}$$

$$M^{\nu\lambda} = \int d^3x\,\mathfrak{M}^{0\nu\lambda}$$

where $\Sigma^{\nu\lambda} = \frac{1}{4}[\gamma^\nu,\gamma^\lambda]$ is the spinor rotation matrix under a Lorentz transformation and adds the spin angular momentum in the last term of $\mathfrak{M}^{\mu\nu\lambda}$. For the space components in particular, we have

$$\mathbf{J} \equiv (M^{23},M^{31},M^{12}) = \int d^3x\,\psi^\dagger\left(\mathbf{r}\times\frac{1}{i}\boldsymbol{\nabla} + \frac{1}{2}\boldsymbol{\sigma}\right)\psi \tag{13.48}$$

which looks familiar as the sum of orbital and spin angular momentum, $\mathbf{J} = \mathbf{L} + \mathbf{S}$.

We can identify one further conservation law for the free Dirac theory when we recall that solutions of the Dirac equation satisfy the condition $(\partial/\partial x^\mu)\,\bar{\psi}\gamma^\mu\psi = 0$ and therefore that

Conserved current $j^\mu = \bar{\psi}\gamma^\mu\psi$

$$Q = \int d^3x\,\psi^\dagger\psi \tag{13.49}$$

is a constant total "charge." This is analogous to the conserved charge (12.63) in the Klein-Gordon theory for non-hermitian fields.

Continuing along the canonical path, we would form the quantum field theory now by imposing the commutator relations (11.39). We know, however, that these lead to many-particle quantum systems which obey the Bose-Einstein statistics. In order to accommodate the exclusion principle in constructing a quantum field theory of the many-electron system, these commutators must now be replaced by anticommutators as in the preceding section.

13.4 Momentum Expansions

The quantization procedure will be carried out in momentum space in terms of the creation and annihilation operators. The development and discussion of the first part of this chapter for the Schrödinger field can then be applied directly, since we again want operators, as in (13.32), which create states in accord with the exclusion principle.

The general wave expansion of a solution of the free Dirac equation (13.40) takes the form, as discussed in Chap. 3,

$$\begin{aligned}\psi(\mathbf{x},t) &= \sum_{\pm s}\int \frac{d^3p}{(2\pi)^{3/2}}\sqrt{\frac{m}{E_p}}\,[b(p,s)u(p,s)e^{-ip\cdot x} + d^\dagger(p,s)v(p,s)e^{ip\cdot x}]\\ \psi^\dagger(\mathbf{x},t) &= \sum_{\pm s}\int \frac{d^3p}{(2\pi)^{3/2}}\sqrt{\frac{m}{E_p}}\,[b^\dagger(p,s)\bar{u}(p,s)\gamma_0 e^{ip\cdot x} + d(p,s)\bar{v}(p,s)\gamma_0 e^{-ip\cdot x}]\end{aligned} \tag{13.50}$$

with $E_p = p_0 = +\sqrt{|\mathbf{p}|^2 + m^2}$. The following useful relations were established in Chap. 3 for the spinors[1] $u(p,s)$ and $v(p,s)$

(*a*) Dirac equation

$$\begin{aligned}(\not{p} - m)u(p,s) &= 0 \qquad & \bar{u}(p,s)(\not{p} - m) &= 0\\ (\not{p} + m)v(p,s) &= 0 \qquad & \bar{v}(p,s)(\not{p} + m) &= 0\end{aligned}$$

[1] For example, (3.16), (3.30). Recall that the notation $u(-p,s)$ means $u(\sqrt{\mathbf{p}^2 + m^2}, -\mathbf{p}, s)$. The sign of energy is always taken positive in spinors. The normalization convention on b, $b^\dagger$ is designed to make quanta carry charge ± 1.

(*b*) Orthogonality

$$
\begin{aligned}
\bar{u}(p,s)u(p,s') &= \delta_{ss'} = -\bar{v}(p,s)v(p,s') \\
u^\dagger(p,s)u(p,s') &= \frac{E_p}{m}\,\delta_{ss'} = v^\dagger(p,s)v(p,s') \\
\bar{v}(p,s)u(p,s') &= 0 = v^\dagger(p,s)u(-p,s')
\end{aligned}
\tag{13.51}
$$

(*c*) Completeness

$$
\sum_{\pm s} [u_\alpha(p,s)\bar{u}_\beta(p,s) - v_\alpha(p,s)\bar{v}_\beta(p,s)] = \delta_{\alpha\beta}
$$

$$
\sum_{\pm s} u_\alpha(p,s)\bar{u}_\beta(p,s) = \left(\frac{\not{p} + m}{2m}\right)_{\alpha\beta} \equiv (\Lambda_+(p))_{\alpha\beta}
$$

$$
-\sum_{\pm s} v_\alpha(p,s)\bar{v}_\beta(p,s) = \left(\frac{m - \not{p}}{2m}\right)_{\alpha\beta} \equiv (\Lambda_-(p))_{\alpha\beta}
$$

Upon second quantization of the Dirac field, the expansion coefficients $b(p,s)$, $b^\dagger(p,s)$, $d(p,s)$, and $d^\dagger(p,s)$ become operators which annihilate and create particles. Since we wish to arrive at an exclusion principle, they are assigned anticommutation relations analogous to (13.26), which read in the continuum notation (13.50),

$$
\begin{aligned}
\{b(p,s),b^\dagger(p',s')\} &= \delta_{ss'}\delta^3(\mathbf{p} - \mathbf{p}') \\
\{d(p,s),d^\dagger(p',s')\} &= \delta_{ss'}\delta^3(\mathbf{p} - \mathbf{p}') \\
\{b(p,s),b(p',s')\} &= \{d(p,s),d(p',s')\} = 0 \\
\{b^\dagger(p,s),b^\dagger(p',s')\} &= \{d^\dagger(p,s),d^\dagger(p',s')\} = 0 \\
\{b(p,s),d(p',s')\} &= \{b(p,s),d^\dagger(p',s')\} = 0 \\
\{d(p,s),b(p',s')\} &= \{d(p,s),b^\dagger(p',s')\} = 0
\end{aligned}
\tag{13.52}
$$

The anticommutation relations for the fields (13.50) can now be derived from (13.52). For example, we find with the aid of (13.51),

$$
\begin{aligned}
&\{\psi_\alpha(\mathbf{x},t),\psi_\beta^\dagger(\mathbf{x}',t)\} \\
&= \sum_{\pm s,\pm s'} \iint \frac{d^3p\,d^3p'}{(2\pi)^3} \sqrt{\frac{m}{E_p}\cdot\frac{m}{E_{p'}}} \cdot \delta^3(\mathbf{p} - \mathbf{p}')\delta_{ss'} \\
&\qquad \times [u_\alpha(p,s)\bar{u}_\tau(p',s')\gamma^0_{\tau\beta}e^{i\mathbf{p}\cdot(\mathbf{x}-\mathbf{x}')} + v_\alpha(p,s)\bar{v}_\tau(p',s')\gamma^0_{\tau\beta}e^{-i\mathbf{p}\cdot(\mathbf{x}-\mathbf{x}')}] \\
&= \int \frac{d^3p}{(2\pi)^3}\frac{1}{2E_p}\{[(\not{p} + m)\gamma^0]_{\alpha\beta}e^{i\mathbf{p}\cdot(\mathbf{x}-\mathbf{x}')} - [(m - \not{p})\gamma^0]_{\alpha\beta}e^{-i\mathbf{p}\cdot(\mathbf{x}-\mathbf{x}')}\} \\
&= \int \frac{d^3p}{(2\pi)^3}\frac{1}{2E_p}e^{i\mathbf{p}\cdot(\mathbf{x}-\mathbf{x}')}2E_p\delta_{\alpha\beta} = \delta^3(\mathbf{x} - \mathbf{x}')\delta_{\alpha\beta}
\end{aligned}
\tag{13.53}
$$

There emerge similarly, in analogy with (13.34)

$$\{\psi(\mathbf{x},t),\psi(\mathbf{x}',t)\} = 0 \qquad \{\psi^\dagger(\mathbf{x},t),\psi^\dagger(\mathbf{x}',t)\} = 0 \tag{13.54}$$

In parallel with the Schrödinger theory we interpret $b^\dagger$ and d as creation operators for electrons. In the Dirac theory, however, we must cope with the negative-energy solutions; d creates a state of negative energy. To exhibit this, we express the energy and momentum operators in terms of the b, d, $b^\dagger$, and $d^\dagger$. Inserting (13.50) in (13.46), we find with the aid of (13.51)

$$\begin{aligned} H = \int d^3x \sum_{\pm s, \pm s'} \iint \frac{d^3p\, d^3p'}{(2\pi)^3} \sqrt{\frac{m^2}{E_pE_{p'}}}\, E_{p'} & \\ \times [b^\dagger(p,s)\bar{u}(p,s)\gamma_0 e^{ip\cdot x} + d(p,s)\bar{v}(p,s)\gamma_0 e^{-ip\cdot x}] & \\ \times [b(p',s')u(p',s')e^{-ip'\cdot x} - d^\dagger(p',s')v(p',s')e^{ip'\cdot x}] & \\ = \sum_{\pm s, \pm s'} \int d^3p\, m[b^\dagger(p,s)b(p,s')u^\dagger(p,s)u(p,s') & \\ - d(p,s)d^\dagger(p,s')v^\dagger(p,s)v(p,s')] & \\ = \sum_{\pm s} \int d^3p\, E_p[b^\dagger(p,s)b(p,s) - d(p,s)d^\dagger(p,s)] & \end{aligned} \tag{13.55}$$

Similarly,

$$\mathbf{P} = \sum_{\pm s} \int d^3p\, \mathbf{p}[b^\dagger(p,s)b(p,s) - d(p,s)d^\dagger(p,s)] \tag{13.56}$$

From these forms for H and $\mathbf{P}$ and from the relations (13.52) it follows that $d(p,s)$ creates a negative-energy particle with $(-E_p,-\mathbf{p})$ and that $b^\dagger(p,s)$ creates a positive-energy particle with $(E_p,\mathbf{p})$. $d^\dagger(p,s)$ and $b(p,s)$ are the corresponding annihilation operators.

We also see that the energy operator in (13.55) is not positive definite. This presents an apparent difficulty because one can always find a state of energy lower than that of any proposed ground state by introducing more particles into negative-energy states. However, as we have discussed at length in Chap. 5, the hole theory interpretation of Dirac removes this difficulty. According to the hole theory, we define the vacuum state as the one obtained by filling all the negative-energy electron states and leaving empty all the positive-energy ones. That we can *in principle* define the vacuum in this way requires use of an exclusion principle and, consequently, of anticommutator relations in formulating the second quantized theory of the Dirac field. Had we attempted to quantize according to the canonical procedure in terms of commutators, we would now find ourselves in fundamental and inextricable difficulties. According to Bose-Einstein statistics, we can always introduce unlimited numbers of particles into any given state; and therefore there would exist no ground state of lowest energy for hamiltonian (13.55). This necessary connection of the anticommutators with the Dirac theory is a particular example of a fundamental theorem in local, Lorentz-invariant field theory, first proved in 1940 by Pauli, according to which particles of half-integer spin must obey Fermi-Dirac statistics and those of integer spin, Bose-Einstein statistics.[1] We return to a more general discussion of this point in Chap. 16.

In view of the hole theory interpretation of the Dirac theory, we now rewrite the energy-momentum four-vector of (13.55) and (13.56) in the form

$$P^\mu = \sum_{\pm s} \int d^3p\, p^\mu[b^\dagger(p,s)b(p,s) + d^\dagger(p,s)d(p,s) - \{d(p,s),d^\dagger(p,s)\}] \tag{13.57}$$

Operating on the vacuum of hole theory, the first two terms vanish, since there are no positive-energy electrons present to be annihilated by $b(p,s)$ and no holes among the negative-energy states to be filled by $d(p,s)$. The last term is just an infinite constant which we discard, since all energies and momenta are measured relative to the vacuum. Formally, this is equivalent to redefining P^μ in terms of normally ordered products of field amplitudes. In normal order the products of field operators are so written that the positive-frequency parts

$$\psi^{(+)} = \sum_{\pm s} \int \frac{d^3p}{(2\pi)^{3/2}} \sqrt{\frac{m}{E_p}}\, b(p,s)u(p,s)e^{-ip\cdot x}$$

$$\bar{\psi}^{(+)} = \overline{\psi^{(-)}} = \sum_{\pm s} \int \frac{d^3p}{(2\pi)^{3/2}} \sqrt{\frac{m}{E_p}}\, d(p,s)\bar{v}(p,s)e^{-ip\cdot x}$$

stand to the right of the negative-frequency parts

$$\psi^{(-)} = \sum_{\pm s} \int \frac{d^3p}{(2\pi)^{3/2}} \sqrt{\frac{m}{E_p}}\, d^\dagger(p,s)v(p,s)e^{ip\cdot x}$$

$$\bar{\psi}^{(-)} = \overline{\psi^{(+)}} = \sum_{\pm s} \int \frac{d^3p}{(2\pi)^{3/2}} \sqrt{\frac{m}{E_p}}\, b^\dagger(p,s)\bar{u}(p,s)e^{ip\cdot x}$$

[1] W. Pauli, *Phys. Rev.*, **58**, 716 (1940), *Ann. Inst. Henri Poincaré*, **6**, 137 (1936). Had we attempted to make the wrong spin and statistics connection in (12.14) and (12.15) for spin-0 quanta, the energy-momentum four-vector would no longer be an operator whose eigenstates describe a physical system, but would be no more than an infinite constant.

For fields which obey anticommutation relations the normal ordering carries with it the instruction to change the sign of the term for each interchange of field amplitudes required to bring the fields into normal order, viz.,

$$:\bar{\psi}_\alpha \psi_\beta: = \bar{\psi}_\alpha^{(+)}\psi_\beta^{(+)} + \bar{\psi}_\alpha^{(-)}\psi_\beta^{(+)} + \bar{\psi}_\alpha^{(-)}\psi_\beta^{(-)} - \psi_\beta^{(-)}\bar{\psi}_\alpha^{(+)} \tag{13.58}$$

With this definition of normal-ordering, a bilinear form such as the energy-momentum four-vector (13.57) is altered only by a c number to

$$P^\mu = \sum_{\pm s} \int d^3p \; p^\mu[b^\dagger(p,s)b(p,s) + d^\dagger(p,s)d(p,s)] \tag{13.59}$$

$b(p,s)$ destroys and $b^\dagger(p,s)$ creates a positive-energy electron with quantum numbers (p,s), and in analogy with the Klein-Gordon theory

$$N^+(p,s) = b^\dagger(p,s)b(p,s)$$

is interpreted as the number operator for positive-energy electrons. The eigenvalues of $N^+(p,s)\,d^3p$ tell how many electrons of spin s are in the momentum interval d^3p. $d^\dagger(p,s)$ destroys a negative-energy electron, which is interpreted in the hole theory as the *creation* of a positron; similarly, $d(p,s)$ destroys a positron, and in (13.59)

$$N^-(p,s) = d^\dagger(p,s)d(p,s)$$

is the number operator for positrons of positive energy. Equation (13.59) is simply the sum

$$P^\mu = \sum_{\pm s} \int d^3p \; p^\mu[N^+(p,s) + N^-(p,s)] \tag{13.60}$$

The charge operator can be similarly expressed in this representation. Inserting (13.50) into (13.49) for the conserved charge, now normal-ordered, we verify by direct calculation that

$$\begin{aligned} Q = \int d^3x \; :\psi^\dagger\psi: &= \sum_{\pm s} d^3p \; :b^\dagger(p,s)b(p,s) + d(p,s)d^\dagger(p,s): \\ &= \sum_{\pm s} \int d^3p \; [N^+(p,s) - N^-(p,s)] \end{aligned} \tag{13.61}$$

The effect of normal-ordering is to dispose of an infinite constant: the total charge of the vacuum of filled negative-energy states. Equations (13.60) and (13.61) show the symmetric appearance of electrons and positrons of equal masses and opposite charges in the quantum theory of the Dirac field. This is the same charge symmetry as established and discussed in the positron theory development of Chap. 5. There the conserved current $\bar{\psi}\gamma^\mu\psi$ was coupled to the electromagnetic potential and the charge (13.61) was specifically identified as an *electric* charge. This identification is based on the classical analogy; however, broader interpretations of the conserved charge are also possible, as we shall discuss in Chap. 15.

Turning next to the angular-momentum operator (13.48), now normal-ordered, we show that the state of a positron at rest and with spin along the z direction, that is,

$$\Psi_{1\,\text{positron}} = d^\dagger(p,s)|0\rangle$$

with $$p = (m,0,0,0) \qquad \text{and} \qquad s = (0,0,0,+1) \tag{13.62}$$

is an eigenstate of $J_z = \int :\psi^\dagger \left[\left(\mathbf{r}\times\frac{1}{i}\nabla\right)_z + \frac{1}{2}\sigma_z\right]\psi: d^3x$ with eigenvalue $+\frac{1}{2}$ and therefore really represents a particle state with spin $+\frac{1}{2}$ along the z axis. We form

$$J_z\,\Psi_{1\,\text{positron}} = [J_z,d^\dagger(p,s)]|0\rangle$$

where we used $J_z|0\rangle = 0$, and compute the commutator:

$$J_z\Psi_{1\,\text{positron}} = -\int d^3x \frac{1}{(2\pi)^{3/2}}\sqrt{\frac{m}{E_p}}\,e^{-ip\cdot x}v^\dagger(p,s)\left[\left(\mathbf{r}\times\frac{1}{i}\nabla\right)_z + \frac{1}{2}\sigma_z\right]\psi(x)|0\rangle$$

Since the orbital angular-momentum operator $L_z = -i(\mathbf{r}\times\nabla)_z$ is hermitian, we may take it to operate to the left, in which case $L_z \to 0$ for a particle at rest.[1] Then, since

$$v(p,s) = \begin{bmatrix}0\\0\\0\\1\end{bmatrix}$$

in the rest frame for our choice (13.62) of s [see (5.7)], we have

$$v^\dagger(p,s)\sigma_z = -v^\dagger(p,s)$$

and

$$\begin{aligned} J_z\Psi_{1\,\text{positron}} &= +\frac{1}{2}\int d^3x \frac{1}{(2\pi)^{3/2}}\,e^{-ip\cdot x}\sqrt{\frac{m}{E_p}}\,v^\dagger(p,s)\psi(x)|0\rangle \\ &= \tfrac{1}{2}d^\dagger(p,s)|0\rangle = +\tfrac{1}{2}\Psi_{1\,\text{positron}} \end{aligned}$$

The quantum formalism of the Dirac field with hole theory thus leads to

[1] To rigorously justify this parts integration, it is better to use packets instead of plane waves.

the desired result that the positron spin is opposite to that of the missing negative-energy electron.

13.5 Relativistic Covariance

The displacement and Lorentz invariance of the quantum theory of the Dirac field is established by verifying that the anticommutation relations (13.53) and (13.54) lead to the Heisenberg equations (11.70) and (11.73):

$$\begin{aligned} i[P^\mu,\psi(x)] &= \frac{\partial\psi(x)}{\partial x_\mu} \\ i[M^{\mu\nu},\psi(x)] &= x^\mu \frac{\partial\psi(x)}{\partial x_\nu} - x^\nu \frac{\partial\psi(x)}{\partial x_\mu} + \frac{1}{4}[\gamma^\mu,\gamma^\nu]\psi(x) \end{aligned} \tag{13.63}$$

The proof of (13.63) with the use of the P^μ and $M^{\mu\nu}$ of (13.46) and (13.47) is straightforward, and it is left as an exercise for the reader.

The equal-time anticommutation relations can be generalized to two different times for the free Dirac field, since we have the explicit solutions (13.50), and can also be put in covariant form. Returning to (13.53) for unequal times we find

$$\begin{aligned} \{\psi_\alpha(\mathbf{x},t),\psi_\beta^\dagger(\mathbf{x}',t')\} &= \int \frac{d^3p}{(2\pi)^3 2E_p} \{[(\not p + m)\gamma^0]_{\alpha\beta} e^{-ip\cdot(x-x')} \\ &\qquad - [(m - \not p)\gamma^0]_{\alpha\beta} e^{ip\cdot(x-x')}\} \\ &= ((i\nabla_x + m)\gamma^0)_{\alpha\beta} i\Delta(x - x') \end{aligned}$$

where $\Delta(x - x')$ is the invariant singular function first encountered in (12.37).

Multiplying by γ^0 gives

$$\{\psi_\alpha(x),\bar\psi_\beta(x')\} = i(i\nabla_x + m)_{\alpha\beta}\,\Delta(x - x') \equiv -iS_{\alpha\beta}(x - x') \tag{13.64}$$

Similarly, we find from (13.54)

$$\{\psi_\alpha(x),\psi_\beta(x')\} = \{\bar\psi_\alpha(x),\bar\psi_\beta(x')\} = 0 \tag{13.65}$$

We can verify the covariance of these relations by applying (11.67), the Lorentz transformation, to the fields:

$$U(a,b)\psi(x)U(a,b)^{-1} = S^{-1}(a)\psi(ax + b)$$

where for a spinor field the matrix S satisfies

$$S^{-1}\gamma^\mu S = a^\mu{}_\nu\gamma^\nu$$

For example, the right-hand side of (13.64) is not an operator and is unchanged if we make a similarity transformation $U(a,b) \cdots U^{-1}(a,b)$ with the unitary operators which Lorentz-transform the states. On the left-hand side we find

$$\begin{aligned} U(a,b)\{\psi_\alpha(x),\bar{\psi}_\beta(x')\}U^{-1}(a,b) &= S^{-1}_{\alpha\tau}(a)\{\psi_\tau(ax+b),\bar{\psi}_\lambda(ax'+b)\}S_{\lambda\beta}(a) \\ &= S^{-1}_{\alpha\tau}(a)(-\nabla\!\!\!/_{ax}+im)_{\tau\lambda}S_{\lambda\beta}(a)\Delta(ax-ax') \\ &= (-\nabla\!\!\!/_x+im)_{\alpha\beta}\Delta(x-x') \end{aligned}$$

since $\Delta(x-x')$ is an invariant function of the coordinate interval and $S^{-1}(a)\nabla\!\!\!/_{ax}S(a) = \nabla\!\!\!/_x$, as already computed in Chap. 2. This verifies covariance of (13.64), since both sides transform under a Lorentz transformation in the same way. For (13.65) this result follows immediately.

As already pointed out in (12.41), $\Delta(x-y)$, and therefore $S(x-y)$, vanishes for space-like intervals, $(x-y)^2 < 0$. From this it follows that, although the fields themselves do not commute, bilinear forms constructed at a point commute with one another for space-like separations of coordinates:

$$\begin{aligned} &[\bar{\psi}_\alpha(x)\psi_\beta(x),\bar{\psi}_\lambda(x')\psi_\tau(x')] \\ &\quad = \bar{\psi}_\alpha(x)\{\psi_\beta(x),\bar{\psi}_\lambda(x')\}\psi_\tau(x') - \{\bar{\psi}_\alpha(x),\bar{\psi}_\lambda(x')\}\psi_\beta(x)\psi_\tau(x') \\ &\qquad + \bar{\psi}_\lambda(x')\bar{\psi}_\alpha(x)\{\psi_\beta(x),\psi_\tau(x')\} - \bar{\psi}_\lambda(x')\{\bar{\psi}_\alpha(x),\psi_\tau(x')\}\psi_\beta(x) \\ &\quad = 0 \qquad \text{for } (x-x')^2 < 0 \end{aligned} \tag{13.66}$$

Since the amplitudes with which we associate a physical meaning—such as charge density or momentum density—are constructed of hermitian bilinear forms and satisfy (13.66), the Dirac theory, like that of the Klein-Gordon field, meets our intuitive demands of microscopic causality.

13.6 The Feynman Propagator

In concluding the discussion of the free Dirac field, we construct the one-particle Green's function corresponding to the Feynman propagator of positron theory. The amplitude to create an electron at the point $x = (\mathbf{x},t)$ is

$$\Psi_\alpha(x) = \psi_\alpha^\dagger(x)|0\rangle$$

where the index $\alpha = 1, 2, 3, 4$ labels the particular spinor component. The β, α spinor component of the amplitude for this electron to propagate to x' $(t' > t)$, where it is destroyed, is then

$$\langle 0|\psi_\beta(x')\psi_\alpha^\dagger(x)|0\rangle\theta(t'-t) \tag{13.67}$$

According to the considerations of Chap. 6, the Feynman propagator does not vanish for $t' < t$ but also describes the propagation of a positive-energy positron created at x' forward in time to the point x, where it is destroyed. As in the case of the Klein-Gordon field, this is accomplished by interchanging the operators in (13.67). The amplitude that a positive-energy positron is created at x' and destroyed at x is

$$\langle 0|\psi_\alpha{}^\dagger(x)\psi_\beta(x')|0\rangle\theta(t - t') \tag{13.68}$$

Both amplitudes (13.67) and (13.68) increase the charge by one unit at the point x' and decrease it at x; their difference forms a Green's function. Defining $S_F(x',x)$ by

$$\begin{aligned}(S_F(x',x)\gamma^0)_{\beta\alpha} = & -i\langle 0|\psi_\beta(x')\psi_\alpha{}^\dagger(x)|0\rangle\theta(t' - t)\\ & + i\langle 0|\psi_\alpha{}^\dagger(x)\psi_\beta(x')|0\rangle\theta(t - t')\end{aligned} \tag{13.69}$$

we find that

$$\begin{aligned}(i\nabla\!\!\!/_{x'} - m)_{\lambda\beta}(S_F(x',x)\gamma^0)_{\beta\alpha} &= \gamma^0_{\lambda\beta}\langle 0|\{\psi_\beta(x'),\psi_\alpha{}^\dagger(x)\}|0\rangle\delta(t' - t)\\ &= \gamma^0_{\lambda\alpha}\delta^4(x' - x)\end{aligned}$$

or

$$(i\nabla\!\!\!/_{x'} - m)S_F(x' - x) = \delta^4(x' - x) \tag{13.70}$$

$S_F(x' - x)$ is identical with the Feynman propagator defined and extensively used in the positron theory discussions of the companion volume. The evaluation of (13.69) gives the same sum over wave functions as in (6.48) for the free Feynman propagator.

In the field theory formalism we have come to the Feynman Green's function by considering positive-energy electrons and positrons always propagating forward in time. The electrons and positrons appear symmetrically in (13.69), the relative minus sign between the terms being dictated by the necessity of forming an anticommutator in (13.70). In the positron theory developed in Chap. 6, we followed along a particle path with the charge, moving forward in time as a positive-energy electron and backward as a negative-energy one. This identification of a positron of positive energy moving forward in time with an electron of negative energy moving backward has already been made in Chap. 5.

The Feynman propagator (13.69) plays a central role in field theory calculations, as it did in the earlier propagator considerations. We may express it more compactly in terms of the time-ordered product introduced in (12.72) for the Klein-Gordon field. In order to accommodate the minus sign in (13.69), we modify the T symbol to include a minus sign for *each* reordering of pairs of fields which are quantized with elementary anticommutators. Thus for two Fermi-

Dirac fields $a(x)$ and $b(x')$

$$T(a(x)b(x')) = a(x)b(x')\theta(t - t') - b(x')a(x)\theta(t' - t)$$
$$= -T(b(x')a(x)) \tag{13.71}$$

The Feynman propagator is

$$S_F(x',x)_{\beta\alpha} = -i\langle 0|T(\psi_\beta(x'),\bar{\psi}_\alpha(x))|0\rangle \tag{13.72}$$

Problems

1. Write the hamiltonian both in the form of n-particle quantum mechanics and in the form of a field theory when two-body interaction potentials are present.

2. Prove that $\mathfrak{M}^{\mu\nu\lambda}$ has the vanishing divergence $\partial/\partial x^\mu\ \mathfrak{M}^{\mu\nu\lambda} = 0$ and, therefore, $M^{\nu\lambda} = \int d^3x\ \mathfrak{M}^{0\nu\lambda}$ is a constant independent of time.

3. Verify (13.54) and (13.56).

4. Prove (13.63). Also, obtain the result in $\mathbf{p}$ space.

14.1 Introduction

Since the electromagnetic field is observable classically, it, above all other fields, should be quantized according to the canonical procedure. It is ironic that of the fields we shall consider it is the most difficult to quantize. Indeed, one of the most common versions of this quantization—that developed by Gupta and Bleuler[1]—abandons the notion we have cherished so far of a positive definite probability. However, a canonical procedure for quantization is possible. It was, in fact, the original procedure applied[2] to the Maxwell field in 1929, but it has the disadvantage of not being manifestly covariant. The Gupta-Bleuler procedure, stimulated twenty years later by the modern methods of covariant calculation, provides a covariant procedure of quantization, although at the cost of a cogent physical interpretation.

We shall follow here the historical procedure of canonical quantization. The Gupta-Bleuler method is discussed in many texts, among them Schweber, Jauch and Rohrlich, and Bogoliubov and Shirkov.[3]

The difficulties of the quantization originate in the use of more variables than there are independent degrees of freedom. One would like to describe the electromagnetic field in terms of the four components of the vector potential A_μ. Although it is the field strengths defined by

$$F_{\mu\nu} = \frac{\partial A_\mu}{\partial x^\nu} - \frac{\partial A_\nu}{\partial x^\mu} \tag{14.1}$$

which have immediate physical significance, we have already seen in our calculations of the companion volume that the potentials A_μ are what occur naturally in the interaction terms and transition amplitudes. However, the four components of A_μ cannot all be treated as independent variables. Therefore, the canonical quantization procedure encounters some difficulties. These are the very same difficulties which must be faced in constructing a classical canonical formulation of the Maxwell field.

Another way of saying this is that the free electromagnetic field when decomposed into plane waves consists only of transverse waves,

[1] S. N. Gupta, *Proc. Phys. Soc. (London)*, **A63**, 681 (1950); K. Bleuler, *Helv. Phys. Acta*, **23**, 567 (1950).

[2] Cf. E. Fermi, *Rev. Mod. Phys.*, **4**, 87 (1932).

[3] S. Schweber, "An Introduction to Relativistic Quantum Field Theory," Harper & Row, Publishers, Incorporated, New York 1961. J. M. Jauch and F. Rohrlich, "The Theory of Photons and Electrons," Addison-Wesley Publishing Company, Inc., Reading, Mass., 1955. N. N. Bogoliubov and D. V. Shirkov, "Introduction to the Theory of Quantized Fields," Interscience Publishers, Inc., New York, 1959.

that is, of waves for which the polarization vector is space-like and orthogonal to the wave vector. This condition of transversality is a restriction on the orientation of the vector potential. Only these two transverse components of the vector potential need be considered dynamical variables and deserve to be quantized. On the other hand, there is no unique invariant way of choosing two independent transverse polarization vectors corresponding to a given wave vector, because there exists always that set of preferred Lorentz frames where the time component of each polarization vector vanishes. It is here that the difficulties with manifest Lorentz covariance begin. This will become only too evident when the details are presented.

In setting up the formalism, we shall abandon *manifest* Lorentz covariance by making a particular choice of photon polarizations. However, we start with the Lorentz-covariant Maxwell field equations, and when the smoke clears, we shall finally be led back to the same covariant rules of calculation derived in Chaps. 7 and 8 on the basis of more intuitive arguments. These rules lead to the same results in all Lorentz frames.

14.2 Quantization

The components of the electromagnetic field strengths **E** and **B** form an antisymmetric tensor of second rank, denoted by $F^{\mu\nu}$, with components

$$F^{\mu\nu} = \mu\downarrow\overset{\nu\,\rightarrow}{\begin{bmatrix} 0 & E_x & E_y & E_z \\ -E_x & 0 & B_z & -B_y \\ -E_y & -B_z & 0 & B_x \\ -E_z & B_y & -B_x & 0 \end{bmatrix}} \tag{14.2}$$

The relation of $F^{\mu\nu}(x)$ to the vector potential $A^\mu(x) = (\Phi,\mathbf{A})$ as given by (14.1) is

$$\mathbf{E} = -\nabla\Phi - \dot{\mathbf{A}} \qquad \mathbf{B} = \nabla\times\mathbf{A} \tag{14.3}$$

This implies the two Maxwell equations

$$\nabla\times\mathbf{E} = -\dot{\mathbf{B}} \qquad \nabla\cdot\mathbf{B} = 0 \tag{14.4}$$

The other two Maxwell equations are expressed by

$$\frac{\partial F^{\mu\nu}}{\partial x^\nu} = 0$$

or

$$\nabla\cdot\mathbf{E} = 0 \qquad \nabla\times\mathbf{B} = \dot{\mathbf{E}} \tag{14.5}$$

in the absence of source charges and currents. Furthermore, it follows that all field components satisfy the wave equation

$$\Box F_{\mu\nu}(x) = 0 \tag{14.6}$$

To any given field strength $F^{\mu\nu}(x)$ there exist many potentials differing from one another by a gauge transformation

$$\tilde{A}_\mu(x) = A_\mu(x) + \frac{\partial \Lambda(x)}{\partial x^\mu} \tag{14.7}$$

where $\Lambda(x)$ may be an arbitrary function of $\mathbf{x}$ and t. Equation (14.7) expresses the gauge freedom in the choice of the vector potential; if $A_\mu(x)$ satisfies (14.1), so does $\tilde{A}_\mu(x)$. At this point we leave open the choice of gauge.

In order to derive (14.5) from a lagrangian by Hamilton's principle, we multiply by an infinitesimal variation $\delta A_\mu(x)$ which vanishes at t_1 and t_2 and integrate over all space-time within the interval (t_1,t_2)

$$\begin{aligned} 0 &= \int_{t_1}^{t_2} d^4x \frac{\partial F^{\mu\nu}(x)}{\partial x^\nu} \delta A_\mu(x) = -\int_{t_1}^{t_2} d^4x\, F^{\mu\nu}\, \delta \frac{\partial A_\mu}{\partial x^\nu} \\ &= -\frac{1}{2} \int_{t_1}^{t_2} d^4x\, F^{\mu\nu}\, \delta F_{\mu\nu} = -\frac{1}{4}\, \delta \int_{t_1}^{t_2} d^4x\, F_{\mu\nu}F^{\mu\nu} \end{aligned} \tag{14.8}$$

A satisfactory lagrangian density for a free Maxwell field is thus

$$\mathfrak{L} = -\frac{1}{4} F_{\mu\nu}F^{\mu\nu} = -\frac{1}{2}\left(\frac{\partial A_\mu}{\partial x^\nu} - \frac{\partial A_\nu}{\partial x^\mu}\right)\frac{\partial A^\mu}{\partial x_\nu} = \frac{1}{2}(E^2 - B^2) \tag{14.9}$$

Equation (14.8) shows that with this $\mathfrak{L}$ Hamilton's principle gives the field equations (14.5) if each of the four components of $A^\mu(x)$ is treated as an independent dynamical degree of freedom.

We now construct conjugate momenta from $\mathfrak{L}$ by the standard prescription:

already a problem. π^0 vanishes.

$$\pi^0 = \frac{\partial \mathfrak{L}}{\partial \dot{A}_0} = 0 \qquad \pi^k = \frac{\partial \mathfrak{L}}{\partial \dot{A}_k} = -\dot{A}^k - \frac{\partial A_0}{\partial x^k} = E^k \tag{14.10}$$

This gives the hamiltonian density

$$\mathfrak{H} = \sum_{k=1}^{3} \pi^k \dot{A}_k - \mathfrak{L} = \tfrac{1}{2}(E^2 + B^2) + \mathbf{E} \cdot \nabla \Phi \tag{14.11}$$

The hamiltonian is simply

$$H = \int d^3x\, \mathfrak{H} = \tfrac{1}{2}\int d^3x\, (E^2 + B^2) \tag{14.12}$$

where the final term is disposed of by an integration by parts and use of the Maxwell equation

$$\nabla \cdot \mathbf{E} = 0$$

Quantization of the Maxwell field is carried out by treating $A^\mu(x)$ as an operator and imposing commutation relations between A^μ and the canonical momenta π^k. We attempt to follow closely the canonical formalism and start with the vanishing equal-time commutators

$$\begin{aligned} [A^\mu(\mathbf{x},t),A^\nu(\mathbf{x}',t)] &= 0 \\ [\pi^k(\mathbf{x},t),\pi^j(\mathbf{x}',t)] &= 0 \\ [\pi^k(\mathbf{x},t),A^0(\mathbf{x}',t)] &= 0 \end{aligned} \tag{14.13}$$

This quantization procedure singles out the scalar potential A_0 at the expense of manifest covariance. Since the momentum $\pi^0(x)$ conjugate to $A^0(x)$ vanishes, $A^0(x)$ commutes with all operators and may be taken to be a pure number (c number) and not an operator, in contrast to the space components $A^k(x)$. It is at this point that we sacrifice manifest covariance and continue with the canonical form of the quantization procedure. In this choice we are armed with the conviction that our starting point, the Maxwell theory, is Lorentz covariant. Although we shall encounter along the way many expressions which are neither Lorentz nor gauge invariant, we shall find at the end that the physical results, namely, transition amplitudes (S-matrix elements), are Lorentz covariant and independent of gauge.

For the equal-time commutators between the potentials $A_j(\mathbf{x}',t)$ and the conjugate momenta $\pi^k(\mathbf{x},t)$ we are led by the canonical procedure to write[1]

$$[\pi^i(\mathbf{x},t),A_j(\mathbf{x}',t)] = -[E^i(\mathbf{x},t),A^j(\mathbf{x}',t)] = -i\delta_{ij}\delta^3(\mathbf{x} - \mathbf{x}') \tag{14.14}$$

However, we must now depart from the straight canonical path because (14.14) is not consistent with the Maxwell equations. Gauss's law

$$\nabla \cdot \mathbf{E} = 0$$

contains no time derivatives and is a constraint on the electric field. Consequently, the divergence with respect to $\mathbf{x}$ of the left-hand side of (14.14) vanishes, which is not true for the δ function on the right-hand side, however. Going into momentum space we see that

$$\sum_{i=1}^{3} \frac{\partial}{\partial x_i} \delta_{ij}\delta^3(\mathbf{x} - \mathbf{x}') = i \int \frac{d^3k}{(2\pi)^3} e^{i\mathbf{k}\cdot(\mathbf{x}-\mathbf{x}')} k_j \tag{14.15}$$

[1] As defined in Eqs. (14.10) and (14.11), π^i is conjugate to $A_i = -A^i$.

Handwritten margin notes (top):
signs:
$B^i = \epsilon_{ijk}\partial_j A^k$ where $\epsilon_{123} = +1$
$\vec{A} = \vec{B}\times\vec{C} \Rightarrow A^i = \epsilon_{ijk}B^j C^k$

write $[\pi^i, A_i]$ $(= [\pi^i, A_j])$
and require
∂_j of it $= 0$
get $\propto k_i$
so, it's $\propto k_i k_j$

To remove the divergence of $\delta_{ij}\delta^3(\mathbf{x} - \mathbf{x}')$, we modify its momentum space expansion by adding on a term proportional to k_j. This term must then be proportional to $k_i k_j$, the only other tensor, aside from δ_{ij}, at our disposal. The coefficient of $k_i k_j$ is determined uniquely from the condition that the divergence vanish, and we obtain the divergenceless, or transverse, "δ function" by the modification

$$\delta_{ij}\delta^3(\mathbf{x} - \mathbf{x}') \to \delta_{ij}^{tr}(\mathbf{x} - \mathbf{x}') \equiv \int \frac{d^3k}{(2\pi)^3} e^{i\mathbf{k}\cdot(\mathbf{x}-\mathbf{x}')}\left(\delta_{ij} - \frac{k_i k_j}{\mathbf{k}^2}\right) \quad (14.16)$$

The commutator condition (14.14) is then replaced by

$$[\pi^i(\mathbf{x},t), A^j(\mathbf{x}',t)] = +i\delta_{ij}^{tr}(\mathbf{x} - \mathbf{x}') \quad (14.17)$$

Notice that this requires $\nabla \cdot \mathbf{A}$ to commute with all operators, since the divergence with respect to $\mathbf{x}'$ of the right-hand side of (14.17) vanishes. That $\nabla \cdot \mathbf{A}$ is a c number already follows from the definition of $\mathbf{E}$ in terms of $\mathbf{A}$ and Φ and from Gauss's law

$$0 = \nabla \cdot \mathbf{E} = -\nabla^2\Phi - \nabla \cdot \dot{\mathbf{A}} \quad (14.18)$$

Since Φ is already known to be a c number, $\nabla \cdot \mathbf{A}$ must also be—except possibly for the zero-frequency mode. in which $\nabla\cdot A = 0$, so $\nabla\cdot A$ is undetermined by (14.18), but by (14.17) is a c-number.

Margin: $\nabla \cdot \vec{A}(k)$ → $\vec{k}\cdot\vec{A}(k)$

Thus the longitudinal part of $\mathbf{A}$ (in Fourier space, the component of $\mathbf{A}$ parallel to the wave vector) and the scalar potential are really not dynamical degrees of freedom. In fact, by an appropriate choice of gauge, $\nabla \cdot \mathbf{A}$ as well as Φ may be set to zero. We carry out this gauge transformation in two steps. First make the transformation

$$A'_\mu = A_\mu - \frac{\partial}{\partial x^\mu}\int_0^t \Phi(\mathbf{x},t')\,dt' \quad (14.19)$$

$\Rightarrow \vec{A}' = \vec{A} + \vec{\nabla}\int_0^t \Phi(x,t')dt'$

which removes the scalar potential.

To remove the longitudinal potential, we seek a $\Lambda(x)$ such that

$$0 = \nabla \cdot \mathbf{A}'' = \nabla \cdot \mathbf{A}' + \nabla^2\Lambda(x)$$

Thus the choice

Margin: since $\nabla^2\frac{1}{r} = -4\pi\delta^3(\vec{r})$

$$\Lambda(x) = \int \frac{d^3x'}{4\pi|\mathbf{x} - \mathbf{x}'|}\nabla' \cdot \mathbf{A}'(\mathbf{x}',t)$$

removes the longitudinal potential, and as a consequence of (14.18) and the vanishing of Φ',

Margin:
$\phi'' = \phi' - \partial_0\int\frac{d^3x'}{4\pi|x-x'|}\nabla'\cdot\vec{A}'(x',t)$
$= \int\frac{d^3x'}{4\pi|x-x'|}\nabla'\cdot\vec{E}'(x',t) = 0$

$$\frac{\partial\Lambda}{\partial t} = 0 \quad \text{and} \quad \Phi'' = \Phi' = 0$$

Margin: since $\vec{E}' = -\nabla\phi' - \dot{\vec{A}}' = -\dot{\vec{A}}'$ and $\nabla'\cdot\vec{E}'(x',t) = 0$

This gauge, for which

$$\Phi = 0 \qquad \nabla \cdot \mathbf{A} = 0 \tag{14.20}$$

is known as the radiation gauge; hereafter, we shall work in this gauge, although we lose manifest Lorentz and gauge covariance. It has the advantage that only the two transverse degrees of freedom of the radiation field appear in the formalism.

14.3 Covariance of the Quantization Procedure

Before accepting this modified quantization procedure as satisfactory, it is necessary to verify that with the commutation relations (14.13) and (14.17) the theory remains invariant under coordinate displacements and spatial rotations. The only effect of a Lorentz transformation should be a change of gauge.

Translational invariance is assured by verifying (11.70) with P_μ given by Noether's theorem (11.49)

$$P^0 = H = \frac{1}{2}\int d^3x :E^2 + B^2: = \frac{1}{2}\int d^3x :\dot{A}^2 + (\nabla \times \mathbf{A})^2:$$

$$\mathbf{P} = \int d^3x :\mathbf{E} \times \mathbf{B}: = -\int d^3x \sum_{i=1}^{3} :\dot{A}_i \nabla A_i: \tag{14.21}$$

where the dots denote normal ordering as in the Klein-Gordon theory. Similarly, to check invariance under spatial rotations, we verify (11.73) for the spatial components $i, j = 1, 2, 3$ with M^{ij} given by (11.57):

$$M^{ij} = \int d^3x : \sum_{r=1}^{3} \dot{A}^r \left(x^i \frac{\partial}{\partial x_j} - x^j \frac{\partial}{\partial x_i}\right) A^r - (\dot{A}^i A^j - \dot{A}^j A^i): \tag{14.22}$$

and with

$$\Sigma_{rs}^{ij} = g^{ir} g_s{}^j - g^i{}_s g^{jr} \tag{14.23}$$

according to (11.54). Finally, a transformation between Lorentz frames in relative motion is generated by

$$M^{0k} = \int d^3x :\left[x^0 \sum_{r=1}^{3} \dot{A}^r \frac{\partial A^r}{\partial x_k} - \frac{x^k}{2}(\dot{A}^2 + (\nabla \times \mathbf{A})^2)\right]: \tag{14.24}$$

Actually, under a Lorentz transformation A_μ does not transform as a four-vector but is supplemented by an additional gauge term.[1]

[1] Despite this, we shall continue to denote the transformation generated by (14.24) a Lorentz transformation.

Under an infinitesimal Lorentz transformation generated by M^{0k} according to (11.72) and (11.73) we find that[1]

$$U(\epsilon)A^{\mu}(x)U^{-1}(\epsilon) = A^{\mu}(x') - \epsilon^{\mu\nu}A_{\nu}(x') + \frac{\partial\Lambda(x',\epsilon)}{\partial x'_{\mu}} \tag{14.25}$$

with $\Lambda(x',\epsilon)$ an operator gauge function. It is clear that such a gauge term is necessary, since if

$$\Phi(x) = A_0(x) = 0$$

it follows that

$$U\Phi(x)U^{-1} = 0 \tag{14.26}$$

for any unitary transformation U. The structure of (14.25) guarantees that the gauge-invariant Maxwell equations are Lorentz covariant. The only additional requirements are that

$$\nabla' \cdot \mathbf{A}'(x') = 0 \tag{14.27}$$

and that the equal-time commutation relations (14.13) and (14.17) be valid also in the transformed frame. These relations may be verified explicitly. Thus two observers O and O' in relative motion who construct in their respective frames a quantum electrodynamics in radiation gauge are assured of a unitary transformation relating the states of O and O'.

The nontrivial calculations verifying (14.21) to (14.27) as well as the establishing of the covariance of the equal-time commutation relations are left as exercises for the reader.

14.4 Momentum Expansions

Expanding the potentials in plane waves and imposing the commutation relations (14.13) and (14.17) leads, as in the Klein-Gordon theory, to the interpretation of the expansion coefficients as creation and annihilation operators. The new feature for the Maxwell theory is that these quanta carry a unit of spin.

In the radiation gauge $\mathbf{A}(\mathbf{x},t)$ is transverse and in the plane-wave expansion

$$\mathbf{A}(\mathbf{x},t) = \int d^3k \sum_{\lambda=1}^{2} \boldsymbol{\varepsilon}(k,\lambda)A(\mathbf{k},\lambda,t)e^{i\mathbf{k}\cdot\mathbf{x}} \tag{14.28}$$

[1] See Prob. 2.

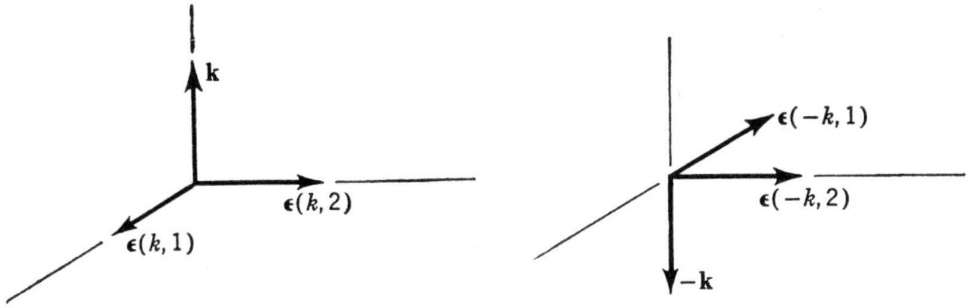

Fig. 14.1 Polarization unit vectors for photons of momentum k and $-k$.

the two unit vectors $\boldsymbol{\varepsilon}(k,\lambda)$ for each $\mathbf{k}$ and for $\lambda = 1, 2$ are orthogonal to $\mathbf{k}$

$$\boldsymbol{\varepsilon}(k,\lambda) \cdot \mathbf{k} = 0 \tag{14.29}$$

in order that $\nabla \cdot \mathbf{A} = 0$. They may also be conveniently chosen to be orthogonal to each other, for each $\mathbf{k}$,

$$\boldsymbol{\varepsilon}(k,\lambda) \cdot \boldsymbol{\varepsilon}(k,\lambda') = \delta_{\lambda\lambda'} \tag{14.30}$$

Then $\boldsymbol{\varepsilon}(k,1)$, $\boldsymbol{\varepsilon}(k,2)$, and $\hat{\mathbf{k}} = \mathbf{k}/|\mathbf{k}|$ form a three-dimensional orthogonal basis system as shown in Fig. 14.1. We also adopt the convention that

$$\boldsymbol{\varepsilon}(-k,1) = -\boldsymbol{\varepsilon}(k,1) \qquad \boldsymbol{\varepsilon}(-k,2) = +\boldsymbol{\varepsilon}(k,2) \tag{14.31}$$

as illustrated. By (14.30), this implies

$$\boldsymbol{\varepsilon}(k,\lambda) \cdot \boldsymbol{\varepsilon}(-k,\lambda') = (-)^\lambda \delta_{\lambda\lambda'} \tag{14.32}$$

From the Maxwell equations, it follows that $\mathbf{A}(x)$ in radiation gauge also satisfies the wave equation

$$\Box \mathbf{A} = 0$$

and we can make the expansion

$$\mathbf{A}(\mathbf{x},t) = \int \frac{d^3k}{\sqrt{2\omega(2\pi)^3}} \sum_{\lambda=1}^{2} \boldsymbol{\varepsilon}(k,\lambda)[a(k,\lambda)e^{-ik\cdot x} + a^\dagger(k,\lambda)e^{ik\cdot x}] \tag{14.33}$$

with $\qquad k_0 = \omega = |\mathbf{k}| \qquad$ and $\qquad k^2 = k_\mu k^\mu = 0$

In the same manner as used in the Klein-Gordon theory, we may invert (14.33) for the amplitudes $a(k,\lambda)$. Using (14.30) and (14.32), we obtain first

$$\omega \int d^3x\, e^{ik\cdot x}\boldsymbol{\varepsilon}(k,\lambda) \cdot \mathbf{A}(x) = \sqrt{\frac{(2\pi)^3\omega}{2}}\,[a(k,\lambda) + (-)^\lambda a^\dagger(-k,\lambda)e^{2i\omega t}]$$

$$i \int d^3x\, e^{ik\cdot x}\boldsymbol{\varepsilon}(k,\lambda) \cdot \dot{\mathbf{A}}(x) = \sqrt{\frac{(2\pi)^3\omega}{2}}\,[a(k,\lambda) - (-)^\lambda a^\dagger(-k,\lambda)e^{2i\omega t}]$$

and then

$$a(k,\lambda) = \int \frac{d^3x\, e^{ik\cdot x}}{\sqrt{2\omega(2\pi)^3}} \boldsymbol{\varepsilon}(k,\lambda)\cdot[\omega\mathbf{A}(x) + i\dot{\mathbf{A}}(x)]$$

$$= i\int \frac{d^3x\, e^{ik\cdot x}}{\sqrt{2\omega(2\pi)^3}} \overleftrightarrow{\partial_0}\boldsymbol{\varepsilon}(k,\lambda)\cdot\mathbf{A}(x)$$

$$a^\dagger(k,\lambda) = -i\int \frac{d^3x\, e^{-ik\cdot x}}{\sqrt{2\omega(2\pi)^3}} \overleftrightarrow{\partial_0}\boldsymbol{\varepsilon}(k,\lambda)\cdot\mathbf{A}(x) \tag{14.34}$$

Computing next the commutation relations for $a(k,\lambda)$ and $a^\dagger(k,\lambda)$ from (14.13) and (14.17), we find, for example,

$$[a(k,\lambda),a^\dagger(k',\lambda')] = \int \frac{d^3x\, d^3x'\, e^{i(k\cdot x-k'\cdot x')}}{2(2\pi)^3\sqrt{\omega\omega'}}(\omega'+\omega)\sum_{i,j=1,2,3}\epsilon_i(k,\lambda)\epsilon_j(k',\lambda')\delta_{ij}^{tr}(\mathbf{x}-\mathbf{x}')$$

Carrying out the integrals with the help of (14.16) and using the orthogonality relations (14.30 and (14.32) on the polarization vectors, we find

$$[a(k,\lambda),a^\dagger(k',\lambda')] = \delta^3(\mathbf{k}-\mathbf{k}')\delta_{\lambda\lambda'} \tag{14.35}$$

In the same manner we also find

$$[a(k,\lambda),a(k',\lambda')] = [a^\dagger(k,\lambda),a^\dagger(k',\lambda')] = 0 \tag{14.36}$$

The expansion coefficients for the two transverse dynamically independent components of the vector potential are thus quantized with the same commutation relations as in the Klein-Gordon theory. We may interpret the $a^\dagger(k,\lambda)$ and $a(k,\lambda)$ as creation and destruction operators of energy ω and momentum $\mathbf{k}$ upon observing that the hamiltonian (14.12) may be written in the momentum representation with the help of (14.3), (14.21), and (14.33) as

$$H = \int \frac{d^3x}{2} :(E^2+B^2): = \int d^3k\, \omega \sum_{\lambda=1}^{2} a^\dagger(k,\lambda)a(k,\lambda) \tag{14.37}$$

Similarly, for the total momentum

$$\mathbf{P} = \int d^3x :\mathbf{E}\times\mathbf{B}: = \int d^3k\, \mathbf{k} \sum_{\lambda=1}^{2} a^\dagger(k,\lambda)a(k,\lambda) \tag{14.38}$$

The vacuum state Φ_0, that is, the state of lowest energy, is an eigenstate of H and $\mathbf{P}$ with eigenvalues zero and satisfies

$$a(k,\lambda)\Phi_0 = 0 \tag{14.39}$$

in complete analogy with (12.17) for the Klein-Gordon theory. The interpretation of $a^\dagger(k,\lambda)$ as a creation operator for a photon of four-momentum k_μ, with $k^2 = k_\mu k^\mu = 0$, is clear if we form the state

$$\Phi_{1,k\lambda} \equiv a^\dagger(k,\lambda)\Phi_0 \equiv a^\dagger(k,\lambda)|0\rangle \tag{14.40}$$

and compute

$$\begin{aligned} P^\mu\Phi_{1,k\lambda} &= \int d^3k'\, k'^\mu \sum_{\lambda'=1}^{2} a^\dagger(k',\lambda')a(k',\lambda')a^\dagger(k,\lambda)|0\rangle \\ &= k^\mu\Phi_{1,k\lambda} \end{aligned} \tag{14.41}$$

14.5 Spin of the Photon

Photons differ from the Klein-Gordon quanta in several ways. Since they satisfy the Einstein condition $k_\mu k^\mu = 0$, they have zero rest mass. In addition, the vector potential $\mathbf{A}(x)$ is real and upon quantization becomes a hermitian operator, so that photons carry no charge but are similar to the neutral mesons which emerge from quantizing the real Klein-Gordon theory. A new feature is the polarization vector $\boldsymbol{\varepsilon}(k,\lambda)$, which labels each photon and which is associated with the spin angular momentum. In particular, the vector character of $\mathbf{A}$ leads to photons of unit spin, while the constraint of transversality removes one degree of freedom. The projection of spin angular momentum along the direction of propagation cannot be zero but only ± 1. To show this, we use the angular-momentum operator (14.22) and compute the three-component of angular momentum of a one-photon state

$$M^{12}\Phi_{1,k\lambda} = [M^{12},a^\dagger(k,\lambda)]|0\rangle \tag{14.42}$$

whose wave vector is along the 3-axis, that is, $k\cdot x = \omega(t - x^3)$. M^{12} consists of two terms, the first of which can be identified as the orbital angular momentum. Its projection along the direction of motion must vanish, as may be verified directly by computing the commutator. There remains only the contribution of the spin terms, which yield with the help of (14.22) and (14.33)

$$[M^{12},a^\dagger(k,\lambda)] = \int \frac{d^3x\, e^{-ik\cdot x}}{\sqrt{2\omega(2\pi)^3}}[\epsilon^1(k,\lambda)\overleftrightarrow{\partial_0}A^2(x) - \epsilon^2(k,\lambda)\overleftrightarrow{\partial_0}A^1(x)] \tag{14.43}$$

using unit vectors $\hat{e}_3 = \frac{\vec{k}}{\omega}$, $\hat{e}_2 = \varepsilon(k,2)$, $\hat{e}_1 = \varepsilon(k,1)$

Using (14.34) and Fig. 14.1, this is just

$$[M^{12},a^\dagger(k,\lambda)] = i\epsilon^1(k,\lambda)a^\dagger(k,2) - i\epsilon^2(k,\lambda)a^\dagger(k,1) \tag{14.44}$$

Use $\varepsilon(k,1) = (1\,0\,0)$
$\varepsilon(k,2) = (0\,1\,0)$
$k = (0\,0\,\omega)$

By forming the linear combinations

$$a_{\mathrm{R}}^{\dagger}(k) = \frac{1}{\sqrt{2}}\,[a^{\dagger}(k,1) + ia^{\dagger}(k,2)]$$
$$a_{\mathrm{L}}^{\dagger}(k) = \frac{1}{\sqrt{2}}\,[a^{\dagger}(k,1) - ia^{\dagger}(k,2)] \tag{14.45}$$

representing right- and left-handed circularly polarized waves, respectively, we find

$$[M^{12},a_{\mathrm{R}}^{\dagger}(k)] = +a_{\mathrm{R}}^{\dagger}(k) \qquad [M^{12},a_{\mathrm{L}}^{\dagger}(k)] = -a_{\mathrm{L}}^{\dagger}(k) \tag{14.46}$$

which shows that a right-circularly polarized photon carries spin 1 along its direction of propagation.

14.6 The Feynman Propagator for Transverse Photons

In order to study the space-time evolution of a state of one transversely polarized photon, we construct the Feynman propagator. In analogy with the discussion of Sec. 12.6, for Klein-Gordon quanta, we form the amplitude for a transverse photon, created with polarization projection μ at x, to propagate forward in time to x' and be destroyed with projection ν:

$$\langle 0|A_\nu(x')A_\mu(x)|0\rangle\theta(t'-t) \tag{14.47}$$

For $t > t'$ we form the amplitude for the photon created at x' with projection ν to be destroyed at x with projection μ.

$$\langle 0|A_\mu(x)A_\nu(x')|0\rangle\theta(t-t') \tag{14.48}$$

The sum of (14.47) and (14.48) defines the Feynman propagator

$$\begin{aligned} iD_F^{\mathrm{tr}}(x',x)_{\nu\mu} &= \langle 0|A_\nu(x')A_\mu(x)|0\rangle\theta(t'-t) + \langle 0|A_\mu(x)A_\nu(x')|0\rangle\theta(t-t') \\ &= \langle 0|T(A_\nu(x')A_\mu(x))|0\rangle \end{aligned} \tag{14.49}$$

where T is the time-ordering operator introduced in (12.72) and (13.71).

We construct the explicit form of $D_F^{\mathrm{tr}}(x',x)_{\nu\mu}$ by expanding the fields in plane waves:

$$\begin{aligned} &D_F^{\mathrm{tr}}(x',x)_{\nu\mu} \\ &= -i\int\frac{d^3k}{2\omega(2\pi)^3}\sum_{\lambda=1,2}\epsilon_\nu(k,\lambda)\epsilon_\mu(k,\lambda)[\theta(t'-t)e^{-ik\cdot(x'-x)} + \theta(t-t')e^{ik\cdot(x'-x)}] \end{aligned} \tag{14.50}$$

In radiation gauge $\epsilon_\nu(k,\lambda)$ has no time component, that is,

$$\epsilon^\nu(k,\lambda) = (0,\boldsymbol{\varepsilon}(k,\lambda))$$

and the transversality condition depends only upon the orientation of the space vector **k** and not on the frequency ω. In the reference system with this gauge we may imitate our earlier propagator discussions and write the Feynman propagator in the form of a four-dimensional integral

$$D_F^{\mathrm{tr}}(x'-x)_{\nu\mu} = \int \frac{d^4k}{(2\pi)^4} \frac{e^{-ik\cdot(x'-x)}}{k^2+i\epsilon} \sum_{\lambda=1}^{2} \epsilon_\nu(k,\lambda)\epsilon_\mu(k,\lambda) \tag{14.51}$$

This Feynman propagator, reminiscent of that encountered first in Chap. 6, and again in Chap. 12, will occur frequently in subsequent calculations. As expressed in (14.51) it is not Lorentz covariant, since the $\epsilon_\nu(k,\lambda)$ are specified with space components only in a particular Lorentz frame. In order to separate out the coordinate dependence explicitly, we introduce a time-like unit vector $\eta^\mu = (1,0,0,0)$ in the frame in which we have carried out the quantization. For a given k^μ, a quartet of independent orthogonal unit vectors is completed by introducing

$$\hat{k}^\mu = \frac{k^\mu - (k\cdot\eta)\eta^\mu}{\sqrt{(k\cdot\eta)^2 - k^2}} \tag{14.52}$$

to join $\epsilon^\mu(k,1)$, $\epsilon^\mu(k,2)$, and η^μ. We may then write

$$\sum_{\lambda=1}^{2} \epsilon_\nu(k,\lambda)\epsilon_\mu(k,\lambda) = -g_{\nu\mu} + \eta_\nu\eta_\mu - \hat{k}_\nu\hat{k}_\mu$$

$$= -g_{\nu\mu} - \frac{k_\nu k_\mu}{(k\cdot\eta)^2 - k^2} + \frac{(k\cdot\eta)(k_\nu\eta_\mu + \eta_\nu k_\mu)}{(k\cdot\eta)^2 - k^2} - \frac{k^2\eta_\nu\eta_\mu}{(k\cdot\eta)^2 - k^2} \tag{14.53}$$

Introducing (14.53) into (14.51) gives

$$D_F^{\mathrm{tr}}(x'-x)_{\nu\mu} = g_{\nu\mu}D_F(x'-x)$$

$$- \int \frac{d^4k}{(2\pi)^4} \frac{e^{-ik\cdot(x'-x)}}{(k^2+i\epsilon)} \frac{k^2\eta_\nu\eta_\mu - (k\cdot\eta)(k_\nu\eta_\mu + \eta_\nu k_\mu) + k_\nu k_\mu}{(k\cdot\eta)^2 - k^2} \tag{14.54}$$

In (14.54), the first term

$$g_{\nu\mu}D_F(x'-x) = \lim_{m^2\to 0} (-g_{\nu\mu})\Delta_F(x'-x,\, m)$$

is just the Feynman propagator used extensively in the calculations of electromagnetic interactions in Chaps. 7, 8, and 9. $\Delta_F(x'-x,\, m)$ is the propagator of a spin-0 boson as encountered in Chap. 9.

All dependence of $D^{\mathrm{tr}}_{F\nu\mu}$ upon the η_μ is in the second term of (14.54). In the calculations of physical amplitudes in field theory, as the discussions of the following chapters show, the propagator $D_F^{\mathrm{tr}}(x' - x)_{\nu\mu}$ always finds itself sandwiched between conserved currents which are the source of the electromagnetic field. The terms proportional to k_μ or k_ν in (14.54) will then vanish as a result of current conservation, as discussed above (7.61) and further illustrated in discussing vacuum polarization (8.9). The remaining η-dependent term in (14.54) is then

$$-\int \frac{d^4k}{(2\pi)^4} \frac{e^{-ik\cdot(x'-x)}}{(k\cdot\eta)^2 - k^2} \eta_\nu\eta_\mu = \frac{-g_{\nu 0}g_{\mu 0}\delta(t' - t)}{4\pi|\mathbf{x}' - \mathbf{x}|} \tag{14.55}$$

in the special coordinate system with $\eta^\mu = (1,0,0,0)$. This has the form of the familiar static Coulomb interaction between two charges. In computing the total interaction between two charges at space-time points x and x', (14.55) is canceled by the Coulomb interaction, which exists in addition to that carried by the pure radiation field. The effective propagator reduces to the Lorentz-covariant first term of (14.54).

This rule, which will be proved in detail in Chap. 17 for general S-matrix elements, restores manifest Lorentz covariance and gauge invariance to the formalism where it is really needed—in the physical scattering amplitudes. The additional gauge-dependent terms in $D_F^{\mathrm{tr}}(x' - x)_{\nu\mu}$ contribute to unobservable quantities such as the renormalization constants Z_1 and Z_2 encountered in Chap. 8, which are not gauge invariant. Their appearance in $D^{\mathrm{tr}}_{F\nu\mu}$ is the price it is necessary to pay in order to quantize the Maxwell field within the canonical framework. These unpleasant terms all disappear, however, when we compute transition rates and cross sections for comparison with experiment.

Problems

1. Compute the commutation relations between field strengths and verify that they vanish for space-like intervals.

Answer:

$$[B_i(x'),B_j(x)] = \left[\delta_{ij}\nabla'\cdot\nabla - \frac{\partial}{\partial x_i'}\frac{\partial}{\partial x_j}\right] iD(x' - x)$$

$$[E_i(x'),B_j(x)] = -\frac{\partial}{\partial x_0'} \sum_{k=1}^{3} \epsilon_{ijk}\frac{\partial}{\partial x^k} iD(x' - x)$$

$$[E_i(x'),E_j(x)] = \left(\delta_{ij}\frac{\partial^2}{\partial x_0\,\partial x_0'} - \frac{\partial}{\partial x^i}\frac{\partial}{\partial x^{j'}}\right) iD(x' - x)$$

2. Verify the energy-momentum and angular-momentum operators constructed in (14.21), (14.22), and (14.24). Verify the transformation of the potentials given in (14.25) for a Lorentz transformation and construct $\Lambda(x',\epsilon)$ and $U(\epsilon)$ explicitly. Complete the proof of covariance of the Maxwell theory quantized in the radiation gauge by establishing (14.27) and the covariance of the equal-time commutators. If $\epsilon_{\mu\nu} = \epsilon_{0k}$

$$\frac{\partial\Lambda(x',\epsilon)}{\partial x'_\mu} = -\epsilon_{0k}\frac{\partial}{\partial x'_\mu}\int\frac{d^3x}{4\pi|\mathbf{x}-\mathbf{x}'|}\frac{\partial A^k}{\partial x_0}$$

By explicit calculation show that M_{ij} and M_{0k} commute with H.

3. Starting with the Maxwell lagrangian, with a (noncovariant) "photon mass" term $-\lambda^2\mathbf{A}^2(x)$ added for later convenience (λ^2 small), quantize the electromagnetic field using only the subsidiary condition $A_0 \equiv 0$. First construct the hamiltonian and show from the Hamilton equations

$$\frac{\partial H}{\partial \pi^k} = \dot{A}_k \qquad \frac{\partial H}{\partial A_k} = -\dot{\pi}^k$$

that as $\lambda \to 0$ one obtains Maxwell's equations, with the possible exception of an extra zero-frequency contribution to Gauss's law:

$$\nabla\cdot\dot{\mathbf{E}} = -\nabla\cdot\ddot{\mathbf{A}}(x) = \lambda^2\nabla\cdot\mathbf{A}(x)$$

Impose canonical commutation relations, for example,

$$i[\pi^k(\mathbf{x},t),A_j(\mathbf{x}',t)] = \delta_{kj}\delta^3(\mathbf{x}-\mathbf{x}')$$

and quantize the longitudinal as well as transverse modes. Show in particular that the dispersion law for these quanta is $\omega^2 = \lambda^2$.

4. (*a*) Solve Maxwell's equations within a volume bounded by two infinite parallel conducting plates separated by distance a. Quantize the electromagnetic field in this region, with due regard to the boundary conditions.

(*b*) Calculate the zero-point energy per unit area E_0, using a cutoff depending only upon frequency, and show that

$$E_0 = C_1 a + C_2 + \frac{\hbar c}{a^3}\frac{B_4}{4!}$$

where C_1 and C_2 are cutoff-dependent quantities and $B_4 = -1/30$ is the fourth Bernoulli number.

(*c*) Show that the force between the (neutral) plates comes from the last term only, and calculate its magnitude. [H. Casimir, *Koninkl. Ned. Akad. Wetenschap., Proc., Ser. B*, 793 (1948); M. Fierz, *Helv. Phys. Acta*, **33**, 855 (1960).]

15

Interacting Fields

15.1 Introduction

A theory of free fields alone has no physical content. The nature of the physical world is revealed to us only through the interactions between fields, to which we now turn.

In constructing general interactions, we shall be motivated by analogy with the electromagnetic field, and we shall first discuss the electromagnetic interactions of a charged particle from the point of view of second quantization.

For nonelectromagnetic interactions, the best clues to the structure of the coupling terms come from experimental observation of various conservation laws. These may be built into the theory by requiring the lagrangian to possess certain symmetries; in particular for each continuous symmetry transformation the Noether theorem of Chap. 11 automatically produces a conserved quantity.

15.2 The Electrodynamic Interaction

We adopted in all our earlier discussions the "minimal" substitution

$$p_\mu \rightarrow p_\mu - eA_\mu \tag{15.1}$$

corresponding to the classical interaction of a point charge, as the prescription for introducing electrodynamic couplings. We may preserve this classical correspondence by imitating our earlier prescription and introducing (15.1) into the lagrangian density for the electron and photon fields, which becomes

$$\mathfrak{L}(x) = \bar{\psi}(x)(i\nabla\!\!\!/ - e_0 A\!\!\!/(x) - m_0)\psi(x) - \tfrac{1}{4}F_{\mu\nu}(x)F^{\mu\nu}(x) \tag{15.2}$$

The lagrangian density (15.2) describes the local interaction of the electron and photon fields at the same point x. From it we deduce the coupled electron-photon field equations by independent variation with respect to the electron and photon field amplitudes

$$(i\nabla\!\!\!/ - m_0)\psi(x) = e_0 A\!\!\!/\psi(x) \tag{15.3a}$$

$$\frac{\partial F^{\mu\nu}(x)}{\partial x^\nu} = e_0\bar{\psi}(x)\gamma^\mu\psi(x) \tag{15.3b}$$

The first equation has the same form as in our earlier study of the one-particle theory. Now, however, the field $A_\mu(x)$ is not taken to be externally prescribed but is included in our dynamical system, since it is coupled back to the electron current in (15.3b). In turn, this electron current is constructed by solving (15.3*a*) for the motion under the influence of the field A_μ.

It is evident from (15.3) that in discussing the coupling between two fields we are up against a nonlinear problem of vast complexity. Even classically this is the case, as found in the study of radiation damping and the runaway solutions for the motions of classical charges under the influence of their own as well as each other's fields.[1] The quantum-mechanical problem certainly is not easier.

The coupling of the fields in (15.3) is already implicit in the calculations of electron-electron scattering, Compton scattering, and electron self-energy in Chaps. 7 and 8. For electron-electron scattering, we computed the motions of two charges under the influence of each other's fields. For Compton scattering we found the modification of the radiation field in the presence of an electron. For the self-energy problem we considered the interaction of the electron with the field A_μ produced by itself.

If we wish to compute the motions of electrons under the influence of an applied external electromagnetic field in addition to the radiation field, we need only add the potential for the applied field $A_\mu^{\text{ext}}(x)$ in (15.1). This leads to the field equations

→ influences the e⁻, but is not influenced by e⁻. Impossible, of course, but useful as an approx. to some experimental conditions.

$$(i\nabla\!\!\!/ - m_0)\psi(x) = e_0[A\!\!\!/(x) + A\!\!\!/^{\text{ext}}(x)]\psi(x)$$

$$\frac{\partial F^{\mu\nu}(x)}{\partial x^\nu} = e_0\bar{\psi}(x)\gamma^\mu\psi(x) \qquad (15.4)$$

contains only A_μ, not A_μ^{ext}

In writing the lagrangian and field equations for the coupled fields we have appended subscripts to the mass m_0 and charge e_0, in anticipation that these will not be the physically observable values m and e for the mass and charge. Indeed, we have already found that m_0 and e_0 are altered when we made the perturbation theory calculations in Chap. 8. There we found that to lowest order in e the corrections to the mass and charge were both logarithmically divergent. We do not enter here into the discussion of whether these renormalization constants $m_0 - m$ and Z_3^{-1}, if calculated exactly, would also be infinite. It is important to realize that the *necessity* of renormalization is in no way connected with the magnitude of the renormalization constants. However, because of the difficulties encountered in perturbation theory, we must proceed with utmost caution to isolate the divergences unambiguously from finite observable physical magnitudes. In fact, the great progress since 1948 has been that quantum electrodynamics has achieved a status of peaceful coexistence with its divergences so that

[1] See, for example, P. A. M. Dirac, *Proc. Roy. Soc.* (*London*), **A167,** 148 (1938), and G. N. Plass, *Rev. Mod. Phys.*, **33,** 37 (1961).

the finite physical amplitudes can be calculated to any desired order of accuracy, as limited in practice only by the labor involved.[1]

In order to extend the formalism from the classical to the quantum domain, we first identify the canonical momenta in order that the equal-time commutation relations may be written down. The interaction term introduced into the lagrangian (15.2) at the suggestion of classical correspondence

$$\mathcal{L}_{\text{int}} = -e_0 \bar{\psi}(x)\gamma_\mu \psi(x) A^\mu(x) \tag{15.5}$$

contains no time derivatives of the fields. The canonical momenta are then unchanged from the free-field expressions and are given by (13.43) and (14.10) if we stay in the radiation, or Coulomb, gauge

$$\nabla \cdot \mathbf{A} = 0 \tag{15.6}$$

because eqn. for Φ
(15.7-8) looks like
Coulomb's potential

as employed in Chap. 14. In this gauge the scalar potential no longer vanishes; from Gauss's law and (15.6)   *← NOTICE, NO ∂_0^2 appears.*

$$\nabla \cdot \mathbf{E}(x) = -\nabla \cdot \dot{\mathbf{A}}(x) - \nabla^2 \Phi(x) = -\nabla^2 \Phi(x) = e_0 \psi^\dagger(x)\psi(x) \neq 0 \tag{15.7}$$

For A^i, the differential eqn. on bot. of pg. 84 can be solved with the usual retarded integral, but an arbitrary free-field $\bar{A}^i$ could be added to it.
Thus, A^i is independent of J^i, whereas A^0 is determined by J^0. [Free field Coulomb Gauge has $A^0 = 0$.]

The scalar potential, however, is still not an independent dynamical variable, but is determined by the instantaneous charge distribution $\rho(\mathbf{x},t) \equiv \psi^\dagger(\mathbf{x},t)\psi(\mathbf{x},t)$

$$A_0(\mathbf{x},t) = \Phi(\mathbf{x},t) = e_0 \int \frac{d^3x'\; \psi^\dagger(\mathbf{x}',t)\psi(\mathbf{x}',t)}{4\pi|\mathbf{x}' - \mathbf{x}|} = e_0 \int \frac{d^3x'\; \rho(\mathbf{x}',t)}{4\pi|\mathbf{x}' - \mathbf{x}|} \tag{15.8}$$

Since the independent variables are the same as those in the free-field theories, we adopt the same canonical commutation relations as in (13.53), (13.54), (14.13), and (14.17):

$$\{\psi_\alpha(\mathbf{x},t),\psi_\beta^\dagger(\mathbf{x}',t)\} = \delta_{\alpha\beta}\delta^3(\mathbf{x} - \mathbf{x}')$$
$$\{\psi_\alpha(\mathbf{x},t),\psi_\beta(\mathbf{x}',t)\} = \{\psi_\alpha^\dagger(\mathbf{x},t),\psi_\beta^\dagger(\mathbf{x}',t)\} = 0$$
$$[\dot{A}_i(\mathbf{x},t),A_j(\mathbf{x}',t)] = -i\delta_{ij}^{\text{tr}}(\mathbf{x} - \mathbf{x}') \qquad i,j = 1,2,3$$
$$[A_i(\mathbf{x},t),A_j(\mathbf{x}',t)] = [\dot{A}_i(\mathbf{x},t),\dot{A}_j(\mathbf{x}',t)] = 0 \tag{15.9}$$

To complete the set, we impose vanishing equal-time commutation relations between the Dirac and Maxwell fields:

see p. 90

$$[\psi_\alpha(\mathbf{x},t),A_i(\mathbf{x}',t)] = 0 \qquad [\psi_\alpha(\mathbf{x},t),\dot{A}_i(\mathbf{x}',t)] = 0 \tag{15.10}$$

We make this choice because the ψ_α and A_i are independent canonical variables. As we have remarked, the scalar potential Φ is not an independent variable, but is determined from ψ by (15.8). It therefore

---

[1] J. Schwinger, *Quantum Electrodynamics*, Dover Publications, Inc., New York, 1958.

satisfies the commutation relations

$$
\begin{aligned}
[\Phi(\mathbf{x},t),A_i(\mathbf{x}',t)] &= [\Phi(\mathbf{x},t),\dot{A}_i(\mathbf{x}',t)] = 0 \\
[\Phi(\mathbf{x},t),\psi_\alpha(\mathbf{x}',t)] &= -\frac{e_0}{4\pi|\mathbf{x}-\mathbf{x}'|}\,\psi_\alpha(\mathbf{x}',t)
\end{aligned}
\tag{15.11}
$$

Passing to the hamiltonian by means of the canonical prescription, we find by repeating the same steps as in the free-field discussions

$$
\mathcal{H} = \frac{\partial \mathcal{L}}{\partial \dot{\psi}}\,\dot{\psi} + \frac{\partial \mathcal{L}}{\partial \dot{A}^k}\,\dot{A}^k - \mathcal{L} \tag{15.12}
$$

which leads to

$$
\mathcal{H} = \psi^\dagger(-i\boldsymbol{\alpha}\cdot\boldsymbol{\nabla} + \beta m_0)\psi + \tfrac{1}{2}(E^2 + B^2) + \mathbf{E}\cdot\boldsymbol{\nabla}\Phi + e_0\bar{\psi}\gamma_\mu\psi A^\mu \tag{15.13}
$$

The hamiltonian is then

$$
H = \int d^3x\,\mathcal{H} = \int d^3x\,\left\{\psi^\dagger(x)\left[\boldsymbol{\alpha}\cdot\left(\frac{1}{i}\boldsymbol{\nabla} - e_0\mathbf{A}\right) + \beta m_0\right]\psi(x) + \frac{1}{2}[E^2(x) + B^2(x)]\right\} \tag{15.14}
$$

where we have partially integrated the $\mathbf{E}\cdot\boldsymbol{\nabla}\Phi$ term to $-e_0\psi^\dagger\psi\Phi$, dropping the inessential surface term.[1]

In this form H is at first sight somewhat surprising, since it reveals explicitly only a coupling of the electron current with the transverse vector potential. What has happened to the electrostatic

[1] The problems faced in building a quantum theory of the coupled electromagnetic and charged Klein-Gordon fields are similar to those already encountered in this chapter for photons and electrons and will not be repeated here. (G. Wentzel, "Quantum Theory of Fields," Interscience Publishers, Inc., New York, 1949.) Indeed, they appear with a vengeance, since the coupling prescription (15.1) introduces interaction terms containing derivatives. The lagrangian density is

$$
\begin{aligned}
\mathcal{L} &= \left[\left(\frac{\partial}{\partial x_\nu} - ie_0A^\nu\right)\varphi^*\right]\left[\left(\frac{\partial}{\partial x^\nu} + ie_0A_\nu\right)\varphi\right] - m_0{}^2\varphi^*\varphi - \frac{1}{4}F_{\mu\nu}F^{\mu\nu} \\
&= \left[\left(\frac{\partial\varphi^*}{\partial x_\nu}\right)\left(\frac{\partial\varphi}{\partial x^\nu}\right) - m_0{}^2\varphi^*\varphi - \frac{1}{4}F_{\mu\nu}F^{\mu\nu}\right] + \left[-ie_0A_\nu\left(\varphi^*\frac{\partial\varphi}{\partial x_\nu} - \left(\frac{\partial\varphi^*}{\partial x_\nu}\right)\varphi\right)\right. \\
&\qquad \left. + e_0{}^2\varphi^*\varphi A_\nu A^\nu\right]
\end{aligned}
$$

Since the second, or interaction, term in $\mathcal{L}$ contains derivatives, the canonical momenta for the Klein-Gordon field are changed to

$$
\begin{aligned}
\pi &\equiv \frac{\partial\mathcal{L}}{\partial\dot{\varphi}} = \dot{\varphi}^* - ie_0A_0\varphi^* \\
\pi^* &\equiv \frac{\partial\mathcal{L}}{\partial\dot{\varphi}^*} = \dot{\varphi} + ie_0A_0\varphi
\end{aligned}
$$

interaction between charges? Actually, it is contained in the electric field energy term $\frac{1}{2}\int E^2\, d^3x$.

To display it, we decompose **E** in terms of its longitudinal and transverse parts

$$\mathbf{E} = \mathbf{E}_l + \mathbf{E}_t \qquad \mathbf{E}_l \equiv -\nabla\Phi \qquad \mathbf{E}_t \equiv -\dot{\mathbf{A}} \tag{15.15}$$

The total field energy then separates into two terms

$$\tfrac{1}{2}\int d^3x\, (E^2 + B^2) = \tfrac{1}{2}\int d^3x\, E_l^2 + \tfrac{1}{2}\int d^3x\, (E_t^2 + B^2) \tag{15.16}$$

where the cross term $\mathbf{E}_l \cdot \mathbf{E}_t$ vanishes by another integration by parts. The first term on the right of (15.16) is the total energy associated with the Coulomb field. Using Gauss's law to express it in terms of the charges gives

$$\frac{1}{2}\int E_l^2\, d^3x = \frac{e_0^2}{8\pi}\int d^3x\, d^3y\, \frac{\psi^\dagger(\mathbf{x},t)\psi(\mathbf{x},t)\psi^\dagger(\mathbf{y},t)\psi(\mathbf{y},t)}{|\mathbf{x}-\mathbf{y}|}$$
$$= \frac{e_0^2}{8\pi}\int d^3x\, d^3y\, \frac{\rho(\mathbf{x},t)\rho(\mathbf{y},t)}{|\mathbf{x}-\mathbf{y}|}$$

The second term in (15.16) is the energy of the transverse radiation field coupled to the current $\mathbf{j} = \psi^\dagger\boldsymbol{\alpha}\psi$, and it has the same structure as the free-field energy.

15.3 Lorentz and Displacement Invariance

The canonical procedure has given us the energy operator H in (15.14). The momentum operator is found in the same way to be

$$\mathbf{P} = \int d^3x\, (-i\psi^\dagger\nabla\psi + \mathbf{E}_t \times \mathbf{B}) \tag{15.17}$$

This coincides with the sum of the free-field momentum operators, (13.46) and (14.21), since the interaction contains no derivatives.

The Heisenberg relations,

$$[P_\mu,\psi(x)] = -i\,\frac{\partial\psi(x)}{\partial x^\mu} \qquad [P_\mu,A^k(x)] = -i\,\frac{\partial A^k(x)}{\partial x^\mu} \tag{15.18}$$

required for a translationally invariant theory, may be verified from the commutation relations (15.9) and (15.10) and from the field equations (15.3). The commutator relations and field equations thus form a mutually consistent set of equations on which to base our theory.

To verify Lorentz invariance, we turn to the angular-momentum tensor which may be constructed from the Noether prescription (11.56) and (11.57). The space components M^{ij} ($i, j = 1, 2, 3$) are just the

Using $\theta_{\sigma\nu}$ instead of $\mathcal{H}$ (from 15.14) in 15.19-20 changes
$\dfrac{M^{0i}}{\mathcal{H}} \rightarrow \dfrac{M^{0i}}{\mathcal{H}} + \int d^3x \left(\frac{i}{2}\right) \psi^\dagger \gamma^i \psi^\dagger \psi^i \psi = \dfrac{M^{0i}}{\theta_{\sigma\sigma}}$ *gives*
exactly $\phi_{5,19}$ for M^{0i} with spin term built into $\theta_{\sigma\sigma}$ part.

sum of the free-particle angular-momentum tensors. Since the commutation relations are also unchanged from the free-particle theory, it follows that the relations (11.72) and (11.73) for three-dimensional rotations are preserved. The interaction introduces additional terms into the generators M^{0k} ($k = 1, 2, 3$) of Lorentz transformations to moving systems. We write

$$M^{0k} = P^k t - \int d^3x \left[x^k \mathcal{3C}(x) - \frac{i}{2} \bar\psi(x)\gamma^k\psi(x) \right] \qquad (15.19)$$

with $\mathcal{3C}$ the hamiltonian density of (15.13). *(15.14)* Under an infinitesimal Lorentz transformation generated by M^{0k}, according to (11.72), we again find that the electromagnetic potentials,

$$A^\mu(x) = (\Phi(x),\mathbf{A}(x))$$

undergo a gauge transformation required to restore the transverse gauge in the new coordinate system.[1] As in (14.25) for the free field, we have

$$U(\epsilon)A^\mu(x)U^{-1}(\epsilon) = A^\mu(x') - \epsilon^{\mu\nu}A_\nu(x') + \frac{\partial\Lambda(x',\epsilon)}{\partial x'_\mu}$$

gauge k-form on $A\mu$

with[2]

$$\Lambda(x,\epsilon) = \epsilon_{0k} \int d^3y \frac{E^k(y) + e_0\rho(y)(y^k - x^k)}{4\pi|\mathbf{x} - \mathbf{y}|} \qquad (15.20)$$

phase k-form on ψ

The Dirac operator $\psi(x)$ undergoes a phase transformation in addition to the Lorentz transformation,

$e^{-ie_0\Lambda}$ *infinitesimal*

$$U(\epsilon)\psi(x)U^{-1}(\epsilon) = [1 - ie_0\Lambda(x,\epsilon)]S_{rs}^{-1}(\epsilon)\psi_s(x') \qquad (15.21)$$

This phase transformation is just the one required for the field equations to transform covariantly under U; for example,

for $\bar\psi$ phase factor is $e^{+ie_0\Lambda}$ so they cancel

$$U(\epsilon)\bar\psi(x)[i\bar\nabla - e_0\bar A(x)]\psi(x)U^{-1}(\epsilon) = \bar\psi(x')[i\bar\nabla_{x'} - e_0\bar A(x')]\psi(x')$$

In order to check the Lorentz invariance of the quantum field *$\bar\psi \psi$* theory formalism completely, it is also necessary to verify the invariance of the equal-time commutation relations[3] (15.9) and (15.10).

The details of these calculations are tedious and not completely trivial; they are left as exercises for the reader. We do not enter into

---

[1] B. Zumino, *J. Math Phys.*, **1**, 1 (1960).

[2] In constructing this form, the total differential $\mathbf{\nabla} \cdot (\mathbf{E}\Phi)$ was removed from $\mathcal{3C}$ in (15.13), while the term $\mathbf{E}_l \cdot \mathbf{E}_l = \mathbf{\nabla} \cdot (\dot{\mathbf{A}}\Phi)$ was not. *Because that's what you get from $\theta_{\sigma\sigma}$.*

[3] We cannot explicitly write covariant expressions for the commutators for time-like separations, as was possible for free fields, since we do not know the exact solutions of the equations of motion.

Either Adding $\nabla\cdot(\mathbf{E}\Phi)$ term or subtracting $\nabla\cdot(\dot{\mathbf{A}}\Phi)$ term from $\mathcal{H}$ in (15.19) adds to the right hand side of 15.20

$$\epsilon_{0j}\,\Phi(x) + \epsilon_{0i}\,\frac{\partial}{\partial x^i}\int d^3y\,\frac{\dot{\Phi}_j\,\Phi(y)}{4\pi|x-y|}$$

them here, assured that our prescription (15.1) has preserved gauge invariance and confident that the physical results are relativistically covariant even though we have chosen to quantize in a special gauge.

15.4 Momentum Expansions

The free-field momentum expansion (13.50) and (14.33) for $\psi(x)$ and $\mathbf{A}(x)$ must be modified in the presence of their interactions, since the fields no longer have the space-time coordinate dependence as solutions of free-wave equations. Indeed, lacking exact solutions of the coupled field equations (15.3), we do not know their complete coordinate development. In this situation it is convenient to make a three-dimensional Fourier expansion of their $\mathbf{x}$ dependence for a particular time, usually taken to be $t = 0$, and to describe their time development by the Heisenberg equations (15.18). We write then by formally integrating (15.18):

$$\psi(\mathbf{x},t) = e^{iHt}\psi(\mathbf{x},0)e^{-iHt} \qquad \mathbf{A}(\mathbf{x},t) = e^{iHt}\mathbf{A}(\mathbf{x},0)e^{-iHt} \tag{15.22}$$

and expand, as in (13.50) and (14.33), in a complete set of free-wave solutions:

$$\begin{aligned}
\psi(\mathbf{x},0) &= \sum_{\pm s} \int \frac{d^3p}{(2\pi)^{3/2}} \sqrt{\frac{m}{E_p}}\, [b(p,s)u(p,s)e^{+i\mathbf{p}\cdot\mathbf{x}} + d^\dagger(p,s)v(p,s)e^{-i\mathbf{p}\cdot\mathbf{x}}] \\
\psi^\dagger(\mathbf{x},0) &= \sum_{\pm s} \int \frac{d^3p}{(2\pi)^{3/2}} \sqrt{\frac{m}{E_p}}\, [b^\dagger(p,s)\bar{u}(p,s)\gamma_0 e^{-i\mathbf{p}\cdot\mathbf{x}} + d(p,s)\bar{v}(p,s)\gamma_0 e^{+i\mathbf{p}\cdot\mathbf{x}}] \\
\mathbf{A}(\mathbf{x},0) &= \int \frac{d^3k}{\sqrt{2\omega(2\pi)^3}} \sum_{\lambda=1}^{2} \boldsymbol{\varepsilon}(k,\lambda)[a(k,\lambda)e^{+i\mathbf{k}\cdot\mathbf{x}} + a^\dagger(k,\lambda)e^{-i\mathbf{k}\cdot\mathbf{x}}] \\
\dot{\mathbf{A}}(\mathbf{x},0) &= \int \frac{d^3k}{\sqrt{2\omega(2\pi)^3}} (-i\omega) \sum_{\lambda=1}^{2} \boldsymbol{\varepsilon}(k,\lambda)[a(k,\lambda)e^{i\mathbf{k}\cdot\mathbf{x}} - a^\dagger(k,\lambda)e^{-i\mathbf{k}\cdot\mathbf{x}}]
\end{aligned} \tag{15.23}$$

with $E_p \equiv \sqrt{|\mathbf{p}|^2 + m^2}$, $\omega = |\mathbf{k}|$, and $u(p,s)$ given by (13.51) and $\boldsymbol{\varepsilon}(k,\lambda)$ by (14.29), etc. The operator expansion coefficients $b(p,s)$, $d(p,s)$, $a(k,\lambda)$ and their hermitian conjugates are assigned the same commutation relations (13.52), (14.35), and (14.36) as for the free-field theory; in addition

$$\begin{aligned}
&[a(k,\lambda),b(p,s)] \\
&\quad = [a^\dagger(k,\lambda),b(p,s)] = [a(k,\lambda),d(p,s)] = [a^\dagger(k,\lambda),d(p,s)] \\
&\quad = 0
\end{aligned} \tag{15.24}$$

This assures the canonical commutation relations, (15.9) and (15.10), at $t = 0$; and hence by (15.22) at all times t.

The fields with their time dependence removed are in the Schrödinger picture and coincide with the Heisenberg fields (15.22) at time $t = 0$.

The formal correspondence between the expansions given in (12.7), (12.57), (13.50), and (14.33) for the free fields and the interacting ones may be summarized by the relations

$$
\begin{array}{ccc}
\textit{Free fields} & & \textit{Interacting fields} \\
a(k,\lambda)e^{-i\omega t} & \longrightarrow & e^{iHt}a(k,\lambda)e^{-iHt} \\
a^\dagger(k,\lambda)e^{+i\omega t} & \longrightarrow & e^{iHt}a^\dagger(k,\lambda)e^{-iHt} \\
b(p,s)e^{-i\omega t} & \longrightarrow & e^{iHt}b(p,s)e^{-iHt}
\end{array}
\tag{15.25}
$$

The operator expansion coefficients, however, no longer retain their simple physical interpretations as creation and destruction operators for single quanta of given definite masses as in (14.41), for instance, for free photons.

15.5 The Self-energy of the Vacuum; Normal Ordering

It is never hard to find trouble in field theory, and a difficulty already appears here. If we take the vacuum expectation value of Gauss's law (15.7), we see that

$$\langle 0|\boldsymbol{\nabla}\cdot\mathbf{E}|0\rangle = e_0\langle 0|\psi^\dagger(x)\psi(x)|0\rangle = e_0\sum_n |\langle 0|\psi^\dagger(x)|n\rangle|^2$$

where the sum is over a complete set of states. Thus even the vacuum carries a charge density which is in fact badly divergent. The physical origin of this charge is clear; it is just the infinite electrostatic charge of the electrons in the negative-energy sea.

This unpleasant situation is easily circumvented. One changes the theory by adding on a uniform external background charge of opposite sign which neutralizes the vacuum. Formally, one does this most conveniently by making the replacement in the preceding equations, including the Maxwell equations, of

$$\bar{\psi}\gamma_\mu\psi \Rightarrow \tfrac{1}{2}[\bar{\psi},\gamma_\mu\psi] \tag{15.26}$$

that is, one antisymmetrizes the electric current under the interchange of the Dirac fields, which amounts to subtracting an infinite c number from the operator current $\bar{\psi}\gamma_\mu\psi$. This change is also identical with normal-ordering the operators in the current. To see this, we refer back to the fields (15.22) and their momentum expansions (15.23) and observe that products of interacting field operators at a common time t can be displaced in time to $t = 0$ with the exponentials $e^{iHt} \cdots e^{-iHt}$ factoring out. These operators obey the same algebra as for noninteracting fields according to (15.24), and the normal ordering can be defined precisely as given for the free-field case; for example

$$:b(p,s)d^\dagger(p',s') + b^\dagger(p,s)b(p',s') + d(p,s)a^\dagger(k,\lambda): \\ = -d^\dagger(p',s')b(p,s) + b^\dagger(p,s)b(p',s') + a^\dagger(k,\lambda)d(p,s)$$

This gives in (15.26)

$$\begin{aligned} \tfrac{1}{2}[\bar{\psi}(x),\gamma_\mu\psi(x)] &= \bar{\psi}(x)\gamma_\mu\psi(x) - \tfrac{1}{2}\{\bar{\psi}(x),\gamma_\mu\psi(x)\} \\ &= :\bar{\psi}(x)\gamma_\mu\psi(x): + \{\bar{\psi}(x)^{(+)},\gamma_\mu\psi(x)^{(-)}\} - \tfrac{1}{2}\{\bar{\psi}(x),\gamma_\mu\psi(x)\} \\ &= :\psi(x)\gamma_\mu\psi(x): + 2\delta^3(0)g_{\mu 0} - 2\delta^3(0)g_{\mu 0} \end{aligned} \qquad (15.27)$$

where the highly singular terms arising in the charge density with $\mu = 0$ are computed and observed to cancel out with the aid of the free-field commutation relations as assigned to the a,b,d, etc. The superscripts $(+)$ and $(-)$ in (15.27) denote the individual terms in (15.23) proportional to a,b,d and $a^\dagger,b^\dagger,d^\dagger$, respectively. As illustrated in (15.25), they reduce to the free-field annihilation and creation operators.

Although it is not yet clear that (15.26) or (15.27) removes the background charge and neutralizes the vacuum, we may recall from the discussion of the noninteracting Dirac field that the vacuum expectation value of a normal product does in fact vanish. When we come later, in Sec. 15.12, to a study of the charge conjugation invariance of quantum electrodynamics, we shall confirm the result

$$\langle 0| :\bar{\psi}(x)\gamma_\mu\psi(x):|0\rangle = 0$$

as a consequence of this symmetry operation. For the present we adopt it for j_μ, writing

$$\begin{aligned} H = \int d^3x\, \mathcal{H}(x) &= \int d^3x\, :\psi^\dagger(x)[\boldsymbol{\alpha}\cdot(-i\nabla - e_0\mathbf{A}(x)) + \beta m_0]\psi(x) \\ &+ \frac{1}{2}[\dot{\mathbf{A}}(x)^2 + (\nabla\times\mathbf{A}(x))^2]: + \frac{e_0^2}{8\pi}\int \frac{d^3x\, d^3y}{|\mathbf{x}-\mathbf{y}|}[:\rho(\mathbf{x},t):\ :\rho(\mathbf{y},t):] \end{aligned} \qquad (15.28)$$

We may furthermore subtract the vacuum expectation value from the hamiltonian in order that energies shall be measured relative to the vacuum. This choice gives

$$\langle 0|H|0\rangle = 0$$

and removes the need of discussing several embarrassing divergences. The first is the zero-point energy of the field oscillators; for example,

$$\langle 0|\dot{\mathbf{A}}^2(x)|0\rangle = \infty$$

as discussed in connection with the Klein-Gordon theory (see (12.46)). In addition, there is an additional divergence in the final term which arises because the charge density $:\rho(x):$ is an operator which does not commute with H, although the total charge $Q \equiv \int :\rho(x):\, d^3x$ does:

$$[Q,H] = 0$$

Consequently, there are fluctuations in the vacuum charge density and a highly divergent Coulomb energy associated with these fluctuations.

The presence of these fluctuation energies of the vacuum raises the question of its very existence. Because of the extreme complexity of the coupled field equations, exact solutions are not available and indeed it has not been possible to show from the formalism that the vacuum state exists, that is, that there exists a lower bound to the energy spectrum of the hamiltonian. However, it is not possible even to begin constructing a sensible physical theory without appealing to the experimental evidence that the vacuum does indeed exist. That this question is nontrivial may be seen by considering a theory in which antiparticles repel particles. It may possibly be energetically favorable for pairs to be continually created, the energy required to do this being provided by the change in potential energy as the particle and antiparticle escape from each other.[1]

It is easy to determine the effect of our normal-ordering prescription for H in (15.28), and the analogous normal ordering of $\mathbf{P}$ and $M_{\mu\nu}$ in (15.17) and (15.19), on the discussion of Lorentz and displacement invariance of quantum electrodynamics given in Sec. 15.3. All that happens is that the resulting field equations appear in normal-ordered form in place of (15.3):

$$(i\nabla\!\!\!/ - m_0)\psi(x) = e_0 :A\!\!\!/\psi(x): \qquad \frac{\partial F^{\mu\nu}(x)}{\partial x^\nu} = e_0 :\bar{\psi}(x)\gamma^\mu\psi(x): \qquad (15.29)$$

Henceforth normal ordering shall be automatically understood to apply. The vacuum subtraction is of no concern, since it amounts to no more than removing a (possibly infinite) c number from the definition of H.

[1] F. J. Dyson, *Phys. Rev.*, **85**, 631 (1952).

15.6 Other Interactions

It is natural to extend the lagrangian formalism developed thus far to treat the interactions of other particles, such as the mesons and nucleons. A straightforward, albeit unimaginative, procedure is to associate with each particle a field which obeys a wave equation consistent with its known properties of spin, mass, and charge. The interactions of these particles among themselves are then modeled after the interaction of the Dirac or scalar particles with the electromagnetic field, and assumed to be local and derivable from a lagrangian density. In addition we require that the interactions be invariant under coordinate displacements and under proper homogeneous Lorentz transformations; that is, the lagrangian density is a scalar under proper Lorentz transformations. If the interactions of some particles appear to be invariant under improper symmetries such as parity, time reversal, or charge conjugation, still further restrictions may be posed. Finally, the conservation laws of nucleon number, lepton number, and electrical charge, as well as the approximate laws of isotopic spin and strangeness conservation, limit still further the form the interaction may take.

We use simplicity as a final although less physical guide in motivating this development. This was done, for instance, in Chap. 10 in discussing meson-nucleon scattering. There, in the charge-independent approximation, the wave equations [(10.33) and (10.34)] were constructed as

$$(i\nabla\!\!\!/ - M_0)\Psi = g_0 i\gamma_5(\boldsymbol{\tau}\cdot\boldsymbol{\phi})\Psi$$
$$(\Box + \mu_0^2)\boldsymbol{\phi} = -g_0\bar{\Psi}i\gamma_5\boldsymbol{\tau}\Psi$$

where the nucleon field has the isotopic components

$$\Psi(x) = \begin{bmatrix}\psi_p(x)\\ \psi_n(x)\end{bmatrix}$$

and the π-meson field, the components

$$\boldsymbol{\phi}(x) = \left(\frac{\varphi_+(x) + \varphi_-(x)}{\sqrt{2}},\, i\,\frac{\varphi_+(x) - \varphi_-(x)}{\sqrt{2}},\, \varphi_0(x)\right)$$

These may now be derived from the lagrangian density

$$\mathcal{L} = \bar{\Psi}(i\nabla\!\!\!/ - M_0)\Psi + \frac{1}{2}\left[\left(\frac{\partial\boldsymbol{\phi}}{\partial x_\mu}\right)\cdot\left(\frac{\partial\boldsymbol{\phi}}{\partial x^\mu}\right) - \mu_0^2\boldsymbol{\phi}\cdot\boldsymbol{\phi}\right] - ig_0\bar{\Psi}\gamma_5\boldsymbol{\tau}\Psi\cdot\boldsymbol{\phi} \tag{15.30}$$

by independent variations with respect to $\Psi(x)$ and $\boldsymbol{\phi}(x)$. We have again appended subscripts to the masses and coupling constants to emphasize that these quantities are not to be identified directly with observed masses and coupling parameters, but will be altered by renormalization effects similar to those occurring in the interactions of photons and electrons as shown in Chap. 8.

We observe at this point the ease with which satisfactory coupled field equations may be derived from $\mathcal{L}$. Replacing the tortuous arguments of Chap. 10 regarding signs [for example (10.17), (10.19), (10.21), (10.24)] is the statement that, as $g_0 \to 0$, $\mathcal{L}$ reduces to the sum of lagrangians for the free particles, the signs of which are determined by the requirement of a positive definite free-particle hamiltonian density. The coupling term in (15.30) is determined uniquely by the requirements that:

1. It contains no derivatives.
2. It is linear in the meson field and bilinear in the nucleon field (implying vertices as drawn in Figs. 10.4 and 10.5)
3. It preserves the rotational invariance in isotopic spin space satisfied by the free lagrangian. (proper & improper X-forms; or 4. It's a true Lorentz scalar (proper + J + P)

We must emphasize again that the lagrangian (15.30) should be at best regarded as a rough guess. Other interaction terms such as

$$\frac{f}{\mu}\,\bar{\Psi}(x)\gamma_5\gamma_\mu\boldsymbol{\tau}\Psi(x)\cdot\frac{\partial\boldsymbol{\phi}(x)}{\partial x_\mu} \qquad \text{or} \qquad g\bar{\Psi}(x)\Psi(x)(\boldsymbol{\phi}\cdot\boldsymbol{\phi})^4$$

might be present. The first term is unpopular because it leads to a theory still divergent in perturbation theory after mass and coupling constant renormalization. The second violates the "principle of maximum simplicity"; if present, it would be in addition to the term in (15.30) but not in place of it, since by itself it cannot lead to processes with single meson production.[1]

One way of removing some of the singular behavior of a derivative coupling theory at high energy is to introduce a form factor which smears out the interaction, writing, for instance,

$$\mathcal{L}'(x) = \int d^4y\, d^4z\, g\bar{\Psi}(z)\gamma_5\gamma_\mu\boldsymbol{\tau}\Psi(y)\cdot\frac{\partial\boldsymbol{\phi}(z)}{\partial z_\mu}F((x-y)^2,(x-z)^2)$$

This is Lorentz invariant and form invariant under displacements. However, it is nonlocal, because local variations in the pion fields at z

[1] "Maximum simplicity" may, however, be subtle. For example, the ("simple") Einstein lagrangian for the gravitational field involves a square root of the determinant of the metric tensor, and from the point of view discussed here it is a mess.

influence the nucleon field at other space-time points, which violates the philosophy to which we have agreed in Chap. 11. In addition, it is very awkward to work with such forms.[1]

The domain of more complicated interaction lagrangians is virtually unexplored. There is as yet a lack of cogent physical motivation for choice of one over another. Such possibilities as above should not be overlooked, however.

Before continuing, we stress that the method of introducing for each particle in nature a separate field and a separate interaction term is dubious and at best phenomenological in character. It is not clear which of the particles in nature might be "bound states" or "excitations" of more fundamental fields.[2] In fact, the present trend of thinking is to attempt to reformulate the theory without the use of the lagrangian, retaining only the fundamental axioms of field theory. Instead of introducing, via a lagrangian, "bare" particles which are then "clothed," that is, gain structure owing to their interactions, one accepts the physical particles which exist and successively "undresses" them by investigating the structures which have the greatest spatial extent. There is considerable optimism that such a program, which is also manifestly phenomenological in approach, is essentially equivalent in physical consequences to the lagrangian formalism.[3]

15.7 Symmetry Properties of Interactions

Experiment and simplicity are our main guides in constructing interactions. In writing lagrangians for physical systems we shall associate a symmetry property of $\mathcal{L}$ with each conservation law observed in the laboratory. The conservation of energy, momentum, and angular momentum was discussed from this point of view in Chap. 11, and the lagrangians we write down will have this translational and Lorentz invariance built in. In addition, the conservation of charge, nucleon number, isotopic spin, etc., may also be associated with "internal" symmetries of the lagrangian. As was demonstrated in Chap. 11 [see (11.58) and following], invariance of $\mathcal{L}$ under a local transformation

[1] For some of the problems arising in nonlocal quantum field theories see M. Chretien and R. E. Peierls, *Proc. Roy. Soc.* (*London*), **A223**, 468 (1954).

[2] Although local fields may be constructed [K. Baumann, *Z. Physik*, **152,** 448 (1958); K. Nishijima, *Progr. Theoret. Phys.* (*Kyoto*), **11,** 995 (1958); and W. Zimmermann, *Nuovo Cimento,* **10,** 567 (1958)] for composite particles, the wave equations they satisfy are undoubtedly quite complicated and the lagrangian formalism awkward at best.

[3] See, for example, G. F. Chew, *Physics,* **1,** 77 (1964).

Q conserved ⟺ $\varphi \to e^{i\alpha Q}\varphi e^{-i\alpha Q} \simeq 1 + i\alpha[Q,\varphi]$ is symmetry ($\mathcal{L}$ invariant under it)

of fields of the form

$$\varphi_r(x) \to \varphi_r(x) - i\epsilon\lambda_{rs}\varphi_s(x) \tag{15.31}$$

where ϵ is an infinitesimal parameter, leads to the conserved current

$$\begin{aligned} J_\mu(x) &= -i \frac{\partial \mathcal{L}}{\partial(\partial\varphi_r(x)/\partial x_\mu)} \lambda_{rs}\varphi_s(x) \\ \frac{\partial J_\mu(x)}{\partial x_\mu} &= 0 \end{aligned} \tag{15.32}$$

and to the conserved "charge"

$$\begin{aligned} Q = \int d^3x\, J_0(x) &= -i \int d^3x \frac{\partial \mathcal{L}}{\partial \dot{\varphi}_r(x)} \lambda_{rs}\varphi_s(x) \\ \frac{\partial Q}{\partial t} &= 0 \end{aligned} \tag{15.33}$$

In this way, for instance, we may find the electromagnetic current of a charged particle. The lagrangian of the electron and photon is invariant under the phase transformation

$$\psi(x) \to \psi(x) - i\epsilon\psi(x) \qquad A_\mu(x) \to A_\mu(x) \tag{15.34}$$

leading to the conserved electromagnetic current according to (15.32),

$$j_\mu(x) = -i \frac{\partial \mathcal{L}}{\partial(\partial\psi/\partial x_\mu)} \psi(x) = \bar{\psi}(x)\gamma_\mu\psi(x) \tag{15.35}$$

with a similar result for a charged Klein-Gordon particle.

In making the transition to quantum mechanics, $j_\mu(x)$ becomes an operator, and it is convenient to normal-order it[1] as in (15.26) and (15.27); that is, $j_\mu(x) \to :\bar{\psi}(x)\gamma_\mu\psi(x):$. This does not change the conservation law, since we are subtracting a constant number (perhaps infinitely large) which has no space-time dependence (since $P|0\rangle = 0$)

$$\langle 0|j_\mu(x)|0\rangle = \langle 0|e^{iP\cdot x} j_\mu(0) e^{-iP\cdot x}|0\rangle = \langle 0|j_\mu(0)|0\rangle$$

As is the case for P_μ and $M_{\mu\nu}$, the conserved quantity Q in (15.33) becomes the *generator* of the desired transformation. Forming the unitary operator

$$U(\epsilon) = e^{i\epsilon Q} \approx 1 + i\epsilon Q \tag{15.36}$$

[1] Henceforth the vacuum subtraction is always assumed to be made in dealing with all conserved quantities.

where ϵ is an infinitesimal parameter and Q is hermitian, we find from (15.31),

$$
\begin{aligned}
U(\epsilon)\varphi_r(x)U^{-1}(\epsilon) &= \varphi_r(x) + i\epsilon[Q,\varphi_r(x)] \\
&= \varphi_r(x) - i\epsilon\lambda_{rs}\varphi_s(x) \qquad (15.37)
\end{aligned}
$$

or

$$[Q,\varphi_r(x)] = -\lambda_{rs}\varphi_s(x)$$

in analogy with (11.69) and (11.70). In canonical field theory with commutation relations (11.39), (15.37) gives

$$Q = -i\int d^3x\ {:}\pi_r(x)\lambda_{rs}\varphi_s(x){:} \qquad (15.38)$$

in agreement with (15.33). For theories quantized with anticommutators we may again check that (15.38) is the generator of the desired symmetry transformation provided λ_{rs} couples together only Fermi fields which satisfy one-time *anticommutation* relations:

$$
\begin{aligned}
\{\psi_{i,\alpha}(\mathbf{x},t),\psi^\dagger_{j,\beta}(\mathbf{x}',t)\} &= \delta_{\alpha\beta}\delta_{ij}\delta^3(\mathbf{x}-\mathbf{x}') \\
\{\psi_{i,\alpha}(\mathbf{x},t),\psi_{j,\beta}(\mathbf{x}',t)\} &= 0 \qquad (15.39) \\
\{\psi^\dagger_{i,\alpha}(\mathbf{x},t),\psi^\dagger_{j,\beta}(\mathbf{x}',t)\} &= 0
\end{aligned}
$$

where α, $\beta = 1, 2, 3, 4$ denote the spinor components and i, $j = 1, 2, \ldots$ denote different Fermi fields such as proton and neutron.

In the following developments we shall always require that different Fermi fields anticommute, according to (15.39), instead of commute with each other, that different Bose fields commute according to (11.39), and that the Fermi fields *commute* with the Bose fields at equal times.[1] For free-field theory this choice of commutation rules is merely a convention of phases. However, for interacting fields one finds difficulty even with the Heisenberg equations of motion without this choice of anticommutators. For example, in our model (15.30) of the meson-nucleon interaction, the hamiltonian is

$$
\begin{aligned}
H = \int d^3x[\Psi^\dagger(-i\boldsymbol{\alpha}\cdot\nabla + \beta M_0)\Psi + \tfrac{1}{2}(\boldsymbol{\pi}\cdot\boldsymbol{\pi} + \nabla\boldsymbol{\phi}\cdot\nabla\boldsymbol{\phi} + \mu_0^2\boldsymbol{\phi}\cdot\boldsymbol{\phi}) \\
+ ig_0\bar{\Psi}\gamma_5\boldsymbol{\tau}\cdot\boldsymbol{\phi}\Psi] \qquad (15.40)
\end{aligned}
$$

and the proton field equation is

$$(i\nabla\!\!\!/ - M_0)\psi_p(x) = ig_0\gamma_5(\sqrt{2}\,\varphi_+(x)\psi_n(x) + \varphi_0(x)\psi_p(x))$$

In order to verify the Heisenberg equation

$$[H,\psi_p(x)] = -i\,\frac{\partial\psi_p(x)}{\partial t}$$

[1] We depart from the canonical rules (11.39) for photons since the components of the electromagnetic potential are not all independent. For these we use (15.9) in the radiation gauge. There is no change in their commutation rules with other fields.

we must assume that $\psi_p(\mathbf{x},t)$ commutes with a bilinear form $\psi_n^\dagger(\mathbf{x},t) \cdots \psi_n(\mathbf{x},t)$ and, as well,

$$\{\psi_p(\mathbf{x},t),\ \varphi_+(\mathbf{x},t)\psi_n(\mathbf{x},t)\} = 0 \tag{15.41}$$

Our choice of commutation relations meets these requirements, as is readily seen. In addition, with this choice the commutation relations (15.39) remain invariant under the transformation (15.37) and the theory formulated in terms of the transformed fields

$$\varphi_r'(x) = U(\epsilon)\varphi_r(x)U^{-1}(\epsilon) \tag{15.42}$$

is unchanged from its original form in terms of the $\varphi_r(x)$.

15.8 Strong Couplings of Pi Mesons and Nucleons

The most interesting applications of these symmetry considerations lie in the realm of strong interactions. Beginning with the π-meson–nucleon interactions, we notice that the model lagrangian (15.30) is invariant under a simultaneous phase change of the neutron and proton fields

Baryon Cons.

$$\Psi \to \Psi - i\epsilon\Psi \tag{15.43}$$

This transformation corresponds to $\lambda_{rs} = \delta_{rs}$ $(r, s = 1, 2)$ in (15.37) and leads to the conserved current, Eq. (15.32),

$$J_\mu^N(x) = \bar{\Psi}\gamma_\mu\Psi = J_\mu^p(x) + J_\mu^n(x) \tag{15.44}$$

and to the constant (15.33)

$$N = \int d^3x\, J_0^N(x) = \int d^3x\, \{\psi_p^\dagger\psi_p + \psi_n^\dagger\psi_n\} = N_p + N_n \tag{15.45}$$

which we call the nucleon number. This conservation law has already been found in (10.39) and (10.40). For a theory with no interactions, N may be interpreted as the number of protons plus neutrons minus the number of antiprotons plus antineutrons. When interactions are turned on, it is no longer possible to compute exactly. However with highly reasonable assumptions about the nature of the states of the interacting system to be discussed on page 143, the same interpretation of N may be made, provided, of course, it is still a constant of the motion.

The model lagrangian (15.30) also has a symmetry operation corresponding to charge conservation. In this transformation only the charged particles undergo a common phase transformation which leaves

$\mathcal{L}$ in (15.30) invariant:

$$
\begin{aligned}
&\psi_p \to \psi_p - i\epsilon\psi_p \\
&\psi_n \to \psi_n \\
&\varphi_+ \to \varphi_+ - i\epsilon\varphi_+ \\
&\varphi_- = \varphi_+^* \to \varphi_- + i\epsilon\varphi_- \\
&\varphi_0 \to \varphi_0
\end{aligned}
\tag{15.46}
$$

In isotopic notation this reads

$$\Psi \to \Psi - i\epsilon\left(\frac{1+\tau_3}{2}\right)\Psi \qquad \boldsymbol{\varphi} \to \boldsymbol{\varphi} - \epsilon(\boldsymbol{\varphi} \times \hat{\boldsymbol{\varphi}}_0) \tag{15.47}$$

where $\hat{\boldsymbol{\varphi}}_0 = (0,0,1)$ as introduced in the isotopic spin formalism in (10.36). The invariance of (15.30) under this transformation may be verified explicitly and leads to the conserved electromagnetic current, by (15.32) and (15.33),

$$j_\mu(x) = \bar{\Psi}\gamma_\mu\left(\frac{1+\tau_3}{2}\right)\Psi - \left(\frac{\partial\boldsymbol{\varphi}}{\partial x^\mu} \times \boldsymbol{\varphi}\right)\cdot\hat{\boldsymbol{\varphi}}_0 \tag{15.48}$$

$$Q = \int j_0(x)\,d^3x = \int d^3x\,[\psi_p^\dagger(x)\psi_p(x) + \varphi_1(x)\dot{\varphi}_2(x) - \varphi_2(x)\dot{\varphi}_1(x)]$$

These coincide with the analogous expressions (10.37) and (10.38) of our propagator discussions.

The invariance of the model $\mathcal{L}$ of (15.30) under rotations in isotopic space leads to the law of conservation of isotopic spin. We leave it as an exercise to verify explicitly that (15.30) is invariant under the rotation

$$\Psi \to \Psi - i\epsilon\,\tfrac{1}{2}\boldsymbol{\tau}\cdot\hat{\boldsymbol{\varphi}}\Psi \qquad \boldsymbol{\varphi}(x) \to \boldsymbol{\varphi}(x) - \epsilon(\boldsymbol{\varphi}(x) \times \hat{\boldsymbol{\varphi}}) \tag{15.49}$$

with the $\hat{\boldsymbol{\varphi}}$ a set of unit vectors, $\hat{\boldsymbol{\varphi}}_1 = (1,0,0)$, $\hat{\boldsymbol{\varphi}}_2 = (0,1,0)$, $\hat{\boldsymbol{\varphi}}_3 = (0,0,1)$. The conserved isotopic spin current inferred from (15.32) and (15.49) is

$$\mathbf{J}_\mu(x) = \tfrac{1}{2}\bar{\Psi}\gamma_\mu\boldsymbol{\tau}\Psi + \left(\boldsymbol{\varphi} \times \frac{\partial\boldsymbol{\varphi}}{\partial x^\mu}\right) \tag{15.50}$$

and the three components of the isotopic spin

$$\mathbf{I} = \int d^3x\,[\tfrac{1}{2}\Psi^\dagger\boldsymbol{\tau}\Psi + (\boldsymbol{\varphi} \times \dot{\boldsymbol{\varphi}})] \tag{15.51}$$

are constants of the motion. Once again these expressions coincide with (10.43) and (10.44). Adding together (15.45) and the third component of (15.51), we find, from (15.48), the relation

$$Q = \frac{N}{2} + I_3 \tag{15.52}$$

as also found in (10.41). We may also verify from (15.51) and the commutation rules that the components of **I** satisfy the angular momentum commutation rules

$$[I_i, I_j] = i I_k \qquad (15.53)$$

so that, as in our propagator discussion of Chap. 10, we may label states by their eigenvalues of I_3 and of I^2. *i.e. choose eigenstates of $I_3 \, \xi \, I^2$ as basis*

By referring back to our model lagrangian (15.30), we have found the symmetry operations (15.43), (15.47), and (15.49), corresponding to laws of nucleon, charge, and isotopic spin conservation. Although the model (15.30) is undoubtedly inadequate for describing the full π-meson–nucleon interaction, when we attempt to develop more general models for comparison with experiment we shall preserve these and other symmetries as verified by observed selection rules.

15.9 Symmetries of Strange Particles

This use of symmetry requirements to ensure the existence of observed constants of the motion is convenient when we introduce lagrangians for the hyperons and K mesons which also participate in the strong interactions. With the discovery that these particles occur in charge multiplets and that certain selection rules govern their strong interactions, it is natural to introduce lagrangians which admit symmetry operations generated by the observed constants of the motion.

We denote by "strange particles" those in addition to leptons, π mesons, and nucleons which, in the absence of the weak interactions, are stable.[1]

These are all illustrated in the energy level diagram in Fig. 15.1 (page 103). As is customary, we have shown just the particles and not the antiparticles in drawing the baryons (N, Σ, Λ, Ξ) and leptons (μ, e, ν, ν'). Experiment, as always, provides the necessary clue to which are the particles and which the antiparticles. In the present case it is the observed rigorous conservation of the total number of baryons minus antibaryons—and analogously for leptons—that fixes Ξ^-, and not its positively charged antiparticle $\overline{\Xi^-}$ as the baryon in the classification of Fig. 15.1. For the mesons, both particles and antiparticles are drawn in the figure. The π^- and K^- are the oppositely charged antiparticles of the π^+ and K^+. The π^0 is its own antiparticle; it carries no charge. The K^0, on the other hand, is distinguishable

---

[1] The Σ^0, like the π^0, is included in this category although it decays electromagnetically $(\Sigma^0 \to \Lambda^0 + \gamma)$; whereas the Ω^- and nuclei are not.

from its antiparticle, the $\bar{K}^0$; for instance, the reaction

$$\pi^- + p \rightarrow \Lambda^0 + K^0 \tag{15.54a}$$

is observed, while

$$\bar{K}^0 + p \rightarrow \Lambda^0 + \pi^+ \tag{15.54b}$$

is not.

With the exception of the Σ^0 and Ξ^0, the spins of the baryons have been measured to be one-half and the spin of the K has been measured to be zero. Assuming spin $\frac{1}{2}$ for all of the baryons, they are described by Dirac equations with appropriate masses, in the absence of interactions, and the π and K mesons are described by Klein-Gordon equations.[1]

The multiplet structure of the mass levels in Fig. 15.1 strongly suggests that the isotopic spin formalism developed for π mesons and nucleons be used for strange particles also. Since all experimental evidence thus far supports a law of isotopic spin conservation in the strong interactions among mesons and baryons, it is useful to introduce the isotopic spin formalism for them. We may then guarantee conservation of isotopic spin by demanding that, in the absence of electromagnetic corrections, the lagrangian be invariant under rotations in isotopic space.

In isotopic space the Σ field is described as a vector, similarly to the π mesons in (15.30):

$$\boldsymbol{\Psi}_\Sigma(x) = \left(\frac{\psi_{\Sigma^+}(x) + \psi_{\Sigma^-}(x)}{\sqrt{2}}, \frac{i[\psi_{\Sigma^+}(x) - \psi_{\Sigma^-}(x)]}{\sqrt{2}}, \psi_{\Sigma^0}(x)\right) \tag{15.55}$$

In a free-field theory $\psi_{\Sigma^+}(x)$ destroys a Σ^+ particle or creates an anti-Σ^+, whereas $\psi_{\Sigma^-}(x)$ destroys a Σ^- or creates an anti-Σ^-. Since hyperons are distinguishable from antihyperons, that is, the Σ^+ differs from the anti-Σ^-, the field $\psi_\Sigma(x)$ is not hermitian, as was the corresponding π-meson isotopic vector. The hermitian conjugate field to (15.55) is given by

$$\boldsymbol{\Psi}_\Sigma^\dagger(x) = \left(\frac{\psi_{\Sigma^+}^\dagger(x) + \psi_{\Sigma^-}^\dagger(x)}{\sqrt{2}}, \frac{-i[\psi_{\Sigma^+}^\dagger(x) - \psi_{\Sigma^-}^\dagger(x)]}{\sqrt{2}}, \psi_{\Sigma^0}^\dagger(x)\right)$$

The Λ is described as a scalar in isotopic space and the cascade Ξ and

[1] See M. Gell-Mann and Y. Ne'eman, "The Eightfold Way," W. A. Benjamin, Inc., New York, 1965, for a discussion of theoretical schemes exploiting the symmetry among the eight baryons and mesons (including the η).

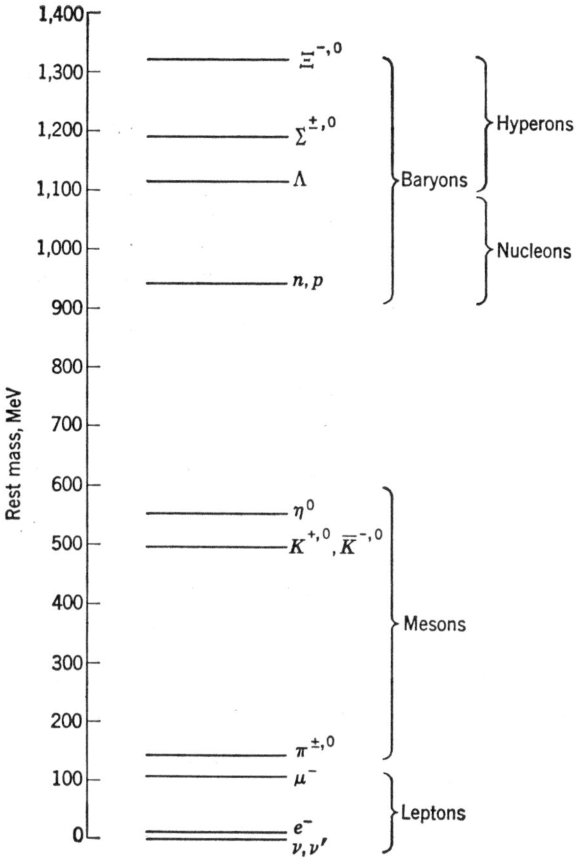

Fig. 15.1 Energy level diagram for lowest lying baryons, mesons, and leptons.

the K meson as isotopic spinors, similar to the nucleon:

$$\Psi_{\Xi}(x) = \begin{bmatrix} \psi_{\Xi^0}(x) \\ \psi_{\Xi^-}(x) \end{bmatrix} \tag{15.56}$$

$$\Phi_K(x) = \begin{bmatrix} \varphi_{K^+}(x) \\ \varphi_{K^0}(x) \end{bmatrix} \tag{15.57}$$

With the above assignments the different fields transform under an infinitesimal rotation about an arbitrary axis $\hat{\mathbf{u}}$ in isotopic space according to

$$\begin{aligned}
\boldsymbol{\phi}_\pi &\to \boldsymbol{\phi}_\pi - \epsilon\boldsymbol{\phi}_\pi \times \hat{\mathbf{u}} \\
\Phi_K &\to \Phi_K - \frac{i\epsilon}{2}\boldsymbol{\tau}\cdot\hat{\mathbf{u}}\Phi_K \\
\Psi_N &\to \Psi_N - \frac{i\epsilon}{2}\boldsymbol{\tau}\cdot\hat{\mathbf{u}}\Psi_N \\
\Psi_\Lambda &\to \Psi_\Lambda \\
\boldsymbol{\Psi}_\Sigma &\to \boldsymbol{\Psi}_\Sigma - \epsilon\boldsymbol{\Psi}_\Sigma \times \hat{\mathbf{u}} \\
\Psi_\Xi &\to \Psi_\Xi - \frac{i\epsilon}{2}\boldsymbol{\tau}\cdot\hat{\mathbf{u}}\Psi_\Xi
\end{aligned} \tag{15.58}$$

The conserved isotopic spin current deduced from (15.58) and (15.32) or (15.37) is the sum of terms [1]

$$\mathbf{J}_\mu(x) = \tfrac{1}{2}\bar{\Psi}_N\gamma_\mu\boldsymbol{\tau}\Psi_N + \tfrac{1}{2}\bar{\Psi}_\Xi\gamma_\mu\boldsymbol{\tau}\Psi_\Xi - i\bar{\boldsymbol{\Psi}}_\Sigma\gamma_\mu \times \boldsymbol{\Psi}_\Sigma + \boldsymbol{\phi}_\pi \times \frac{\partial\boldsymbol{\phi}_\pi}{\partial x^\mu} + \tfrac{1}{2}i\left(\Phi_K^\dagger\boldsymbol{\tau}\frac{\partial\Phi_K}{\partial x^\mu} - \frac{\partial\Phi_K^\dagger}{\partial x^\mu}\boldsymbol{\tau}\Phi_K\right) \tag{15.59}$$

and the conserved total isotopic spin is

$$\mathbf{I} = \int \mathbf{J}_0(x)\, d^3x$$

These are the generalizations of (15.50) and (15.51) for nucleons and π mesons alone; for the three-component, in particular, we have

$$\begin{aligned}
I_3 = \int d^3x\, [&\tfrac{1}{2}\psi_p^\dagger\psi_p - \tfrac{1}{2}\psi_n^\dagger\psi_n + \tfrac{1}{2}\psi_{\Xi^0}^\dagger\psi_{\Xi^0} - \tfrac{1}{2}\psi_{\Xi^-}^\dagger\psi_{\Xi^-} + \psi_{\Sigma^+}^\dagger\psi_{\Sigma^+} - \psi_{\Sigma^-}^\dagger\psi_{\Sigma^-} \\
&+ (\varphi_{1_\pi}\dot{\varphi}_{2_\pi} - \varphi_{2_\pi}\dot{\varphi}_{1_\pi}) \\
&+ \tfrac{1}{2}i(\varphi_{K^+}^\dagger\dot{\varphi}_{K^+} - \dot{\varphi}_{K^+}^\dagger\varphi_{K^+} - \varphi_{K^0}^\dagger\dot{\varphi}_{K^0} + \dot{\varphi}_{K^0}^\dagger\varphi_{K^0})] \\
= \tfrac{1}{2}(N_p - N_n &+ N_{\Xi^0} - N_{\Xi^-} + N_{K^+} - N_{K^0}) + N_{\Sigma^+} - N_{\Sigma^-} + N_{\pi^+}
\end{aligned} \tag{15.60}$$

We again find that $\mathbf{I}$ has the commutation properties of an angular momentum, (15.53), so that I_3 and I^2 may be used simultaneously as quantum numbers for states of the system.

Isotopic spin conservation is an approximate conservation law of the strong interactions and is violated by electromagnetic corrections. For example, the small mass differences among members of a given multiplet in Fig. 15.1 are attributed to electromagnetic effects and are neglected in writing a lagrangian in the charge-independent approximation for which (15.58) is a symmetry operation.

The nucleon number (15.45) is no longer conserved when we consider strange-particle reactions, such as (15.54*a*). It is replaced, however, by an exact, or absolute, conservation law on the total baryon number. That is, the total number of baryons minus antibaryons is conserved in any reaction; or, in the language of Feynman graphs,

[1] This is true if there are no derivative coupling terms in $\mathcal{L}$.

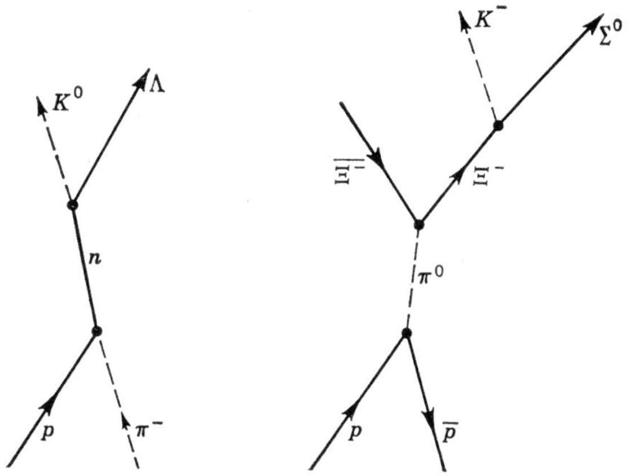

Fig. 15.2 Processes conserving the total number of baryons minus antibaryons.

baryon lines continue through interaction vertices, rather than originating or terminating at them, as illustrated in Fig. 15.2. This suggests that $\mathcal{L}$ be invariant under the simultaneous phase transformation of the baryon fields

$$\begin{aligned} \Psi_N &\to \Psi_N - i\epsilon\Psi_N \\ \Psi_\Lambda &\to \Psi_\Lambda - i\epsilon\Psi_\Lambda \\ \boldsymbol{\Psi}_\Sigma &\to \boldsymbol{\Psi}_\Sigma - i\epsilon\boldsymbol{\Psi}_\Sigma \\ \Psi_\Xi &\to \Psi_\Xi - i\epsilon\Psi_\Xi \end{aligned} \tag{15.61}$$

leading to the constant baryon number

$$B = N_N + N_\Lambda + N_\Sigma + N_\Xi \tag{15.62}$$

as a generalization of (15.45).

Finally, we associate the requirement that charge be conserved with the invariance of $\mathcal{L}$ under simultaneous phase transformations of charged particles only, as in (15.47). Thus $\mathcal{L}$ should be invariant under the transformation, with $\hat{\mathbf{u}}_3 \equiv (0, 0, 1)$ in isotopic space

$$\begin{aligned} \boldsymbol{\phi}_\pi &\to \boldsymbol{\phi}_\pi - \epsilon\boldsymbol{\phi}_\pi \times \hat{\mathbf{u}}_3 \\ \Phi_K &\to \Phi_K - \frac{i\epsilon}{2}(1 + \tau_3)\Phi_K \\ \Psi_N &\to \Psi_N - \frac{i\epsilon}{2}(1 + \tau_3)\Psi_N \\ \Psi_\Lambda &\to \Psi_\Lambda \\ \boldsymbol{\Psi}_\Sigma &\to \boldsymbol{\Psi}_\Sigma - \epsilon\boldsymbol{\Psi}_\Sigma \times \hat{\mathbf{u}}_3 \\ \Psi_\Xi &\to \Psi_\Xi - \frac{i\epsilon}{2}(-1 + \tau_3)\Psi_\Xi \end{aligned} \tag{15.63}$$

leading to the constant

$$
\begin{aligned}
Q &= \int d^3x\,[\psi_p^\dagger\psi_p + \psi_{\Sigma^+}^\dagger\psi_{\Sigma^+} - \psi_{\Sigma^-}^\dagger\psi_{\Sigma^-} - \psi_{\Xi^-}^\dagger\psi_{\Xi^-} \\
&\qquad + (\varphi_{1_\pi}\dot\varphi_{2_\pi} - \varphi_{2_\pi}\dot\varphi_{1_\pi}) + i(\varphi_{K^+}^\dagger\dot\varphi_{K^+} - \dot\varphi_{K^+}^\dagger\varphi_{K^+})] \\
&= N_p + N_{\Sigma^+} - N_{\Sigma^-} - N_{\Xi^-} + N_{\pi^+} + N_{K^+}
\end{aligned}
\tag{15.64}
$$

The transformation (15.63) can be considered to be composed of a rotation about the 3-axis in isotopic space, plus a phase transformation of the K, N, and Ξ particles given by

$$
\Phi_K \to \Phi_K - \frac{i\epsilon}{2}\Phi_K \qquad \Psi_N \to \Psi_N - \frac{i\epsilon}{2}\Psi_N \qquad \Psi_\Xi \to \Psi_\Xi + \frac{i\epsilon}{2}\Psi_\Xi \tag{15.65}
$$

Since $\mathfrak{L}$ is invariant under rotations around the 3-axis in isotopic space, leading to the constant (15.60), the transformation (15.65) must also be a symmetry operation leading to conservation of what is called "hypercharge" Y:

$$
Y = N_K + N_N - N_\Xi \tag{15.66}
$$

Recalling (15.60) and (15.64), one finds the relation

$$
Q = \frac{Y}{2} + I_3 \tag{15.67}
$$

Closely related to Y is the "strangeness" S defined by

$$
S = Y - B \tag{15.68}
$$

where B is the baryon number (15.62); it must also be conserved in strong interactions. Since the constants S, Y, B, I_3, and Q are all proportional to the number operators for the various particles, we may assign a quantum number to each particle in terms of which to discuss additive conservation laws. The accompanying table lists these quantum numbers for the baryons and mesons; the quantum numbers of the antibaryons are the negative of those of the baryons. The conservation laws of hypercharge Y, Eq. (15.66), and of strangeness S, Eq. (15.68), emerge solely as consequences of conservation of Q, I_3, and

	Q	I_3	B	Y	S
π	1, 0, −1	1, 0, −1	0	0	0
K	1, 0	½, −½	0	1	1
N	1, 0	½, −½	1	1	0
Λ	0	0	1	0	−1
Σ	1, 0, −1	1, 0, −1	1	0	−1
Ξ	0, −1	½, −½	1	−1	−2

B. They lead to the experimentally observed selection rule of associated production; that is, a K meson, for example, can be produced only in association with a Λ or Σ, as in (15.54*a*), while (15.54*b*) violates S or Y conservation by two units.[1]

We have so far considered only the strong interactions. If we accept the minimal prescription

$$p_\mu \rightarrow p_\mu - eA_\mu$$

for introducing electromagnetic interactions, we still retain B, Y, and S as constants of the motion, since the coupling is diagonal in baryon number and hypercharge, or strangeness. Therefore, by (15.67), conservation of charge Q implies conservation of I_3; that is, the electromagnetic interaction terms introduced into $\mathcal{L}$ are invariant under rotations about the 3-axis in isotopic space. However, the full symmetry under rotations in isotopic space is destroyed by the electromagnetic transitions such as photoproduction processes,

$$\begin{aligned} \gamma + p &\rightarrow p + \pi^0 \\ &\rightarrow n + \pi^+ \end{aligned}$$

which lead to final π-meson–nucleon states of both $I = \frac{1}{2}$ and $\frac{3}{2}$.

The weak interactions violate I_3 conservation and, therefore, by (15.67) and (15.68), strangeness or hypercharge conservation as well. These weak couplings account for the instability of the strange particles via, for instance,

$$K^0 \rightarrow \pi^+ + \pi^-$$

Only charge Q and baryon number B, along with the lepton numbers L_e and L_μ (see Prob. 15.22), remain as absolutely conserved quantum numbers to the present limits of experimental tests.

15.10 Improper Symmetries

The symmetry operations considered so far have all been generated by infinitesimal transformations. There are, in addition to these, "improper," or discrete, transformations which cannot be generated by a succession of infinitesimal steps but which provide further useful selection rules and information on the nature of interaction terms. These are the space inversion or parity transformation $\mathcal{P}$, the time-

[1] A. Pais, *Phys. Rev.*, **86,** 633 (1952); M. Gell-Mann, *Phys. Rev.*, **92,** 833 (1953); K. Nishijima, *Progr. Theoret. Phys. (Kyoto)*, **12,** 107 (1954).

reversal transformation $\mathfrak{T}$, and the charge conjugation transformation $\mathcal{C}$.

15.11 Parity

To define the meaning of the parity transformation, we add to the lagrangian density a term representing the interaction, generally electromagnetic, of the quantum system with the measuring apparatus:

$$\mathcal{L} \rightarrow \mathcal{L} - j_\mu(x) A^\mu_{\text{ext}}(x) \tag{15.69}$$

where $A^\mu_{\text{ext}}(x)$ is to be treated as a classical prescribed external field which interacts with the current operator $j_\mu(x)$ of the system. If we invert the measuring apparatus, that is, consider a new physical system for which the external fields applied in preparing and analyzing the states of the system are given by

$$\tilde{A}^\mu_{\text{ext}}(x) = (A^0_{\text{ext}}(-\mathbf{x},t), -\mathbf{A}_{\text{ext}}(-\mathbf{x},t)) = A_\mu^{\text{ext}}(-\mathbf{x},t) \tag{15.70}$$

the dynamics of the new system is the same as that of the original one, provided parity is conserved. In particular, if the action

$$\tilde{J} = \int d^4x \, [\mathcal{L} - j_\mu(x) \tilde{A}^\mu_{\text{ext}}(x)]$$

of the new system is related to the action J of the original system by a unitary transformation $\mathcal{P}$, the equations of motion are unchanged. This is guaranteed if $\mathcal{P}$ has the property

$$\mathcal{P}\mathcal{L}(\mathbf{x},t)\mathcal{P}^{-1} = \mathcal{L}(-\mathbf{x},t) \tag{15.71}$$

and, for the electromagnetic current,

$$\mathcal{P} j_\mu(\mathbf{x},t)\mathcal{P}^{-1} = j^\mu(-\mathbf{x},t) \tag{15.72}$$

In addition, $\mathcal{P}$ must leave the commutation relations invariant, in which case the new system and the original one satisfy identical dynamical laws and we say that parity is conserved.

We first consider free-particle theories and construct explicitly the operator $\mathcal{P}$ for $\mathcal{L}$ a free-particle lagrangian in (15.71). For the free Klein-Gordon theory of Chap. 12 [see (12.2), (12.4), and (12.63)], the condition

$$\mathcal{P}\varphi(\mathbf{x},t)\mathcal{P}^{-1} = \pm\varphi(-\mathbf{x},t) \tag{15.73}$$

clearly satisfies (15.71) and (15.72) and leaves the commutation relations invariant.

The choice of the + or − sign in (15.73) defines what is called the "intrinsic parity" of the particle described by this field; the + sign is chosen for scalar particles and the − sign for pseudoscalar ones like the π meson appearing in (15.30). It is a specific rule of transformation of the field which creates the particle by operating on the vacuum state

$$\varphi(x)|0\rangle \tag{15.74}$$

and is determined when interactions are introduced between different particles.[1] The intrinsic parity differs from the orbital parity associated with the particle wave function in a state of given orbital angular momentum. The wave function $\mathcal{f}_l(x)$ for a particle formed from the vacuum, by (15.74), in an angular-momentum state l, denoted $|n = 1; l\rangle$, has the property

$$\begin{aligned}\mathcal{f}_l(\mathbf{x},t) = \langle n = 1; l|\varphi(\mathbf{x},t)|0\rangle &= (-)^l\langle n = 1; l|\varphi(-\mathbf{x},t)|0\rangle \\ &= (-)^l\mathcal{f}_l(-\mathbf{x},t)\end{aligned} \tag{15.75}$$

which is no more than a statement of the evenness or oddness of $\mathcal{f}_l(x)$. On the other hand, the parity of the state $|n = 1; l\rangle$ relative to the vacuum which by convention is even, that is,

$$\mathcal{P}|0\rangle = |0\rangle \tag{15.76}$$

is found by considering

$$\begin{aligned}\langle n = 1; l|\mathcal{P}\varphi(\mathbf{x},t)|0\rangle &= \langle n = 1; l|\mathcal{P}\varphi(\mathbf{x},t)\mathcal{P}^{-1}|0\rangle \\ &= \pm\,\langle n = 1; l|\varphi(-\mathbf{x},t)|0\rangle \\ &= \pm(-)^l\mathcal{f}_l(\mathbf{x},t)\end{aligned} \tag{15.77}$$

and is the product of the intrinsic parity, ±, and the orbital parity, $(-)^l$. Thus a pseudoscalar π meson in a p state has even parity.

In terms of the momentum-space expansions, (12.7) or (12.57), the parity transformation (15.73) satisfies[2]

$$\mathcal{P}a(k)\mathcal{P}^{-1} = \pm a(-k) \qquad \mathcal{P}a^\dagger(k)\mathcal{P}^{-1} = \pm a^\dagger(-k) \tag{15.78}$$

Acting on a momentum eigenstate, $\mathcal{P}$ produces a new state with all momenta $\mathbf{k}_1, \ldots, \mathbf{k}_n$ replaced by their negatives $-\mathbf{k}_1, \ldots, -\mathbf{k}_n$ but with all other quantum numbers, such as charge or numbers of particles, remaining unchanged.

[1] An arbitrary phase $e^{i\varphi}$ may be given in (15.73), but it is not discussed here for lack of physical interest.

[2] The transformation of $k \to -k$ in the argument of the creation and annihilation operators applies only to the direction of the space components of the momenta.

Equations (15.73) or (15.78) determining $\mathcal{P}$ are most readily solved in momentum space. Introducing

$$\mathcal{P} = e^{iP} \tag{15.79}$$

we rewrite (15.78) as

$$\mathcal{P}a(k)\mathcal{P}^{-1} = a(k) + i[P,a(k)] + \frac{i^2}{2!}[P, [P,a(k)]] + \cdots + \frac{i^n}{n!}[P, [. . .[P,a(k)],] . . .] + \cdots = -a(-k) \tag{15.80}$$

where we have chosen the minus sign on the right-hand side for a pseudoscalar field. This form suggests the choice

$$[P,a(k)] = \frac{\lambda}{2}[a(k) \pm a(-k)] \tag{15.81}$$

with λ and the sign to be determined. Then

$$[P,[P,a(k)]] = \tfrac{1}{2}\lambda^2 [a(k) \pm a(-k)]$$

and by (15.80),

Uses λ - continuous corresponds to P + 3-D Rotation?

$$\begin{aligned}\mathcal{P}a(k)\mathcal{P}^{-1} &= a(k) \\ &\quad + \tfrac{1}{2}\left[i\lambda + \frac{(i\lambda)^2}{2!} + \cdots + \frac{(i\lambda)^n}{n!} + \cdots\right][a(k) \pm a(-k)] \\ &= \tfrac{1}{2}[a(k) \mp a(-k)] + \tfrac{1}{2}e^{i\lambda}[a(k) \pm a(-k)]\end{aligned} \tag{15.82}$$

Choosing the plus sign in (15.81) and setting $\lambda = \pi$, we succeed in solving (15.82) for a pseudoscalar field. From (15.81) we readily find P_{ps} to be[1]

$$P_{ps} = -\frac{\pi}{2}\int d^3k\, [a^\dagger(k)a(k) + a^\dagger(k)a(-k)] = P_{ps}^\dagger$$

and

$$\mathcal{P}_{ps} = \exp\left\{\frac{-i\pi}{2}\int d^3k\, [a^\dagger(k)a(k) + a^\dagger(k)a(-k)]\right\} \tag{15.83}$$

For the scalar field one finds in the same way, by using a + sign on the right-hand side of (15.80),

$$\mathcal{P}_s = \exp\left\{\frac{-i\pi}{2}\int d^3k\, [a^\dagger(k)a(k) - a^\dagger(k)a(-k)]\right\} \tag{15.84}$$

[1] P. Federbush and M. Grisaru, *Nuovo Cimento*, **9**, 1058 (1958). Notice that P and $\mathcal{P}$ are nonlocal operators in coordinate space, as is required in order to reflect a particle at point **x** to point $-$**x**.

The unitarity of $\mathcal{P}$ is guaranteed by the hermiticity of P in (15.83) and (15.84), and the convention (15.76) that the vacuum be an even-parity eigenstate is guaranteed by the normal ordering of the operators in P.

For interacting fields, we can similarly construct a parity operator $\mathcal{P}$ which satisfies (15.73). We must then check to see whether (15.71) and (15.72) remain valid so that $\mathcal{P}$, as a symmetry operator of the theory, commutes with the hamiltonian

$$[\mathcal{P},H] = 0 \tag{15.85}$$

and is a constant of the motion.

In order to construct in general a $\mathcal{P}$ satisfying (15.73), we first construct a $\mathcal{P}_0$ which satisfies (15.73) at $t = 0$. This is carried out just as for the free fields, because, from (15.23), the expansion coefficients of the interacting fields at $t = 0$ satisfy the same commutation algebra as the free fields. Thus $\mathcal{P}_0$ is given by the free-field solution, (15.83) or (15.84), with the expansion coefficients of the interacting field at $t = 0$ replacing the free-field creation and destruction operators. To obtain $\mathcal{P}$, one then uses the hamiltonian operator to displace in time. That is, if

$$\mathcal{P}_0\varphi(\mathbf{x},0)\mathcal{P}_0^{-1} = \pm\varphi(-\mathbf{x},0) \tag{15.86}$$

then

$$\mathcal{P}(t) = e^{iHt}\mathcal{P}_0e^{-iHt} \tag{15.87}$$

satisfies (15.73) for a general time t. If $\mathcal{P}$ is a symmetry operation and (15.85) is valid, we have immediately from (15.87) that

$$\mathcal{P}(t) = \mathcal{P}(0) = \mathcal{P}_0$$

For the free Dirac field the fundamental conditions (15.71) and (15.72) and the invariance of the commutation relations are satisfied by the choice

$$\mathcal{P}\psi(\mathbf{x},t)\mathcal{P}^{-1} = \gamma_0\psi(-\mathbf{x},t) \tag{15.88}$$

That the Dirac equation is invariant under this parity transformation is shown in (2.33); invariance of the anticommutators (13.53) and (13.54) is readily verified. Again it is convenient to work directly in terms of the expansion coefficients in constructing $\mathcal{P}$ for a Dirac field, and we write

$$\begin{aligned}\int\frac{d^3p}{(2\pi)^{3/2}}\sqrt{\frac{m}{E_p}}\sum_{\pm s}\,&[\mathcal{P}b(p,s)\mathcal{P}^{-1}u(p,s)e^{-iE_pt+i\mathbf{p}\cdot\mathbf{x}} + \mathcal{P}d^\dagger(p,s)\mathcal{P}^{-1}v(p,s)e^{+iE_pt-i\mathbf{p}\cdot\mathbf{x}}]\\ = \int\frac{d^3p}{(2\pi)^{3/2}}\sqrt{\frac{m}{E_p}}\sum_{\pm s}\,&[b(p,s)\gamma_0u(p,s)e^{-iE_pt-i\mathbf{p}\cdot\mathbf{x}} + d^\dagger(p,s)\gamma_0v(p,s)e^{+iE_pt+i\mathbf{p}\cdot\mathbf{x}}]\end{aligned} \tag{15.89}$$

Changing **p** to **−p** and using the spinor properties

$$\gamma_0\, u(-p,s) = u(p,s) \qquad \gamma_0\, v(-p,s) = -v(p,s) \tag{15.90}$$

in the right-hand side of (15.89) we obtain the conditions

$$\mathcal{P} b(p,s)\mathcal{P}^{-1} = b(-p,s) \qquad \mathcal{P} d^\dagger(p,s)\mathcal{P}^{-1} = -d^\dagger(-p,s) \tag{15.91}$$

with similar ones for $b^\dagger$ and d.

Since $b^\dagger$ operating on a vacuum creates a one-electron or baryon state and $d^\dagger$ creates a positron or antibaryon, we see by (15.91) that an electron has the opposite parity from a positron in the same orbital state. With the arbitrary phase choice given in (15.88) the electron state transforms under the parity operator as a scalar and the positron as a pseudoscalar, according to (15.91). An electron-positron pair in a relative s state has an odd intrinsic parity independent of the phase conventions. Similarly, the parity of a baryon-antibaryon pair in a relative s state is odd if we are dealing with a theory for which $\mathcal{P}$ is a symmetry operation as in (15.85): → sandwich $\mathcal{P}\mathcal{P}^{-1}$ in between op's and use $\mathcal{P}|0\rangle = |0\rangle$

$$\mathcal{P} \int d^3p\, f(\mathbf{p}^2) b^\dagger(p,s) d^\dagger(p,s)|0\rangle = -\int d^3p\, f(\mathbf{p}^2) b^\dagger(p,s) d^\dagger(p,s)|0\rangle \tag{15.92}$$

To construct an explicit expression for the parity operator for a Dirac field, we need only repeat the steps (15.79) to (15.84), finding

$$\mathcal{P}_{\text{Dirac}} = \exp\,(iP_{\text{Dirac}})$$

$$P_{\text{Dirac}} = -\frac{\pi}{2}\int d^3p[b^\dagger(p,s)b(p,s) - b^\dagger(p,s)b(-p,s) + d^\dagger(p,s)\, d(p,s) + d^\dagger(p,s)d(-p,s)] \tag{15.93}$$

Another consequence of the appearance of γ_0 in the parity operator is that $\bar\psi\gamma_5\psi$ is a pseudoscalar; that is,

$\gamma_0\gamma_5\gamma_0 = -\gamma_5$

$$\mathcal{P}\bar\psi(\mathbf{x},t)\gamma_5\psi(\mathbf{x},t)\mathcal{P}^{-1} = -\bar\psi(-\mathbf{x},t)\gamma_5\psi(-\mathbf{x},t) \tag{15.94}$$

This is important for meson-nucleon interactions. For instance, the model lagrangian (15.30) satisfies (15.71) under the combined parity operation

$$\mathcal{P} = \mathcal{P}_p\mathcal{P}_n\mathcal{P}_{\pi^+}\mathcal{P}_{\pi^0}\mathcal{P}_{\pi^-}$$

applied to all nucleons and mesons if (and only if) the pseudoscalar operator (15.83) is chosen for the mesons as required by experiment.

The intrinsic parity of the electromagnetic field is determined by correspondence arguments, since it is coupled to classical currents. Therefore, for photons we specify

$$\mathcal{P}\mathbf{A}(\mathbf{x},t)\mathcal{P}^{-1} = -\mathbf{A}(-\mathbf{x},t) \tag{15.95}$$

This leaves the Maxwell lagrangian invariant. $\mathcal{P}$ is constructed by imitating the steps taken for the Klein-Gordon field.

15.12 Charge Conjugation

The symmetry operation of charge conjugation is associated with the interchange of the role of particle and antiparticle. In the special application to electrons in Chap. 5 this amounted to reversing the sign of the electric charge and of the electromagnetic field. This interpretation is retained here. In terms of the lagrangian (15.69), which includes the external fields A_μ^{ext} of the measuring apparatus, this leads to the requirement, for a theory which is charge conjugation invariant, that there exist a unitary operator $\mathcal{C}$ such that

$$\mathcal{C}\mathcal{L}(x)\mathcal{C}^{-1} = \mathcal{L}(x) \qquad \mathcal{C}j_\mu(x)\mathcal{C}^{-1} = -j_\mu(x) \tag{15.96}$$

where $j_\mu(x)$ is the electromagnetic current. $\mathcal{C}$ must also change electrically neutral particles, described by non-hermitian fields, such as K^0, n, and Λ to their antiparticles $\overline{K^0}$, $\bar{n}$, and $\bar{\Lambda}$, in order that the conservation laws of strangeness, nucleon number, and isotopic spin be invariant under $\mathcal{C}$. For photons and π^0 mesons, described by hermitian fields, the particle is not distinguished from the antiparticle, and under $\mathcal{C}$ the hermitian field can at most change by a factor -1. In the case of the electromagnetic field, $\mathbf{A}(x)$ must transform under $\mathcal{C}$ according to

$$\mathcal{C}\mathbf{A}(x)\mathcal{C}^{-1} = -\mathbf{A}(x) \tag{15.97}$$

in order to leave the $\mathbf{j}(x) \cdot \mathbf{A}(x)$ term in the lagrangian invariant. In momentum space this leads directly to

$$\mathcal{C}a^\dagger(k,\lambda)\mathcal{C}^{-1} = -a^\dagger(k,\lambda) \tag{15.98}$$

and imitating the procedure developed in constructing the parity operator, we find

$$\mathcal{C} = \exp\left[i\pi \int d^3k \sum_{\lambda=1}^{2} a^\dagger(k,\lambda)a(k,\lambda)\right] \tag{15.99}$$

In the presence of interactions, $\mathcal{C}$ in (15.99) may not be a symmetry operator of the theory. In this case we construct $\mathcal{C}$, as in (15.86) and (15.87) for $\mathcal{P}$, by forming

$$\mathcal{C}(t) = e^{iHt}\mathcal{C}_0 e^{-iHt} \tag{15.100}$$

where $\mathcal{C}_0$ satisfies (15.97) at $t = 0$ and is given by (15.99) in terms of the expansion coefficients of the interacting photon field at $t = 0$. For free-field theory the vacuum $|0\rangle$ is nondegenerate and therefore an eigenstate of the symmetry operator $\mathcal{C}$. We follow the analogous convention to (15.76) for $\mathcal{P}$ of choosing the phase for the unitary operator $\mathcal{C}$ so that

$$\mathcal{C}|0\rangle = +|0\rangle \tag{15.101}$$

that is, the vacuum is an even eigenstate. For an n-photon state the eigenvalue of $\mathcal{C}$ according to (15.97) and this choice is $(-1)^n$. This is denoted as the *charge parity* of a state. In the presence of interactions we retain these conventions when further assuming the vacuum state to be nondegenerate.

For π^0 mesons the choice of sign associated with the transformation $\mathcal{C}$ is dictated by the observation of the decay $\pi^0 \to 2\gamma$. If charge conjugation invariance holds in strong and electromagnetic couplings, the π^0 must be even under $\mathcal{C}$ if it is to couple with the state of two photons, which by (15.97) is even.

Turning next to the charged π-meson fields, it is natural to consider φ and φ^*, which create the minus and plus charge eigenstates. Since $\mathcal{L}(x) \to \mathcal{L}(x)$ and $j_\mu(x) \to -j_\mu(x)$ under the transformation $\varphi \rightleftarrows \varphi^*$, we search for a $\mathcal{C}$ which has the property, up to an arbitrary phase,

$$\mathcal{C}\varphi(x)\mathcal{C}^{-1} = \varphi^*(x) \qquad \mathcal{C}\varphi^*(x)\mathcal{C}^{-1} = \varphi(x) \tag{15.102}$$

In terms of the creation and destruction operators (12.57) for π^+ and π^-, this reads

$$\mathcal{C}a_+^\dagger(k)\mathcal{C}^{-1} = a_-^\dagger(k) \qquad \mathcal{C}a_-^\dagger(k)\mathcal{C}^{-1} = a_+^\dagger(k) \tag{15.103}$$

Alternatively, in terms of the hermitian fields $\varphi_1(x)$ and $\varphi_2(x)$ of (12.52)

$$\mathcal{C}\varphi_1(x)\mathcal{C}^{-1} = \varphi_1(x) \qquad \mathcal{C}\varphi_2(x)\mathcal{C}^{-1} = -\varphi_2(x) \tag{15.104}$$

and $C\varphi_3 C^{-1} = +\varphi_3$

or in momentum space,

$$\mathcal{C}a_1(k)\mathcal{C}^{-1} = a_1(k) \qquad \mathcal{C}a_2(k)\mathcal{C}^{-1} = -a_2(k) \tag{15.105}$$

which shows that $\mathcal{C}$ is a reflection operator in isotopic space about the (1,3) plane. We see that eigenstates of $\mathcal{C}$, as is clear physically, must contain equal numbers of π^+ and π^- mesons and hence are electrically neutral. Formally, this is verified from the observation that

$$\mathcal{C}Q = -Q\mathcal{C} \tag{15.106}$$

To construct $\mathcal{C}$ for the complete π-meson field $(\varphi_1,\varphi_2,\varphi_3)$, we recall from (15.104) that $\mathcal{C}$ fails to commute only with φ_2. Following the procedure employed for the photon, we find

$$\mathcal{C} = \exp\left[i\pi \int d^3k\, a_2^\dagger(k)a_2(k)\right] \tag{15.107}$$

In the general case when it is not a constant of the motion we construct $\mathcal{C}$ as in (15.100).

Similar considerations apply to the K mesons. The remarks above may be transcribed without change to the K^+ and K^- fields. Since for K^+ mesons $Q_{K^+} = S_{K^+} = Y_{K^+}$ by (15.64), (15.66), and (15.68), it follows from (15.105) that $\mathcal{C}$ anticommutes with S and Y. In addition, the K^0 and $\bar{K}^0$ must also transform into each other under $\mathcal{C}$; this transformation is accomplished in the same way as for K^+ and K^-, the only difference being that Q now vanishes. However, it is still true that $\mathcal{C}$ anticommutes with S and Y. Since $\varphi^*_{K^0}$ and φ_{K^0} create the K^0 and $\bar{K}^0$ which are eigenstates of S, the hermitian linear combinations[1]

$$\varphi_{K_1} = \frac{1}{\sqrt{2}}\,(\varphi_{K^0} - \varphi^*_{K^0}) \qquad \varphi_{K_2} = \frac{1}{\sqrt{2}}\,(\varphi_{K^0} + \varphi^*_{K^0}) \tag{15.108}$$

in analogy with the φ_1 and φ_2 of (15.104) for charged π mesons, create the states of even and odd charge conjugation. These states are important in the theory of the weak decays of the neutral K mesons.[2]

A discussion of charge conjugation for a Dirac particle has already been given in the companion volume. The free Dirac equation is invariant under the replacement

$$\psi(x) \rightarrow C\bar{\psi}^T(x) \tag{15.109}$$

where C is a 4×4 matrix with the property

$$C\gamma_\mu C^{-1} = -\gamma_\mu^T \qquad \text{or} \qquad C_{\alpha\beta}\gamma^\mu_{\beta\lambda}C^{-1}_{\lambda\tau} = -\gamma^\mu_{\tau\alpha} \tag{15.110}$$

In Eq. (5.6) we chose

$$C = i\gamma^2\gamma^0 = -C^{-1} = -C^\dagger = -C^T \tag{15.111}$$

in the representation with $\gamma_0 = \gamma_0{}^T$ and $\gamma_2 = \gamma_2{}^T$. In field theory we seek a unitary operator $\mathcal{C}$ which generates the transformation (15.109):

$$\begin{aligned} &\mathcal{C}\psi_\alpha(x)\mathcal{C}^{-1} = C_{\alpha\beta}\bar{\psi}_\beta(x) = (C\gamma^0)_{\alpha\beta}\psi^\dagger_\beta(x) \\ \text{and} \qquad &\mathcal{C}\bar{\psi}_\alpha(x)\mathcal{C}^{-1} = -\psi_\beta(x)C^{-1}_{\beta\alpha} \end{aligned} \tag{15.112}$$

with the matrix C given by (15.110) and (15.111), up to an arbitrary and uninteresting phase factor. The matrix order in (15.112) is explicitly shown by the indices and must be carefully watched.

It is readily verified that under $\mathcal{C}$ the commutation relations (13.53) and (13.54)—as well as the Dirac equation—are invariant and that $\mathcal{L}$ in (13.42) changes only by an inessential total divergence.

[1] The K^0 is known to have odd intrinsic parity. Thus according to (15.108), the K_1 is even under the combined operation $\mathcal{C}\mathcal{P}$ and the K_2 odd.

[2] See *Proc.* 1964 *Intern. Conf. High-energy Phys.* (*Dubna*).

Acting upon $\bar{\psi}(x)\gamma_\mu\psi(x)$, $\mathcal{C}$ gives

$$\begin{aligned}\mathcal{C}\bar{\psi}(x)\gamma^\mu\psi(x)\mathcal{C}^{-1} &= -\psi_\alpha(x)C^{-1}_{\alpha\beta}\gamma^\mu_{\beta\lambda}C_{\lambda\tau}\bar{\psi}_\tau(x) \\ &= \psi_\alpha(x)\gamma^\mu_{\tau\alpha}\bar{\psi}_\tau(x) \qquad (15.113)\end{aligned}$$

However, we have already seen in Sec. 15.5 that the identification of $\bar{\psi}(x)\gamma_\mu\psi(x)$ with the current leads to difficulties unless one antisymmetrizes (or, equivalently, normal-orders) the fermion field operators as in (15.26). Identifying

$$j_\mu(x) = \tfrac{1}{2}[\bar{\psi}(x),\gamma_\mu\psi(x)]$$

it follows directly from (15.113) that

$$\mathcal{C}j_\mu(x)\mathcal{C}^{-1} = -j_\mu(x) \qquad (15.114)$$

and therefore the electromagnetic current is odd under $\mathcal{C}$ as demanded in (15.96). It also follows that if the vacuum is nondegenerate, it is an eigenstate of $\mathcal{C}$, and $\langle 0|j_\mu(x)|0\rangle = 0$.

To construct $\mathcal{C}$, we go into momentum space, recalling from Chap. 5, p. 69, that the electron and positron spinors are related by

$$\begin{aligned}(C\gamma^0)_{\alpha\beta}u^\dagger_\beta(p,s) &= v_\alpha(p,s)e^{i\varphi(p,s)} \\ (C\gamma^0)_{\alpha\beta}v^\dagger_\beta(p,s) &= u_\alpha(p,s)e^{i\varphi(p,s)}\end{aligned}$$

Carrying out the momentum expansion of (15.112), we find

$$\begin{aligned}\mathcal{C}b(p,s)\mathcal{C}^{-1} &= d(p,s)e^{i\varphi(p,s)} \\ \mathcal{C}d^\dagger(p,s)\mathcal{C}^{-1} &= b^\dagger(p,s)e^{i\varphi(p,s)}\end{aligned}$$

which shows that the charge conjugation transformation interchanges the particle and antiparticle operators, in accord with its definition. The explicit construction of $\mathcal{C}$ follows along a path similar to that taken earlier. First it is convenient to break $\mathcal{C}$ into a product of two unitary transformations

$$\mathcal{C} = \mathcal{C}_2\mathcal{C}_1 \qquad (15.115)$$

and to choose $\mathcal{C}_1$ to remove the phase factor φ:

$$\begin{aligned}\mathcal{C}_1 b(p,s)\mathcal{C}_1^{-1} &= e^{i\varphi(p,s)}b(p,s) \\ \mathcal{C}_1 d^\dagger(p,s)\mathcal{C}_1^{-1} &= e^{i\varphi(p,s)}d^\dagger(p,s)\end{aligned}$$

By explicit construction we find

$$\mathcal{C}_1 = \exp\left\{-i\int d^3p\sum_{\pm s}\varphi(p,s)[b^\dagger(p,s)b(p,s) - d^\dagger(p,s)d(p,s)]\right\} \qquad (15.116)$$

$\mathcal{C}_2$ is then constructed by the same techniques as for the parity operator (15.84) and found to be

$$\mathcal{C}_2 = \exp\left\{\frac{i\pi}{2}\int d^3p \sum_{s=1}^{2} [b^\dagger(p,s) - d^\dagger(p,s)][b(p,s) - d(p,s)]\right\} \quad (15.117)$$

If $\mathcal{C}$ is not a constant of the motion, (15.116) and (15.117) remain valid at $t = 0$ only and, as before, $\mathcal{C}$ is constructed at general times t by (15.100).

We leave as an exercise the demonstration that the introduction of π-meson–nucleon couplings according to the lagrangian model (15.30) does not destroy charge-conjugation invariance, provided the $\mathcal{C}$ transformation is applied simultaneously to all fields appearing in $\mathcal{L}$. For electromagnetic couplings, the symmetry is preserved; it was, in fact, the way the symmetry was introduced.

We consider the decay of positronium as an interesting illustration of the application of charge conjugation invariance to the prediction of selection rules. In the same way as for the neutral K mesons, we can form positronium eigenstates which are even or odd under $\mathcal{C}$. To form positronium, let us first construct a free electron-positron pair from the vacuum and superpose states of different spin and momentum to represent the initial positronium state of a given angular momentum

$$\Psi_{e^+e^-} = \int d^3p\, d^3p' \sum_{s,s'} \mathfrak{F}(p,s;p',s')b^\dagger(p,s)d^\dagger(p',s')|0\rangle \quad (15.118)$$

Although (15.118) is not the exact state in the presence of electromagnetic couplings, it has the same symmetry properties as the true physical state because of the invariance of the electromagnetic interaction under $\mathcal{C}$. Therefore, we need only consider what kinds of amplitudes $\mathfrak{F}(p,s;p's')$ correspond to states even under $\mathcal{C}$, which are observed to decay into two photons, and what are the odd states decaying to three photons. Applying $\mathcal{C}$ to (15.118), we find, using the anticommutation algebra for the $b^\dagger$ and $d^\dagger$ operators,[1]

$$\begin{aligned}\mathcal{C}\Psi_{e^+e^-} &= \int d^3p\, d^3p' \sum_{s,s'} \mathfrak{F}(p,s;p',s')d^\dagger(p,s)b^\dagger(p',s')|0\rangle \\ &= -\int d^3p\, d^3p' \sum_{s,s'} \mathfrak{F}(p',s';p,s)b^\dagger(p,s)d^\dagger(p',s')|0\rangle\end{aligned}$$

Evidently a state *even* under the interchange of electron and positron

$$\mathfrak{F}(p,s;p',s') = +\mathfrak{F}(p',s';p,s)$$

[1] We have omitted phases $\eta(p,s)$. For justification of this, see Prob. 23.

is *odd* under $\mathcal{C}$ and vice versa. This means that the $^3S_1$ triplet state of positronium decays with three-photon emission and the $^1S_0$ singlet state decays into two γ rays.

A symmetric state of boson and antiboson is even under $\mathcal{C}$, because the minus sign coming from the anticommutation relations is missing. As a general rule, it is convenient to remember that the eigenvalue under charge conjugation of a particle-antiparticle pair is $+1$ if the particles are in a state allowed for two identical particles (even for bosons, odd for fermions). For an odd eigenfunction the situation is just reversed.

15.13 Time Reversal

The time-reversal transformation changes the direction of time from t to $t' = -t$. In constructing this transformation for the one-particle Dirac theory we found that it is a symmetry operation if it includes in addition to the instruction to replace t by t' the command to take the complex conjugate and multiply the wave function by a matrix $T = i\gamma^1\gamma^3$ in the representation where only γ^2 is imaginary. In field theory we seek an operator $\mathfrak{I}$ which transforms physical states developing in time t to states as would be viewed on backward running film with $t' = -t$. The quantum conditions (11.70) make it clear that $\mathfrak{I}$, as in the one-particle theory, will not be a linear operator. Consider, for example

$$[H,\varphi_r(\mathbf{x},t)] = -i\,\frac{\partial\varphi_r(\mathbf{x},t)}{\partial t} \tag{15.119}$$

If we seek a unitary operator $\mathfrak{U}$ which leaves the action invariant and which transforms $\varphi_r(\mathbf{x},t)$ to $W_{rs}\varphi_s(\mathbf{x},t') = \mathfrak{U}\varphi_r(\mathbf{x},t)\mathfrak{U}^{-1}$, we find

$$[\mathfrak{U}H\mathfrak{U}^{-1},\varphi_s(\mathbf{x},t')] = +\,i\,\frac{\partial\varphi_s(\mathbf{x},t')}{\partial t'} \tag{15.120}$$

In order to restore (15.119), it is necessary to have $\mathfrak{U}$ transform H to $-H$ in (15.120). This is unacceptable on physical grounds, since the eigenvalues of H must be positive relative to the vacuum state, before and after the transformation. Faced with this situation, we settle, as in (5.14), for a nonunitary $\mathfrak{I}$, obtained by appending to the unitary $\mathfrak{U}$ the instruction K to *take the complex conjugate of all c numbers.*[1]

[1] The operation of taking a complex conjugate is nonlinear; $\mathfrak{I}$ is called an antilinear, or antiunitary, operator. See E. P. Wigner, *Göttinger Nachr.*, **31**, 546 (1932), W. Pauli (ed.), "Niels Bohr and the Development of Physics," McGraw-Hill Book Company, New York, 1955, and G. Lüders, *Ann. Phys.* (*N.Y.*), **2**, 1 (1957).

Then if

$$\mathfrak{T} = \mathfrak{U}K \qquad \text{and} \qquad \mathfrak{T}H\mathfrak{T}^{-1} = H \tag{15.121}$$

(15.119) will be invariant under $\mathfrak{T}$. In terms of the lagrangian density (15.69), including interaction with an external field, the theory will be time-reversal invariant provided a $\mathfrak{T}$ exists such that the commutation relations are invariant:

$$\mathfrak{T}\mathfrak{L}(\mathbf{x},t)\mathfrak{T}^{-1} = \mathfrak{L}(\mathbf{x},-t) \tag{15.122}$$

and

$$\mathfrak{T}j_\mu(\mathbf{x},t)\mathfrak{T}^{-1} = j^\mu(\mathbf{x},-t) \tag{15.123}$$

In (15.123) the electromagnetic currents are reversed while the charges are unchanged under time reversal. This is required by classical correspondence, since for the external electromagnetic fields

$$A_\mu(\mathbf{x},t) \to A^\mu(\mathbf{x},-t) \tag{15.124}$$

under time reversal; and therefore by (15.123),

$$j_\mu(\mathbf{x},t)A^\mu(\mathbf{x},t) \to +\ j_\mu(\mathbf{x},-t)A^\mu(\mathbf{x},-t)$$

Thus $\mathfrak{T}$ changes the action according to

$$\mathfrak{T}J(t_2,t_1)\mathfrak{T}^{-1} = \int_{t_1}^{t_2} \mathfrak{L}(\mathbf{x},-t)\, d^3x\, dt = \int_{-t_2}^{-t_1} d^3x\, dt\, \mathfrak{L}(x) = J(-t_1,-t_2) \tag{15.125}$$

$J(-t_1,-t_2)$ differs from $J(t_2,t_1)$ only by a translation in time, which is also a symmetry operation of the theory. Therefore, (15.122) and (15.123) are satisfactory criteria for time-reversal invariance.

We turn to a construction of $\mathfrak{T}$ for the various free fields discussed so far, starting with the electromagnetic field. According to (15.123),

$$\mathfrak{T}\mathbf{A}(\mathbf{x},t)\mathfrak{T}^{-1} = -\mathbf{A}(\mathbf{x},-t) \tag{15.126}$$

since the currents producing the field are reversed. This transformation satisfies (15.122) for the Maxwell lagrangian in transverse gauge as in (14.9). It also leaves invariant the equal-time commutation relations (14.13) and (14.17), owing to the presence of K. We go into momentum space to construct $\mathfrak{T}$. Inserting the expansion (14.33) into (15.126), we find

$$\begin{aligned}
\mathfrak{T}\mathbf{A}(\mathbf{x},t)\mathfrak{T}^{-1} &= \int \frac{d^3k}{\sqrt{(2\pi)^3 2\omega}} \sum_{\lambda=1}^{2} \boldsymbol{\varepsilon}(k,\lambda)[\mathfrak{U}a(k,\lambda)\mathfrak{U}^{-1}e^{i\omega t - i\mathbf{k}\cdot\mathbf{x}} \\
&\qquad + \mathfrak{U}a^\dagger(k,\lambda)\mathfrak{U}^{-1}e^{-i\omega t + i\mathbf{k}\cdot\mathbf{x}}] \\
= -\mathbf{A}(\mathbf{x},-t) &= -\int \frac{d^3k}{\sqrt{(2\pi)^3 2\omega}} \sum_{\lambda=1}^{2} \boldsymbol{\varepsilon}(k,\lambda)[a(k,\lambda)e^{i\omega t + i\mathbf{k}\cdot\mathbf{x}} \\
&\qquad + a^\dagger(k,\lambda)e^{-i\omega t - i\mathbf{k}\cdot\mathbf{x}}]
\end{aligned} \tag{15.127}$$

With the convention used in (14.31)

$$\boldsymbol{\varepsilon}(k,1) = -\boldsymbol{\varepsilon}(-k,1) \qquad \boldsymbol{\varepsilon}(k,2) = +\boldsymbol{\varepsilon}(-k,2)$$

we have

$$\mathfrak{U}a(k,1)\mathfrak{U}^{-1} = +a(-k,1) \qquad \mathfrak{U}a(k,2)\mathfrak{U}^{-1} = -a(-k,2) \tag{15.128}$$

To solve (15.128), we refer back to the solutions for the parity operators of the scalar and pseudoscalar fields (15.78), (15.83), and (15.84) and find

$$\mathfrak{U} = \exp\left\{\frac{-i\pi}{2}\int d^3k\,[a^\dagger(k,1)a(k,1) - a^\dagger(k,1)a(-k,1) + a^\dagger(k,2)a(k,2) + a^\dagger(k,2)a(-k,2)]\right\} \tag{15.129}$$

For the free hermitian Klein-Gordon field, the criteria for a satisfactory $\mathfrak{T}$, (15.122) and (15.123), are satisfied by the choice

$$\mathfrak{T}\varphi(\mathbf{x},t)\mathfrak{T}^{-1} = \pm\varphi(\mathbf{x},-t) \tag{15.130}$$

For the charged field, the requirement that $\mathfrak{T}$ transform $j_\mu(\mathbf{x},t)$ to $j^\mu(\mathbf{x},-t)$ suggests the choice

$$\mathfrak{T}\varphi(\mathbf{x},t)\mathfrak{T}^{-1} = \pm\varphi(\mathbf{x},-t) \tag{15.131}$$

Thus for the three hermitian components of the π-meson field, we take

$$\mathfrak{T}\begin{bmatrix}\varphi_1(\mathbf{x},t)\\ \varphi_2(\mathbf{x},t)\\ \varphi_3(\mathbf{x},t)\end{bmatrix}\mathfrak{T}^{-1} = \pm\begin{bmatrix}+\varphi_1(\mathbf{x},-t)\\ -\varphi_2(\mathbf{x},-t)\\ +\varphi_3(\mathbf{x},-t)\end{bmatrix} \tag{15.132}$$

with the overall phase arbitrary. $\mathfrak{T}$ may be constructed by following the analogous steps used for the Maxwell field.

For the Dirac theory we seek an operator $\mathfrak{T}$ such that

$$\mathfrak{T}\psi_\alpha(\mathbf{x},t)\mathfrak{T}^{-1} = T_{\alpha\beta}\psi_\beta(\mathbf{x},-t) \tag{15.133}$$

which must satisfy (15.122) and (15.123), and leave invariant the anticommutators (13.53) and (13.54). We readily verify that these criteria are satisfied if we take T to be the same matrix found in (5.15) for the one-particle theory, namely:

$$T = i\gamma^1\gamma^3 \qquad T\gamma_\mu T^{-1} = \gamma_\mu{}^T = \gamma^{\mu *}$$

with

$$T = T^\dagger = T^{-1} = -T^* \tag{15.134}$$

in the representation in which only γ^2 is imaginary; for instance

$$\begin{aligned}
\mathfrak{I} j_\mu(\mathbf{x},t)\mathfrak{I}^{-1} &= \psi^\dagger(\mathbf{x},-t)T^{-1}(\gamma_0\gamma_\mu)^*T\psi(\mathbf{x},-t) - \langle 0|\bar\psi\gamma_\mu\psi|0\rangle \\
&= \bar\psi(\mathbf{x},-t)\gamma^\mu\psi(\mathbf{x},-t) - \langle 0|\bar\psi\gamma_\mu\psi|0\rangle \\
&= j^\mu(\mathbf{x},-t)
\end{aligned}$$

checks (15.123).

The property (15.133) differs from one-particle theory, where $\psi(\mathbf{x},t) \to T\psi^*(\mathbf{x},-t)$ upon the instruction of complex conjugation. In field theory the analogous transformation $\psi \to T\psi^\dagger$ is unacceptable, since this would transform, for instance, a state of an electron at rest into a positron state.

Going into momentum space to construct $\mathfrak{I}$, (15.133) becomes

$$\begin{aligned}
\int \frac{d^3p}{\sqrt{(2\pi)^3}} \sqrt{\frac{m}{E}} \sum_{\pm s} [\mathfrak{U}b(p,s)\mathfrak{U}^{-1}u^*(p,s)\, e^{iEt-i\mathbf{p}\cdot\mathbf{x}} \\
+ \mathfrak{U}d^\dagger(p,s)\mathfrak{U}^{-1}v^*(p,s)\, e^{-iEt+i\mathbf{p}\cdot\mathbf{x}}] \\
= \int \frac{d^3p}{\sqrt{(2\pi)^3}} \sqrt{\frac{m}{E}} \sum_{\pm s} [b(p,s)Tu(p,s)\, e^{iEt+i\mathbf{p}\cdot\mathbf{x}} \\
+ d^\dagger(p,s)Tv(p,s)\, e^{-iEt-i\mathbf{p}\cdot\mathbf{x}}]
\end{aligned} \tag{15.135}$$

It follows from our discussion of time reversal in the one-particle theory [see (5.16)] that

$$Tu(p,s) = u^*(-p,-s)e^{i\alpha_+(p,s)} \qquad Tv(p,s) = v^*(-p,-s)e^{i\alpha_-(p,s)} \tag{15.136}$$

where the α's are phase factors which depend upon the spin state. By applying T again to (15.136), one finds, since $T^2 = 1$,

$$\alpha_\pm(p,s) = \pi + \alpha_\pm(-p,-s) \tag{15.137}$$

Using (15.136), we satisfy (15.135) provided

$$\begin{aligned}
\mathfrak{U}b(p,s)\mathfrak{U}^{-1} &= -b(-p,-s)e^{i\alpha_+(p,s)} \\
\mathfrak{U}d^\dagger(p,s)\mathfrak{U}^{-1} &= -d^\dagger(-p,-s)e^{i\alpha_-(p,s)}
\end{aligned} \tag{15.138}$$

The transformation $\mathfrak{U}$ is most easily found by breaking it into the product of two unitary transformations

$$\mathfrak{U} = \mathfrak{U}_2\mathfrak{U}_1 \tag{15.139}$$

$\mathfrak{U}_1$ may be chosen to remove the phase factors

$$\begin{aligned}
\mathfrak{U}_1 b(p,s)\mathfrak{U}_1^{-1} &= e^{i\alpha_+(p,s)}b(p,s) \\
\mathfrak{U}_1 d^\dagger(p,s)\mathfrak{U}_1^{-1} &= e^{i\alpha_-(p,s)}d^\dagger(p,s)
\end{aligned} \tag{15.140}$$

and is explicitly constructed to be

$$\mathfrak{U}_1 = \exp\left\{-i\int d^3p \sum_{\pm s} [\alpha_+(p,s)b^\dagger(p,s)b(p,s) - \alpha_-(p,s)d^\dagger(p,s)d(p,s)]\right\} \tag{15.141}$$

$\mathfrak{U}_2$ then satisfies

$$\mathfrak{U}_2 b(p,s)\mathfrak{U}_2^{-1} = -b(-p,-s) \quad \mathfrak{U}_2 d^\dagger(p,s)\mathfrak{U}_2^{-1} = -d^\dagger(-p,-s) \tag{15.142}$$

and using the same techniques as for the parity operator (15.83) is found to be

$$\mathfrak{U}_2 = \exp\left\{-i\frac{\pi}{2}\int d^3p \sum_{\pm s} [b^\dagger(p,s)b(p,s) + b^\dagger(p,s)b(-p,-s) - d^\dagger(p,s)d(p,s) - d^\dagger(p,s)d(-p,-s)]\right\} \tag{15.143}$$

From (15.138) we see that the time-reversed state of one free electron or positron with energy-momentum $E_p,\mathbf{p}$, and spin $\mathbf{s}$ is a state of the same particle, but with eigenvalues $E_p, -\mathbf{p}, -\mathbf{s}$ for reversed spin and momentum but again positive energy. The corresponding wave functions for these states are related by[1]

$$\begin{aligned}
\psi_{E_p,\mathbf{p},\mathbf{s}}(\mathbf{x},t) &= \langle 0\,|\psi(\mathbf{x},t)|\,1 \text{ electron}; \mathbf{p},\mathbf{s}\rangle \\
&= \langle K0|K\psi(\mathbf{x},t)\ 1 \text{ electron}; \mathbf{p},\mathbf{s}\rangle^* \\
&= \langle K0|\mathfrak{U}^{-1}\mathfrak{U}K\psi(\mathbf{x},t)\ 1 \text{ electron}; \mathbf{p},\mathbf{s}\rangle^* \\
&= \langle 0\,|\mathfrak{T}\psi(\mathbf{x},t)\mathfrak{T}^{-1}|\ \mathfrak{T}\ (1 \text{ electron}; \mathbf{p},\mathbf{s})\rangle^* \\
&= -e^{i\alpha_+(p,s)}T^*\langle 0\,|\psi(\mathbf{x},-t)|\ 1 \text{ electron}; -\mathbf{p},-\mathbf{s}\rangle^* \\
&= e^{i\alpha_+(p,s)}T\psi^*_{E_p,-\mathbf{p},-\mathbf{s}}(\mathbf{x},-t)
\end{aligned} \tag{15.144}$$

Equation (15.144) shows that the time-reversed *wave functions* are related, as in the one-particle theory, to each other's complex conjugates.

In the case of interacting fields, we may borrow the operators we have constructed, although we have only the expansions of the field operators at time $t = 0$. Since the commutation relations at $t = 0$ are unchanged from free-field commutators, we may construct a $\mathfrak{T}_0$ which

[1] Note that we explicitly separate the nonlinear operator K in the second line by taking the complex conjugate. Also $T^* = -T$ by (15.134).

preserves, for instance, all the relations of the form

$$
\begin{aligned}
\mathfrak{T}_0\psi_\alpha(\mathbf{x},0)\mathfrak{T}_0^{-1} &= T_{\alpha\beta}\psi_\beta(\mathbf{x},0)\\
\mathfrak{T}_0\varphi(\mathbf{x},0)\mathfrak{T}_0^{-1} &= \pm\varphi(\mathbf{x},0)\\
\mathfrak{T}_0\dot{\varphi}(\mathbf{x},0)\mathfrak{T}_0^{-1} &= \mp\dot{\varphi}(\mathbf{x},0)\\
\mathfrak{T}_0\mathbf{A}(\mathbf{x},0)\mathfrak{T}_0^{-1} &= -\mathbf{A}(\mathbf{x},0)\\
\mathfrak{T}_0\dot{\mathbf{A}}(\mathbf{x},0)\mathfrak{T}_0^{-1} &= +\dot{\mathbf{A}}(\mathbf{x},0)
\end{aligned}
\tag{15.145}
$$

found for the free-field operators. Here $\mathfrak{T}_0$ is identical with that constructed for the free fields, with the operators $a^\dagger$, b, d, etc. replaced by the expansion coefficients of (15.23) evaluated at $t = 0$. Writing[1]

$$
\mathfrak{T} = e^{-iHt}\mathfrak{T}_0 e^{-iHt} \tag{15.146}
$$

to form $\mathfrak{T}$ at arbitrary times t, we find, for instance,

$$
\begin{aligned}
\mathfrak{T}\psi_\alpha(\mathbf{x},t)\mathfrak{T}^{-1} &= e^{-iHt}\mathfrak{T}_0\psi_\alpha(\mathbf{x},0)\mathfrak{T}_0^{-1}e^{iHt}\\
&= e^{-iHt}T_{\alpha\beta}\psi_\beta(\mathbf{x},0)e^{iHt}\\
&= T_{\alpha\beta}\psi_\beta(\mathbf{x},-t)
\end{aligned}
$$

as desired [see (15.133)].

Similarly, we find

$$
\mathfrak{T}\varphi(\mathbf{x},t)\mathfrak{T}^{-1} = \pm\varphi(\mathbf{x},-t) \qquad \text{and} \qquad \mathfrak{T}\mathbf{A}(\mathbf{x},t)\mathfrak{T}^{-1} = -\mathbf{A}(\mathbf{x},-t)
$$

If the $\mathfrak{T}$ so constructed also satisfies (15.122) and (15.123), it is a symmetry operation of the theory, in which case

$$
[\mathfrak{T},H] = 0 \tag{15.147}
$$

and (15.146) simplifies to

$$
\mathfrak{T} = \mathfrak{T}_0
$$

15.14 The $\mathfrak{T}\mathcal{C}\mathcal{P}$ Theorem

We can readily verify that the electromagnetic interactions and the π-meson–nucleon interactions introduced in (15.30) are invariant under the separate symmetry operations of $\mathfrak{T}$, $\mathcal{C}$, and $\mathcal{P}$ which we have explicitly constructed for general interactions. It is, of course, possible to modify the interaction terms by a judicious insertion of a few i's and γ_5's here and there and in this way destroy $\mathfrak{T}$, $\mathcal{C}$, and $\mathcal{P}$ as symmetry operations without affecting invariance under proper Lorentz transformations and displacements. It is remarkable, however, that the

[1] In the analogous construction for the time dependence of $\mathcal{P}$ and $\mathcal{C}$ we displaced time $t \to 0 \to t$. Here for time reversal we displace $t \to 0 \to -t$, which explains the different sign of the exponential factor.

product of $\mathcal{T}$, $\mathcal{C}$, and $\mathcal{P}$ remains a symmetry operation provided only that:

1. The theory, which in the present context means a local theory for which there exists an appropriately normal-ordered hermitian lagrangian density, is covariant under proper Lorentz transformations.

2. The theory is quantized with the usual connection between spin and statistics; for example, Klein-Gordon and Maxwell fields are quantized with commutation relations leading to Bose-Einstein statistics, and Dirac fields obey anticommutation relations leading to the exclusion principle.

This is the $\mathcal{TCP}$ theorem of Lüders and Zumino, Pauli, and Schwinger.[1] We construct a proof for interacting Klein-Gordon, Maxwell, and Dirac fields by showing that with the successive applications of $\mathcal{T}(t)$, $\mathcal{C}(t)$, and $\mathcal{P}(t)$, evaluated at a common time t, the hamiltonian H satisfies

$$\mathcal{PCT}H\mathcal{T}^{-1}\mathcal{C}^{-1}\mathcal{P}^{-1} = H \tag{15.148}$$

Since our assumptions are based upon the properties of the lagrangian density, it is more convenient to start by considering it and to work our way back to (15.148). We shall first show that

$$\mathcal{PCT}\mathcal{L}(\mathbf{x},t)\mathcal{T}^{-1}\mathcal{C}^{-1}\mathcal{P}^{-1} = \mathcal{L}(-\mathbf{x},-t)$$

with $\mathcal{P}$, $\mathcal{C}$, $\mathcal{T}$ evaluated at time t.

By the Lorentz invariance of $\mathcal{L}$, we mean that it is an hermitian operator built up out of scalars formed from products of $\varphi_r(x)$ and $A_\mu(x)$ and their derivatives $\partial/\partial x^\mu$ and, in addition, of bilinear forms of spinor fields or their derivatives $\bar{\psi}^A\Gamma\psi^B$, which transform as tensors. The indices A and B label the internal degree of freedom (p,e^-,ν, etc.), and Γ is one of the sets of matrices $1, i\gamma_5,\gamma_\mu,\gamma_5\gamma_\mu,\sigma_{\mu\nu}$. The action of $\mathcal{PCT}$ on a scalar hermitian field φ_r is

$$\mathcal{PCT}\varphi_r(\mathbf{x},t)\mathcal{T}^{-1}\mathcal{C}^{-1}\mathcal{P}^{-1} = \pm\varphi_r(-\mathbf{x},-t) \tag{15.149}$$

where the $\pm$ sign is arbitrary, depending upon the choice of signs in (15.132). In terms of operators producing charge eigenstates, (15.149) reads

$$\mathcal{PCT}\varphi(\mathbf{x},t)\mathcal{T}^{-1}\mathcal{C}^{-1}\mathcal{P}^{-1} = \pm\varphi^*(-\mathbf{x},-t)$$
$$\mathcal{PCT}\varphi^*(\mathbf{x},t)\mathcal{T}^{-1}\mathcal{C}^{-1}\mathcal{P}^{-1} = \pm\varphi(-\mathbf{x},-t)$$

[1] G. Lüders, *Ann. Phys.* (*N.Y.*), **2,** 1 (1957).

Putting together $\mathcal{P}$, $\mathcal{C}$, and $\mathcal{T}$ for the spinor fields [(15.88), (15.112), and (15.133)], we find

$$\mathcal{P}\mathcal{C}\mathcal{T}\psi_\alpha^A(\mathbf{x},t)\mathcal{T}^{-1}\mathcal{C}^{-1}\mathcal{P}^{-1} = -i(\gamma^0\gamma_5)_{\alpha\beta}\bar{\psi}_\beta{}^A(-\mathbf{x},-t)$$
$$= +i\gamma^5_{\alpha\beta}\psi_\beta^{A\dagger}(-\mathbf{x},-t)$$

and

$$\mathcal{P}\mathcal{C}\mathcal{T}\bar{\psi}_\alpha{}^A(\mathbf{x},t)\mathcal{T}^{-1}\mathcal{C}^{-1}\mathcal{P}^{-1} = -i\psi_\beta{}^A(-\mathbf{x},-t)(\gamma_5\gamma_0)_{\beta\alpha} \tag{15.150}$$

For the bilinear forms in the spinor fields there follows from (15.150)

$$\mathcal{P}\mathcal{C}\mathcal{T}\bar{\psi}_\alpha{}^A(\mathbf{x},t)\Gamma_{\alpha\beta}\psi_\beta{}^B(\mathbf{x},t)\mathcal{T}^{-1}\mathcal{C}^{-1}\mathcal{P}^{-1}$$
$$= -\psi_\lambda{}^A(-\mathbf{x},-t)(\gamma_5\gamma_0\Gamma^*\gamma_0\gamma_5)_{\lambda\tau}\bar{\psi}_\tau{}^B(-\mathbf{x},-t)$$
$$= -\psi_\lambda{}^A(-\mathbf{x},-t)\Gamma'_{\tau\lambda}\bar{\psi}_\tau{}^B(-\mathbf{x},-t) \tag{15.151}$$

where $\Gamma' = +\Gamma$ for $\Gamma = 1,\ i\gamma_5,\ \sigma_{\mu\nu}$

$\Gamma' = -\Gamma$ for $\Gamma = \gamma_\mu, \gamma_5\gamma_\mu$

Retaining earlier assumptions that independent Fermi fields ψ^A and ψ^B anticommute with each other and that products of spinor fields always appear in $\mathcal{L}$ in normal order, we arrive at

$$\mathcal{P}\mathcal{C}\mathcal{T}:\bar{\psi}_\alpha{}^A(\mathbf{x},t)\Gamma_{\alpha\beta}\psi_\beta{}^B(\mathbf{x},t):\mathcal{T}^{-1}\mathcal{C}^{-1}\mathcal{P}^{-1}$$
$$= +:\bar{\psi}_\tau{}^B(-\mathbf{x},-t)\Gamma'_{\tau\lambda}\psi_\lambda{}^A(-\mathbf{x},-t): \tag{15.152}$$

In fact only for the vector case $\Gamma = \gamma^\mu$ with identical Fermi fields $A = B$ is the normal-ordering needed to proceed from (15.151) to (15.152), since by (15.39)

$$-\psi_\lambda{}^A(-\mathbf{x},-t)\Gamma'_{\tau\lambda}\bar{\psi}_\tau{}^B(-\mathbf{x},-t)$$
$$= +\bar{\psi}_\tau{}^B(-\mathbf{x},-t)\Gamma'_{\tau\lambda}\psi_\lambda{}^A(-\mathbf{x},-t) - \delta_{AB}\delta^3(0)\operatorname{Tr}(\gamma_0\Gamma')$$

But it was just in terms of this vector current discussed in Sec. 15.5 that we defined $\mathcal{P}$, $\mathcal{C}$, and $\mathcal{T}$ in (15.72), (15.96), and (15.123) such that

$$\mathcal{P}\mathcal{C}\mathcal{T}j^\mu(\mathbf{x},t)\mathcal{T}^{-1}\mathcal{C}^{-1}\mathcal{P}^{-1} = -j^\mu(-\mathbf{x},-t) = -:\bar{\psi}(-\mathbf{x},-t)\gamma^\mu\psi(-\mathbf{x},-t): \tag{15.153}$$

For the electromagnetic field we find, upon putting together $\mathcal{P}$, $\mathcal{C}$, and $\mathcal{T}$ from (15.95), (15.97), and (15.126),

$$\mathcal{P}\mathcal{C}\mathcal{T}\mathbf{A}(\mathbf{x},t)\mathcal{T}^{-1}\mathcal{C}^{-1}\mathcal{P}^{-1} = -\mathbf{A}(-\mathbf{x},-t) \tag{15.154}$$

Applying $\mathcal{T}$, $\mathcal{C}$, and $\mathcal{P}$ to the constraint equation (15.8) for $A_0(\mathbf{x},t)$ gives

$$\mathcal{P}\mathcal{C}\mathcal{T}A_0(\mathbf{x},t)\mathcal{T}^{-1}\mathcal{C}^{-1}\mathcal{P}^{-1} = -\frac{e_0}{4\pi}\int\frac{d^3y}{|\mathbf{x}-\mathbf{y}|}j_0(-\mathbf{y},-t) = -A_0(-\mathbf{x},-t)$$

as may be checked readily from (15.152). Therefore, we have

$$\mathcal{P}\mathcal{C}\mathcal{T}A_\mu(x)\mathcal{T}^{-1}\mathcal{C}^{-1}\mathcal{P}^{-1} = -A_\mu(-x) \tag{15.155}$$

Finally, if derivatives appear, we note that

$$\frac{\partial}{\partial x_\mu} = -\frac{\partial}{\partial(-x_\mu)} \tag{15.156}$$

We may summarize these results by noticing that $\mathcal{P}\mathcal{C}\mathcal{T}$ effects the following changes:

1. All coordinates x_μ are changed into $x'_\mu = -x_\mu$; thus

$$\frac{\partial}{\partial x'_\mu} = -\frac{\partial}{\partial x_\mu}$$

2. Hermitian scalar fields $\varphi_r(x)$ are transformed into $+\varphi_r(x')$, with the arbitrary phase fixed here to be $+1$. The electromagnetic field $A_\mu(x)$ is transformed into $-A_\mu(x')$.
3. All even-rank tensors including bilinear forms of the Fermi fields or their derivatives are transformed into their hermitian conjugates, and all odd-rank ones are transformed into the negatives of their hermitian conjugates.
4. All other *c* numbers are replaced by their complex conjugates.

Because $\mathcal{L}$ is a scalar, all tensor indices in any given term of $\mathcal{L}$ are contracted and there are an even number of minus signs associated with condition 3; they may therefore be ignored. The net effect of the instructions is equivalent to the instruction to take the hermitian conjugate of $\mathcal{L}$. The question of ordering of factors is eliminated by the restriction that $\mathcal{L}$ be normal-ordered. It is here that the connection of spin and statistics enters, since the operation of normal ordering introduces a minus sign, (13.58), for anticommuting spin-½ fields and a plus sign, (12.25), for the Bose fields of spin 0 and 1. Consequently, we find for a hermitian $\mathcal{L}$ that

$$\mathcal{P}\mathcal{C}\mathcal{T}\mathcal{L}(x)\mathcal{T}^{-1}\mathcal{C}^{-1}\mathcal{P}^{-1} = \mathcal{L}(x') = \mathcal{L}(-\mathbf{x},-t) \tag{15.157}$$

The transition from $\mathcal{L}$ to the hamiltonian density is given as usual by

$$\mathcal{H}(x) = -\mathcal{L}(x) + \sum_r :\pi_r(x)\dot{\varphi}_r(x): \tag{15.158}$$

where the sum extends over all fields, Fermi or Bose, appearing in $\mathcal{L}$. Since $\mathcal{T}$, $\mathcal{C}$, and $\mathcal{P}$ were constructed to leave the commutation relations invariant

$$[\pi_r(\mathbf{x},t),\varphi_s(\mathbf{x}',t)] = -i\delta^3(\mathbf{x}-\mathbf{x}')\delta_{rs}$$

for Bose fields, or

$$\{\pi_r(\mathbf{x},t),\varphi_s(\mathbf{x}',t)\} = +i\delta^3(\mathbf{x}-\mathbf{x}')\delta_{rs}$$

for Fermi fields, we know that

C.R.
since K is all that matters in RHS of C.R. and $-\eta_r^*\eta_r = -1$

$$\mathcal{P}\mathcal{C}\mathcal{T}\pi_r(\mathbf{x},t)\mathcal{T}^{-1}\mathcal{C}^{-1}\mathcal{P}^{-1} = -\eta_r^*\pi_r(-\mathbf{x},-t)$$

if

$$\mathcal{P}\mathcal{C}\mathcal{T}\varphi_r(\mathbf{x},t)\mathcal{T}^{-1}\mathcal{C}^{-1}\mathcal{P}^{-1} = \eta_r\varphi_r(-\mathbf{x},-t) \tag{15.159}$$

also

$$\mathcal{P}\mathcal{C}\mathcal{T}\dot{\varphi}_r(\mathbf{x},t)\mathcal{T}^{-1}\mathcal{C}^{-1}\mathcal{P}^{-1} = -\eta_r\frac{\partial}{\partial(-t)}\varphi_r(-\mathbf{x},-t) \equiv -\eta_r\dot{\varphi}_r(-\mathbf{x},-t)$$

We conclude that

$$\mathcal{P}\mathcal{C}\mathcal{T}\mathcal{H}(\mathbf{x},t)\mathcal{T}^{-1}\mathcal{C}^{-1}\mathcal{P}^{-1} = \mathcal{H}(-\mathbf{x},-t)$$

and therefore (15.148) follows and the $\mathcal{T}\mathcal{C}\mathcal{P}$ theorem is proved.[1]

Problems

1. Show that $\Box\mathbf{A} = e_0\mathbf{j}_{tr}$ in radiation gauge and construct the transverse current operator.

2. Show that to lowest order in e_0 the vacuum charge density, before normal-ordering, is cubically divergent.

3. Verify that $[Q,H] = 0$, showing constancy of the total charge in quantum electrodynamics.

4. Verify the Heisenberg relations (15.18).

5. Verify the rules in (15.20) and (15.21) for the transformations of the field operators under a Lorentz transformation.

6. Complete the proof of Lorentz invariance of quantum electrodynamics in the radiation gauge by verifying invariance of the equal-time commutation relations (15.9) and (15.10).

7. By choosing the gauge $A_0 = 0$ and adding a small photon mass term, as in Chap. 14, Prob. 3, quantize the charged scalar field. (The lagrangian is given in the footnote, page 87, and Appendix B.)

8. Repeat Prob. 7 for a charged Dirac particle.

9. Verify the Heisenberg equation of motion for a charge-independent meson-nucleon interaction given by the hamiltonian (15.40), using (15.41). What hap-

[1] An elegant theorem on the connection between $\mathcal{T}\mathcal{C}\mathcal{P}$, spin and statistics, and locality (weak local commutativity) has been established within the axiomatic framework. For a discussion of the approach see R. F. Streator and A. S. Wightman, "PCT, Spin and Statistics, and All That," W. A. Benjamin, Inc., New York, 1964.

pens if one attempts to construct a quantum theory with this H if the proton and neutron fields are specified to *commute* with each other?

10. Verify the nucleon number, isotopic spin, and charge symmetries of the model lagrangian (15.30) and construct the constants from Noether's theorem.

11. Write a general lagrangian $\mathcal{L}$ which is bilinear in the eight baryon fields, linear in the boson fields with no derivative couplings, and invariant under the baryon number, isotopic spin, and charge or strangeness transformations. Introduce electromagnetic interactions and compute the conserved currents.

12. Verify (15.75) for the orbital parity by expanding the field in spherical waves and using properties of spherical harmonics.

13. Verify (15.93) for the form of the parity operator for a Dirac field.

14. Construct the parity operator for the electromagnetic field and verify the invariance of the lagrangian and commutators of quantum electrodynamics under a parity transformation.

15. Verify invariance of the Dirac equation and of the commutation relations (13.53) and (13.54) under charge conjugation. Construct $\mathcal{C}$ in (15.115).

16. Verify that vector and tensor currents formed from bilinear forms in the Dirac fields are odd and that scalar, axial, and pseudoscalar are even under $\mathcal{C}$. Discuss possible CP-invariant β-decay interactions.

17. Construct the $\mathcal{C}$ transformation for the model lagrangian (15.30).

18. The combination known as $G = e^{i\pi I_y}\mathcal{C}$ is useful for classifying particles. Determine how the baryons and mesons transform under G. In particular, show that a state of n pions with $Q = 0$ has G parity $(-)^n$.

19. Construct $\mathcal{T}$ for the π-meson field satisfying (15.132).

20. What becomes of a state of one right circularly polarized photon under time reversal?

21. Verify the construction of $\mathcal{T}$ for a Dirac field in (15.141) and (15.143).

22. Construct a phenomenological weak interaction lagrangian for leptons, nucleons, and mesons consistent with the ideas discussed in Chap. 10. Discuss the conservation laws and the symmetries of $\mathcal{L}$ they imply.

23. What is the explicit dependence of the phases $\phi(p,s)$ in the charge conjugation transformation on p. 116 on the momentum p and spin s? We omitted these phases in discussing positronium states on p. 117. Justify their neglect by constructing $\mathcal{F}$ for given L and S, using the angular-momentum operator.

16

Vacuum Expectation Values and the S Matrix

16.1 Introduction

It is the goal of quantum field theory as a physical theory to describe the dynamics of interacting particles which are observed in nature. We have already seen how the properties of free particles emerge from the application of the formal quantization procedure to classical fields and how symmetries are protected and constants of the motion identified in constructing interaction lagrangians. The task remaining is that of constructing and studying, as well as evaluating, general matrix elements which describe the dynamical behavior of interacting particles.

We are interested both in the amplitudes for single particles to propagate in space-time—that is, the one-particle Green's functions—and in the transition amplitudes for interacting particles between different initial and final states, that is, the S matrix. One of the main results will be to reconstruct from the quantum field theory formalism the Feynman rules of calculation developed in the companion volume.

16.2 Properties of Physical States

The problem of constructing exact solutions of the coupled nonlinear equations for interacting fields, such as (15.4), has so far proved too formidable for solution. Before turning to various approximation procedures, let us study how far one can go in determining the properties of exact states Φ and of propagators on the basis of invariance arguments alone. In particular, displacement and Lorentz invariance play important roles and are symmetries common to all theories of interest here.

To begin with we choose the states Φ to be eigenstates of energy and momentum. This is possible because the existence of a conserved energy-momentum four-vector P^μ is guaranteed by the assumed displacement invariance as discussed in Chap. 11. We work in the Heisenberg picture, specifying Φ by the eigenvalues of P^μ and of all other mutually commuting constants of the motion.

In addition, we put certain conditions on the eigenspectrum of P^μ on physical grounds. In the absence of an exact solution for the assumed P^μ these requirements are, of course, unproved. We assume that:

1. The eigenvalues of energy and momentum all lie within the forward light cone

$$P^2 = P_\mu P^\mu \geq 0 \qquad P^0 \geq 0 \tag{16.1}$$

2. There exists a nondegenerate Lorentz-invariant ground state of lowest energy. This is the vacuum state

$$\Phi_0 \equiv |0\rangle$$

and by convention the zero of energy is so chosen that

$$P^0|0\rangle = 0 \tag{16.2}$$

From (16.1) it follows that

$$\mathbf{P}|0\rangle = 0$$

also. Lorentz invariance of the vacuum (16.2) assures that $|0\rangle$ appears as the vacuum state to observers in all Lorentz systems.

3. There exist stable single-particle states

$$\Phi_{1(i)} \equiv |P^{(i)}\rangle$$

with $P^{(i)}_\mu P^{(i)\mu} = m_i^2$ for each stable particle of mass m_i.

Ignoring temporarily "infrared" complications associated with zero-mass states of photons and neutrinos, we add a fourth requirement:

4. The vacuum and single-particle states form a discrete spectrum in P^μ. This is illustrated by Fig. 16.1, which shows the energy-momentum spectrum of, say, π mesons. The π meson is, of course, not a stable particle in nature; its observed half-life for decay, primarily via

$$\pi^+ \rightarrow \mu^+ + \nu'$$

is $\sim 2 \times 10^{-8}$ sec, as already discussed in Chap. 10. This is a very long half-life relative to the natural frequency $\hbar/m_\pi c^2 \sim 5 \times 10^{-24}$ sec, and to a first approximation with neglect of the weak couplings we treat the π meson as stable, as in our earlier propagator discussions. We associate with it a field φ in constructing the lagrangian and energy-momentum four-vector with the eigenspectrum of Fig. 16.1 for π-meson states.

In the same way we shall associate a field $\varphi(x)$ or $\psi(x)$ with each discrete (or nearly discrete) state appearing in the spectrum of P_μ. The spirit of this approach is quite similar to that of perturbation theory. Having introduced fields for the stable particles into a lagrangian, we assume that their mutual interactions do not violently change the spectrum of states from the form, for instance, of Fig. 16.1. This is evidently a very strong assumption and a critical limitation to this approach, since it automatically excludes bound states. Haag, Nishijima, and Zimmermann,[1] with an axiomatic approach, have

[1] K. Nishijima, *Phys. Rev.*, **111**, 995 (1958).

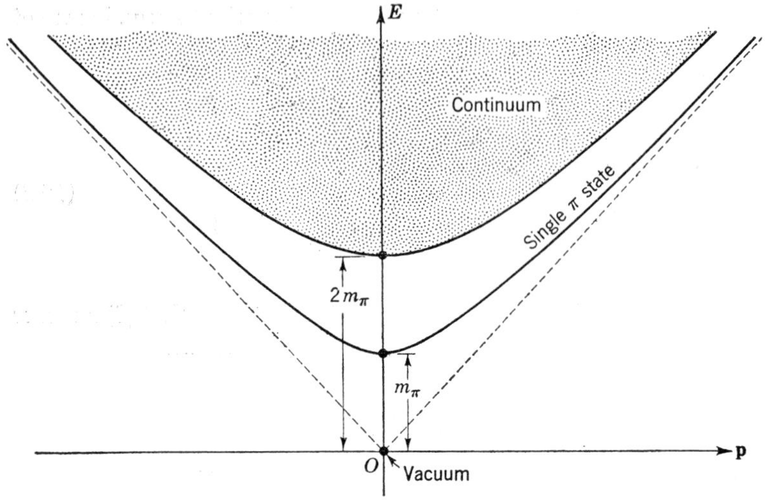

Fig. 16.1 Energy-momentum spectrum of π mesons.

recently made progress in showing how to construct a local field defined at each space-time point x and associated with a stable bound state of two given elementary fields—for example, how to represent a deuteron formed from a pair of nucleon fields.[1] However, the precise relation between such a field and the interaction terms in the lagrangian is not clear.[2]

16.3 Construction of In-fields and In-states; the Asymptotic Condition

Since we are primarily concerned with scattering problems, we want to construct first the states which give a simple description of the physical system at the initial time $t \to -\infty$. At this time the particles in a scattering event have not yet interacted with each other and propagate under the influence of their self-interactions only. Therefore, to start

[1] The prescription is simple. If the one-particle "composite" state α is spin 0 and coupled to spinless fields $A(x)$ and $B(x)$, the field operator φ_α may be chosen to be

$$\varphi_\alpha(x) = \lim_{\substack{\epsilon^2<0 \\ \epsilon\to 0}} \frac{A(x+\epsilon)B(x-\epsilon)}{\sqrt{2E_\alpha(2\pi)^3}\langle 0|A(x+\epsilon)B(x-\epsilon)|\alpha\rangle}$$

[2] See in this connection the comments of S. Weinberg and A. Salam, *Proc. 1962 Intern. Conf. High-energy Phys., CERN, Geneva,* 683–687.

building a theory of interactions, we seek operators creating independent particle states with each particle propagating with its physical mass. In the spirit of our discussion of the preceding paragraph we assume that these states for all particle numbers and momenta form a complete set.

For the case of noninteracting particles described by free-field lagrangians the one-particle states were generated from the vacuum by one of the creation operators appearing in the free-field expansion; repeated applications of n creation operators led to the n-particle states. The particle interpretation was derived from the spectrum of the energy-momentum operator P^μ and the algebra of the creation and annihilation operators as derived from the assumed commutation or anticommutation relations. For the interacting fields we seek now the corresponding operators creating states of single physical particles. For simplicity, we discuss these operators first for the hermitian meson field $\varphi(x)$ satisfying the wave equation

not m, but m_0

$$(\Box + m_0^2)\, \varphi(x) = j(x) \tag{16.3}$$

and the equal-time commutation relations $[\varphi(x), \varphi(y)] = i \, z \, \Delta(x-y)$
to be consistent
with (16.4) & (16.5b)

$$[\varphi(\mathbf{x},t),\varphi(\mathbf{y},t)] = [\pi(\mathbf{x},t),\pi(\mathbf{y},t)] = 0 \qquad \text{with } (16.4) \text{ & } (16.5b)$$
$$[\pi(\mathbf{x},t),\varphi(\mathbf{y},t)] = -i\delta^3(\mathbf{x} - \mathbf{y}) \tag{16.4}$$

The current $j(x)$ is a scalar operator constructed from the fields interacting with $\varphi(x)$ locally at x. If for simplicity we exclude derivative coupling terms from $j(x)$, we have

$$\pi(x) = \dot{\varphi}(x)$$

Beyond these assumptions, $j(x)$ is of arbitrary form. It might express the coupling with nucleon sources or, perhaps, a self-coupling of the form

$$j(x) = \lambda\varphi^3(x)$$

in which case the field equation is derivable from the simple lagrangian density

$$\mathcal{L} = \frac{1}{2}\left(\frac{\partial\varphi}{\partial x_\mu}\frac{\partial\varphi}{\partial x^\mu} - m_0^2\varphi^2 + \frac{1}{2}\lambda\varphi^4\right)$$

Let us denote by $\varphi_{\text{in}}(x)$ the operator which we are seeking in order to construct the states of single physical particles. $\varphi_{\text{in}}(x)$ will be formed by a suitable functional of the exact $\varphi(x)$ and any other fields present in $j(x)$, and its existence will be shown by explicit construction. To ensure that $\varphi_{\text{in}}(x)$ has the desired interpretation of an operator which creates a free physical meson from the vacuum, we assign to it the following properties which were explored for free fields in the preceding chapters:

1. $\varphi_{in}(x)$ transforms under coordinate displacements and Lorentz transformations in the same way as the corresponding $\varphi(x)$. This choice guarantees the covariance of the one-particle states formed by $\varphi_{in}(x)$. For displacements, in particular, we have

$$[P^\mu, \varphi_{in}(x)] = -i \frac{\partial \varphi_{in}(x)}{\partial x_\mu} \qquad (16.5)$$

2. The space-time development of $\varphi_{in}(x)$ is described by a free-particle Klein-Gordon equation with the physical mass m:

$$(\Box + m^2)\varphi_{in}(x) = 0 \qquad (16.6)$$

And assume
it amplitudes
to do so
= 1. See
below (16.16)
below (16.37).

It follows from these two defining equations that $\varphi_{in}(x)$ creates the physical one-particle state from the vacuum. To confirm this, consider an arbitrary eigenstate

$$P^\mu|n\rangle = p_n^\mu|n\rangle \qquad (16.7)$$

and form the following matrix element with the vacuum state $|0\rangle$:

$$-i\frac{\partial}{\partial x_\mu}\langle n|\varphi_{in}(x)|0\rangle = \langle n|[P^\mu, \varphi_{in}(x)]|0\rangle = p_n^\mu\langle n|\varphi_{in}(x)|0\rangle$$

Iterating this operation gives, by (16.6),

$$(\Box + m^2)\langle n|\varphi_{in}(x)|0\rangle = (m^2 - p_n{}^2)\langle n|\varphi_{in}(x)|0\rangle = 0 \qquad (16.8)$$

Therefore, the only states produced from the vacuum by $\varphi_{in}(x)$ are those with $p_n{}^2 = m^2$, that is, the one-particle states of mass m. The Fourier expansion of $\varphi_{in}(x)$ is the same as that given for free fields in Eq. (12.7), which are also solutions of a Klein-Gordon equation (16.6):

$$\varphi_{in}(x) = \int d^3k \, [a_{in}(k)f_k(x) + a_{in}^\dagger(k)f_k^*(x)]$$

with
$$f_k(x) \equiv \frac{1}{\sqrt{(2\pi)^3 2\omega_k}} e^{-ik\cdot x} \qquad (16.9)$$

for
commutators,
see prob. #1
and
$$\omega_k \equiv \sqrt{k^2 + m^2} \equiv k^0$$

as in (12.7), and

$$a_{in}(k) = i\int d^3x \, f_k^*(x)\overleftrightarrow{\partial}_0\varphi_{in}(x)$$

as in (12.9). The operator coefficients $a_{in}(k)$ satisfy the commutation relations obtained from (16.5):

$$[P^\mu, a_{in}(k)] = -k^\mu a_{in}(k) \qquad [P^\mu, a_{in}^\dagger(k)] = +k^\mu a_{in}^\dagger(k) \qquad (16.10)$$

By repeated application of $a_{\text{in}}^{\dagger}(k)$ to the vacuum, we can build up general n-particle eigenstates. By (16.10) and the assumption of a unique stable vacuum of zero energy,

$$P^{\mu}a_{\text{in}}^{\dagger}(k_1) \cdots a_{\text{in}}^{\dagger}(k_N)|0\rangle \equiv P^{\mu}|k_1 \cdots k_N \text{ in}\rangle$$
$$= \sum_{i=1}^{N} k_i^{\mu} a_{\text{in}}^{\dagger}(k_1) \cdots a_{\text{in}}^{\dagger}(k_N)|0\rangle \qquad (16.11)$$
$$a_{\text{in}}(k)|0\rangle = 0$$

and

$$\langle p_1 \cdots p_M \text{ in}|k_1 \cdots k_N \text{ in}\rangle = 0$$ orthogonality

unless $M = N$ and the set $(p_1, \ldots, p_M)$ coincides with $(k_1, \ldots, k_N)$. As remarked before, the set of states formed with all possible choices of number N and momenta k_i^{μ} is assumed to be complete.

To express the $\varphi_{\text{in}}(x)$ in terms of the fields $\varphi(x)$, we rewrite (16.3) in terms of the physical mass m by adding a mass counterterm

$$\delta m^2 \varphi(x) \equiv (m^2 - m_0^2)\varphi(x) \qquad (16.12)$$

to both sides of the equation:

$$(\square + m^2)\varphi(x) = j(x) + \delta m^2 \varphi(x) \equiv \tilde{j}(x) \qquad (16.13)$$

The current $\tilde{j}(x)$ is now treated as the source giving rise to the scattered waves. Removing these scattered waves from $\varphi(x)$ leaves just the free waves propagating with mass m as given by $\varphi_{\text{in}}(x)$ and suggests that we write[1]

$$\sqrt{Z}\,\varphi_{\text{in}}(x) = \varphi(x) - \int d^4y\,\Delta_{\text{ret}}(x - y; m)\tilde{j}(y) \qquad (16.14)$$

where

$$\Delta_{\text{ret}}(x - y; m) = 0 \qquad \text{for } x_0 < y_0 \qquad (16.15)$$

is the retarded Green's function (see Appendix C) which satisfies

$$(\square_x + m^2)\Delta_{\text{ret}}(x - y; m) = \delta^4(x - y) \qquad (16.16)$$

As defined in (16.14), $\varphi_{\text{in}}(x)$ satisfies the two conditions (16.5) and (16.6) required for the in-fields. Since $\tilde{j}(y)$ in (16.13) is a scalar operator, we have, for example,

$$\begin{aligned}\sqrt{Z}\,\varphi_{\text{in}}(x + a) &= \varphi(x + a) - \int d^4y\,\Delta_{\text{ret}}(x + a - y)\,\tilde{j}(y)\\ &= e^{iP\cdot a}\varphi(x)e^{-iP\cdot a} - \int d^4y'\,\Delta_{\text{ret}}(x - y')\,\tilde{j}(y' + a)\\ &= e^{iP\cdot a}[\varphi(x) - \int d^4y'\,\Delta_{\text{ret}}(x - y')\,\tilde{j}(y')]e^{-iP\cdot a}\\ &= e^{iP\cdot a}\sqrt{Z}\,\varphi_{\text{in}}(x)e^{-iP\cdot a}\end{aligned}$$

as required.

The constant $\sqrt{Z}$ in (16.14) is included ~~to permit~~ so that we can assume normalization of $\varphi_{\text{in}}(x)$ to unit amplitude for its matrix element to create one-particle

[1] Z is often denoted by Z_3; we reserve this notation for the photon.

and not assume the same for $\varphi(x)$

states from the vacuum and will be evaluated shortly. The form of (16.14) is reasonable and tempts us to suggest that as $x_0 \to -\infty$ the interaction term vanishes according to (16.15) so that

$$\varphi(x) \overset{?}{\to} \sqrt{Z}\,\varphi_{\text{in}}(x) \qquad \text{as } x_0 \to -\infty \tag{16.17}$$

and in accord with our intuitive notion of causality the field operator reduces to the incident free wave.

Asymptotic conditions similar to (16.17) are commonly applied to wave functions in one-particle quantum mechanics and were used extensively in the propagator discussions of Chaps. 6 and 9. We used there the adiabatic hypothesis to isolate initial and final free-particle states from the interaction. Alternatively, we may achieve the same isolation by constructing wave packets to represent localized solutions which do not overlap the interaction region before or after the scattering.

Equation (16.17), on the other hand, is an *operator* statement and leads to contradictions arising from the impossibility of isolating $\varphi(x)$ from the operator $\jmath(y)$, which includes all self-interactions at $x_0 \to -\infty$. It is the matrix elements of the field operators that are the quantum field theory analogues of the "wave functions," and it is to these and not to the operators directly that we must apply the asymptotic condition. The correct asymptotic condition as stated by Lehmann, Symanzik, and Zimmermann takes the following form.[1] We let $|\alpha\rangle$ and $|\beta\rangle$ be any two normalizable states and define the field operator $\varphi^f(t)$ by smearing $\varphi(x)$ over a space-like region according to

$$\varphi^f(t) = i\int d^3x\, f^*(\mathbf{x},t)\overleftrightarrow{\partial_0}\varphi(\mathbf{x},t) \tag{16.18}$$

with $f(\mathbf{x},t)$ is an arbitrary *normalizable* solution of the Klein-Gordon equation:

$$(\Box + m^2)f(x) = 0 \tag{16.19}$$

The asymptotic condition then states[2]

$$\lim_{t\to-\infty} \langle\alpha|\varphi^f(t)|\beta\rangle = \sqrt{Z}\,\langle\alpha|\varphi^f_{\text{in}}|\beta\rangle \tag{16.20}$$

[1] H. Lehmann, K. Symanzik, and W. Zimmermann, *Nuovo Cimento*, 1, 1425 (1955).

[2] Equation (16.20) differs from (16.17) by the very important ordering of operations. Equation (16.20) requires that we construct normalized packet states and form the matrix element before going to the limit $t \to -\infty$. See W. Zimmermann, *Nuovo Cimento*, **10**, 597 (1958), and also O. W. Greenberg, doctoral dissertation, Princeton University, Princeton, N.J., 1956.

where φ_{in}^f, defined by

$$\varphi_{\text{in}}^f = i\int d^3x\, f^*(\mathbf{x},t)\overleftrightarrow{\partial}_0\varphi_{\text{in}}(\mathbf{x},t) \tag{16.21}$$

is independent of time t by Green's theorem, (16.6) and (16.19).

Equation (16.20), or the "weak asymptotic condition," is what will always be meant as the precise statement when we write (16.17). It expresses the initial conditions on the localized wave packets representing the incoming particles. The states (16.11) formed with the in-field creation operators

$$\begin{aligned} |k_1 \cdots k_n \text{ in}\rangle &= a_{\text{in}}^\dagger(k_1)|k_2 \cdots k_n \text{ in}\rangle \\ &= a_{\text{in}}^\dagger(k_1)a_{\text{in}}^\dagger(k_2) \cdots a_{\text{in}}^\dagger(k_n)|0\rangle \end{aligned} \tag{16.22}$$

are always to be understood as limits of normalizable states formed by the φ_{in}^f of (16.21) for the packets $f^*(\mathbf{x},t)$ approaching the monochromatic plane waves $f_k^*(x)$ of (16.9). A complete set of in-states and the asymptotic condition are all that we need to specify at $t \to -\infty$ the initial dynamics in a scattering experiment.

Provided $\sqrt{Z}$ is nonzero, Eq. (16.14) defines the field operator we sought for constructing the initial scattering states. Since we are here working with a mathematical idealization of a physical theory in terms of local lagrangians and field operators, we cannot guarantee that Z is not zero owing to wild behavior of the theory at infinite energies. This indeed may be the case if, as discussed earlier, the present formalism is the local limit of some unknown, more elaborate theory. With this reservation in mind we proceed formally to derive a general expression for Z, the square of the amplitude for $\varphi(x)$ to produce a one-particle state from the vacuum, from our initial assumptions on the spectrum of P^μ, (16.1) and (16.2), and from canonical commutation relations (16.4).

16.4 Spectral Representation for the Vacuum Expectation Value of the Commutator and the Propagator for a Scalar Field

In order to obtain an expression for Z, we construct the general form of the vacuum expectation value of the commutator of two fields:

$$i\Delta'(x,x') \equiv \langle 0|[\varphi(x),\varphi(x')]|0\rangle \tag{16.23}$$

The commutator itself cannot be given because, in contrast with the free-field theory, Chap. 12, we cannot solve the field equations here.

However, the general form of (16.23) can be derived on the basis of invariance arguments and the assumptions on the spectrum of P^μ.

Inserting a complete set of eigenstates (16.7) between the two field operators in (16.23) and using displacement invariance of the theory (16.5) to write

$$\langle n|\varphi(x)|m\rangle = \langle n|e^{iP\cdot x}\varphi(0)e^{-iP\cdot x}|m\rangle$$
$$= e^{i(p_n - p_m)\cdot x}\langle n|\varphi(0)|m\rangle \qquad (16.24)$$

we have $\quad \sum_{n=?}^{\infty} \int \frac{d^3k_1 \cdots d^3k_n}{n!} \quad$ see H. W.

$$\Delta'(x,x') = -i\sum_n \langle 0|\varphi(0)|n\rangle\langle n|\varphi(0)|0\rangle(e^{-ip_n\cdot(x-x')} - e^{ip_n\cdot(x-x')})$$
$$= \Delta'(x - x') \qquad (16.25)$$

It is convenient to group together all states corresponding to the same eigenvalue p_n. Therefore, introducing

$$1 = \int d^4q \; \delta^4(p_n - q)$$

we rewrite (16.25) as

$$\Delta'(x - x') = \frac{-i}{(2\pi)^3}\int d^4q \left[(2\pi)^3 \sum_n \delta^4(p_n - q)|\langle 0|\varphi(0)|n\rangle|^2 \right]$$
$$\times (e^{-iq\cdot(x-x')} - e^{iq\cdot(x-x')}) \qquad (16.26)$$

The quantity within the brackets is the spectral amplitude[1] $\rho(q)$,

$$\rho(q) \equiv (2\pi)^3 \sum_n \delta^4(p_n - q)|\langle 0|\varphi(0)|n\rangle|^2 \qquad (16.27)$$

$\rho(q) = \delta(q^2 - m^2)\theta(q_0)$
for free fields
(chap. 12)

and $\theta(q_0)$
or $\varepsilon(q_0)$

and measures the contribution to Δ' from all states with energy-momentum eigenvalue q. We may next appeal to Lorentz invariance of the sum in (16.27) to conclude that $\rho(q)$ is a scalar function of q^2 only. Explicitly we have, using Lorentz invariance of the vacuum

$$U(a)|0\rangle = |0\rangle$$

and of the scalar field $\varphi(0)$

$$U(a)\varphi(0)U^{-1}(a) = \varphi(0)$$
$$\rho(q) = (2\pi)^3 \sum_n \delta^4(p_n - q)|\langle 0|\varphi(0)|U(a)n\rangle|^2 \qquad (16.28)$$

where a is the matrix of coefficients for a proper Lorentz transformation. Lorentz invariance of the δ function, verified by looking at its Fourier expansion, allows us to write

$$\delta^4(p_n - q) = \delta^4[(p_n - q)a^{-1}] \qquad (16.29)$$

[1] In terms of unrenormalized fields. who?

$$\Delta'(x) = \frac{-i}{(2\pi)^3}\int d^4q \; \rho(q)\left[e^{-iqx} - e^{iqx} \right]$$

Finally, carrying out the sum over the complete set labeled by $|m\rangle = |U(a)n\rangle$, with eigenvalues

$$p_m{}^\mu = \langle m|P^\mu|m\rangle = \langle n|U^{-1}(a)P^\mu U(a)|n\rangle = (p_n a^{-1})^\mu \quad (16.30)$$

we arrive at the result

$$\rho(q) = (2\pi)^3 \sum_m \delta^4(p_m - qa^{-1})|\langle 0|\varphi(0)|m\rangle|^2 = \rho(qa^{-1})$$

Since $\rho(q)$ vanishes outside the forward light cone owing to the assumed conditions on the spectrum of P_μ, we may write

NOT THE SAME

$$\rho(q) = \bar{\rho}(q^2)\theta(q_0) \quad (16.31)$$

where $\rho(q^2)$ vanishes for $q^2 < 0$ and is real and positive semidefinite for $q^2 \geq 0$. Equation (16.26) can then be put into the form of a weighted integral over the mass parameter of the free-field commutator function

$$\Delta'(x - x') = \frac{-i}{(2\pi)^3} \int d^4q \, \rho(q^2)\theta(q_0)(e^{-iq\cdot(x-x')} - e^{iq\cdot(x-x')})$$

problem when $q^2 = 0$　$\int d\sigma^2 \, \delta(\sigma^2) = ?$　*= 1 here*

$$= \frac{-i}{(2\pi)^3} \int_0^\infty d\sigma^2 \, \rho(\sigma^2) \int d^4q \, \delta(q^2 - \sigma^2)\epsilon(q_0)e^{-iq\cdot(x-x')}$$

I think

$$= \int_0^\infty d\sigma^2 \, \rho(\sigma^2)\Delta(x - x', \sigma) \quad (16.32)$$

plays role of mass

$\theta(q_0)$, $\epsilon(q_0)$, and the invariant Δ function with mass parameter σ are as defined in Appendix C.

We refer to (16.32) as a spectral representation for the vacuum expectation value of the commutator. It was derived for quantum electrodynamics by Källén in 1952 and for the present case by Lehmann in 1954.[1] The above derivation goes through with essentially no change for the various Green's functions of the theory, and in particular

$$\Delta_F'(x - x') = -i\langle 0|T(\varphi(x)\varphi(x'))|0\rangle = \int_0^\infty d\sigma^2 \, \rho(\sigma^2)\Delta_F(x - x', \sigma) \quad (16.33)$$

or in momentum space

$$\Delta_F'(p) = \int_0^\infty d\sigma^2 \, \rho(\sigma^2) \frac{1}{p^2 - \sigma^2 + i\epsilon}$$

with the same weight function ρ as in (16.32).

---

[1] G. Källén, *Helv. Phys. Acta*, **25**, 417 (1952); H. Lehmann, *Nuovo Cimento*, **11**, 342 (1954). See also A. Wightman (unpublished, 1953) as quoted in S. Schweber, "An Introduction to Relativistic Quantum Field Theory," Harper & Row, Publishers, Incorporated, New York, 1961. We have freely interchanged orders of integration and summation. For the justification for this, see the references.

Although we cannot explicitly evaluate the infinite sum for the spectral amplitude in (16.27), we can, by separating out the contribution of the one-particle states, prove the condition

$$0 \leq Z < 1 \tag{16.34}$$

and establish finally the contradiction with (16.17) as an *operator statement.*

By the assumptions on the spectrum of P^μ illustrated in Fig. 16.1, $\rho(\sigma^2)$ for $\sigma^2 = m^2$ can be computed by considering only the one-particle matrix element $\langle 0|\varphi(x)|p\rangle$ in (16.27). By (16.14) this is given by

$$\langle 0|\varphi(x)|p\rangle = \sqrt{Z}\,\langle 0|\varphi_{\text{in}}(x)|p\rangle + \int d^4y\,\Delta_{\text{ret}}(x-y;m)\langle 0|\jmath(y)|p\rangle \tag{16.35}$$

The second term in (16.35) vanishes by (16.13)

$$\begin{aligned}\langle 0|\jmath(y)|p\rangle &= \langle 0|(\Box + m^2)\varphi(y)|p\rangle \\ &= (\Box + m^2)e^{-ip\cdot y}\langle 0|\varphi(0)|p\rangle \\ &= (m^2 - p^2)\langle 0|\varphi(y)|p\rangle = 0\end{aligned} \tag{16.36}$$

Therefore

$$\langle 0|\varphi(x)|p\rangle = \sqrt{Z}\,\langle 0|\varphi_{\text{in}}(x)|p\rangle \tag{16.37}$$

By definition, $\varphi_{\text{in}}(x)$ is normalized so that its matrix elements to create one-particle states from the vacuum have unit amplitude, coinciding with the free-particle result. By (16.9) and (16.11),

$$\begin{aligned}\langle 0|\varphi_{\text{in}}(x)|p\rangle &= \int d^3k\,\frac{e^{-ik\cdot x}}{\sqrt{(2\pi)^3 2\omega_k}}\langle 0|a_{\text{in}}(k)|p\rangle \\ &= \frac{e^{-ip\cdot x}}{\sqrt{(2\pi)^3 2\omega_p}}\end{aligned} \tag{16.38}$$

and $\sqrt{Z}$ in (16.37) represents the amplitude for creating the one-particle state from the vacuum via $\varphi(x)$.

The contribution of the one-particle states to the sum in (16.27) is, by (16.37) and (16.38),

$$(2\pi)^3 \int d^3p\,\delta^4(p-q)\,\frac{Z}{(2\pi)^3 2\omega_p} = Z\delta(q^2 - m^2)\theta(q_0) \tag{16.39}$$

Separating this in the sum over states for the spectral amplitude, we may write in place of (16.32),

$$\Delta'(x-x') = Z\Delta(x-x';m) + \int_{m_1^2}^{\infty} d\sigma^2\,\rho(\sigma^2)\Delta(x-x';\sigma) \tag{16.40}$$

where the threshold m_1^2 now coincides with the square of the mass of the lightest continuum state beyond the discrete one-particle term which contributes to $\rho(\sigma^2)$; in Fig. 16.1, for example, $m = m_\pi$ and $m_1^2 = 4m_\pi^2$.

Taking a time derivative of (16.40) and then setting $t' = t$ produces the constraint (16.34) on Z that we want. Using the canonical commutation relations (16.4) along with the definitions (16.23) of Δ' and (12.42) for Δ, we find

$$\begin{aligned}\lim_{t'\to t}\left(i\frac{\partial}{\partial t}\Delta'(x-x')\right) &= \langle 0|[\dot\varphi(\mathbf{x},t),\varphi(\mathbf{x}',t)]|0\rangle = -i\delta^3(\mathbf{x}-\mathbf{x}') \\ &= \lim_{t'\to t}\left(i\frac{\partial}{\partial t}\Delta(x-x';\sigma)\right)\end{aligned} \tag{16.41}$$

so that

$$1 = Z + \int_{m_1^2}^{\infty}\rho(\sigma^2)\,d\sigma^2 \tag{16.42}$$

Together with the condition (16.31) that $\rho(\sigma^2)$ is never negative, this assures that

$$0 \le Z < 1 \tag{16.43}$$

as claimed in (16.34) if the integral in (16.42) exists and the calculation makes any sense. The limit $Z = 1$ is excluded if there is any coupling to the continuum states. This is in accord with intuition. We expect that the amplitude for $\varphi(x)$ to produce the one-particle states out of the vacuum is reduced to less than its value of unity for the free-field case, since it can produce continuum states as well. However, Z cannot be zero if (16.14) is to define $\varphi_{\text{in}}(x)$ and permit the construction of states as in (16.20) and (16.22). It is uncomfortable, then, to find that individual terms in the perturbation expansion violate this requirement.

In the presence of interactions, so that $Z < 1$, we find the contradiction with the strong asymptotic condition (16.17). Assuming its validity and repeating the steps in (16.41), we find

$$\begin{aligned}\lim_{t\to-\infty}\langle 0|[\dot\varphi(\mathbf{x},t),\varphi(\mathbf{x}',t)]|0\rangle &= -i\delta^3(\mathbf{x}-\mathbf{x}') \\ &\overset{?}{=} \lim_{t\to-\infty} Z\langle 0|[\dot\varphi_{\text{in}}(\mathbf{x},t),\varphi_{\text{in}}(\mathbf{x}',t)]|0\rangle\end{aligned} \tag{16.44}$$

whereas (16.8) and (16.38) imply

$$\langle 0|[\varphi_{\text{in}}(x),\varphi_{\text{in}}(x')]|0\rangle = i\Delta(x-x') \tag{16.45}$$

By taking a time derivative of (16.45) and comparing with (16.44), we would conclude that Z is unity. In this case the loose interchange of limiting procedures has evidently been unjustified.

16.5 The Out-fields and Out-states

Just as we have reduced the dynamics at $t \to -\infty$ to that of free particles with the aid of the in-fields, so we can do the same thing at $t \to +\infty$ by suitably defining out-fields $\varphi_{\text{out}}(x)$. We want such a simple description of the physical system at $t \to +\infty$, since this is the situation in the final state of a scattering problem.

The $\varphi_{\text{out}}(x)$ may be constructed by proceeding along lines closely parallel to the development for $\varphi_{\text{in}}(x)$. They are defined to satisfy, in analogy with (16.5) and (16.6),

$$[P^\mu, \varphi_{\text{out}}(x)] = -i\frac{\partial \varphi_{\text{out}}(x)}{\partial x_\mu} \tag{16.46}$$

$$(\Box + m^2)\varphi_{\text{out}}(x) = 0 \tag{16.47}$$

and therefore $\varphi_{\text{out}}(x)$ produces from the vacuum one-particle states only, as in (16.8). Also, from the expansion analogous to (16.9),

$$\varphi_{\text{out}}(x) = \int d^3k\,[a_{\text{out}}(k)f_k(x) + a^\dagger_{\text{out}}f_k^*(x)] \tag{16.48}$$

we find

$$[P^\mu, a_{\text{out}}(k)] = -k^\mu a_{\text{out}}(k) \qquad [P^\mu, a^\dagger_{\text{out}}(k)] = k^\mu a^\dagger_{\text{out}}(k) \tag{16.49}$$

in analogy with (16.10).

Now, in contrast with (16.20), we want an asymptotic condition in the form

$$\lim_{t\to+\infty} \langle\alpha|\varphi^f(t)|\beta\rangle = \sqrt{Z}\,\langle\alpha|\varphi^f_{\text{out}}|\beta\rangle \tag{16.50}$$

or simply

$$\varphi(x) \to \sqrt{Z}\,\varphi_{\text{out}}(x) \qquad \text{as } t \to +\infty \tag{16.51}$$

where (16.51) is understood as meaning weak operator convergence (16.50), as discussed for $\varphi_{\text{in}}(x)$. This suggests the definition

$$\sqrt{Z}\,\varphi_{\text{out}}(x) = \varphi(x) - \int d^4y\,\Delta_{\text{adv}}(x-y;m)\,j(y) \tag{16.52}$$

in place of (16.14). $\Delta_{\text{adv}}(x-y;m)$ is the advanced Green's function.

$$\begin{aligned}(\Box_x + m^2)\Delta_{\text{adv}}(x-y;m) &= \delta^4(x-y)\\ \Delta_{\text{adv}}(x-y;m) &= 0 \qquad x_0 - y_0 > 0\end{aligned} \tag{16.53}$$

The normalizing constant $\sqrt{Z}$ is again introduced so that $\varphi_{\text{out}}(x)$ has unit amplitude to produce one-particle states from the vacuum and therefore, by (16.36) to (16.38), is identical with the $\sqrt{Z}$ in (16.14):

$$\begin{aligned}\langle 0|\varphi(x)|p\rangle &= \sqrt{Z}\langle 0|\varphi_{\text{out}}(x)|p\rangle \\ &= \sqrt{Z}\langle 0|\varphi_{\text{in}}(x)|p\rangle \\ &= \sqrt{Z}\frac{1}{\sqrt{(2\pi)^3 2\omega_p}}\, e^{-ip\cdot x}\end{aligned} \tag{16.54}$$

From (16.54) it follows that the vacuum expectation values of the commutators of $\varphi_{\text{in}}(x)$ and $\varphi_{\text{out}}(x)$ are simply those for the free fields

$$\begin{aligned}\langle 0|[\varphi_{\text{in}}(x),\varphi_{\text{in}}(y)]|0\rangle &= i\Delta(x-y) \\ \langle 0|[\varphi_{\text{out}}(x),\varphi_{\text{out}}(y)]|0\rangle &= i\Delta(x-y)\end{aligned} \tag{16.55}$$

The proof that these commutators are in fact c numbers and can be written without taking vacuum expectation values on the left-hand side,

$$[\varphi_{\text{in}}(x),\varphi_{\text{in}}(y)] = [\varphi_{\text{out}}(x),\varphi_{\text{out}}(y)] = i\Delta(x-y) \tag{16.56}$$

has been given by Zimmermann and is left as an exercise.[1]

16.6 The Definition and General Properties of the S Matrix

We now have all the necessary properties of the $\varphi_{\text{in}}(x)$ and $\varphi_{\text{out}}(x)$ and all the formal machinery at hand for defining and studying the transition amplitudes, or S-matrix elements, of experimental interest. We start from an initial state of the system with n noninteracting (that is, spatially separated) physical particles with quantum numbers $p_1, \ldots, p_n$, denoted by

$$|p_1 \cdots p_n \text{ in}\rangle = |\alpha \text{ in}\rangle \tag{16.57}$$

The label p will denote, in addition to the momentum, all internal quantum numbers such as charge and strangeness characterizing the particle. This is possible since according to its definition (16.14) $\varphi_{\text{in}}(x)$ has the same transformation properties under an internal symmetry transformation as $\varphi(x)$. More precisely, if

$$[Q,\varphi_r(x)] = -\lambda_{rs}\varphi_s(x)$$

then

$$[Q,\tilde{j}_r(x)] = (\square + m^2)[Q,\varphi_r(x)] = -\lambda_{rs}\tilde{j}_s(x)$$

and therefore from (16.14)

$$[Q,\varphi_r^{\text{in}}(x)] = -\lambda_{rs}\varphi_s^{\text{in}}(x)$$

[1] Zimmermann, *op. cit.*

According to this we can interpret the constants of the interacting system in terms of the quantum numbers of free initial (or final) particle states. For example, the constants constructed in Chap. 15 for the strongly coupled particles can all be applied directly to the description of the free particles in the in- and out-states constructed here.

The element of the S matrix for a transition from such an initial state to a final state in which m particles emerge with $p_1' \cdots p_m'$, denoted by

$$|p_1' \cdots p_m' \text{ out}\rangle = |\beta \text{ out}\rangle \tag{16.58}$$

is given by the probability amplitude

$$S_{\beta\alpha} = \langle\beta \text{ out}\,|\alpha \text{ in}\rangle \tag{16.59}$$

Eq. (16.59) defines the $\beta\alpha$ element of the S matrix.

It is instructive to compare (16.59) with the definition of the S matrix in nonrelativistic propagator theory. In Eq. (6.16), the (f,i) element of the S matrix was expressed as

$$S_{fi} = \lim_{t\to\infty} \int d^3x\, \varphi_f^*(\mathbf{x},t)\Psi_i^+(\mathbf{x},t) = \lim_{t\to\infty} (\varphi_f(\mathbf{x},t),\Psi_i^+(\mathbf{x},t)) \tag{16.60}$$

$\Psi_i^+(\mathbf{x},t)$ is the exact scattering solution to the Schrödinger equation (6.14), with the in boundary condition of reducing to the incident plane wave at $t \to -\infty$. S_{fi} is just the amplitude of the projection of Ψ_i^+ at $t \to +\infty$ on a given free final state, $\varphi_f(\mathbf{x},t)$. To rewrite S_{fi} in the form analogous to (16.59), we introduce $\Psi_f^-(\mathbf{x},t)$, the exact scattering solution to the Schrödinger equation with the out boundary condition of approaching, at $t \to +\infty$, the free wave $\varphi_f(\mathbf{x},t)$ with the quantum numbers f of the final state. $\Psi_f^-(\mathbf{x},t)$ consists of the free wave plus a superposition of spherical waves converging, in the past, on the scatterer and disappearing as $t \to +\infty$. It is a solution of the Schrödinger equation but with the retarded Green's function in (6.14) replaced by an advanced one:

$$\Psi_f^-(x') = \varphi_f(x') + \int d^4x_1' G_0^{\text{adv}}(x';x_1')V(x_1')\Psi_f^-(x_1')$$

with

$$G_0^{\text{adv}}(\mathbf{x}',t';\mathbf{x}_1',t_1') = 0 \qquad \text{for } t' > t_1' \tag{16.61}$$

Then $\Psi_f^-(\mathbf{x}',t') \to \varphi_f(\mathbf{x}',t')$ as $t' \to +\infty$ and in (16.60) gives

$$\begin{aligned} S_{fi} &= \lim_{t\to\infty} (\Psi_f^-(\mathbf{x},t),\Psi_i^+(\mathbf{x},t)) \\ &= \lim_{t\to\infty} \int d^3x\, \Psi_f^{-*}(\mathbf{x},0)e^{iHt}e^{-iHt}\Psi_i^+(\mathbf{x},0) \\ &= (\Psi_f^-(\mathbf{x},0),\Psi_i^+(\mathbf{x},0)) \end{aligned} \tag{16.62}$$

$\Psi(\mathbf{x},0)$ is the wave function in the Heisenberg representation with the time dependence removed, and (16.62) is the analogue of (16.59) in terms of the wave functions with the in and out boundary conditions.

From (16.59) we have, by assumption of the completeness of the in- and out-states, all matrix elements of an operator S which transforms the in- to the out-states:

$$(a) \quad \langle \beta \text{ in}|S = \langle \beta \text{ out}| \quad (b) \quad \langle \beta \text{ out}|S^{-1} = \langle \beta \text{ in}| \tag{16.63}$$

From this follows

$$S_{\beta\alpha} = \langle \beta \text{ in}|S|\alpha \text{ in}\rangle$$

The S matrix is of central interest to us because its matrix elements express transition amplitudes and are closely related to physical measurements. A number of important properties of S follow from the initial assumptions on the spectrum of states and from the properties of the $\varphi_{\text{in}}(x)$ and of the $\varphi_{\text{out}}(x)$. They may now be enumerated.

1. Stability of the vacuum state requires $|S_{00}| = 1$, or

$$\langle 0 \text{ in}|S = \langle 0 \text{ out}| = e^{i\varphi_0}\langle 0 \text{ in}| \qquad \text{Assumption 10.7 about}$$

The vacuum state is by assumption unique and the phase φ_0 may be put to zero, so that

$$\langle 0 \text{ out}| = \langle 0 \text{ in}| = \langle 0| \tag{16.64}$$

and so $S_{00} = 1$.

2. Stability of the one-particle state also requires

$$\langle p \text{ in}|S|p \text{ in}\rangle = \langle p \text{ out}|p \text{ in}\rangle = \langle p \text{ in}|p \text{ in}\rangle = 1 \qquad \text{property of } \varphi_{in} \text{ and } \varphi_{out} \tag{16.65}$$

since $|p \text{ in}\rangle = |p \text{ out}\rangle = |p\rangle$ according to (16.54).

3. S transforms the in- to the out-fields according to

$$\varphi_{\text{in}}(x) = S\varphi_{\text{out}}(x)S^{-1} \qquad (\text{Just another way to say } 16.59) \tag{16.66}$$

To show this, we consider the matrix element

$$\langle \beta \text{ out}|\varphi_{\text{out}}(x)|\alpha \text{ in}\rangle = \langle \beta \text{ in}|S\varphi_{\text{out}}(x)|\alpha \text{ in}\rangle \qquad \text{property of } \varphi_{in} \text{ & } \varphi_{out} \text{ S'} + \text{def. of S}$$

Now $\langle \beta \text{ out}|\varphi_{\text{out}}(x)$ is an out-state, and we may write, by (16.63), $\langle \beta \text{ out}|\varphi_{\text{out}}(x) = \langle \beta \text{ in}|\varphi_{\text{in}}(x)S$, from which it follows that

$$\langle \beta \text{ in}|\varphi_{\text{in}}(x)S|\alpha \text{ in}\rangle = \langle \beta \text{ in}|S\varphi_{\text{out}}(x)|\alpha \text{ in}\rangle$$

By completeness of the in-states, we arrive at (16.66).

4. The S matrix defined here is unitary. From (16.63)

$$S^\dagger|\alpha \text{ in}\rangle = |\alpha \text{ out}\rangle \qquad \text{adjoint of (16.63)(a)} \tag{16.67}$$

Consequently, using (16.63) again

$$\langle\beta \text{ in}|SS^\dagger|\alpha \text{ in}\rangle = \langle\beta \text{ out}|\alpha \text{ out}\rangle = \delta_{\beta\alpha} \tag{16.68}$$

and

$$SS^\dagger = S^\dagger S = 1 \tag{16.69}$$

5. S is translation and Lorentz invariant,[1] that is,

$$U(a,b)SU^{-1}(a,b) = S \tag{16.70}$$

where $U(a,b)$ defined in (11.66), (11.69), and (11.72) is the unitary operator generating the transformation

$$x'_\mu = b_\mu + a_\mu{}^\nu x_\nu$$

To prove (16.70), we introduce (16.66) into the transformation equation (11.67) for field operators under $U(a,b)$

$$\begin{aligned}\varphi_{\text{in}}(ax + b) &= U(a,b)\varphi_{\text{in}}(x)U^{-1}(a,b) = US\varphi_{\text{out}}(x)S^{-1}U^{-1} \\ &= USU^{-1}\varphi_{\text{out}}(ax + b)US^{-1}U^{-1}\end{aligned} \tag{16.71}$$

But $\varphi_{\text{in}}(ax + b) = S\varphi_{\text{out}}(ax + b)S^{-1}$, and therefore

$$S = U(a,b)SU^{-1}(a,b)$$

is Lorentz invariant.

16.7 The Reduction Formula for Scalar Fields

Having these general properties of S and spurred on by interest in its matrix elements, since $|S_{\beta\alpha}|^2$ measures the probability of experimentally observed transitions between in-states α and out-states β, we face up to the highly nontrivial task of actually computing $S_{\beta\alpha}$.

Until 1954 the only systematic approach to a calculation of the S matrix was the perturbation theory expansion in powers of the interaction current $\jmath(x)$ in (16.13). Progress since then has come from the developments initiated primarily by Low[2] and by Lehmann, Symanzik, and Zimmermann[3] (LSZ), who showed how to bring to the fore some of the general information contained in S without resorting to weak-coupling perturbation expansions. They achieved this by applying the asymptotic conditions (16.20) and (16.50) to express matrix elements of physical interest in terms of vacuum expectation values of

[1] Bear in mind that this does not suffice when electromagnetic interactions are present, since a gauge transformation must accompany each Lorentz transformation to reestablish radiation gauge in the new frame of reference. We must therefore establish in this case that the S matrix is gauge invariant.

[2] F. Low, *Phys. Rev.*, **97**, 1392 (1955).

[3] Lehmann, Symanzik, and Zimmermann, *op. cit.*

the field operators. We have already seen an example of the advantage of working with vacuum expectation values in the developments leading to (16.40) for the field commutator. We there achieved a compact, general form for $\Delta'(x - x')$ by invoking Lorentz invariance and other general properties of the theory.

In what is to follow we shall find the vacuum expectation values of products of field operators somewhat more tractable to work with than the matrix elements in the form (16.59). On the one hand, it is possible to expand the field operators $\varphi(x)$ directly in a perturbation series and to construct in this way an expansion of the S-matrix elements in terms of vacuum expectation values of products of the free-wave in-field operators; rules of calculation are found for these expressions and are just the Feynman rules of the earlier propagator approach. On the other hand, invariance arguments such as invoked above in studying $\Delta'(x - x')$ are most readily identified and used when we study matrix elements of the Heisenberg operators φ taken between unique, invariant vacuum states. With the aid of these, considerable progress beyond the perturbation methods has been achieved.

To this end we proceed to develop step by step the general "reduction technique" of LSZ which extracts the information from the physical states in (16.59) and displays it in products of field operators sandwiched between vacuum states. Consider the S-matrix element

$$S_{\beta,\alpha p} = \langle \beta \text{ out} | \alpha p \text{ in} \rangle \tag{16.72}$$

where β denotes the emerging particles in the out-state $|\beta \text{ out}\rangle$ and $|\alpha p \text{ in}\rangle$ is the in-state corresponding to an assemblage α of in particles plus an additional incoming particle with momentum p.

Using the asymptotic condition, we want to extract particle p from the in-state, introducing in place of it a suitable field operator. Using (16.22), (16.9), and (16.48), we write[1]

$$\begin{aligned} \langle \beta \text{ out} | \alpha p \text{ in} \rangle &= \langle \beta \text{ out} | a_{\text{in}}^\dagger(p) | \alpha \text{ in} \rangle \\ &= \langle \beta \text{ out} | a_{\text{out}}^\dagger(p) | \alpha \text{ in} \rangle + \langle \beta \text{ out} | a_{\text{in}}^\dagger(p) - a_{\text{out}}^\dagger(p) | \alpha \text{ in} \rangle \\ &= \langle \beta - p \text{ out} | \alpha \text{ in} \rangle - i \langle \beta \text{ out} | \int d^3x \, f_p(x) \overleftrightarrow{\partial}_0 [\varphi_{\text{in}}(x) - \varphi_{\text{out}}(x)] | \alpha \text{ in} \rangle \end{aligned} \tag{16.73}$$

Here $|\beta - p \text{ out}\rangle$ represents an out-state with particle p, if present, removed from the set β; if p is not included in β, the first term of (16.73) is absent. If $|\alpha p \text{ in}\rangle$ represents an initial two-particle scattering state,

[1] Rigorously, we must work with normalizable states and replace $a_{\text{in}}(p)$ by $\varphi_{\text{in}}{}^f$ of (16.21). In practical applications, however, we generally resort, with due caution, to the simple plane-wave solutions.

including momentum.
Hence "forward scattering"
All satisfy
$(\Box + m^2)(f) = 0$
for Yout

then $\langle \beta - p \text{ out}|\alpha \text{ in}\rangle$ contributes only to the forward elastic scattering in which the projectile and target particles preserve their quantum numbers. The terms on the right-hand side of (16.73) are time-independent by Green's theorem, and the asymptotic condition (16.20) and (16.50) permits the replacement of $\varphi_{\text{in}}(\mathbf{x},x_0)$ by the field $(1/\sqrt{Z})\,\varphi(\mathbf{x},x_0)$ in the limit $x_0 \to -\infty$ and of $\varphi_{\text{out}}(\mathbf{x},x_0)$ by $(1/\sqrt{Z})\,\varphi(\mathbf{x},x_0)$ as $x_0 \to +\infty$; that is

$$\langle \beta \text{ out}|\alpha p \text{ in}\rangle = \langle \beta - p \text{ out}|\alpha \text{ in}\rangle$$
$$+ \frac{i}{\sqrt{Z}}\,(\lim_{x_0 \to +\infty} - \lim_{x_0 \to -\infty})\int d^3x\, f_p(\mathbf{x},x_0)\overleftrightarrow{\partial}_0\langle \beta \text{ out}|\varphi(\mathbf{x},x_0)|\alpha \text{ in}\rangle \quad (16.74)$$

This accomplishes the first step in the reduction procedure. For a more convenient and covariant form, we incorporate the time limits in (16.74) into a four-dimensional volume integral by the identity

see also
pu. 25

$$(\lim_{x_0 \to +\infty} - \lim_{x_0 \to -\infty})\int d^3x\, g_1(x)\overleftrightarrow{\partial}_0 g_2(x) = \int_{-\infty}^{\infty} d^4x\, \frac{\partial}{\partial x_0}[g_1(x)\overleftrightarrow{\partial}_0 g_2(x)]$$
$$= \int_{-\infty}^{\infty} d^4x\left[g_1(x)\frac{\partial^2}{\partial x_0^2}g_2(x) - \frac{\partial^2 g_1(x)}{\partial x_0^2}g_2(x)\right] \quad (16.75)$$

Introducing (16.75) into (16.74), using the property that $f_p(x)$ satisfies the Klein-Gordon equation

$$\frac{\partial^2 f_p(x)}{\partial x_0^2} = (\nabla^2 - m^2)f_p(x) \quad (16.76)$$

and integrating ∇^2 by parts[1] onto $\varphi(\mathbf{x},t)$, we obtain the desired form

$$\langle \beta \text{ out}|\alpha p \text{ in}\rangle = \langle \beta - p \text{ out}|\alpha \text{ in}\rangle$$
$$+ \frac{i}{\sqrt{Z}}\int d^4x\, f_p(x)(\Box + m^2)\langle \beta \text{ out}|\varphi(x)|\alpha \text{ in}\rangle \quad (16.77)$$

The above procedure can now be repeated until we remove all particles from the states and are left with a vacuum expectation value of a product of field operators. For example, let us remove an out particle p' from the assemblage $\beta = \gamma p'$ in (16.77). Repeating the steps (16.74), with appropriate hermitian conjugates, we find

$$\langle \beta \text{ out}|\varphi(x)|\alpha \text{ in}\rangle$$
$$= \langle \gamma \text{ out}|\varphi(x)|\alpha - p' \text{ in}\rangle + \langle \gamma \text{ out}|a_{\text{out}}(p')\varphi(x) - \varphi(x)a_{\text{in}}(p')|\alpha \text{ in}\rangle$$
$$= \langle \gamma \text{ out}|\varphi(x)|\alpha - p' \text{ in}\rangle$$
$$- i\int d^3y\langle \gamma \text{ out}|\varphi_{\text{out}}(y)\varphi(x) - \varphi(x)\varphi_{\text{in}}(y)|\alpha \text{ in}\rangle\overleftrightarrow{\partial}_{y_0}f_{p'}^*(y) \quad (16.78)$$

---

[1] No surface terms arise from these partial integrations, by the usual assumption that the physical system is localized in space.

(¹) not necessary. See p. 25

The asymptotic condition again permits the replacement of the in- and out-fields by $(1/\sqrt{Z})\varphi(y)$ at $y_0 \to -\infty$ and $+\infty$, respectively, in the matrix element in (16.78), which can then be written in terms of the time-ordered product (12.72),

$$\langle \gamma \text{ out}|\varphi_{\text{out}}(y)\varphi(x) - \varphi(x)\varphi_{\text{in}}(y)|\alpha \text{ in}\rangle$$

$$= \frac{1}{\sqrt{Z}} \left(\lim_{y_0 \to +\infty} - \lim_{y_0 \to -\infty} \right) \langle \gamma \text{ out}|T(\varphi(y)\varphi(x))|\alpha \text{ in}\rangle \quad (16.79)$$

Finally, with the aid of (16.75) and (16.76), we arrive at

$$\langle \gamma p' \text{ out}|\varphi(x)|\alpha \text{ in}\rangle = \langle \gamma \text{ out}|\varphi(x)|\alpha - p' \text{ in}\rangle$$
$$+ \frac{i}{\sqrt{Z}} \int d^4y \langle \gamma \text{ out}|T(\varphi(y)\varphi(x))|\alpha \text{ in}\rangle \overleftarrow{(\Box_y + m^2)} f^*_{p'}(y) \quad (16.80)$$

The road is now clear to apply this reduction technique to remove all particles from the states until we arrive at the vacuum expectation value of a product of field operators:

$$\langle p_1 \cdots p_n \text{ out}|q_1 \cdots q_m \text{ in}\rangle$$
$$= \left(\frac{i}{\sqrt{Z}}\right)^{m+n} \prod_{i=1}^{m} \int d^4x_i \prod_{j=1}^{n} \int d^4y_j \, f_{q_i}(x_i) \overrightarrow{(\Box_{x_i} + m^2)}$$
$$\times \langle 0|T(\varphi(y_1) \cdots \varphi(y_n)\varphi(x_1) \cdots \varphi(x_m))|0\rangle \overleftarrow{(\Box_{y_j} + m^2)} f^*_{p_j}(y_j)$$
$$\text{for all } p_i \neq q_j \quad (16.81)$$

In writing (16.81) we have, for simplicity, assumed all $p_i \neq q_j$ and dropped the forward-scattering terms appearing in (16.77) and (16.80); these present no problem, since they too can be further reduced by successively applying the same technique.

Equation (16.81) serves as a cornerstone for all calculations of scattering amplitudes in modern quantum field theory. We remark, as a preview of things to come, that $\langle 0|T(\varphi(z_1) \cdots \varphi(z_r))|0\rangle$ represents the sum of all Feynman graphs with r particles created or destroyed at $(z_1 \cdots z_r)$, as illustrated by Fig. 16.2. It is the complete r-particle Green's function. The factors $(\Box_i + m^2)$ in (16.81) remove the propagators for the external legs leading into the interaction blob in the diagram. To see this schematically, note that $\Box_i + m^2$ becomes $m^2 - p_i^2$ in momentum space, canceling the corresponding propagator factor $i/(p_i^2 - m^2)$. The reduction formula (16.81) states that the S-matrix element is just the Green's function for the $r = n + m$ particles with the external legs removed and with the external momenta put onto the mass shell $p_i^2 = q_j^2 = m^2$. We explore this further in the following chapters.

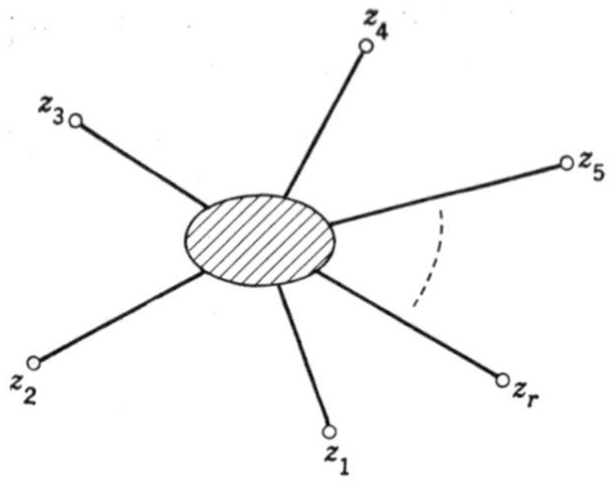

Fig. 16.2 The complete r-particle Green's function.

The close analogy of (16.81) with the S matrix constructed in propagator theory can be brought to the fore by looking back at (6.30) and converting the volume integrals there to four-dimensional form. Using the retarded property of the Green's function there, we rewrite (6.30), with the aid of the free Schrödinger equation

$$\left(i\frac{\partial}{\partial t} + \frac{1}{2m}\nabla^2\right)\varphi = 0$$

$$\begin{aligned}
S_{fi} &= -i \lim_{t'\to\infty} \int d^3x' \int d^4x\ \varphi_f^*(\mathbf{x}',t')\frac{\partial}{\partial t}[G(x';x)\varphi_i(\mathbf{x},t)] \\
&= \lim_{t'\to\infty} \int d^3x' \int d^4x\ \varphi_f^*(x')G(x';x)\overleftarrow{\left(-i\frac{\partial}{\partial t} + \frac{\nabla_x^2}{2m}\right)}\varphi_i(x) \\
&= -i\int d^4x' \int d^4x\ \varphi_f^*(x')\overrightarrow{\left(i\frac{\partial}{\partial t'} + \frac{\nabla_{x'}^2}{2m}\right)}G(x';x)\overleftarrow{\left(-i\frac{\partial}{\partial t} + \frac{\nabla_x^2}{2m}\right)}\varphi_i(x)
\end{aligned} \tag{16.82}$$

Comparing with (16.81), we see that aside from the nonrelativistic replacement

$$(\Box + m^2) \to \left(i\frac{\partial}{\partial t} + \frac{\nabla^2}{2m}\right)$$

the general r-particle Green's function has been replaced by the exact retarded one-particle Green's function $G(x';x)$ for motion in an applied external field. We leave it as an exercise to show that a similar result is obtained from (6.56) in positron theory, with now the exact Feynman propagator appearing.

16.8 In- and Out-fields and Spectral Representation for the Dirac Theory

The main problems encountered in extending the formalism developed so far only for scalar fields to spinor and electromagnetic fields are primarily formal ones. Spin complicates the algebra in the Dirac case, and the gauge problem occurs in the Maxwell theory. However, the basic ideas which have already been laid out in discussing the in- and out-fields, in establishing the spectral form for the vacuum expectation value of the commutator, and in developing the reduction formula may be applied here also.[1]

The Dirac equation is written in terms of the physical mass m by adding a mass counterterm to the source in analogy with (16.13):

$$(i\nabla\!\!\!/ - m)\psi(x) = \mathfrak{j}(x) \qquad \mathfrak{j}(x) \equiv j(x) - (m - m_0)\psi(x) \tag{16.83}$$

The fields $\psi(x)$ are assigned the familiar equal-time anticommutation relations

$$\{\psi_\alpha(\mathbf{x},t),\psi_\beta(\mathbf{y},t)\} = 0 \qquad \{\psi_\alpha(\mathbf{x},t),\psi_\beta^\dagger(\mathbf{y},t)\} = \delta^3(\mathbf{x}-\mathbf{y})\delta_{\alpha\beta}$$

as discussed in (15.9).

The in-field $\psi_{\text{in}}(x)$ is introduced and defined as before as the inhomogeneous term in the integral equation corresponding to (16.83), that is

$$\sqrt{Z_2}\,\psi_{\text{in}}(x) = \psi(x) - \int d^4y\, S_{\text{ret}}(x - y, m)\,\mathfrak{j}(y) \tag{16.84}$$

where $S_{\text{ret}}(x)$ is the retarded Green's function

$$\begin{aligned}(i\nabla\!\!\!/_x - m)S_{\text{ret}}(x - y, m) &= \delta^4(x - y)\\ S_{\text{ret}}(x - y) &= 0 \qquad x_0 < y_0\end{aligned} \tag{16.85}$$

The constant $\sqrt{Z_2}$ is again determined from the normalization of $\psi_{\text{in}}(x)$ to unit amplitude for its matrix element to produce one-particle states from the vacuum.

As defined above, $\psi_{\text{in}}(x)$ satisfies the free wave equation with mass m

$$(i\nabla\!\!\!/ - m)\psi_{\text{in}}(x) = 0 \tag{16.86}$$

with the same transformation properties as $\psi(x)$; in particular

$$[P_\mu,\psi_{\text{in}}(x)] = -i\,\frac{\partial\psi_{\text{in}}(x)}{\partial x^\mu} \tag{16.87}$$

[1] H. Lehmann, *op. cit.;* M. Gell-Mann and F. Low, *Phys. Rev.*, **95,** 1300 (1954).

Take $(\Box_x^2+m^2)$ both sides. Use $S_{ret}' = -(i\not\partial_x+m)\Delta_{ret}$, $(\Box_x^2+m^2)\Delta_r = \delta^4(x-y)$
$i\not\partial_x \to -i\not\partial_y$ on $\delta^4(x-y)$, $i\not p \to (i\not\partial_y+m)\bar{J}(y)$. Integrate $\int d^4y \to$
$(i\not\partial_x+m)J(y) = (i\not\partial_x+m)(i\not\partial_x-m)\psi$
$= -(\Box^2+m^2)\psi$ ✓

From (16.84) it follows that each spinor component of $\psi_{\text{in}}(x)$ satisfies the Klein-Gordon equation and therefore, as in (16.8), $\psi_{\text{in}}(x)$ produces only one-particle states of mass m from the vacuum.

The Fourier expansion of $\psi_{\text{in}}(x)$ is that of a free Dirac field and may be written, as in Chap. 13,

$$\psi_{\text{in}}(x) = \int d^3p \sum_{\pm s} [b_{\text{in}}(p,s)U_{ps}(x) + d_{\text{in}}^{\dagger}(p,s)V_{ps}(x)] \qquad (16.88)$$

where

$$U_{ps}(x) = \frac{1}{(2\pi)^{3/2}}\sqrt{\frac{m}{E_p}}\,u(p,s)e^{-ip\cdot x}$$

$$V_{ps}(x) = \frac{1}{(2\pi)^{3/2}}\sqrt{\frac{m}{E_p}}\,v(p,s)e^{ip\cdot x}$$

with $E_p \equiv \sqrt{|\mathbf{p}|^2 + m^2}$. Inverting (16.88) and its hermitian conjugate for the operator coefficients gives

$$\begin{aligned} b_{\text{in}}(p,s) &= \int d^3x\; U_{ps}^{\dagger}(x)\psi_{\text{in}}(x) \\ b_{\text{in}}^{\dagger}(p,s) &= \int d^3x\; \psi_{\text{in}}^{\dagger}(x)U_{ps}(x) \\ d_{\text{in}}(p,s) &= \int d^3x\; \psi_{\text{in}}^{\dagger}(x)V_{ps}(x) \\ d_{\text{in}}^{\dagger}(p,s) &= \int d^3x\; V_{ps}^{\dagger}(x)\psi_{\text{in}}(x) \end{aligned} \qquad (16.89)$$

for commutation relations, see 16.121 and prob. #1

These satisfy commutation relations derived from (16.87), in analogy with (16.10):

$$\begin{aligned} [P^\mu, b_{\text{in}}(p,s)] &= -p^\mu\, b_{\text{in}}(p,s) \\ [P^\mu, b_{\text{in}}^{\dagger}(p,s)] &= +p^\mu\, b_{\text{in}}^{\dagger}(p,s) \\ [P^\mu, d_{\text{in}}(p,s)] &= -p^\mu\, d_{\text{in}}(p,s) \\ [P^\mu, d_{\text{in}}^{\dagger}(p,s)] &= +p^\mu\, d_{\text{in}}^{\dagger}(p,s) \end{aligned} \qquad (16.90)$$

These coincide with the free-field results and show that we can build general n-particle states by repeated application of $b_{\text{in}}^{\dagger}$ and $d_{\text{in}}^{\dagger}$ upon the vacuum, in analogy with (16.11). We again assume, as always, that the vacuum is a unique stable state of zero energy and that a complete set of states is generated by repeated application of the in-field operators.

The asymptotic condition, as for the scalar field, is

$$\psi(x) \to \sqrt{Z_2}\,\psi_{\text{in}}(x) \qquad \text{as } t \to -\infty$$

in the sense of weak operator convergence (16.20); that is, for normalizable states

$$\lim_{t\to-\infty} \langle\alpha|\psi^f(t)|\beta\rangle = \sqrt{Z_2}\,\langle\alpha|\psi_{\text{in}}^f|\beta\rangle \qquad (16.91)$$

ψ_{in}^f is defined by the first of Eqs. (16.89) with $U_{ps}^{\dagger}(x)$ replaced by a localized packet. $\psi^f(t)$ is defined by the same equation with $\psi_{\text{in}}(x)$ replaced by $\psi(x)$.

We may similarly introduce out-fields to provide a simple description of the physical system at $t \to +\infty$. Instead of (16.84) we write

$$\sqrt{Z_2}\,\psi_{\text{out}}(x) = \psi(x) - \int d^4y\, S_{\text{adv}}(x - y)\,\tilde{\jmath}(y) \qquad (16.92)$$

where

$$\begin{aligned} (i\nabla_x - m)S_{\text{adv}}(x - y) &= \delta^4(x - y) \\ S_{\text{adv}}(x - y) &= 0 \qquad x_0 > y_0 \end{aligned} \qquad (16.93)$$

Equations (16.86) to (16.90) remain valid with replacement of in by out. The asymptotic condition is now

$$\lim_{t \to +\infty} \langle\alpha|\psi^f(t)|\beta\rangle = \sqrt{Z_2}\,\langle\alpha|\psi^f_{\text{out}}|\beta\rangle \qquad (16.94)$$

The equation determining Z_2, the probability that $\psi(x)$ forms a one-particle state of mass m from the vacuum, can be constructed in analogy with the scalar theory by considering the spectral representation for

$$S'_{\alpha\beta}(x,x') = i\langle 0|\{\psi_\alpha(x),\bar{\psi}_\beta(x')\}|0\rangle \qquad (16.95)$$

Inserting a complete set of states

$$P^\mu|n\rangle = p^\mu_n|n\rangle$$

and displacing the fields to the origin leads to

$$S'_{\alpha\beta}(x,x') = S'_{\alpha\beta}(x - x') = i\sum_n [\langle 0|\psi_\alpha(0)|n\rangle\langle n|\bar{\psi}_\beta(0)|0\rangle e^{-ip_n\cdot(x-x')} + \langle 0|\bar{\psi}_\beta(0)|n\rangle\langle n|\psi_\alpha(0)|0\rangle e^{ip_n\cdot(x-x')}] \qquad (16.96)$$

As before, we introduce the spectral amplitude by grouping together in the sum over n all states of given four-momentum q

$$\rho_{\alpha\beta}(q) = (2\pi)^3 \sum_n \delta^4(p_n - q)\langle 0|\psi_\alpha(0)|n\rangle\langle n|\bar{\psi}_\beta(0)|0\rangle \qquad (16.97)$$

and set out to construct its general form from invariance arguments. $\rho(q)$ is a 4×4 matrix and may be expanded in terms of the 16 linearly independent products of the γ matrices:

$$\rho_{\alpha\beta}(q) = \rho(q)\delta_{\alpha\beta} + \rho_\mu(q)\gamma^\mu_{\alpha\beta} + \rho_{\mu\nu}(q)\sigma^{\mu\nu}_{\alpha\beta} + \tilde{\rho}(q)\gamma^5_{\alpha\beta} + \tilde{\rho}_\mu(\gamma^\mu\gamma^5)_{\alpha\beta} \qquad (16.98)$$

The structure of the coefficients of the matrices is strongly limited by the requirement of Lorentz covariance.[1] The terms in (16.98) must

[1] For a field coupled to the radiation field we must always accompany a Lorentz transformation by a gauge transformation to restore the transverse gauge used in the quantization. Because $S'(x,x')$ is not gauge invariant [cf. Eq. (15.21)], the considerations of this section will not apply. See Prob. 8.

transform as dictated by the defining equation (16.97). Recall that under a Lorentz transformation, the field operators satisfy

$$\begin{aligned} U(a)\psi_\alpha(0)U^{-1}(a) &= S^{-1}_{\alpha\lambda}(a)\psi_\lambda(0) \\ U(a)\bar{\psi}_\alpha(0)U^{-1}(a) &= \bar{\psi}_\lambda(0)S_{\lambda\alpha}(a) \end{aligned} \tag{16.99}$$

according to page 62. The matrix S is defined in Chap. 2 by

$$S^{-1}\gamma^\mu S = a^\mu{}_\nu\gamma^\nu \tag{16.100}$$

Inserting into (16.97) and using Lorentz invariance of the vacuum as assumed at the outset, $U|0\rangle = |0\rangle$, we find

$$\rho_{\alpha\beta}(q) = \sum_n (2\pi)^3\delta^4(p_n - q)S^{-1}_{\alpha\lambda}(a)S_{\delta\beta}(a)\langle 0|\psi_\lambda(0)|U(a)n\rangle\langle U(a)n|\bar{\psi}_\delta(0)|0\rangle \tag{16.101}$$

Again taking advantage of Lorentz invariance of $\delta^4(p_n - q)$, as in (16.29), to rewrite the sum over the complete set of states labeled by $|m\rangle = |U(a)n\rangle$, we may rewrite (16.101) as

$$\rho_{\alpha\beta}(q) = S^{-1}_{\alpha\lambda}(a)\sum_m (2\pi)^3\delta^4(p_m - qa^{-1})\langle 0|\psi_\lambda(0)|m\rangle\langle m|\bar{\psi}_\delta(0)|0\rangle S_{\delta\beta}(a)$$

or simply

$$\rho(q) = S^{-1}(a)\rho(qa^{-1})S(a) \tag{16.102}$$

Equation (16.102), together with the general expansion of $\rho(q)$ given in (16.98), determines the structure of the coefficients ρ, ρ_μ, etc. For instance, if (16.98) is inserted into (16.102), it follows that

$$\rho(q) = \rho(qa^{-1})$$

that is, ρ transforms as a Lorentz scalar. Similarly,

$$\rho_\mu(q) = a_\mu{}^\nu\rho_\nu(qa^{-1})$$

transforms as a Lorentz four-vector, and so on. Since $\rho_{\alpha\beta}$ is a function only of q, and vanishes outside the forward light cone, it follows that ρ and $\tilde{\rho}$ depend only upon q^2, $\rho_\mu(q)$ and $\tilde{\rho}_\mu(q)$ are scalar functions of q^2 multiplied by q_μ, and $\rho_{\mu\nu}$ is proportional to $q_\mu q_\nu$. In this way the form (16.98) is limited to

$$\rho_{\alpha\beta}(q) = \rho_1(q^2)q_{\alpha\beta} + \rho_2(q^2)\delta_{\alpha\beta} + \tilde{\rho}_1(q^2)(q\gamma^5)_{\alpha\beta} + \tilde{\rho}_2(q^2)\gamma^5_{\alpha\beta} \tag{16.103}$$

To reduce the form of $\rho_{\alpha\beta}(q)$ further, we must require invariance of the theory under the parity transformation $\mathcal{P}$, which has the property

$$\mathcal{P}\psi_\alpha(0)\mathcal{P}^{-1} = \gamma^0_{\alpha\lambda}\psi_\lambda(0) \tag{16.104}$$

Inserting (16.104) into (16.97) and carrying out the steps analogous to those used for proper Lorentz transformations, we arrive at the analogue of (16.102):

$$\rho_{\alpha\beta}(\mathbf{q},q_0) = \gamma^0_{\alpha\lambda}\rho_{\lambda\delta}(-\mathbf{q},q_0)\gamma^0_{\delta\beta} \tag{16.105}$$

Inserting (16.103) into (16.105) one finds

$$\tilde{\rho}_1 = \tilde{\rho}_2 = 0 \tag{16.106}$$

owing to the extra sign change arising from γ_0 anticommuting with γ_5. Thus (16.103) reduces to the final form

$$\rho_{\alpha\beta}(q) = \rho_1(q^2)q_{\alpha\beta} + \rho_2(q^2)\delta_{\alpha\beta} \tag{16.107}$$

Our following discussion is based on the form of (16.107). $\mathcal{P}$ is, of course, not a symmetry operation in nature when effects of the weak couplings are included. A more complete discussion should be based on the form in (16.103), and, similarly, in (16.84) and (16.92) the renormalization factor Z_2 should be recognized as a matrix of form $a + b\gamma_5$. We refer the interested reader to the literature.[1]

The spectral amplitude for the second term in (16.96) can be related directly to (16.97) with the aid of $\mathcal{P}\mathcal{C}\mathfrak{T}$ invariance of the theory. Introducing the nonlinear operation K of taking the complex conjugate which is contained in the time-reversal operator $\mathfrak{T} = \mathfrak{U}K$, we write

$$\begin{aligned}\langle 0|\bar{\psi}_\beta(x')\psi_\alpha(x)|0\rangle &= \langle K0|K\bar{\psi}_\beta(x')\psi_\alpha(x)0\rangle^* \\ &= \langle K\bar{\psi}_\beta(x')\psi_\alpha(x)0|K0\rangle\end{aligned}$$

Introducing now $1 = (\mathcal{P}\mathcal{C}\mathfrak{U})^{-1}(\mathcal{P}\mathcal{C}\mathfrak{U})$ and using the $\mathcal{P}\mathcal{C}\mathfrak{T}$ invariance of the vacuum state

$$\mathcal{P}\mathcal{C}\mathfrak{T}|0\rangle \equiv \theta|0\rangle = |0\rangle$$

leads to

$$\langle 0|\bar{\psi}_\beta(x')\psi_\alpha(x)|0\rangle = \langle \theta\bar{\psi}_\beta(x')\theta^{-1}\theta\psi_\alpha(x)\theta^{-1}0|0\rangle \tag{16.108}$$

Recalling from (15.150) how the Dirac field transforms under $\mathcal{P}\mathcal{C}\mathfrak{T}$

$$\begin{aligned}\theta\psi_\alpha(x)\theta^{-1} &= -i(\gamma_0\gamma_5\bar{\psi}(-x))_\alpha = i(\gamma_5\psi^\dagger(-x))_\alpha \\ \theta\bar{\psi}_\beta(x')\theta^{-1} &= -i(\psi(-x')\gamma_5\gamma_0)_\beta\end{aligned}$$

we obtain, using $\gamma_5 = \gamma_5{}^T$,

$$\langle 0|\bar{\psi}_\beta(x')\psi_\alpha(x)|0\rangle = -(\gamma^5)_{\alpha\tau}\langle 0|\psi_\tau(-x)\bar{\psi}_\lambda(-x')|0\rangle(\gamma^5)_{\lambda\beta} \tag{16.109}$$

[1] K. Sekine, *Nuovo Cimento*, **11,** 87 (1959); K. Hiida, *Phys. Rev.*, **132,** 1239 (1963); **134,** B174 (1964).

Inserting (16.109) along with (16.107) into (16.96) gives finally

$$S'_{\alpha\beta}(x - x') = i \int \frac{d^4q}{(2\pi)^3}\, \theta(q_0)\big([\not{q}\rho_1(q^2) + \rho_2(q^2)]_{\alpha\beta} e^{-iq\cdot(x-x')}$$
$$- \{\gamma_5[\not{q}\rho_1(q^2) + \rho_2(q^2)]\gamma_5\}_{\alpha\beta} e^{iq\cdot(x-x')}\big)$$
$$= i \int \frac{d^4q}{(2\pi)^3}\, \theta(q_0)[i\nabla_x \rho_1(q^2) + \rho_2(q^2)]_{\alpha\beta}$$
$$\times (e^{-iq\cdot(x-x')} - e^{iq\cdot(x-x')}) \qquad (16.110)$$

Since ρ vanishes for space-like q^2, we may also write this as an integral over the mass spectrum by introducing

$$\rho(q^2) = \int_0^\infty \rho(M^2)\delta(q^2 - M^2)\, dM^2$$

We find

$$S'_{\alpha\beta}(x - x') = -\int dM^2\, [i\rho_1(M^2)\nabla_x + \rho_2(M^2)]_{\alpha\beta}\Delta(x - x'; M)$$
$$= \int dM^2 \{\rho_1(M^2) S_{\alpha\beta}(x - x'; M)$$
$$+ [M\rho_1(M^2) - \rho_2(M^2)]_{\alpha\beta}\Delta(x - x'; M)\} \qquad (16.111)$$

where the invariant Δ and S functions are as given in Appendix C.

The above derivation of the spectral representation goes through unchanged for the vacuum expectation value of the time-ordered product of the Dirac fields; it is necessary only to replace the S and Δ in (16.111) by the Feynman propagators S_F and Δ_F:

$$S'_{F\alpha\beta}(x - x') = +\int dM^2\, [i\rho_1(M^2)\nabla_x + \rho_2(M^2)]_{\alpha\beta}\Delta_F(x - x'; M) \qquad (16.112)$$

or in momentum space

$$S'_F(p) = \int_0^\infty dM^2\, [\not{p}\rho_1(M^2) + \rho_2(M^2)] \frac{1}{p^2 - M^2 + i\epsilon}$$

Similar forms can be written for the other Green's functions as well.

Comparing (16.111) with (16.32), we see that in the Dirac theory with the spin degree of freedom, the spectral form is given in terms of integrals over two instead of one unknown scalar function. The properties of ρ_1 and ρ_2 analogous to those of $\rho(q^2)$ are

(i)	$\rho_1(M^2)$ and $\rho_2(M^2)$ are real	
(ii)	$\rho_1(M^2) \geq 0$	(16.113)
(iii)	$M\rho_1(M^2) - \rho_2(M^2) \geq 0$	

To prove (i), we take the complex conjugate of (16.97):

$$\begin{aligned}\rho^*_{\alpha\beta}(q) &= \sum_n (2\pi)^3\delta^4(p_n - q)\langle n|\bar\psi_\tau(0)|0\rangle\gamma^0_{\tau\alpha}\gamma^0_{\beta\lambda}\langle 0|\psi_\lambda(0)|n\rangle \\ &= [\gamma_0\rho(q)\gamma_0]_{\beta\alpha} \\ &= [\rho_1(q^2)q^* + \rho_2(q^2)]_{\alpha\beta}\end{aligned}$$

To prove (ii), we form the trace; from (16.97) and (16.107)

$$\begin{aligned}\mathrm{Tr}\,\gamma_0\rho(q) = 4q_0\rho_1(q^2) &= \sum_n (2\pi)^3\delta^4(p_n - q)\sum_{\alpha=1}^{4}\langle 0|\psi_\alpha(0)|n\rangle\langle n|\psi^\dagger_\alpha(0)|0\rangle \\ &= \sum_n (2\pi)^3\delta^4(p_n - q)\sum_{\alpha=1}^{4}|\langle 0|\psi_\alpha(0)|n\rangle|^2 \geq 0\end{aligned}$$

Since $q_0 > 0$, $\rho_1(q^2) > 0$.

(iii) is proved similarly by forming the absolute square of the operator $(i\nabla\!\!\!/ - M)\psi$. Demonstration of this is left as an exercise.

We may now extract from the spectral amplitudes the contribution coming from the discrete one-particle states, as done for the scalar field, and obtain a condition on Z_2 analogous to (16.42). Forming the matrix element of (16.83) from the vacuum to a one-particle state, $|ps\rangle$, gives

$$\begin{aligned}(i\nabla\!\!\!/ - m)\langle 0|\psi(x)|ps\rangle &= (p\!\!\!/ - m)\langle 0|\psi(0)|ps\rangle e^{-ip\cdot x} \\ &= \langle 0|\tilde{j}(0)|ps\rangle e^{-ip\cdot x}\end{aligned} \qquad (16.114)$$

Again appealing to known proper Lorentz transformation properties, we recall from (16.99) that $\langle 0|\psi(0)|ps\rangle$ transforms as a spinor wave function of momentum p and spin s:

$$\begin{aligned}\langle 0|\psi_\alpha(0)|ps\rangle &= \langle 0|U(a)\psi_\alpha(0)U^{-1}(a)|U(a)ps\rangle \\ &= S^{-1}_{\alpha\beta}(a)\langle 0|\psi_\beta(0)|p's'\rangle\end{aligned}$$

and may therefore be written[1]

$$\langle 0|\psi(0)|ps\rangle = au(p,s) + bv(p,-s) = (a + b\gamma_5)u(p,s) \qquad (16.115)$$

Imposing the condition that the theory is to be invariant under the parity transformation removes the b term in (16.115). By (15.88) and (15.91)

$$\begin{aligned}\langle 0|\psi(0)|p_0,-\mathbf{p},s\rangle &= (a + b\gamma_5)u(-p,s) = \langle 0|\mathcal{P}\psi(0)\mathcal{P}^{-1}|ps\rangle \\ &= \gamma_0(a + b\gamma_5)u(p,s) = (a - b\gamma_5)u(-p,s)\end{aligned} \qquad (16.116)$$

[1] See Prob. 9.

where the identification $u(-p,s) = \gamma_0 u(p,s)$ follows from the Dirac equation.[1] Therefore, $b = 0$ and from (16.114), it follows that

$$\langle 0|\bar{j}(0)|ps\rangle = a(\not{p} - m)u(p,s) = 0 \tag{16.117}$$

and

$$\begin{aligned}\langle 0|\psi(x)|ps\rangle &= \sqrt{Z_2}\,\langle 0|\psi_{\text{in}}(x)|ps\rangle \\ &= \frac{\sqrt{Z_2}}{(2\pi)^{3/2}}\sqrt{\frac{m}{E_p}}\,u(p,s)e^{-ip\cdot x}\end{aligned} \tag{16.118}$$

$$\langle 0|\psi(x)|ps\rangle = \sqrt{Z_2}\,\langle 0|\psi_{\text{out}}(x)|ps\rangle \tag{16.119}$$

Because in- and out-fields produce only one-particle states from the vacuum with matrix elements given by (16.118), it follows in analogy with (16.55) that

$$\langle 0|\{\psi_\alpha^{\text{in}}(x),\bar{\psi}_\beta^{\text{in}}(y)\}|0\rangle = -iS_{\alpha\beta}(x-y) \tag{16.120}$$

In analogy with (16.56) the anticommutators are c numbers and we can write

$$\{\psi_\alpha^{\text{in}}(x),\bar{\psi}_\beta^{\text{in}}(y)\} = \{\psi_\alpha^{\text{out}}(x),\bar{\psi}_\beta^{\text{out}}(y)\} = -iS_{\alpha\beta}(x-y) \tag{16.121}$$

Demonstration of this is left as an exercise.

Using (16.118), we separate the contribution of the one-particle state to the spectral amplitude (16.97), which is

$$\begin{aligned}Z_2\int\frac{d^3p}{(2\pi)^3}\sum_{\pm s}(2\pi)^3\delta^4(p-q)\frac{m}{E_p}\,u_\alpha(p,s)\bar{u}_\beta(p,s) \\ = Z_2(\not{q}+m)_{\alpha\beta}\delta(q^2-m^2)\theta(q_0)\end{aligned}$$

Inserting into (16.111), we find

$$\begin{aligned}S'_{\alpha\beta}(x-x') &= Z_2S_{\alpha\beta}(x-x',m) \\ &\quad - \int_{m_1^2}^{\infty} dM^2\,[i\rho_1(M^2)\nabla\!\!\!/_x + \rho_2(M^2)]_{\alpha\beta}\Delta(x-x',M)\end{aligned} \tag{16.122}$$

where the spectral integral starts at the threshold m_1 of the continuum spectrum.

At $t = t'$ the left-hand side of (16.122) is known from the assumed one-time anticommutation relations for the fields $\psi(x)$

$$\begin{aligned}S'_{\alpha\beta}(\mathbf{x}-\mathbf{x}',0) &= i\langle 0|\{\psi_\alpha(\mathbf{x},t),\bar{\psi}_\beta(\mathbf{x}',t)\}|0\rangle \\ &= i\gamma^0_{\alpha\beta}\delta^3(\mathbf{x}-\mathbf{x}')\end{aligned} \tag{16.123}$$

[1] As before, $u(-p,s) \equiv u(p_0, -\mathbf{p},s)$.

We find then as a condition on the magnitude of Z_2

$$1 = Z_2 + \int_{m_1{}^2}^{\infty} dM^2\, \rho_1(M^2) \tag{16.124}$$

or by (16.113),

$$0 \le Z_2 < 1 \tag{16.125}$$

in analogy with (16.34).

In deriving this condition on Z_2, the probability of forming a one-Dirac-particle state from the vacuum, we have relied heavily on Lorentz invariance of the theory. Since a gauge change must accompany each Lorentz transformation if electromagnetic couplings are present, and since $S'(x - x')$ is not gauge invariant, condition (16.125) is not valid in quantum electrodynamics and Z_2, a gauge-dependent number, has no simple physical interpretation.

16.9 The Reduction Formula for Dirac Fields

The general properties of the S matrix which were discussed in Sec. 16.6 apply when spin-½ as well as spin-0 particles are present in the in- and out-states. The reduction technique developed for scalar fields in Sec. 16.7 may be extended to matrix elements between states with Dirac particles present with only minor modifications in technical details.

From (16.88), (16.90), and (16.121) it is apparent that we construct n-particle in- (and out-) states just as in the free Dirac theory by repeated application to the vacuum of

$$b_{\text{in}}^\dagger(p,s) = \int \psi_{\text{in}}^\dagger(x) U_{ps}(x)\, d^3x \qquad d_{\text{in}}^\dagger(\bar{p},\bar{s}) = \int V_{\bar{p}\bar{s}}^\dagger(x) \psi_{\text{in}}(x)\, d^3x \tag{16.126}$$

A general in-state, for example, with the indicated quantum numbers is written

$$\begin{aligned} |(\bar{p}_k\bar{s}_k) \;.\;.\;.\; (\bar{p}_1\bar{s}_1);\, (p_js_j) \;.\;.\;.\; (p_1s_1);\, q_1 \;.\;.\;.\; q_n \text{ in}\rangle \\ = d_{\text{in}}^\dagger(\bar{p}_k\bar{s}_k) \;.\;.\;.\; d_{\text{in}}^\dagger(\bar{p}_1\bar{s}_1) b_{\text{in}}^\dagger(p_js_j) \;.\;.\;.\; b_{\text{in}}^\dagger(p_1s_1) \\ \times\, a_{\text{in}}^\dagger(q_1) \;.\;.\;.\; a_{\text{in}}^\dagger(q_n)|0\rangle \end{aligned} \tag{16.127}$$

where the convention adopted here is to list the arguments of the fermion fields reading from right to left in the order in which they are created in (16.127), with particles (p_is_i) preceding the antiparticles $(\bar{p}_i\bar{s}_i)$. This convention fixes the signs of the states and takes care of the bookkeeping questions arising from the anticommutation algebra of the fermion operators.

To start the reduction procedure, we remove a Dirac particle (ps) from the in-state, replacing it by the matrix element of a field operator.

Applying the asymptotic condition (16.91) and (16.94), we repeat the steps of (16.73) and (16.74), writing

$$\begin{aligned}
&\langle\beta\ \text{out}|(ps),\ \alpha\ \text{in}\rangle \\
&\quad = \langle\beta - (ps)\ \text{out}|\alpha\ \text{in}\rangle + \langle\beta\ \text{out}|b_{\text{in}}^{\dagger}(ps) - b_{\text{out}}^{\dagger}(ps)|\alpha\ \text{in}\rangle \\
&\quad = \langle\beta - (ps)\ \text{out}|\alpha\ \text{in}\rangle + \int d^3x\ \langle\beta\ \text{out}|[\psi_{\text{in}}^{\dagger}(x) - \psi_{\text{out}}^{\dagger}(x)]|\alpha\ \text{in}\rangle U_{ps}(x) \\
&\quad = \langle\beta - (ps)\ \text{out}|\alpha\ \text{in}\rangle \\
&\qquad - \frac{1}{\sqrt{Z_2}}\int d^4x \langle\beta\ \text{out}|\ \frac{\partial}{\partial x^0}\,(\bar{\psi}(x)\gamma^0 U_{ps}(x))|\alpha\ \text{in}\rangle
\end{aligned} \tag{16.128}$$

Since $U_{ps}(x)$ is a solution of the free Dirac equation, we introduce

$$\gamma_0 \frac{\partial}{\partial x^0}\, U_{ps}(x) = (-\boldsymbol{\gamma}\cdot\boldsymbol{\nabla} - im)U_{ps}(x)$$

and partially integrate the $-\boldsymbol{\gamma}\cdot\boldsymbol{\nabla}$ onto the $\bar{\psi}(x)$ in (16.128). This gives for the second term

$$-\frac{i}{\sqrt{Z_2}}\int d^4x\ \langle\beta\ \text{out}|\bar{\psi}(x)|\alpha\ \text{in}\rangle(-i\overleftarrow{\nabla\!\!\!/} - m)U_{ps}(x) \tag{16.129}$$

In a similar way, removing an antiparticle from the in-state leads to

$$\frac{i}{\sqrt{Z_2}}\int d^4x\ \bar{V}_{\bar{p}\bar{s}}(x)(i\overrightarrow{\nabla\!\!\!/} - m)\langle\beta\ \text{out}|\psi(x)|\alpha\ \text{in}\rangle \tag{16.130}$$

removing a particle from the out-state leads to

$$\frac{-i}{\sqrt{Z_2}}\int d^4x\ \bar{U}_{p's'}(x)(i\overrightarrow{\nabla\!\!\!/} - m)\langle\beta\ \text{out}|\psi(x)|\alpha\ \text{in}\rangle \tag{16.131}$$

and removing an antiparticle from the out-state leads to

$$\frac{+i}{\sqrt{Z_2}}\int d^4x\ \langle\beta\ \text{out}|\bar{\psi}(x)|\alpha\ \text{in}\rangle(-i\overleftarrow{\nabla\!\!\!/} - m)V_{\bar{p}'\bar{s}'}(x) \tag{16.132}$$

The pair of expressions (16.129) and (16.132) shows the close formal connection between the amplitude describing the interaction of an incident particle (electron with ps) in the in-state and the amplitude for the interaction of an emerging antiparticle (positron with $\bar{p}'\bar{s}'$) in the out-state; between the two expressions it is necessary only to replace $U_{ps}(x)$ by $-V_{\bar{p}'\bar{s}'}(x)$, that is, $u(p,s)e^{-ip\cdot x}$ by $-v(\bar{p}',\bar{s}')e^{-i(-\bar{p}')\cdot x}$. This is the field theoretic statement of the result of the propagator theory developed in Chap. 6, and which instructed us to calculate positron processes by propagating negative-energy electrons backward in time. There is a similar correspondence in (16.130) and (16.131) between the outgoing electron and incoming positron.

To continue with the reduction process until all particles are removed from the state vectors and we arrive at a vacuum expectation value of a product of fields, we form time-ordered products as in (16.79). Removing, for example, a Dirac particle from the in-state in a matrix element containing both scalar and spinor fields gives

$$\begin{aligned}\langle\beta\ \text{out}|T(x_1 \cdots z_p)_{\alpha_1 \cdots \beta_p}|(ps)\alpha\ \text{in}\rangle \\ \equiv \langle\beta\ \text{out}|T(\varphi(x_1) \cdots \varphi(x_n)\psi_{\alpha_1}(y_1) \cdots \psi_{\alpha_m}(y_m) \\ \times \bar\psi_{\beta_1}(z_1) \cdots \bar\psi_{\beta_p}(z_p))|(ps)\alpha\ \text{in}\rangle \\ = (-)^{m+p}\langle\beta - (ps)\ \text{out}|T(x_1 \cdots z_p)_{\alpha_1 \cdots \beta_p}|\alpha\ \text{in}\rangle \\ + \langle\beta\ \text{out}|T(x_1 \cdots z_p)_{\alpha_1 \cdots \beta_p}b^\dagger_{\text{in}}(p,s)|\alpha\ \text{in}\rangle \\ - \langle\beta\ \text{out}|(-)^{m+p}b^\dagger_{\text{out}}(p,s)T(x_1 \cdots z_p)_{\alpha_1 \cdots \beta_p}|\alpha\ \text{in}\rangle \end{aligned} \quad (16.133)$$

The sign $(-)^{m+p}$ is governed by the number of sign changes dictated by the definition of time ordering for fermion fields [see Eq. (13.71)]:

$$T(\psi_\alpha(x)\psi_\beta(y)) = \psi_\alpha(x)\psi_\beta(y)\theta(x_0 - y_0) - \psi_\beta(y)\psi_\alpha(x)\theta(y_0 - x_0)$$

Inserting the asymptotic condition in the second term and imitating the earlier discussion leading to (16.129), we find

$$-\frac{i}{\sqrt{Z_2}}\int d^4x\, \langle\beta\ \text{out}|T(\varphi(x_1) \cdots \psi_{\alpha_1}(y_1) \cdots \bar\psi_{\beta_p}(z_p)\bar\psi_\lambda(x))|\alpha\ \text{in}\rangle \\ \times (-i\overleftarrow{\nabla}_x - m)_{\lambda\tau}U_{ps}(x)_\tau \quad (16.134)$$

The corresponding expression to remove an antiparticle from the in-state is

$$\frac{i}{\sqrt{Z_2}}\int d^4x\, \bar V_{\bar p\bar s}(x)_\tau(i\overrightarrow{\nabla}_x - m)_{\tau\lambda} \\ \times (-)^{m+p}\langle\beta\ \text{out}|T(\psi_\lambda(x)\varphi(x_1) \cdots \bar\psi_{\beta_p}(z_p))|\alpha\ \text{in}\rangle \quad (16.135)$$

The expression to remove a particle from the out-state is

$$-\frac{i}{\sqrt{Z_2}}\int d^4x\, \bar U_{p's'}(x)_\tau(i\overrightarrow{\nabla}_x - m)_{\tau\lambda} \\ \times \langle\beta\ \text{out}|T(\psi_\lambda(x)\varphi(x_1) \cdots \bar\psi_{\beta_p}(z_p))|\alpha\ \text{in}\rangle \quad (16.136)$$

And the expression to remove an antiparticle from the out-state is

$$\frac{i}{\sqrt{Z_2}}\int d^4x\, \langle\beta\ \text{out}|T(\varphi(x_1) \cdots \bar\psi_{\beta_p}(z_p)\bar\psi_\lambda(x))|\alpha\ \text{in}\rangle(-)^{m+p} \\ \times (-i\overleftarrow{\nabla}_x - m)_{\lambda\tau}V_{\bar p'\bar s'}(x)_\tau \quad (16.137)$$

In this way we eventually arrive at the vacuum expectation value

$$\langle 0|T(\varphi(x_1) \cdots \psi(y_1) \cdots \bar\psi(z_1) \cdots)|0\rangle \quad (16.138)$$

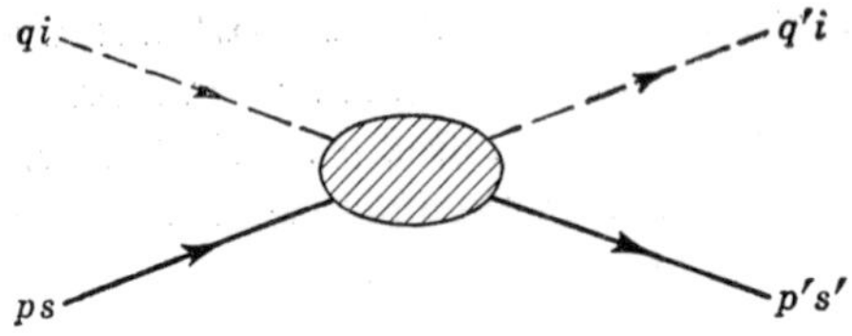

Fig. 16.3 Meson-proton scattering with kinematics as shown.

which, as will be shown in the next chapter, stands for the sum of all Feynman diagrams with lines representing scalar particles entering or leaving at x, fermions entering or antifermions leaving at z_i, and antifermions entering or fermions leaving at y_i. In the reduction formula the Klein-Gordon and Dirac operators lop off the legs of the external particles and put them on their mass shells, with the factors

$$\int d^4x\, f_p(x)(\square_x + m^2) \qquad \int d^4x\, \bar{U}_{p's'}(x)(i\nabla\!\!\!/_x - m) \qquad \text{etc.}$$

The factors $i/\sqrt{Z}$, $-i/\sqrt{Z_2}$, $i/\sqrt{Z_2}$ for each boson, fermion, and antifermion, respectively, renormalize their wave functions appropriately so that the final result may be equated directly to the transition amplitude.

As an illustrative example of considerable interest, consider the scattering of a meson of type i by a proton, with the kinematics as shown in Fig. 16.3. Applying the reduction procedure to the S matrix gives

$$\begin{aligned} S_{fi} &= \langle q';p's' \text{ out}|q;ps \text{ in}\rangle \\ &= \delta_{fi} + \frac{1}{Z_2 Z}\int d^4x\, d^4x'\, d^4z\, d^4z'\, f_{q'}^*(x')\overrightarrow{(\square_{x'} + \mu^2)} \\ &\quad \times [\bar{U}_{p's'}(z')\overrightarrow{(i\nabla\!\!\!/_{z'} - m)}]_\sigma \langle 0|T(\psi_\sigma(z')\bar{\psi}_\tau(z)\varphi_i(x)\varphi_i(x'))|0\rangle \\ &\quad \times [\overleftarrow{(-i\nabla\!\!\!/_z - m)}U_{ps}(z)]_\tau \overleftarrow{(\square_x + \mu^2)} f_q(x) \end{aligned} \tag{16.139}$$

In (16.139), μ and m denote the meson and proton masses, respectively; $\delta_{fi} \neq 0$ only if the in- and out-states coincide, that is, for forward scattering. In carrying through practical calculations, plane waves (16.9) and (16.88) are inserted

$$\begin{aligned} f_q(x) &= \frac{1}{(2\pi)^{3/2}\sqrt{2\omega_q}}\, e^{-iq\cdot x} \\ U_{ps}(x) &= \frac{1}{(2\pi)^{3/2}}\sqrt{\frac{m}{E_p}}\, u(p,s)e^{-ip\cdot x} \end{aligned} \tag{16.140}$$

as the limiting cases of normalizable packet solutions; with this continuum normalization

$$\delta_{fi} = \delta^3(\mathbf{q}' - \mathbf{q})\, \delta^3(\mathbf{p}' - \mathbf{p})\, \delta_{s's}$$

16.10 In- and Out-states and the Reduction Formula for Photons

The final extension of the formal developments of this chapter to the radiation field must be made separately owing to the complications arising from the noncovariant gauge choice and quantization procedure applied to the Maxwell theory.

In the radiation gauge the vector potential is transverse and satisfies the wave equation

$$\Box \mathbf{A} = e_0 \mathbf{j}^{\text{tr}} \tag{16.141}$$

where

$$e_0 \mathbf{j}^{\text{tr}} \equiv e_0 \mathbf{j} + \frac{\partial \mathbf{E}_l}{\partial t}$$

with

$$\nabla \cdot \mathbf{E} = e_0 \rho = e_0 j_0 \qquad \text{and} \qquad \nabla \cdot \mathbf{j}^{\text{tr}} = 0$$

defines the transverse current source. The longitudinal part of the vector potential vanishes in this gauge, and the scalar potential is determined from Gauss's law

$$\nabla^2 A_0 = e_0 \rho$$

We then have two dynamically independent components of the vector potential which obey wave equations (16.141) and one-time commutation relations [see (15.9)]

$$\begin{aligned} [A_i(\mathbf{x},t), A_k(\mathbf{x}',t)] &= [\dot{A}_i(\mathbf{x},t), \dot{A}_k(\mathbf{x}',t)] = 0 \\ [A_i(\mathbf{x},t), \dot{A}_k(\mathbf{x}',t)] &= i\delta_{ik}^{\text{tr}}(\mathbf{x} - \mathbf{x}') \end{aligned} \tag{16.142}$$

These are similar to the canonical theory for scalar fields, differing only in that the mass is zero and that the field is transverse and therefore only the transverse part of the δ function, as defined in (14.16), appears.

In analogy with the discussion for the scalar field, we introduce transverse in- (and out-) fields which are assigned properties as in (16.5) and (16.6):

$$[P^\mu, \mathbf{A}_{\text{in}}(x)] = -i \frac{\partial}{\partial x_\mu} \mathbf{A}_{\text{in}}(x) \qquad \Box \mathbf{A}_{\text{in}}(x) = 0 \tag{16.143}$$

From this it follows, as in (16.8), that $\mathbf{A}_{\text{in}}(x)$ creates only[1] the one-photon states, with $P_n^2 = 0$, from the vacuum. No mass counterterm is introduced into (16.143), since the physical incident and emerging photons are massless and obey the Einstein condition $k_\mu k^\mu = 0$.

The Fourier expansion of $\mathbf{A}_{\text{in}}(x)$ is the same as that for free fields:

$$\mathbf{A}_{\text{in}}(x) = \int d^3k \sum_{\lambda=1}^{2} [a_{\text{in}}(k,\lambda)\mathbf{A}_{k,\lambda}(x) + a_{\text{in}}^\dagger(k,\lambda)\mathbf{A}_{k,\lambda}^*(x)]$$

with
$$\mathbf{A}_{k,\lambda}(x) = \frac{1}{\sqrt{(2\pi)^3 2k_0}} e^{-ik\cdot x}\boldsymbol{\varepsilon}(k,\lambda) \tag{16.144}$$

and, upon inversion,

$$\begin{aligned} a_{\text{in}}(k,\lambda) &= i\int d^3x\, \mathbf{A}_{k,\lambda}^*(x) \cdot \overleftrightarrow{\partial_0}\mathbf{A}_{\text{in}}(x) \\ &= -i\int d^3x\, A_{k,\lambda}^*(x)_\mu \overleftrightarrow{\partial_0} A_{\text{in}}(x)^\mu \end{aligned} \tag{16.145}$$

Repeated application of $a_{\text{in}}^\dagger(k,\lambda)$ onto the vacuum state builds up the general n-photon in-states, as in (16.10) and (16.11). Similar statements apply to the out-fields and -states. As in the earlier discussions, these are assumed to be complete. Continuing to imitate the discussion of the scalar fields (16.14) and (16.52), we write for the relation of $\mathbf{A}_{\text{in}}(x)$ and $\mathbf{A}_{\text{out}}(x)$ to $\mathbf{A}(x)$

$$\begin{aligned} \sqrt{Z_3}\,\mathbf{A}_{\text{in}}(x) &= \mathbf{A}(x) - e_0\int d^4y\, D_{\text{ret}}(x-y)\mathbf{j}^{\text{tr}}(y) \\ \sqrt{Z_3}\,\mathbf{A}_{\text{out}}(x) &= \mathbf{A}(x) - e_0\int d^4y\, D_{\text{adv}}(x-y)\mathbf{j}^{\text{tr}}(y) \end{aligned} \tag{16.146}$$

D_{ret} and D_{adv} are the $m \to 0$ limit of the corresponding Green's functions Δ_{ret} and Δ_{adv}, and $\sqrt{Z_3}$ is introduced so that the in and out matrix elements from the vacuum to the one-photon state are normalized to 1.

So far, all is completely analogous to the scalar theory discussion. The asymptotic condition, always in the sense of the weak operator convergence (16.20), is

$$\begin{aligned} \mathbf{A}(\mathbf{x},t) &\to \sqrt{Z_3}\,\mathbf{A}_{\text{in}}(\mathbf{x},t) \qquad \text{as } t \to -\infty \\ \mathbf{A}(\mathbf{x},t) &\to \sqrt{Z_3}\,\mathbf{A}_{\text{out}}(\mathbf{x},t) \qquad \text{as } t \to +\infty \end{aligned} \tag{16.147}$$

[1] We ignore here the infrared difficulty raised by the possibility of states of more than one photon of ~ 0 frequency. This problem is handled separately in all practical calculations, as we saw in Chap. 8, and it will be further discussed in Chap. 17.

In developing the reduction formula for photons, there is only the one minor change from the scalar field result (16.81)

$$\frac{1}{\sqrt{Z}} f_{q_i}(x_i) (\overrightarrow{\Box_{x_i} + m^2}) \langle 0| \cdots \varphi(x_i) \cdots |0\rangle$$
$$\to \frac{1}{\sqrt{Z_3}} \mathbf{A}_{k_i,\lambda_i}(x_i) \overrightarrow{\Box_{x_i}} \cdot \langle 0| \cdots \mathbf{A}(x_i) \cdots |0\rangle$$
$$= -\frac{1}{\sqrt{Z_3}} A^{\mu}_{k_i,\lambda_i}(x_i) \overrightarrow{\Box_{x_i}} \langle 0| \cdots A_\mu(x_i) \cdots |0\rangle \tag{16.148}$$

For example, if a photon $k'\lambda'$ is removed from an out-state, (16.80) is replaced by

$$\langle \gamma(k'\lambda') \text{ out}|\varphi(x)|\alpha \text{ in}\rangle = \langle \gamma \text{ out}|\varphi(x)|\alpha - (k'\lambda') \text{ in}\rangle$$
$$+ \frac{-i}{\sqrt{Z_3}} \int d^4y \, \langle \gamma \text{ out}|T(A_\mu(y)\varphi(x)|\alpha \text{ in}\rangle \overleftarrow{\Box_y} A^{\mu *}_{k',\lambda'}(y) \tag{16.149}$$

The additional minus sign in (16.149), not shared by (16.80), comes from the space-like nature of the polarization unit vector

$$\epsilon_\mu \epsilon^\mu = -\boldsymbol{\varepsilon} \cdot \boldsymbol{\varepsilon} = -1$$

Thus far the lack of explicit covariance of our quantization procedure has introduced no new problems. That the S matrix is gauge invariant and therefore by (16.70) Lorentz invariant in the presence of electromagnetic interactions will be shown, term by term, from a series expansion in powers of the strengths of the interaction currents in the following chapter. To low orders in the interaction, this has already been established by explicit calculation in the propagator approach in Chaps. 7 and 8.

Here we show only that the normalization constant Z_3 is gauge invariant, as are the rest of the terms in the defining equation (16.146).

As a preview of the general renormalization discussion, we remark that this is an important property, not shared by the Z_2 and Z of charge-bearing fermion and boson fields. Recall from Chap. 8 that we found, when all the pieces had been put together in the calculation of the vertex to order e^3, that the only cutoff-dependent constant remaining there was the Z_3 that renormalized the photon wave function and the charge according to $e = \sqrt{Z_3}\, e_0$. The Z_3 there coincides, as we shall see, with the photon Z_3 here, and the gauge invariance of the numerical value of the physical charge depends on the gauge invariance of Z_3.

16.11 Spectral Representation for Photons

We again imitate the discussion of the scalar field and turn to the vacuum expectation value of the commutator

$$iD'_{ij}(x,x')^{\text{tr}} \equiv \langle 0|[A_i(x),A_j(x')]|0\rangle \qquad (16.150)$$

We have some difficulty in establishing a spectral representation for D'^{tr}_{ij} owing to the lack of explicit Lorentz invariance of our treatment of the Maxwell theory. Consider first

$$\begin{aligned}\langle 0|A_i(x)A_j(x')|0\rangle &= \sum_n e^{-ip_n\cdot(x-x')}\langle 0|A_i(0)|n\rangle\langle n|A_j(0)|0\rangle \\ &= \int \frac{d^4q}{(2\pi)^3}\,\theta(q_0)e^{-iq\cdot(x-x')}\rho_{ij}(q) \qquad (16.151)\end{aligned}$$

which defines the spectral amplitude in the by-now-familiar way:

$$\rho_{ij}(q) \equiv \sum_n \langle 0|A_i(0)|n\rangle\langle n|A_j(0)|0\rangle(2\pi)^3\delta^4(p_n - q) \qquad (16.152)$$

Also as in (16.37) we may separate out the one-photon state

$$\langle 0|\mathbf{A}(x)|p\lambda\rangle = \sqrt{Z_3}\,\langle 0|\mathbf{A}_{\text{in}}(x)|p\lambda\rangle \qquad (16.153)$$

This gives

$$\rho_{ij}(q) = Z_3\delta(q^2)\sum_{\lambda=1}^{2}\epsilon_i(q,\lambda)\epsilon_j(q,\lambda) + \pi_{ij}(q) \qquad (16.154)$$

The second term of the commutator (16.150) can also be put in terms of this spectral amplitude if we invoke the $\mathfrak{TCP}$ invariance of electrodynamics. Under the $\theta = \mathfrak{TCP}$ transformation (15.154),

$$\theta\mathbf{A}(x)\theta^{-1} = -\mathbf{A}(-x) \qquad \text{and} \qquad \theta|0\rangle = |0\rangle$$

This gives, upon writing $\mathfrak{T} = \mathfrak{U}K$, where K denotes complex conjugation,

$$\begin{aligned}\langle 0|A_j(x')A_i(x)|0\rangle &= \langle K0|KA_j(x')A_i(x)0\rangle^* \\ &= \langle 0|\theta A_j(x')\theta^{-1}\theta A_i(x)\theta^{-1}|0\rangle^* \\ &= \langle 0|A_i(-x)A_j(-x')|0\rangle \qquad (16.155)\end{aligned}$$

In writing the last form we used the identity $\langle A|B\rangle^* = \langle B|A\rangle$ together with the hermiticity of the field amplitude. This shows that $\rho_{ij}(q)$ is symmetric in its indices and, together with (16.152), that it is a real function of q:

$$\rho_{ij}(q) = \rho_{ji}(q) = \rho_{ij}(q)^* \qquad (16.156)$$

Inserting in (16.150) and using (16.151) and (16.154), we find

$$D'_{ij}(x - x')^{\text{tr}} = Z_3 D_{ij}(x - x')^{\text{tr}} - i \int \frac{d^4q}{(2\pi)^3}\, \theta(q_0)(e^{-iq\cdot(x-x')} - e^{iq\cdot(x-x')})\pi_{ij}(q) \qquad (16.157)$$

where

$$D_{ij}(x - x')^{\text{tr}} = -i \int \frac{d^4q}{(2\pi)^3}\, \theta(q_0)\delta(q^2)(e^{-iq\cdot(x-x')} - e^{iq\cdot(x-x')}) \left(\delta_{ij} - \frac{q_i q_j}{|\mathbf{q}|^2}\right)$$

To explore the properties of $\pi_{ij}(q)$, we follow the method of Evans and Fulton,[1] who found it convenient to consider first, instead of (16.152), the gauge-invariant Lorentz tensor

$$J_{\mu\nu}(q) = \sum_n \langle 0|j_\mu(0)|n\rangle\langle n|j_\nu(0)|0\rangle(2\pi)^3\delta^4(p_n - q) \qquad (16.158)$$

The current operators are constructed from the fields that are the sources of the electromagnetic field, and, in addition to being gauge-invariant four-vectors, satisfy a differential current conservation law

$$\frac{\partial j_\mu(x)}{\partial x_\mu} = 0 \qquad (16.159)$$

For example, for a Dirac electron

$$j_\mu(x) = \bar{\psi}(x)\gamma_\mu\psi(x) - \langle 0|\bar{\psi}(x)\gamma_\mu\psi(x)|0\rangle \qquad (16.160)$$

It follows from (16.158) that $J_{\mu\nu}$ is a Lorentz tensor of rank 2 and is unchanged by the gauge transformation which must accompany a Lorentz transformation to the new frame of reference. The continuity equation leads to a further limitation on the form of $J_{\mu\nu}(q)$. From (16.159), it follows that

$$p_n{}^\mu\langle 0|j_\mu(x)|n\rangle = 0 \qquad (16.161)$$

In (16.158) this means that

$$q^\mu J_{\mu\nu} = q^\nu J_{\mu\nu} = 0 \qquad (16.162)$$

From the two requirements that $J_{\mu\nu}(q)$ transform as a second-rank tensor and have zero four-divergence, we conclude, by arguments similar to those used for the vacuum polarization tensor in (8.16), that $J_{\mu\nu}$ has the structure

$$J_{\mu\nu}(q) = \left(-g_{\mu\nu} + \frac{q_\mu q_\nu}{q^2}\right) J(q^2) \qquad (16.163)$$

[1] L. Evans and T. Fulton, *Nucl. Phys.*, **21,** 492 (1960).

It is now only necessary to evaluate $J_{\mu\nu}(q)$ in a particular Lorentz frame using radiation gauge for the potentials in order to relate $J_{\mu\nu}(q)$ to the spectral function $\rho_{ij}(q)$ of interest to us. In particular, we are interested only in the transverse parts of the current, according to the definition of $\rho_{ij}(q)$ in (16.152) and the wave equation (16.141):

$$e_0^2 \sum_n \langle 0|j_i^{\text{tr}}(0)|n\rangle\langle n|j_j^{\text{tr}}(0)|0\rangle(2\pi)^3\delta^4(p_n - q)$$
$$= (q^2)^2\rho_{ij}(q) = (q^2)^2\pi_{ij}(q) \quad (16.164)$$

where the last form follows from (16.154) and the identity $x^2\delta(x) = 0$.

The general form of $\pi_{ij}(q)$, as given in (16.164), is limited by the requirement of invariance under three-dimensional rotations:

$$\pi_{ij}(q) = \delta_{ij}\pi(q_0,|\mathbf{q}|^2) - \frac{q_iq_j}{|\mathbf{q}|^2}\tilde{\pi}(q_0,|\mathbf{q}|^2) \quad (16.165)$$

The requirement (16.141)

$$\nabla \cdot \mathbf{j}^{\text{tr}}(x) = 0$$

leads to the conditions

$$q_i\pi_{ij}(q) = q_j\pi_{ij}(q) = 0$$

Consequently,

$$\pi_{ij}(q) = \left(\delta_{ij} - \frac{q_iq_j}{|\mathbf{q}|^2}\right)\pi(q_0,|\mathbf{q}|^2) \quad (16.166)$$

In order to connect $\pi_{ij}(q)$ to $J_{\mu\nu}(q)$, we observe that the *transverse* parts of π_{ij} and $J_{\mu\nu}$ are very simply related. We compare the two quantities

$$\epsilon_i\epsilon_j\pi_{ij}(q) \qquad \text{and} \qquad \epsilon_\mu\epsilon_\nu J^{\mu\nu}(q)$$

where

$$\epsilon_\mu = (0,\boldsymbol{\varepsilon}) \qquad \boldsymbol{\varepsilon} \cdot \mathbf{q} = -\epsilon_\mu q^\mu = 0 \quad (16.167)$$

in the special quantization frame. Since, with the aid of (16.141),

$$e_0\langle 0|\mathbf{j}^{\text{tr}}(x)|n\rangle = e_0\langle 0|\mathbf{j}(x)|n\rangle - \langle 0|\nabla\dot{\Phi}(x)|n\rangle$$

we conclude that

$$\epsilon_i\langle 0|j_i^{\text{tr}}(0)|n\rangle = \epsilon_i\langle 0|j_i(0)|n\rangle \quad (16.168)$$

for any state $|n\rangle$ such that

$$\boldsymbol{\varepsilon} \cdot \mathbf{p}_n = 0$$

According to (16.167) and the momentum δ function in (16.164), we are interested in matrix elements satisfying this condition. Therefore, with (16.158), (16.164), and (16.168) we now find the relation of $\pi_{ij}(q)$

to $J_{\mu\nu}(q)$

$$\begin{aligned}\epsilon_i\epsilon_j\pi_{ij}(q) &= \frac{e_0^2}{q^4}\sum_n \langle 0|\boldsymbol{\varepsilon}\cdot\mathbf{j}^{\rm tr}(0)|n\rangle\langle n|\boldsymbol{\varepsilon}\cdot\mathbf{j}^{\rm tr}(0)|0\rangle(2\pi)^3\delta^4(q-p_n)\\ &= \frac{e_0^2}{q^4}\sum_n \langle 0|\boldsymbol{\varepsilon}\cdot\mathbf{j}(0)|n\rangle\langle n|\boldsymbol{\varepsilon}\cdot\mathbf{j}(0)|0\rangle(2\pi)^3\delta^4(q-p_n)\\ &= \frac{e_0^2}{q^4}\,\epsilon_\mu\epsilon_\nu J^{\mu\nu}(q)\end{aligned} \tag{16.169}$$

Using the general forms (16.163) and (16.166), we conclude that

$$\begin{aligned}\pi_{ij}(q) &= \left(\delta_{ij}-\frac{q_iq_j}{|\mathbf{q}|^2}\right)\pi(q_0,|\mathbf{q}|^2)\\ &= e_0^2\left(\delta_{ij}-\frac{q_iq_j}{|\mathbf{q}|^2}\right)\frac{J(q^2)}{q^4}\end{aligned} \tag{16.170}$$

With (16.170) we have succeeded in expressing $\pi_{ij}(q)$ in terms of a scalar gauge-invariant amplitude $J(q^2)$. Inserting this into (16.157), imitating an earlier step in (16.32) by writing

$$J(q^2) = \int dM^2\,\delta(q^2-M^2)J(M^2)$$

and introducing the notation

$$\begin{aligned}\Delta_{ij}(x-x',M^2)^{\rm tr} = -i\int\frac{d^4k}{(2\pi)^3}\,&\theta(k_0)\delta(k^2-M^2)\\ &\times\left(\delta_{ij}-\frac{k_ik_j}{|\mathbf{k}|^2}\right)(e^{-ik\cdot(x-x')}-e^{ik\cdot(x-x')})\end{aligned} \tag{16.171}$$

for extension to arbitrary mass M of the transverse $D_{ij}{}^{\rm tr}$ function of (16.157), we arrive at a compact form for the spectral representation:

$$D'_{ij}(x-x')^{\rm tr} = Z_3D_{ij}(x-x')^{\rm tr} + \int dM^2\,\Delta_{ij}(x-x',M^2)^{\rm tr}\Pi(M^2) \tag{16.172}$$

with

$$\Pi(M^2) = \frac{e_0^2J(M^2)}{M^4} \tag{16.173}$$

In momentum space the Feynman propagator is

$$D'_{F_{ij}}(q)^{\rm tr} = \left(\frac{Z_3}{q^2+i\epsilon}+\int_0^\infty\frac{dM^2\,\Pi(M^2)}{q^2-M^2+i\epsilon}\right)\left\{\delta_{ij}-\frac{q_iq_j}{|\mathbf{q}|^2}\right\}$$

To express Z_3 in terms of an integral over the spectral weight $\Pi(M^2)$ as done in (16.42) for the scalar field, we take the time derivative of (16.171) at $t = t'$ and use the commutators (16.142) and the definition

(16.150) to find

$$1 = Z_3 + \int_{M_1^2}^{\infty} dM^2\, \Pi(M^2) \qquad (16.174)$$

The gauge and Lorentz invariance of Z_3 is now obvious. The weight function $\Pi(M^2)$ is always nonnegative, as is clear from (16.169) and (16.170):

$$0 \leq \epsilon_i \epsilon_j \pi_{ij}(q) = \frac{e_0^2 J(q^2)}{q^4} = \pi(q^2)$$

Therefore

$$0 \leq Z_3 = 1 - \int_{M_1^2}^{\infty} dM^2\, \Pi(M^2) < 1 \qquad (16.175)$$

and the probability to produce one photon from the vacuum with $\mathbf{A}(x)$ is bounded between 0 and 1 according to (16.175), in analogy with the results found in (16.34) and (16.125) for spin-0 and -½ fields in the absence of electromagnetic couplings.[1]

16.12 Connection between Spin and Statistics[2]

On the basis of the spectral representations established in this chapter we can discuss in some detail the connection between spin and statistics for spin-0 and -½ bosons or fermions which was mentioned in Chap. 15. It has been proved for a local field theory such as we have been discussing, which satisfies the requirements of Lorentz covariance and has a unique ground state, that integer spin fields must be quantized as Bose fields and half-integer ones as Fermi fields if the condition of microscopic causality is to be satisfied. According to the condition of microscopic causality, local densities $\mathcal{O}(x)$ of observable operator quantities

$$\mathcal{O} \equiv \int d^3x\, \mathcal{O}(\mathbf{x},t)$$

do not interfere and therefore are required to commute for space-like separations; that is

$$[\mathcal{O}(x),\mathcal{O}(y)] = 0 \qquad \text{for } (x-y)^2 < 0 \qquad (16.176)$$

Here we shall show that this requirement is incompatible with quantization of spin-0 Klein-Gordon fields with anticommutators and of spin-½ Dirac fields with commutators.

The observables in these theories, such as charge or current densities, are generally constructed from quadratic forms in the field amplitudes. It is not difficult to perform the algebraic manipulations show-

[1] For implications when $Z_3 = 0$ see J. Schwinger, *Phys. Rev.*, **125**, 397 (1962).
[2] See R. Streator and A. Wightman, *op. cit.*

ing that for the bilinear form

$$\mathcal{O}(x) \equiv \varphi_a(x)\,\varphi_b(x)$$

(16.176) is valid provided that the field amplitudes commute or anticommute for all space-like separations. More precisely, the general condition for validity of (16.176) is that

$$[\varphi_r(x),\varphi_s(y)] = 0 \qquad (x-y)^2 < 0 \tag{16.177a}$$

or

$$\{\varphi_r(x),\varphi_s(y)\} = 0 \qquad (x-y)^2 < 0 \tag{16.177b}$$

Here $\varphi_r(x)$ and $\varphi_s(y)$ denote combinations of φ and/or φ^* for Klein-Gordon fields or different spinor components of ψ and $\bar{\psi}$ for Dirac fields.

For the Klein-Gordon field, (16.177*a*) is satisfied in the presence of interactions on grounds of Lorentz invariance and the canonical commutation relations for equal times. If on the other hand we attempt to quantize the Klein-Gordon field in terms of anticommutators as Fermi fields, we arrive at a contradiction with (16.177) when we consider the vacuum expectation value

$$\langle 0|\{\varphi_r(x),\varphi_s(y)\}|0\rangle \equiv \Delta_1'(x-y) \tag{16.178}$$

and use the same invariance arguments as applied to (16.23) and leading to (16.32). The only difference is that the minus sign in (16.26) becomes a plus sign so that

$$\Delta_1'(x-y) = Z\Delta_1(x-y;m^2) + \int_{m_1^2} d\sigma^2\,\rho(\sigma^2)\Delta_1(x-y;\sigma^2)$$

with

$$\Delta_1(x-y) \equiv \int \frac{d^3k}{(2\pi)^3 2\omega_k}\,(e^{-ik\cdot(x-y)} + e^{ik\cdot(x-y)})$$

$\Delta_1(x-y)$ is the symmetric counterpart of $\Delta(x-y)$, obeying the Klein-Gordon equation, but *not* vanishing for space-like intervals $(x-y)^2 < 0$. Indeed, for large separations such that $-(x-y)^2 > \dfrac{1}{m^2}$

$$\Delta_1(\mathbf{x},t,m^2) \sim \frac{\exp\left(-m\sqrt{|\mathbf{x}|^2 - t^2}\right)}{|\mathbf{x}|^2 - t^2}$$

and therefore for large $\mathbf{x}$

$$\Delta_1'(\mathbf{x},0) \sim \frac{Ze^{-m|\mathbf{x}|}}{|\mathbf{x}|^2} + \int_{m_1^2}^{\infty} d\sigma^2\rho(\sigma^2)\,\frac{e^{-\sigma|\mathbf{x}|}}{|\mathbf{x}|^2}$$

Therefore, (16.177*b*) cannot be valid and the requirement of microscopic causality is violated. This establishes the connection between spin and statistics.

If one attempts to quantize a Fermi field with commutators, the same phenomenon occurs. The change in relative sign between the two terms corresponding to the two different orderings of the operators changes the Δ function in the spectral representation (16.111) to a Δ_1 function, and again a contradiction with microscopic causality arises. Since fields of spin higher than $\frac{1}{2}$ can be considered as being built up of products of spin-$\frac{1}{2}$ fields, this establishes the connection in general.

It is worthwhile observing that, if one quantizes, say, a Bose field with anticommutators, the violation of microscopic causality is sizable only at distances comparable to the Compton wavelength of the particle involved, generally $\sim 10^{-13}$ cm. The experimental agreement with this extremely significant prediction of local field theory then strongly supports the general notions of local field theory, at least at distances comparable to the Compton wavelengths of the elementary particles.

Problems

1. By considering the expression

$$\int d^4x \int d^4y\, f_\alpha^*(x) f_\beta^*(y)(\Box_x + m^2)(\Box_y + m^2) T(\varphi(x)\varphi(y))$$

show upon interchange of order of integration that

$$[\varphi_{\rm in}(x),\varphi_{\rm in}(y)] = [\varphi_{\rm out}(x),\varphi_{\rm out}(y)]$$

Then, by considering the same integral operator on a trilinear product of fields $\varphi(x)\varphi(y)\varphi(z)$, show that

$$[[\varphi_{\rm in}(x),\varphi_{\rm in}(y)],\varphi(z)] = 0$$

and thus establish (16.56). W. Zimmermann, *Nuovo Cimento*, **10,** 567 (1958), has justified the interchange of integration order. Establish in a similar way (16.121).

2. Show that the S matrix (6.56) may be expressed directly in terms of the Feynman propagator S_F', analogously to (16.82).

3. Calculate the mass shift δm^2 of a scalar field in terms of the spectral weight function $\rho(p^2)$ in (16.27). Assume everything is finite.

4. Show that $S = \theta_{\rm in}^{-1}\theta$, with $\theta = \mathcal{H}\mathcal{C}\mathcal{P}$.

5. Show that $\langle 0|[\varphi_{\rm in}(x),\varphi_{\rm out}(y)]|0\rangle = +i\Delta(x - y, m)$.

6. Establish properties 1 to 5 of the S matrix, Sec. 16.6, when spinor and nonhermitian scalar fields are present.

7. Prove property (iii), Eq. (16.113).

8. Construct the spectral representation of an electron in quantum electrodynamics and discuss properties of the weight functions that appear.

9. Establish (16.115). *Hint:* see Sec. 3.1.

10. Establish the spin-statistics connection for fermions without assuming parity conservation [as is implied in (16.111)].

17.1 Introduction

At present there are two general approaches known for computing transition amplitudes and matrix elements of physical interest in order to bring relativistic quantum field theory to grips with laboratory observations. The first approach is to make a systematic expansion in powers of the coupling parameters which measure the strengths of the interactions. In this way the interacting fields $\varphi(x)$ are developed in terms of the known in-fields $\varphi_{\text{in}}(x)$ which satisfy free wave equations and commutation relations. Such an expansion of the S matrix will lead right back to the Feynman graphs and rules of calculation—as well as the divergent integrals—which emerged from our propagator considerations in Chaps. 7, 8, and 9. The second approach is to extend the technique discussed in the preceding chapter for constructing vacuum expectation values of the field commutators to vacuum expectation values of three and more fields. Approximation schemes based on analytic function theory which are not limited to weak-coupling strengths as in the first, or perturbation, approach may be developed. This development will be discussed in the following chapter; here we turn to the formulation of covariant perturbation theory.

Our task, then, is to expand transition amplitudes and matrix elements in a power series in the interaction strength and obtain rules for computing the terms in the expansion. In doing this we wish to express the interacting fields $\varphi(x)$ in terms of the asymptotic fields $\varphi_{\text{in}}(x)$ which possess known properties. Fundamental to perturbation theory is the assumption that the spectrum of exact states is in one-to-one correspondence with the "unperturbed" states, in this case the asymptotic in- and out-states. In particular, we assume that for each field $\varphi(x)$ in the lagrangian there is an in-field $\varphi_{\text{in}}(x)$. The analogous requirement exists in nonrelativistic potential theory; for a perturbation theory of the scattering process to converge, there must exist no bound states in the potential.

It is very doubtful that perturbation theory converges in the realm of the strong interaction processes involving nuclear particles. We may be optimistic and assume it is applicable to electrodynamic processes for which the expansion parameter is $\alpha = 1/137$. Indeed, we have already found in Chaps. 7 and 8 that perturbation methods lead to impressive quantitative agreement with measurements of electrodynamic processes. The applicability of perturbation theory to weak interactions remains at this time an open question.

The integral form of the field equations, as in (16.14), (16.84), and (16.146), may serve as the basis for the expansion of the fields in terms of in-fields, and a satisfactory perturbation theory may be constructed. What is necessary is to expand the current operator $j(x)$ on the right-hand side in an in-field series. However, this procedure does not lead directly to Feynman graphs,[1] owing to the presence of $\Delta_{\text{ret}}(x)$ in the expansion instead of $\Delta_F(x)$. Motivated by the wish to reconstruct Feynman's propagator theory from a field-theoretic starting point, Dyson[2] constructed a perturbation theory based upon the U matrix, to which we now turn.

17.2 The U Matrix

The equal-time commutation relations for the interacting fields $\varphi(\mathbf{x},t)$ and conjugate momenta $\pi(\mathbf{x},t)$ are identical[3] with those satisfied by the in-fields[4] $\varphi_{\text{in}}(\mathbf{x},t)$ and $\pi_{\text{in}}(\mathbf{x},t)$. Moreover, these fields form complete operator sets since, according to our assumptions, a complete set of states can be formed from the vacuum by repeated application of φ or of φ_{in}. Therefore, because the fields φ and φ_{in} are in one-to-one correspondence according to our assumption of the applicability of perturbation theory, they may be related[5] by a unitary transformation[6] $U(t)$:

$$\varphi(\mathbf{x},t) = U^{-1}(t)\varphi_{\text{in}}(\mathbf{x},t)U(t) \tag{17.1}$$

$$\pi(\mathbf{x},t) = U^{-1}(t)\pi_{\text{in}}(\mathbf{x},t)U(t) \tag{17.2}$$

The dynamics of the operator $U(t)$ can now be found, since equations of motion have already been given for both $\varphi(x)$ and $\varphi_{\text{in}}(x)$.

[1] F. J. Dyson, *Phys. Rev.*, **82**, 428 (1951).

[2] F. J. Dyson, *Phys. Rev.*, **75**, 486, 1736 (1949).

[3] We exclude theories with derivative couplings from consideration here. The electromagnetic interactions of charged spin-0 and spin-1 particles are examples of theories with derivative couplings. For discussions of problems they raise for developing perturbation expansions see P. T. Matthews, *Phys. Rev.*, **76**, 684, 1489 (1949), F. Rohrlich, *Phys. Rev*, **80**, 666 (1950), and T. D. Lee and C. N. Yang, *Phys. Rev.*, **128**, 885 (1962).

[4] By $\varphi(x)$ we mean a generic name for any boson or fermion field. Specific properties of these fields such as spin and isospin will not play a crucial role in the formalism.

[5] Although this is a theorem in ordinary quantum mechanics, the proof breaks down for systems with a nondenumerable number of degrees of freedom. See R. Haag, *Kgl. Danske Videnskab. Selskab. Mat.-Fys. Medd.*, **29**(12) (1955), and Prob. 2. Here we *assume* the existence of $U(t)$.

[6] U may be covariantly defined on a general space-like surface instead of at constant t, but to no great advantage. For the general formalism see J. Schwinger, *Phys. Rev.*, **74**, 1439 (1948), **75**, 651 (1949), and **76**, 790 (1949); S. Tomonoga, *Progr. Theoret. Phys.* (*Kyoto*), **1**, 27 (1946).

In particular, it follows from the defining relations for the in-fields, which satisfy free-field commutation relations and field equations, that

$$\begin{aligned} \frac{\partial \varphi_{\text{in}}(x)}{\partial t} &= i[H_{\text{in}}(\varphi_{\text{in}},\pi_{\text{in}}),\varphi_{\text{in}}] \\ \frac{\partial \pi_{\text{in}}(x)}{\partial t} &= i[H_{\text{in}}(\varphi_{\text{in}},\pi_{\text{in}}),\pi_{\text{in}}] \end{aligned} \tag{17.3}$$

where $H_{\text{in}}(\varphi_{\text{in}},\pi_{\text{in}})$ is the hamiltonian for a free field described in terms of the physical mass m. In addition, for the exact Heisenberg fields, displacement invariance ensures that

$$\frac{\partial \varphi(x)}{\partial t} = i[H(\varphi,\pi),\varphi(x)] \tag{17.4}$$

which also implies

$$\frac{\partial \pi(x)}{\partial t} = i[H(\varphi,\pi),\pi(x)] \tag{17.5}$$

Therefore, we find

$$\begin{aligned} \dot{\varphi}_{\text{in}}(x) &= \frac{\partial}{\partial t} U(t)\varphi(\mathbf{x},t)U^{-1}(t) \\ &= [\dot{U}(t)U^{-1}(t),\varphi_{\text{in}}(x)] + i[H(\varphi_{\text{in}},\pi_{\text{in}}),\varphi_{\text{in}}(x)] \\ &= \dot{\varphi}_{\text{in}}(x) + [\dot{U}U^{-1} + iH_I(\varphi_{\text{in}},\pi_{\text{in}}),\varphi_{\text{in}}(x)] \end{aligned} \tag{17.6}$$

and

$$\dot{\pi}_{\text{in}}(x) = \dot{\pi}_{\text{in}}(x) + [\dot{U}U^{-1} + iH_I(\varphi_{\text{in}},\pi_{\text{in}}),\pi_{\text{in}}(x)]$$

where

$$H_I(\varphi_{\text{in}},\pi_{\text{in}}) = H(\varphi_{\text{in}},\pi_{\text{in}}) - H_{\text{in}}(\varphi_{\text{in}},\pi_{\text{in}}) \equiv H_I(t) \tag{17.7}$$

is the interaction term written in terms of in-fields and with its explicit time dependence displayed in the last form, which appears frequently in what follows. $H_I(t)$ contains the mass counterterm which was introduced in Chap. 7.[1] From (17.6) it follows that

$$i\dot{U}(t)U^{-1}(t) = H_I(t) + E_0(t)$$

[1] For the specific model of the self-coupled scalar field with the lagrangian

$$\mathcal{L}(\varphi) = \frac{1}{2}:\left[\frac{\partial\varphi}{\partial x_\mu}\frac{\partial\varphi}{\partial x^\mu} - \mu_0{}^2\varphi^2 + \frac{1}{2}\lambda_0\varphi^4\right]:$$

we find

$$\mathcal{L}_{\text{in}}(\varphi_{\text{in}}) = \frac{1}{2}:\left[\frac{\partial\varphi_{\text{in}}}{\partial x_\mu}\frac{\partial\varphi_{\text{in}}}{\partial x^\mu} - \mu^2\varphi_{\text{in}}^2\right]:$$

and

$$\mathcal{L}_I(\varphi_{\text{in}}) = -\mathcal{H}_I(\varphi_{\text{in}}) = \frac{1}{2}:\left[\frac{1}{2}\lambda_0\varphi_{\text{in}}^4 + (\mu^2 - \mu_0{}^2)\varphi_{\text{in}}^2\right]:$$

For fermions the mass counterterm in $\mathcal{H}_I(\varphi_{\text{in}})$ takes the form

$$(m_0 - m):\bar{\psi}_{\text{in}}\psi_{\text{in}}:$$

where $E_0(t)$ commutes with $\varphi_{\text{in}}(\mathbf{x},t)$ and $\pi_{\text{in}}(\mathbf{x},t)$ and is therefore a c-number function of time.

Defining

$$H_I'(t) = H_I(t) + E_0(t)$$

we construct U by solving

$$i\,\frac{\partial U(t)}{\partial t} = H_I'(t)U(t) \tag{17.8}$$

A solution for U in terms of the in-fields serves as the basis for the perturbation expansion. With the aid of U we can write the vacuum expectation values of products of field operators—to which all S-matrix elements have been reduced—as an infinite series of products of in-fields, which can be computed from their known free-field properties.

We continue by integrating (17.8). An initial condition is required, and this can be conveniently applied by turning our attention to the operator

$$U(t,t') \equiv U(t)U^{-1}(t') \tag{17.9}$$

which is also a solution of (17.8) and reduces to the unit operator[1] at $t = t'$:

$$\begin{aligned} i\,\frac{\partial U(t,t')}{\partial t} &= H_I'(t)U(t,t') \\ U(t,t) &= 1 \end{aligned} \tag{17.10}$$

Equation (17.10) is reminiscent of that encountered in the Dirac form of time-dependent perturbation theory in nonrelativistic quantum mechanics and has the same form of solution. The integral form of (17.10) is

$$U(t,t') = 1 - i\int_{t'}^{t} dt_1\, H_I'(t_1)U(t_1,t')$$

and has the expansion

$$\begin{aligned} U(t,t') = 1 - i\int_{t'}^{t} dt_1\, H_I'(t_1) + (-i)^2 \int_{t'}^{t} dt_1\, H_I'(t_1) \int_{t'}^{t_1} dt_2\, H_I'(t_2) \\ + \cdots + (-i)^n \int_{t'}^{t} dt_1 \int_{t'}^{t_1} dt_2 \cdots \int_{t'}^{t_{n-1}} dt_n\, H_I'(t_1) \cdots H_I'(t_n) \\ + \cdots \end{aligned} \tag{17.11}$$

Each product of interaction terms is in time-ordered form, since $t_1 \geq t_2 \cdots \geq t_n$, and we may therefore write $T(H_I'(t_1) \cdots H_I'(t_n))$ for

[1] We shall often use $H_I(t)$ as shorthand for $H_I(\varphi_{\text{in}}(\mathbf{x},t))$.

$H_I'(t_1) \cdots H_I'(t_n)$ in (17.11) without changing anything,

$$U(t,t') = 1 + \sum_{n=1}^{\infty} (-i)^n \int_{t'}^{t} dt_1 \int_{t'}^{t_1} dt_2 \cdots \int_{t'}^{t_{n-1}} dt_n \times T(H_I'(t_1) \cdots H_I'(t_n)) \qquad (17.12)$$

The time-ordered expression is symmetric upon interchange of any of the coordinates $t_1, \ldots, t_n$, since each interchange in the order of appearance of the $H_I'(t_i)$ involves exchanging the positions of an even number of fermion fields; hence there are an even number of minus signs. We make use of this symmetry to symmetrize the integration interval with respect to the n indices. For $n = 2$ we recognize

$$\int_{t'}^{t} dt_1 \int_{t'}^{t_1} dt_2\, T(H_I'(t_1)H_I'(t_2)) = \int_{t'}^{t} dt_2 \int_{t'}^{t_2} dt_1\, T(H_I'(t_1)H_I'(t_2))$$
$$= \frac{1}{2} \int_{t'}^{t} dt_1 \int_{t'}^{t} dt_2\, T(H_I'(t_1)H_I'(t_2))$$

For general n we can similarly carry out the $n!$ permutations of the n indices and extend the integration region over the n-dimensional cube from t' to t. Each of the $n!$ permuted time-ordered regions contributes equally, and we therefore write

$$U(t,t') = 1 + \sum_{n=1}^{\infty} \frac{(-i)^n}{n!} \int_{t'}^{t} dt_1 \cdots \int_{t'}^{t} dt_n\, T(H_I'(t_1) \cdots H_I'(t_n))$$
$$\equiv T\left(\exp\left[-i \int_{t'}^{t} H_I'(t)\, dt\right]\right) = T\left(\exp\left[-i \int_{t'}^{t} d^4x\, \mathcal{H}_I(\varphi_{\text{in}}(x))\right]\right) \qquad (17.13)$$

where the exponential form is defined by (17.13) as the symbolic summary of the time-ordered series with which it coincides when expanded in a power series in the interaction.

A useful multiplicative rule for the U operators is

$$U(t,t') = U(t,t'')U(t'',t') \qquad (17.14)$$

as may be seen either from the definition (17.9) or from (17.13). A special case of (17.14) is

$$U(t,t') = U^{-1}(t',t) \qquad (17.15)$$

17.3 Perturbation Expansion of Tau Functions and the S Matrix

We are now in a position to use (17.1) and (17.13) to express S-matrix elements in terms of vacuum expectation values of in-fields, which can be computed from their known free-field properties. With the reduction technique of the preceding chapter the calculation of S-matrix elements has been reduced to one basic ingredient—the vacuum expectation values of time-ordered products of Heisenberg fields $\varphi(x)$

$$\tau(x_1, \ldots, x_n) = \langle 0|T(\varphi(x_1) \cdots \varphi(x_n))|0\rangle \tag{17.16}$$

Expressing these in terms of the in-fields, using (17.1) and (17.9), we find

$$\begin{aligned}\tau(x_1, \ldots, x_n) &= \langle 0|T(U^{-1}(t_1)\varphi_{\text{in}}(x_1)U(t_1,t_2)\varphi_{\text{in}}(x_2)U(t_2,t_3) \cdots \\ &\qquad U(t_{n-1},t_n)\varphi_{\text{in}}(x_n)U(t_n))|0\rangle \\ &= \langle 0|T(U^{-1}(t)U(t,t_1)\varphi_{\text{in}}(x_1)U(t_1,t_2) \cdots \\ &\qquad U(t_{n-1},t_n)\varphi_{\text{in}}(x_n)U(t_n,-t)U(-t))|0\rangle\end{aligned}$$

where t is a reference time which we shall allow to approach ∞; in this limit t is later than all the t_i and $-t$ is earlier. We may then extract $U^{-1}(t)$ and $U(t)$ from the time-ordered product and write in the symbolic shorthand of (17.13)

$$\begin{aligned}&\tau(x_1, \ldots, x_n) \\ &\quad = \langle 0|U^{-1}(t)T\left(\varphi_{\text{in}}(x_1) \cdots \varphi_{\text{in}}(x_n) \exp\left[-i\int_{-t}^{t} H_I'(t')\,dt\right]\right) U(-t)|0\rangle\end{aligned} \tag{17.17}$$

Equation (17.17) expresses the τ functions, and therefore the S matrix, in terms of the in-fields except for the operators $U^{-1}(t)$ and $U(-t)$, which we now remove by showing that the vacuum $|0\rangle$ is the eigenstate of these operators in the limit $t \to \infty$. To accomplish this, we consider an arbitrary in-state $|\alpha p \text{ in}\rangle$ containing a particle p together with anything else α. We then write, using (16.9) and (16.73) for the case that p is a Klein-Gordon particle,

$$\begin{aligned}&\langle p\alpha \text{ in}|U(-t)|0\rangle = \langle \alpha \text{ in}|a_{\text{in}}(p)U(-t)|0\rangle \\ &\quad = -i\int d^3x\, f_p^*(\mathbf{x},-t')\left(\frac{\overrightarrow{\partial}}{\partial t'} - \frac{\overleftarrow{\partial}}{\partial t'}\right)\langle \alpha \text{ in}|\varphi_{\text{in}}(\mathbf{x},-t')U(-t)|0\rangle\end{aligned} \tag{17.18}$$

A similar form results from (16.126) and (16.145) for fermions and photons. Now by using (17.1), (17.18) becomes

$$\begin{aligned}&\langle p\alpha \text{ in}|U(-t)|0\rangle \\ &\quad = -i\int d^3x\, f_p^*(\mathbf{x},-t')\overleftrightarrow{\partial_0'}\langle \alpha \text{ in}|U(-t')\varphi(\mathbf{x},-t')U^{-1}(-t')U(-t)|0\rangle\end{aligned}$$

where
$$\overleftrightarrow{\partial_0'} \equiv \frac{\overrightarrow{\partial}}{\partial t'} - \frac{\overleftarrow{\partial}}{\partial t'}$$

as defined earlier. This approaches, as $t = t' \to \infty$,

$$\sqrt{Z_3}\,\langle\alpha\ \text{in}|U(-t)a_{\text{in}}(p)|0\rangle + i\int d^3x\, f_p^*(\mathbf{x},-t) \times \langle\alpha\ \text{in}|\dot U(-t)\varphi(\mathbf{x},-t) + U(-t)\varphi(\mathbf{x},-t)\dot U^{-1}(-t)U(-t)|0\rangle \quad (17.19)$$

according to the asymptotic condition (16.20). Evidently

$$a_{\text{in}}(p)|0\rangle = 0$$

there being no incoming particles in the vacuum state, and the first term of (17.19) vanishes. A short calculation verifies that the second term also vanishes:

$$\dot U\varphi + U\varphi\dot U^{-1}U = \dot U U^{-1}\varphi_{\text{in}}U + \varphi_{\text{in}}U\dot U^{-1}U = [\dot U U^{-1},\varphi_{\text{in}}]U = -i[H_I,\varphi_{\text{in}}]U = 0$$

Here we have used (17.1), (17.8), and our assumption of no derivative couplings in H_I; the common time argument $-t$ was suppressed.

We therefore conclude that

$$\langle\alpha p\ \text{in}|U(-t)|0\rangle \to 0 \qquad \text{as } t \to \infty$$

for all in-states αp containing a particle. From this it follows that

$$U(-t)|0\rangle = \lambda_-|0\rangle \qquad \text{when } t \to \infty \tag{17.20}$$

In a similar way we can show that

$$U(t)|0\rangle = \lambda_+|0\rangle \qquad \text{when } t \to \infty$$

The constants λ_- and λ_+ appear in (17.17) in the form, for $t \to \infty$,

$$\begin{aligned}\lambda_-\lambda_+^* &= \langle 0|U^{-1}(t)|0\rangle\langle 0|U(-t)|0\rangle \\ &= \langle 0|U(-t)U^{-1}(t)|0\rangle = \langle 0|U(-t,t)|0\rangle \\ &= \langle 0|T\left(\exp\left[i\int_{-t}^{t} dt'\, H_I'(t')\right]\right)|0\rangle \\ &= \langle 0|T\left(\exp\left[-i\int_{-t}^{t} dt'\, H_I'(t')\right]\right)|0\rangle^{-1}\end{aligned}$$

The τ function (17.17) may thus be rewritten as

$$\tau(x_1, \ldots, x_n) = \frac{\langle 0|T\left(\varphi_{\text{in}}(x_1)\cdots\varphi_{\text{in}}(x_n)\exp\left[-i\int_{-t}^{t} dt'\, H_I'(t')\right]\right)|0\rangle}{\langle 0|T\left(\exp\left[-i\int_{-t}^{t} dt'\, H_I'(t')\right]\right)|0\rangle} \tag{17.21}$$

for $t \to \infty$. Finally, let us cancel the c-number factor defined above (17.8)

$$\exp\left[-i\int_{-t}^{t} dt'\, E_0(t')\right]$$

between numerator and denominator, replacing $H'_I(t)$ by $H_I(t)$, and take the limit $t \to \infty$:

$$\tau(x_1, \ldots, x_n) = \frac{\langle 0|T\left(\varphi_{\text{in}}(x_1) \cdots \varphi_{\text{in}}(x_n) \exp\left[-i\int_{-\infty}^{\infty} dt\, H_I(t)\right]\right)|0\rangle}{\langle 0|T\left(\exp\left[-i\int_{-\infty}^{\infty} dt\, H_I(t)\right]\right)|0\rangle}$$

$$= \frac{\sum_{m=0}^{\infty} \frac{(-i)^m}{m!} \int_{-\infty}^{\infty} d^4y_1 \cdots d^4y_m \, \langle 0|T(\varphi_{\text{in}}(x_1) \cdots \varphi_{\text{in}}(x_n)\mathcal{H}_I(\varphi_{\text{in}}(y_1)) \cdots \mathcal{H}_I(\varphi_{\text{in}}(y_m)))|0\rangle}{\sum_{m=0}^{\infty} \frac{(-i)^m}{m!} \int_{-\infty}^{\infty} d^4y_1 \cdots d^4y_m \langle 0|T(\mathcal{H}_I(\varphi_{\text{in}}(y_1)) \cdots \mathcal{H}_I(\varphi_{\text{in}}(y_m)))|0\rangle} \tag{17.22}$$

Equation (17.22) is the fundamental result. The S matrix in (16.81) or (16.139), for example, is first expressed in terms of τ functions and then by (17.22) as an expansion completely in terms of in-field operators. It is essentially an algebraic problem to proceed now to the Feynman graphs and rules of calculation.

17.4 Wick's Theorem

To reduce a term in (17.22) to integrals we can calculate, we shall attempt to move the destruction operators step by step to the right and the creation operators to the left; they will then vanish when standing next to the vacuum state. This program of normal-ordering a time-ordered product, which gives rise to the Feynman amplitude, was first developed in 1949 by Dyson[1] and later extended by Wick,[2] who stated and proved the following theorem:

$$\begin{aligned}
T(\varphi_{\text{in}}(x_1) \cdots \varphi_{\text{in}}(x_n)) &= :\varphi_{\text{in}}(x_1) \cdots \varphi_{\text{in}}(x_n): \\
&+ [\langle 0|T(\varphi_{\text{in}}(x_1)\varphi_{\text{in}}(x_2))|0\rangle :\varphi_{\text{in}}(x_3) \cdots \varphi_{\text{in}}(x_n): + \text{permutations}] \\
&+ [\langle 0|T(\varphi_{\text{in}}(x_1)\varphi_{\text{in}}(x_2))|0\rangle\langle 0|T(\varphi_{\text{in}}(x_3)\varphi_{\text{in}}(x_4))|0\rangle :\varphi_{\text{in}}(x_5) \cdots \varphi_{\text{in}}(x_n): \\
&\qquad + \text{permutations}] \\
&+ \cdots \\
&+ \begin{cases} [\langle 0|T(\varphi_{\text{in}}(x_1)\varphi_{\text{in}}(x_2))|0\rangle \cdots \langle 0|T(\varphi_{\text{in}}(x_{n-1})\varphi_{\text{in}}(x_n))|0\rangle \\ \qquad + \text{permutations } (n \text{ even})] \\ [\langle 0|T(\varphi_{\text{in}}(x_1)\varphi_{\text{in}}(x_2))|0\rangle \cdots \langle 0|T(\varphi_{\text{in}}(x_{n-2})\varphi_{\text{in}}(x_{n-1}))|0\rangle\varphi_{\text{in}}(x_n) \\ \qquad + \text{permutations } (n \text{ odd})] \end{cases}
\end{aligned} \tag{17.23}$$

[1] F. J. Dyson, *op. cit.*
[2] G. C. Wick, *Phys. Rev.*, **80**, 268 (1950).

The vacuum expectation values, or *contractions*, arise from (anti) commuting the fields to arrange them in the normal order, and they are the field-theoretic expressions for the Feynman propagators, as we have seen in the discussions of free fields. Recall that normal products of operators

$$:\varphi_{\text{in}}(x_1) \cdots \varphi_{\text{in}}(x_n):$$

are formed by decomposing each operator into its positive-frequency and negative-frequency parts:

$$\varphi_{\text{in}}(x) = \varphi_{\text{in}}{}^{(+)}(x) + \varphi_{\text{in}}{}^{(-)}(x) \tag{17.24}$$

where $\varphi_{\text{in}}{}^{(+)}(x)$ contains the destruction and $\varphi_{\text{in}}{}^{(-)}(x)$ the creation operator. All creation operators are then placed to the left of all destruction operators, with a minus sign introduced for each permutation of Fermi fields required to arrive at the normal ordering. To be specific, we may write

$$:\varphi_{\text{in}}(x_1) \cdots \varphi_{\text{in}}(x_n): = \sum_{A,B} \delta_p \prod_{i \text{ in } A} \varphi_{\text{in}}{}^{(-)}(x_i) \prod_{j \text{ in } B} \varphi_{\text{in}}{}^{(+)}(x_j) \tag{17.25}$$

where the sum is over all sets A and B of the n indices, each index appearing once, and δ_p denotes the sign of the permutation of the Fermi fields. The vacuum expectation value of a normal product always vanishes because the destruction operators act to the right on the vacuum and the creation operators act to the left:

$$\varphi_{\text{in}}{}^{(+)}|0\rangle = \langle 0|\varphi_{\text{in}}{}^{(-)} = 0 \tag{17.26}$$

The utility of Dyson's program of reducing time-ordered to normal-ordered products is based on this property, since by (17.22) it is only necessary to compute vacuum expectation values in evaluating τ functions and S-matrix elements. It follows from (17.23) that

1. If n is odd, $\langle 0|T(\varphi_{\text{in}}(x_1) \cdots \varphi_{\text{in}}(x_n))|0\rangle = 0$
2. If n is even,

$$\langle 0|T(\varphi_{\text{in}}(x_1) \cdots \varphi_{\text{in}}(x_n))|0\rangle = \sum_{\text{permutations}} \delta_p \langle 0|T(\varphi_{\text{in}}(x_1)\varphi_{\text{in}}(x_2))|0\rangle \times \cdots \langle 0|T(\varphi_{\text{in}}(x_{n-1})\varphi_{\text{in}}(x_n))|0\rangle \tag{17.27}$$

which expresses the S matrix in terms of known Feynman propagators for free particles with their physical masses. This is the result we want.

The Wick theorem from which our result (17.27) followed may be proved by induction. Theorem (17.23) is certainly true for $n = 1$; for $n = 2$ we may readily verify it also. Consider first a single hermitian boson field φ_{in}. We observe that

$$T(\varphi_{\text{in}}(x_1)\varphi_{\text{in}}(x_2)) = :\varphi_{\text{in}}(x_1)\varphi_{\text{in}}(x_2): + (c \text{ number})$$

since to go from time ordering to normal ordering means only an interchange of the order of various pairs of creation and destruction operators; the commutators left over after such interchange are c numbers. To obtain the c number, we take the vacuum expectation value and use the fact that the vacuum expectation value of a normal product of operators vanishes:

$$T(\varphi_{\text{in}}(x_1)\varphi_{\text{in}}(x_2)) = :\varphi_{\text{in}}(x_1)\varphi_{\text{in}}(x_2): + \langle 0|T(\varphi_{\text{in}}(x_1)\varphi_{\text{in}}(x_2))|0\rangle \quad (17.28)$$

The corresponding argument goes through in the same way for Fermi fields:

$$T(\psi_{\text{in}}(x_1)\bar{\psi}_{\text{in}}(x_2)) = :\psi_{\text{in}}(x_1)\bar{\psi}_{\text{in}}(x_2): + \langle 0|T(\psi_{\text{in}}(x_1)\bar{\psi}_{\text{in}}(x_2))|0\rangle \quad (17.29)$$

Proceeding by induction, we suppose (17.23) is true for some n and establish it for $n + 1$. We consider

$$T(\varphi_{\text{in}}(x_1) \cdots \varphi_{\text{in}}(x_{n+1}))$$

and choose t_{n+1} as the earliest time; thus

$$\begin{aligned} T(\varphi_{\text{in}}(x_1) \cdots \varphi_{\text{in}}(x_{n+1})) &= T(\varphi_{\text{in}}(x_1) \cdots \varphi_{\text{in}}(x_n))\varphi_{\text{in}}(x_{n+1}) \\ &= :\varphi_{\text{in}}(x_1) \cdots \varphi_{\text{in}}(x_n):\varphi_{\text{in}}(x_{n+1}) \\ &\quad + \sum_{\text{perm}} \langle 0|T(\varphi_{\text{in}}(x_1)\varphi_{\text{in}}(x_2))|0\rangle \\ &\quad \times :\varphi_{\text{in}}(x_3) \cdots \varphi_{\text{in}}(x_n):\varphi_{\text{in}}(x_{n+1}) + \cdots \end{aligned} \quad (17.30)$$

To recast this into the form of the Wick expansion, it is necessary to find the rule for introducing $\varphi_{\text{in}}(x_{n+1})$ into the n-fold normal product. We do this using (17.25) and (17.26):

$$\begin{aligned} &:\varphi_{\text{in}}(x_1) \cdots \varphi_{\text{in}}(x_n):\varphi_{\text{in}}(x_{n+1}) \\ &= \sum_{A,B} \delta_p \prod_{i \text{ in } A} \varphi_{\text{in}}{}^{(-)}(x_i) \prod_{j \text{ in } B} \varphi_{\text{in}}{}^{(+)}(x_j)[\varphi_{\text{in}}{}^{(+)}(x_{n+1}) + \varphi_{\text{in}}{}^{(-)}(x_{n+1})] \\ &= \sum_{A,B} \delta_{p'} \prod_{i \text{ in } A} \varphi_{\text{in}}{}^{(-)}(x_i) \prod_{j \text{ in } B} \varphi_{\text{in}}{}^{(+)}(x_j)\varphi_{\text{in}}{}^{(+)}(x_{n+1}) \\ &+ \sum_{A,B} \delta_{p'} \prod_{i \text{ in } A} \varphi_{\text{in}}{}^{(-)}(x_i)\varphi_{\text{in}}{}^{(-)}(x_{n+1}) \prod_{j \text{ in } B} \varphi_{\text{in}}{}^{(+)}(x_j) \\ &+ \sum_{A,B} \delta_{p'} \prod_{i \text{ in } A} \varphi_{\text{in}}{}^{(-)}(x_i) \sum_{k \text{ in } B} \prod_{\substack{j \text{ in } B \\ j \neq k}} \varphi_{\text{in}}{}^{(+)}(x_j)\langle 0|\varphi_{\text{in}}{}^{(+)}(x_k)\varphi_{\text{in}}{}^{(-)}(x_{n+1})|0\rangle \end{aligned} \quad (17.31)$$

$\delta_{p'}$ is the sign of the permutation appropriate to the ordering of the factors in that particular term of (17.31). What has happened in (17.31) is that in absorbing $\varphi_{\text{in}}(x_{n+1})$ into the normal product we have left behind a string of terms (the last line) which involve a commutator (or anticommutator) of $\varphi_{\text{in}}{}^{(-)}(x_{n+1})$ with the $\varphi_{\text{in}}{}^{(+)}(x_k)$ of the set B. This commutator has been replaced by the indicated vacuum expectation value. We again reduce such vacuum expectation values to Feynman propagators with the aid of (17.26) and of the prior specification that t_{n+1} is the earliest time:

$$\begin{aligned}\langle 0|\varphi_{\text{in}}{}^{(+)}(x_k)\varphi_{\text{in}}{}^{(-)}(x_{n+1})|0\rangle &= \langle 0|\varphi_{\text{in}}(x_k)\varphi_{\text{in}}(x_{n+1})|0\rangle \\ &= \langle 0|T(\varphi_{\text{in}}(x_k)\varphi_{\text{in}}(x_{n+1}))|0\rangle\end{aligned}$$

Finally, we rewrite (17.31) as

$$\begin{aligned}:\varphi_{\text{in}}(x_1)\cdots\varphi_{\text{in}}(x_n):\varphi_{\text{in}}(x_{n+1}) &= :\varphi_{\text{in}}(x_1)\cdots\varphi_{\text{in}}(x_{n+1}): \\ &+ \sum_k \delta_p:\varphi_{\text{in}}(x_1)\cdots\varphi_{\text{in}}(x_{k-1})\varphi_{\text{in}}(x_{k+1})\cdots\varphi_{\text{in}}(x_n): \\ &\times \langle 0|T(\varphi_{\text{in}}(x_k)\varphi_{\text{in}}(x_{n+1}))|0\rangle\end{aligned}$$

and check that with this result (17.30) takes the form of (17.23) for $n+1$, thus proving Wick's theorem. There is no substitute at this point for the reader's writing out the cases $n = 3$ and $n = 4$ in detail in order to convince himself of this result.

In applying Wick's theorem to (17.22) we recall that the interaction hamiltonian $\mathcal{H}_I(\varphi_{\text{in}}(y))$ is already normal-ordered.[1] Therefore, in working out the time-ordered products in (17.22) there appear no contraction terms containing two field amplitudes at the same coordinate y from the same interaction term $\mathcal{H}_I$. These start out in normal order in (17.22), and evidently

$$T:\varphi_{\text{in}}(y)\varphi_{\text{in}}(y): = :\varphi_{\text{in}}(y)\varphi_{\text{in}}(y): \tag{17.32}$$

17.5 Graphical Representation

Three classes of nonvanishing contraction terms appear in the reduction of (17.22); for a hermitian Klein-Gordon field (12.74),

$$\langle 0|T(\varphi_{\text{in}}(x)\varphi_{\text{in}}(y))|0\rangle = i\Delta_F(x-y,\mu^2) = i\int\frac{d^4k}{(2\pi)^4}\frac{e^{-ik\cdot(x-y)}}{k^2-\mu^2+i\epsilon} \tag{17.33a}$$

[1] The static Coulomb interaction term in (15.28) is an exception. See in particular Sec. 17.9. However, there are still no contraction terms at the same point.

or a complex field (12.70)

$$\langle 0|T(\varphi_{\text{in}}(x)\varphi^*_{\text{in}}(y))|0\rangle = i\Delta_F(x-y,\mu^2) \tag{17.33b}$$

for a Dirac field (13.72)

$$\begin{aligned}\langle 0|T(\psi_\alpha^{\text{in}}(x)\bar\psi_\beta^{\text{in}}(y))|0\rangle &= iS_F(x-y,m)_{\alpha\beta} \\ &= i\int \frac{d^4p}{(2\pi)^4}\frac{e^{-ip\cdot(x-y)}(\not p + m)}{p^2-m^2+i\epsilon}\end{aligned} \tag{17.33c}$$

for the electromagnetic field (14.51) and (14.53)

$$\begin{aligned}\langle 0|T(A_\mu^{\text{in}}(x)A_\nu^{\text{in}}(y))|0\rangle &= iD_F{}^{\text{tr}}(x-y)_{\mu\nu} \\ = i\int \frac{d^4k}{(2\pi)^4}\frac{e^{-ik\cdot(x-y)}}{k^2+i\epsilon}\Bigg[-g_{\mu\nu} &- \frac{k_\mu k_\nu}{(k\cdot\eta)^2-k^2} + \frac{(k\cdot\eta)(k_\mu\eta_\nu+\eta_\mu k_\nu)}{(k\cdot\eta)^2-k^2} \\ &- \frac{k^2\eta_\mu\eta_\nu}{(k\cdot\eta)^2-k^2}\Bigg]\end{aligned} \tag{17.33d}$$

where $\eta = (1,0,0,0)$ in the Lorentz frame in which quantization was carried out. We represent these propagators graphically as in the Feynman theory, with the pictorial correspondence shown in Fig. 17.1. In writing the propagators we generally suppress the mass argument except when explicitly needed.

With these lines to represent each of the contractions appearing in the Dyson-Wick expansion (17.23) of the τ function (17.22), we may give each of the terms in (17.22) a graphical representation. Since the interaction hamiltonians contain products of field operators at the same point, the propagators associated with contractions involving these operators will be "tied together" at these points, which are called *vertices*.

To illustrate these remarks with a specific example, we return to the self-coupled scalar field, with

$$\mathcal{H}_I(\varphi_{\text{in}}(x)) = -\tfrac{1}{4}\lambda_0 : \varphi_{\text{in}}^4(x) : + \tfrac{1}{2}(\mu_0^2-\mu^2) : \varphi_{\text{in}}^2(x) : \tag{17.34}$$

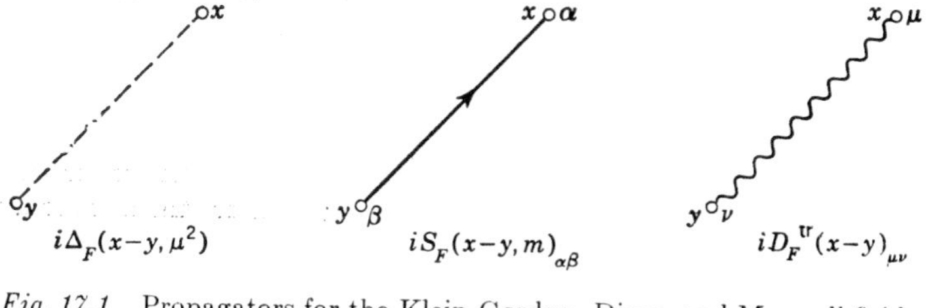

Fig. 17.1 Propagators for the Klein-Gordon, Dirac, and Maxwell fields.

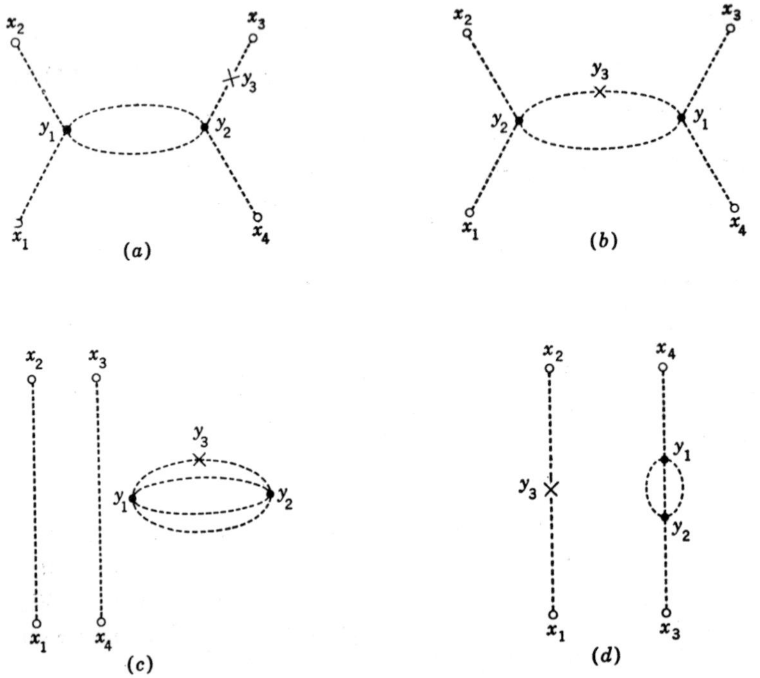

Fig. 17.2 Typical graphs contained in Eq. (17.35).

A contribution to the expansion of $\tau(x_1,x_2,x_3,x_4)$ which is second order in λ_0 and first order in the mass correction comes from a term such as

$$\frac{\lambda_0{}^2\delta\mu^2}{32}\langle 0|T(\varphi_{\text{in}}(x_1)\varphi_{\text{in}}(x_2)\varphi_{\text{in}}(x_3)\varphi_{\text{in}}(x_4):\varphi_{\text{in}}^4(y_1)::\varphi_{\text{in}}^4(y_2)::\varphi_{\text{in}}^2(y_3):)|0\rangle \tag{17.35}$$

The reduction of this expression leads to graphs with the typical form illustrated in Fig. 17.2.

In these pictures we represent the interaction by a four-line vertex for the λ_0 term and a two-line vertex for the mass counterterm (see Fig. 17.3); each line leaving the vertex represents the contraction of

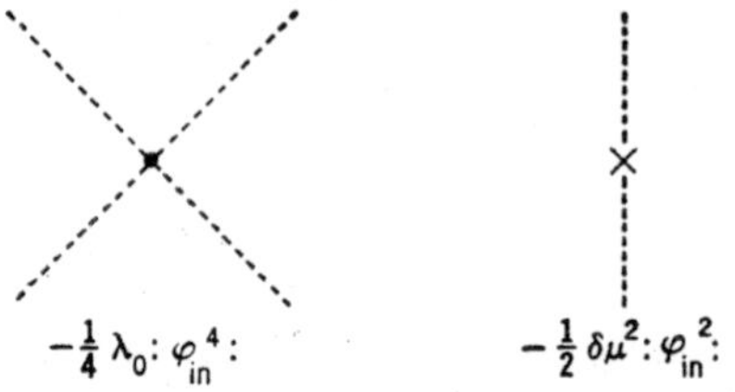

Fig. 17.3 Vertices for meson-meson scattering and for the mass counterterm.

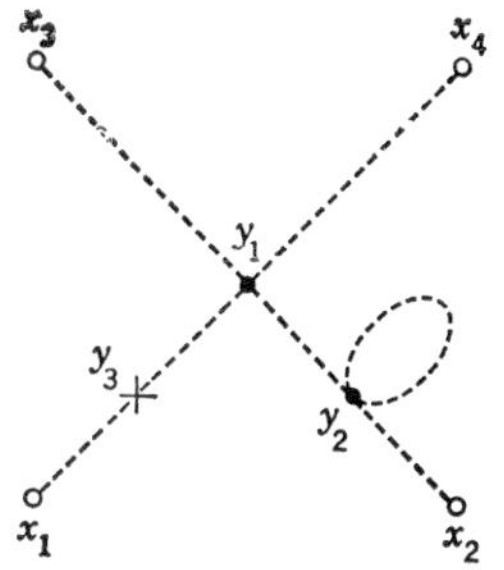

Fig. 17.4 Self-energy "tadpole" not occurring in Eq. (17.35).

each of the fields appearing in the interaction term $\mathcal{H}_I$ with some other field. A typical term which does *not* occur in the expansion is drawn in Fig. 17.4. It is excluded because $\mathcal{H}_I$ has been normal-ordered, and contractions involving two fields at the same interaction vertex do not occur.

The pictures show the general correspondence with Feynman graphs of terms such as (17.35) in the expansion of a τ function. To obtain the rules which were found in propagator theory for Feynman graphs, we need only solve a combinatorial problem, that is, find the number of times the same graph (differing only by a labeling of vertices) appears in the Dyson-Wick decomposition (17.23). For example, for the diagrams associated with the reduction of (17.35), there are 3! ways of permuting the arguments y_1, y_2, and y_3 of the vertex points; each such permuted graph gives an equal contribution to τ. In general there are $m!$ permutations of the $y_1, \ldots, y_m$ which cancel out the $1/m!$ in (17.22). Also, in the diagrams illustrated in Fig. 17.2 there are 4! ways of permuting the four field operators in $\frac{1}{4}\lambda_0 : \varphi_{\text{in}}^4 :$ with the four propagators converging at a λ_0 vertex and 2! ways of associating the two field operators with the self-mass vertices $-\frac{1}{2}\delta\mu^2 : \varphi_{\text{in}}^2 :$. The general rule is therefore to multiply each interaction term in which a given field amplitude appears to the rth power by $r!$.

17.6 Vacuum Amplitudes

Some graphs associated with the expansion of the denominator of (17.22) are illustrated in Fig. 17.5. These vacuum bubbles also appear in the expansion of the numerator, as in Fig. 17.2*c*, and, as we shall see, cancel out the denominator.

Graphs associated with the numerator differ from those associated with the denominator in that they possess *external lines*, that is, those

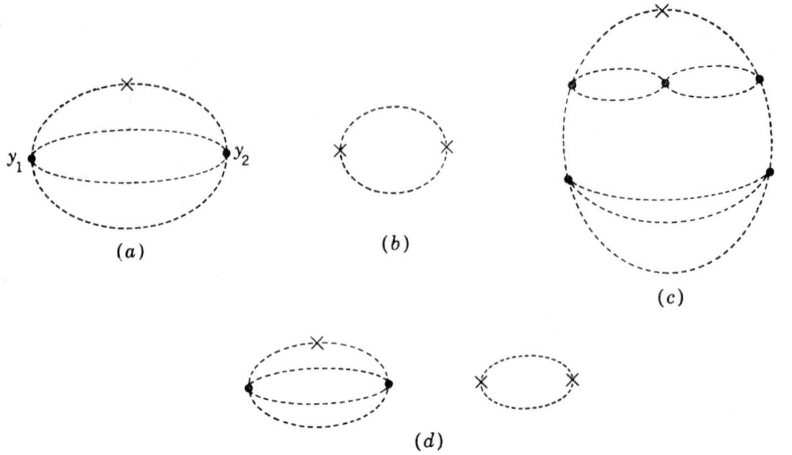

Fig. 17.5 Vacuum bubbles.

contractions associated with the field operators $\varphi_{\text{in}}(x_1) \cdots \varphi_{\text{in}}(x_n)$ appearing in $\tau(x_1, \ldots, x_n)$. We call a *disconnected part* a subgraph occurring in the expansion which is not connected in any way with any external line. A graph which has no disconnected part we shall call *connected.* Any graph appearing in the numerator can be separated uniquely into a connected and disconnected part; the contribution to the τ function can likewise be so separated.

For all graphs which have as their connected part a contribution which is of sth order in the interaction $\mathcal{H}_I$, the numerator of the τ function takes the form

$$\sum_{p=0}^{\infty} \frac{(-i)^p}{p!} \int d^4y_1 \cdots d^4y_p \langle 0|T(\varphi_{\text{in}}(x_1) \cdots \varphi_{\text{in}}(x_n)\mathcal{H}_I(y_1) \cdots \mathcal{H}_I(y_s))|0\rangle_c$$
$$\times \frac{p!}{s!(p-s)!} \langle 0|T(\mathcal{H}_I(y_{s+1}) \cdots \mathcal{H}_I(y_p))|0\rangle \quad (17.36)$$

where the subscript c on the vacuum expectation value denotes that connected parts only are included. The combinatorial factor

$$\binom{p}{s} = \frac{p!}{s!(p-s)!}$$

records the number of ways of extracting s terms $\mathcal{H}_I$ from a set of p and allows us to rewrite (17.36) as

$$\frac{(-i)^s}{s!}\int d^4y_1 \cdots d^4y_s\langle 0|T(\varphi_{\text{in}}(x_1)\cdots\varphi_{\text{in}}(x_n)\mathcal{H}_I(y_1)\cdots\mathcal{H}_I(y_s))|0\rangle_c$$
$$\times\sum_{r=0}^{\infty}\frac{(-i)^r}{r!}\int d^4z_1\cdots d^4z_r\langle 0|T(\mathcal{H}_I(z_1)\cdots\mathcal{H}_I(z_r))|0\rangle \tag{17.37}$$

Equation (17.37) now has the form of an sth-order connected graph times an infinite series over vacuum bubbles as illustrated in Fig. 17.5, which just cancels the denominator of (17.22). In general, we write

$$\tau(x_1, \ldots, x_n) = \frac{\sum_i G_i(x_1, \ldots, x_n)}{\sum_k D_k} = \frac{\sum_i G_i^c(x_1, \ldots, x_n)\sum_k D_k}{\sum_k D_k}$$
$$= \sum_i G_i^c(x_1, \ldots, x_n) \tag{17.38}$$

where G_i^c is the contribution of a connected graph and D_k that of the disconnected part. In words, (17.38) means that we may ignore all disconnected parts in calculating the τ function; it is simply the sum of the contributions of all *connected* Feynman graphs.

17.7 Spin and Isotopic Spin; Pi-nucleon Scattering

For most calculations of physical interest we shall be concerned with interaction hamiltonians $\mathcal{H}_I$ containing several kinds of fields with indices labeling spin and isotopic spin states. The π-nucleon interaction introduced in Chap. 15 has the form, including mass counterterms,

$$\mathcal{H}_I = g:\bar{\psi}i\gamma_5\boldsymbol{\tau}\cdot\boldsymbol{\phi}\psi: - \tfrac{1}{2}\delta\mu^2:\boldsymbol{\phi}\cdot\boldsymbol{\phi}: - \delta M:\bar{\psi}\psi: \tag{17.39}$$

To illustrate the bookkeeping on the various indices, let us consider the lowest-order graphs in π-nucleon scattering. The relevant τ function for this process appears in (16.139) and is

$$\langle 0|T(\psi(z_2)_{\alpha r}\bar{\psi}(z_1)_{\beta s}\varphi_i(x_1)\varphi_j(x_2))|0\rangle \tag{17.40}$$

where (α,β) and (r,s) refer to nucleon spin and isotopic spin indices, respectively, and (i,j) to the isotopic indices of the π mesons. The

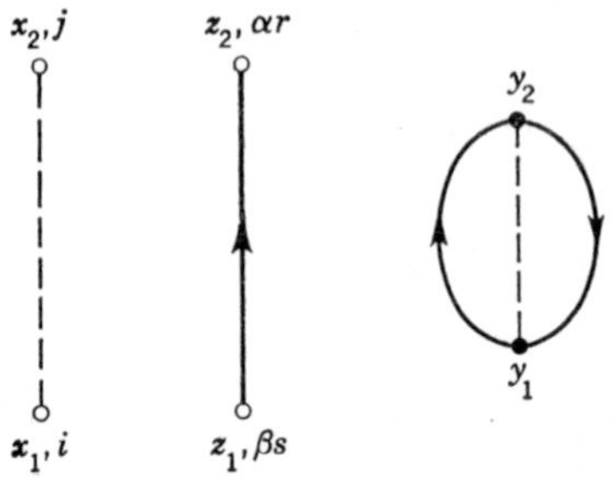

Fig. 17.6 Disconnected graph not contained in π-nucleon amplitude, Eq. (17.41).

lowest-order contribution to the scattering is provided by

$$\tau^{(2)}_{ij,\alpha\beta,rs}(x_1,x_2,z_1,z_2) = \frac{(-i)^2}{2!}\int d^4y_1\, d^4y_2$$
$$\times \langle 0|T(\psi^{\text{in}}_{\alpha r}(z_2)\bar{\psi}^{\text{in}}_{\beta s}(z_1)\varphi^{\text{in}}_i(x_1)\varphi^{\text{in}}_j(x_2)\mathcal{H}_I(y_1)\mathcal{H}_I(y_2))|0\rangle_c \qquad (17.41)$$

with the subscript c indicating that only connected graphs are to be included, according to (17.38). Thus the graph shown in Fig. 17.6 is excluded. The mass counterterms in (17.39) are each of order g^2 and can operate only once in a g^2 calculation. They serve to modify the self-energy contributions to (17.40) corresponding to the graphs shown in Fig. 17.7.

This kind of mass renormalization has been encountered in Chap. 8 in connection with radiative corrections to electron scattering. Thus the graphs of Fig. 17.8 will ensure (to order g^2) that a physical meson or nucleon propagates with the observed mass μ or M. In our present calculation we shall restrict our attention to nonforward scattering; the graphs of Fig. 17.7 and Fig. 17.8 do not contribute.

We turn to the remaining contributions in the reduction of (17.41), corresponding to the graphs in Fig. 17.9. These change the quantum

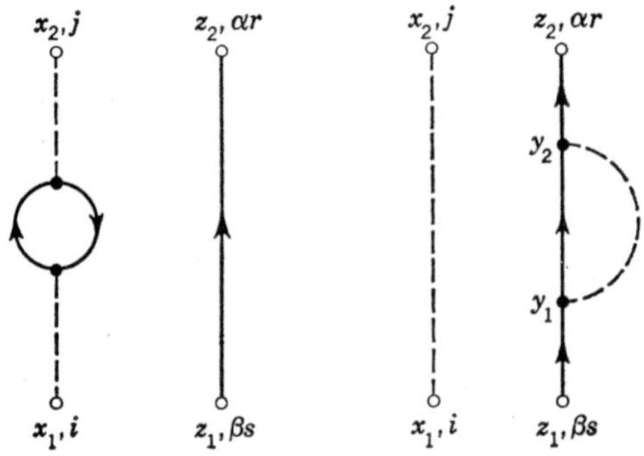

Fig. 17.7 Self-energy contribution to π-nucleon amplitude.

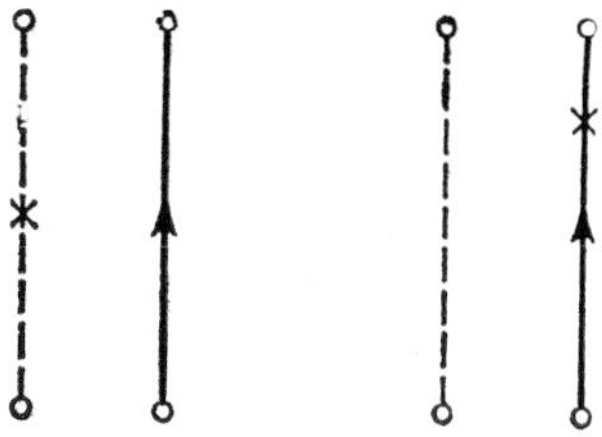

Fig. 17.8 Mass counterterms in π-nucleon amplitude.

numbers of the pion and nucleon between initial and final states and are represented by the following diagrams and matrix elements:

$$\tau^{(2a)}_{ij,\alpha\beta,rs}(x_1,x_2,z_1,z_2) = \frac{(-i)^2}{2!}\,2!g^2 \int d^4y_1\,d^4y_2\,(\tau_j\tau_i)_{rs} i\Delta_F(y_1 - x_1)$$
$$\times\, i\Delta_F(x_2 - y_2)$$
$$\times\,[iS_F(z_2 - y_2)i\gamma_5 iS_F(y_2 - y_1)i\gamma_5 iS_F(y_1 - z_1)]_{\alpha\beta} \quad (17.42a)$$
$$\tau^{(2b)}_{ij,\alpha\beta,rs}(x_1,x_2,z_1,z_2) = \frac{(-i)^2}{2!}\,2!g^2 \int d^4y_1\,d^4y_2\,(\tau_i\tau_j)_{rs} i\Delta_F(y_2 - x_1)$$
$$\times\, i\Delta_F(x_2 - y_1)$$
$$\times\,[iS_F(z_2 - y_2)i\gamma_5 iS_F(y_2 - y_1)i\gamma_5 iS_F(y_1 - z_1)]_{\alpha\beta} \quad (17.42b)$$

There is no substitute for experience! The reader is urged to work out (17.42) in detail to obtain the necessary confidence in the factors and the spin and isotopic spin matrix orderings.

In order to evaluate a transition amplitude, we want to insert the τ function (17.42) into the reduction formula (16.139) relating the τ

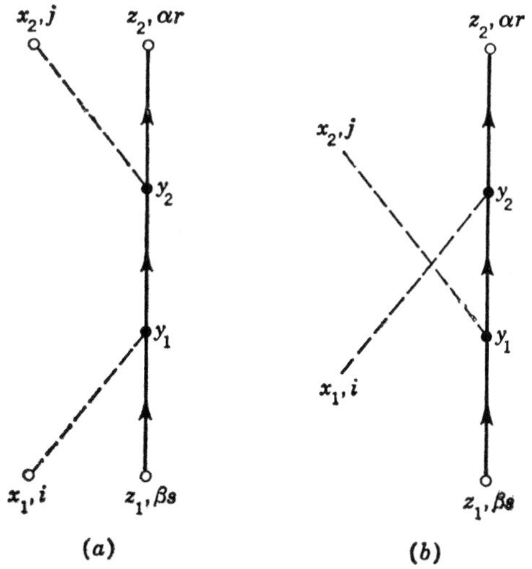

Fig. 17.9 Second-order π-nucleon scattering, Eq. (17.42).

function to the S matrix. Ignoring the forward-scattering terms,

$$S^{(2)}(p_2,q_2;p_1,q_1) = \left(\frac{i}{\sqrt{Z_3}}\right)^2 \left(\frac{-i}{\sqrt{Z_2}}\right)^2 \int d^4x_1\, d^4x_2\, d^4z_1\, d^4z_2$$
$$\times f_{q_2}^*(x_2)\hat{\phi}_{2j}^*[\bar{U}_{p_2s_2}(z_2)(i\overrightarrow{\nabla}_{z_2} - M)]_\alpha(\Box_{x_2} + \mu^2)\chi_r^\dagger(2)\tau^{(2)}_{ij,\alpha\beta,rs}(x_1,x_2,z_1,z_2)$$
$$\times \chi_s(1)[(-i\overleftarrow{\nabla}_{z_1} - M)U_{p_1s_1}(z_1)]_\beta(\overleftarrow{\Box}_{x_1} + \mu^2)\hat{\phi}_{1i}f_{q_1}(x_1) \qquad (17.43)$$

where the f_q are normalized boson wave functions of momentum q, the $U_{ps}(z)$ are normalized fermion wave functions, and the $\hat{\phi}_i$ and χ_r are isotopic wave functions.

In the spirit of the perturbation approach we set $Z_2 = Z_3 = 1$, since we are only calculating to order g^2. The effect of the Klein-Gordon and Dirac operators is to remove the propagators of the external legs, since

$$(\Box_x + \mu^2)\Delta_F(x - y) = -\delta^4(x - y)$$
$$(i\nabla_x - M)S_F(x - y) = S_F(x - y)(-i\overleftarrow{\nabla}_y - M) = \delta^4(x - y) \qquad (17.44)$$

The propagators are thus replaced by the wave functions of the external particles. Carrying out the δ-function integrations and inserting the wave functions simplifies (17.43) to

$$S^{(2)}(p_2,q_2;p_1,q_1) = -g^2 \int \frac{d^4y_1\, d^4y_2}{(2\pi)^6\sqrt{2\omega_1 2\omega_2}} \sqrt{\frac{M^2}{E_1E_2}}$$
$$\times \left\{ \begin{array}{c} e^{i(q_2+p_2)\cdot y_2}[\chi^\dagger(2)\boldsymbol{\tau}\cdot\hat{\boldsymbol{\phi}}_2^*\boldsymbol{\tau}\cdot\hat{\boldsymbol{\phi}}_1\chi(1)] \\ \times [\bar{u}(p_2,s_2)i\gamma_5 iS_F(y_2 - y_1)i\gamma_5 u(p_1,s_1)]e^{-i(p_1+q_1)\cdot y_1} \\ + e^{i(p_2-q_1)\cdot y_2}[\chi^\dagger(2)\boldsymbol{\tau}\cdot\hat{\boldsymbol{\phi}}_1\boldsymbol{\tau}\cdot\hat{\boldsymbol{\phi}}_2^*\chi(1)] \\ \times [\bar{u}(p_2,s_2)i\gamma_5 iS_F(y_2 - y_1)i\gamma_5 u(p_1,s_1)]e^{-i(p_1-q_2)\cdot y_1} \end{array} \right\} \qquad (17.45)$$

A Fourier transformation into momentum space, as carried out frequently in Chaps. 7 to 10, leads directly to the expression written down in (10.54) for this amplitude.

The crossing symmetry of the π-nucleon scattering amplitude noted in (10.54) and (10.55) follows immediately from the reduction formula and the form of (17.40). Evidently (17.40) is invariant under the interchange $i \leftrightarrow j$, $x_1 \leftrightarrow x_2$. Thus if $\hat{\phi}_{1i} \leftrightarrow \hat{\phi}_{2i}^*$ and $f_{q_1}(x_1) \leftrightarrow f_{-q_2}(x_1)$ in the reduction formula (17.43), the S matrix remains unchanged. The first important application to strong-coupling problems was made by Gell-Mann and Goldberger[1] to show that the limit of π-N scattering as $q_1 = q_2 \to 0$ is isotopic spin independent. In this limit S is unchanged by the interchange of isotopic indices i and

[1] M. Gell-Mann and M. L. Goldberger, unpublished.

j and, according to charge independence, has the form $S_{ij} = \delta_{ij}S$. The details of this argument we leave as an exercise. A second application of crossing symmetry is the Pomeranchuk theorem,[1]

$$\lim_{E\to\infty} \sigma_{\text{tot}}(A + B \to A + B) = \lim_{E\to\infty} \sigma_{\text{tot}}(\bar{A} + B \to \bar{A} + B)$$

We shall discuss this theorem in connection with the forward-scattering dispersion relations in Chap. 18.

17.8 Pi-Pi Scattering

As an additional example of the Dyson-Wick reduction procedure we consider a perturbation calculation of π-π scattering, according to the interaction (17.39). The lowest-order contribution is of order g^4 and illustrated in Fig. 17.10. Graphs such as illustrated in Fig. 17.11 contribute only to the forward-scattering amplitude; these we shall ignore. For the graphs of Fig. 17.10, we may set $Z_3 = 1$ and ignore mass counterterms, and the S-matrix element is, for nonforward scattering,

$$S(q_3,q_4;q_1,q_2) = \frac{1}{(2\pi)^6\sqrt{2\omega_1 2\omega_2 2\omega_3 2\omega_4}} \times \int d^4x_1 \cdots d^4x_4\{\exp i[q_3\cdot x_3 + q_4\cdot x_4 - q_1\cdot x_1 - q_2\cdot x_2]\} \times \hat{\phi}_{1i}\hat{\phi}_{2j}\hat{\phi}^*_{3k}\hat{\phi}^*_{4l}K_{x_1}K_{x_2}K_{x_3}K_{x_4}\tau_{ijkl}(x_1,x_2,x_3,x_4) \qquad (17.46)$$

Again the $\hat{\phi}_i$ are isotopic wave functions and we define

$$K_x = \Box_x + \mu^2$$

To this order,

$$\begin{aligned}\tau^{(4)}_{ijkl}(x_1,x_2,x_3,x_4) &= \frac{(-ig)^4}{4!}4!\int d^4y_1 \cdots d^4y_4\langle 0|T(\varphi_i(x_1)\varphi_r(y_1))|0\rangle \\ &\times \langle 0|T(\varphi_j(x_2)\varphi_s(y_2))|0\rangle\langle 0|T(\varphi_k(x_3)\varphi_t(y_3))|0\rangle\langle 0|T(\varphi_l(x_4)\varphi_u(y_4))|0\rangle \\ &\times \langle 0|T(:\bar{\psi}(y_1)i\gamma_5\tau_r\psi(y_1)::\bar{\psi}(y_2)i\gamma_5\tau_s\psi(y_2): \\ &\times :\bar{\psi}(y_3)i\gamma_5\tau_t\psi(y_3)::\bar{\psi}(y_4)i\gamma_5\tau_u\psi(y_4):)|0\rangle \\ &= -g^4\int d^4y_1\,d^4y_2\,d^4y_3\,d^4y_4\,\Delta_F(x_1-y_1)\Delta_F(x_2-y_2) \\ &\times \Delta_F(x_3-y_3)\Delta_F(x_4-y_4)\,\text{Tr}\,\tau_i\tau_j\tau_k\tau_l \\ &\times \text{Tr}\,i\gamma_5 iS_F(y_1-y_2)i\gamma_5 iS_F(y_2-y_3)i\gamma_5 iS_F(y_3-y_4)i\gamma_5 iS_F(y_4-y_1) \\ &+ 5 \text{ terms obtained by permuting external lines}\end{aligned} \qquad (17.47)$$

The minus sign is worthy of attention. It occurs because in order to contract $\psi(y_4)$ with $\bar{\psi}(y_1)$, it is necessary to make an odd number of

[1] Y. Pomeranchuk, *J.E.T.P. (USSR)*, **34,** 725 (1958).

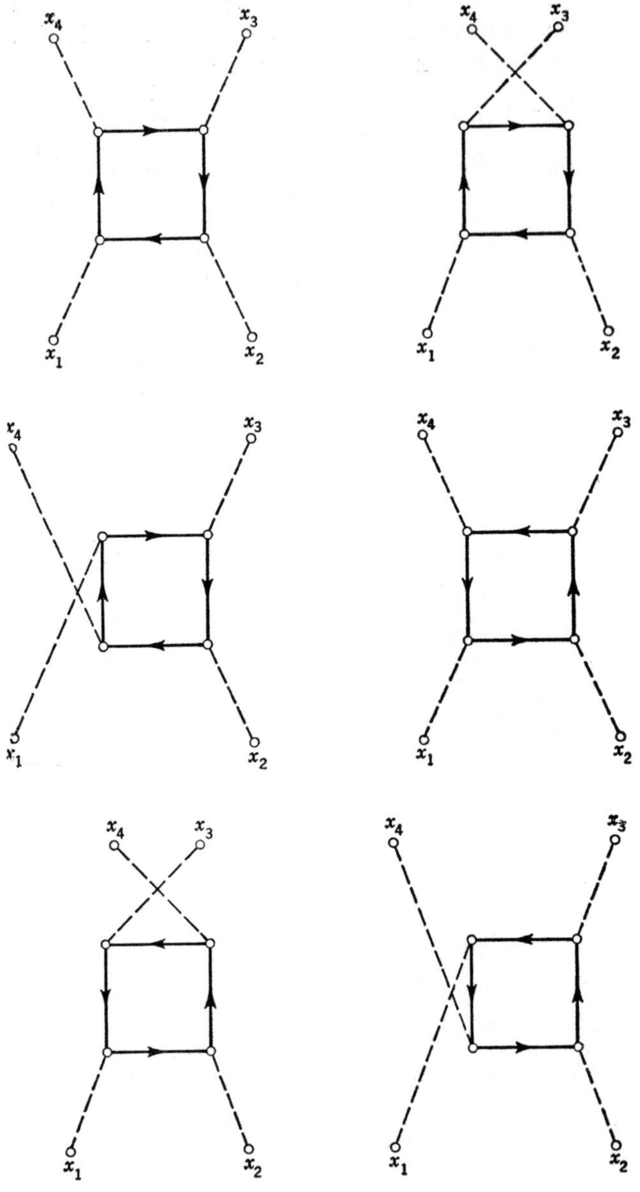

Fig. 17.10 Fourth-order π-π scattering.

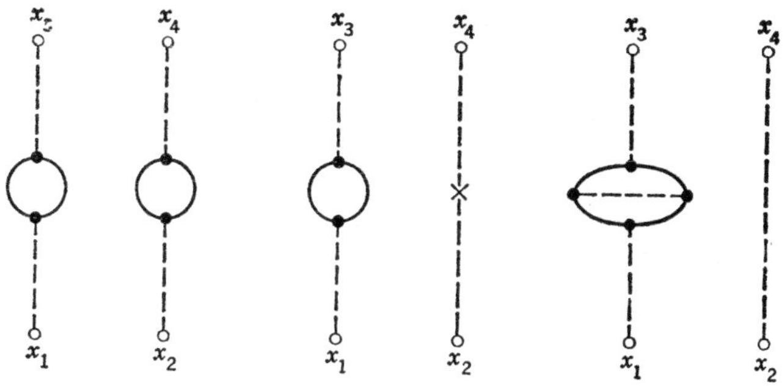

Fig. 17.11 Fourth-order contributions to forward π-π scattering

permutations. This is a general rule; for each closed fermion loop in a graph multiply by (-1). The proof is the same as above. For a general closed loop, as in Fig. 17.12, the Wick product in question is

$$\begin{aligned}
&+\langle 0|T(\bar\psi(y_1)\Gamma\psi(y_1)\bar\psi(y_2)\Gamma\psi(y_2)\cdots\bar\psi(y_n)\Gamma\psi(y_n))|0\rangle \\
&= -\langle 0|T(\psi(y_n)\bar\psi(y_1)\Gamma\psi(y_1)\bar\psi(y_2)\Gamma\cdots\psi(y_{n-1})\bar\psi(y_n)\Gamma)|0\rangle \\
&= (-)\ \mathrm{Tr}\ iS_F(y_n-y_1)\Gamma iS_F(y_1-y_2)\Gamma\cdots iS_F(y_{n-1}-y_n)\Gamma \\
&\qquad + \text{permutations} \qquad (17.48)
\end{aligned}$$

Putting together (17.46) and (17.47) and going into momentum space yields

$$S(q_3,q_4;q_1,q_2) = -\frac{i(2\pi)^4\delta^4(q_3+q_4-q_1-q_2)}{(2\pi)^6\sqrt{16\omega_1\omega_2\omega_3\omega_4}}\,\mathfrak{M}$$

with

$$\begin{aligned}
\mathfrak{M} = -ig^4\ \mathrm{Tr}\ \boldsymbol{\tau}\cdot\hat{\boldsymbol{\phi}}_1\boldsymbol{\tau}\cdot\hat{\boldsymbol{\phi}}_2\boldsymbol{\tau}\cdot\hat{\boldsymbol{\phi}}_3^*\boldsymbol{\tau}\cdot\hat{\boldsymbol{\phi}}_4^* \int \frac{d^4p}{(2\pi)^4}\ \mathrm{Tr}\ i\gamma_5\frac{i}{\not p+\not q_1-M+i\epsilon}i\gamma_5 \\
\times \frac{i}{\not p+\not q_1+\not q_2-M+i\epsilon}i\gamma_5\frac{i}{\not p+\not q_4-M+i\epsilon}i\gamma_5\frac{i}{\not p-M+i\epsilon} \\
+ 5\ \text{permutations} \qquad (17.49)
\end{aligned}$$

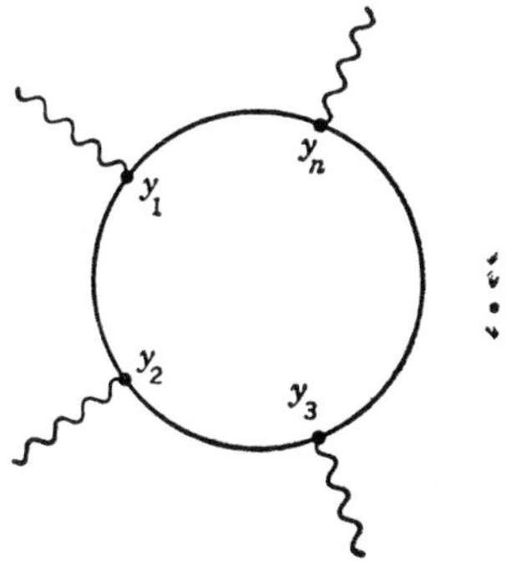

Fig. 17.12 Closed fermion loop in a general graph.

Checking with Appendix B we find that (17.49) agrees precisely with the expression written down directly from the rules.

Unfortunately, the integral over the internal momentum p diverges logarithmically.[1] This infinity calls for an additional counterterm in the lagrangian of the meson interactions. The simplest form of such a term is

$$-\mathcal{L}_I = \mathcal{H}_I(\varphi) = \tfrac{1}{4}\,\delta\lambda\!:\!(\boldsymbol{\phi}\cdot\boldsymbol{\phi})^2\!: \tag{17.50}$$

where $\delta\lambda$ is an infinite constant designed to remove the divergence in (17.49).

The finite contributions to π-π scattering can then be evaluated after subtracting the infinite counterterms. It is a lengthy task to evaluate the integral and traces in (17.49) and not especially rewarding owing to the lack of convergence of the perturbation series ($g^2/4\pi = 15$). However, several properties of the scattering amplitude can be established directly on the basis of its symmetries:

1. $\mathfrak{M}$ is a scalar function of the invariants

$$s = (q_1 + q_2)^2 \qquad t = (q_1 - q_3)^2 \qquad u = (q_1 - q_4)^2 \tag{17.51}$$

where
$$s + t + u = 4\mu^2$$

and the range of values for physical scatterings $q_1 + q_2 \to q_3 + q_4$ is $s \geq 4\mu^2$, $t \leq 0$, $u \leq 0$.

2. In the approximation of charge independence $\mathfrak{M}$ has the isotopic structure

$$\mathfrak{M} = (\hat{\boldsymbol{\phi}}_1 \cdot \hat{\boldsymbol{\phi}}_2)(\hat{\boldsymbol{\phi}}_3^* \cdot \hat{\boldsymbol{\phi}}_4^*)A(s,t,u) + (\hat{\boldsymbol{\phi}}_1 \cdot \hat{\boldsymbol{\phi}}_3^*)(\hat{\boldsymbol{\phi}}_2 \cdot \hat{\boldsymbol{\phi}}_4^*)B(s,t,u) + (\hat{\boldsymbol{\phi}}_1 \cdot \hat{\boldsymbol{\phi}}_4^*)(\hat{\boldsymbol{\phi}}_2 \cdot \hat{\boldsymbol{\phi}}_3^*)C(s,t,u) \tag{17.52}$$

3. $\mathfrak{M}$ possesses crossing symmetry, that is, it is invariant under interchange of any of the external particles

$$q_2 \leftrightarrow -q_3,\ \hat{\boldsymbol{\phi}}_2 \leftrightarrow \hat{\boldsymbol{\phi}}_3^*;\ s \leftrightarrow t$$
$$q_2 \leftrightarrow -q_4,\ \hat{\boldsymbol{\phi}}_2 \leftrightarrow \hat{\boldsymbol{\phi}}_4^*;\ s \leftrightarrow u$$
$$q_3 \leftrightarrow q_4,\quad \hat{\boldsymbol{\phi}}_3^* \leftrightarrow \hat{\boldsymbol{\phi}}_4^*;\ t \leftrightarrow u$$

This follows from the structure of (17.46) and the symmetry of the τ function and implies

$$A(s,t,u) = B(t,s,u) = C(u,t,s) = A(s,u,t) = B(u,s,t) = C(t,u,s) \tag{17.53}$$

[1] No such infinity appears in quantum electrodynamics because of gauge invariance, which imposes additional conditions on the amplitude for the scattering of light by light and leads to a finite result. See Sec. 19.10 for further discussion of this point.

Finally, it is often useful to decompose S into isotopic spin channels; we exhibit here the combinations of isotopic invariants P_I which correspond to definite isotopic spin states for the reaction $1 + 2 \to 3 + 4$:

$$\begin{aligned} P_0 &= \tfrac{1}{3}(\hat{\phi}_1 \cdot \hat{\phi}_2)(\hat{\phi}_3^* \cdot \hat{\phi}_4^*) \\ P_1 &= \tfrac{1}{2}[(\hat{\phi}_1 \cdot \hat{\phi}_3^*)(\hat{\phi}_2 \cdot \hat{\phi}_4^*) - (\hat{\phi}_1 \cdot \hat{\phi}_4^*)(\hat{\phi}_2 \cdot \hat{\phi}_3^*)] \\ P_2 &= \tfrac{1}{2}[(\hat{\phi}_1 \cdot \hat{\phi}_3^*)(\hat{\phi}_2 \cdot \hat{\phi}_4^*) + (\hat{\phi}_1 \cdot \hat{\phi}_4^*)(\hat{\phi}_2 \cdot \hat{\phi}_3^*)] \\ &\qquad - \tfrac{1}{3}(\hat{\phi}_1 \cdot \hat{\phi}_2)(\hat{\phi}_3^* \cdot \hat{\phi}_4^*) \end{aligned} \tag{17.54}$$

Defining

$$P_I = \hat{\phi}_{1i}\hat{\phi}_{2j}\hat{\phi}_{3k}^*\hat{\phi}_{4l}^* P_I(ijkl) \tag{17.55}$$

we may verify the orthogonality properties

$$\sum_{m,n=1}^{3} P_I(ijmn)P_{I'}(mnkl) = P_I(ijkl)\delta_{II'} \tag{17.56}$$

We leave as an exercise the detailed arguments leading to the construction of these operators. We then have

$$\mathfrak{M}(q_3,q_4;q_1,q_2) = A_0P_0 + A_1P_1 + A_2P_2$$

$$\text{with} \quad A_0 = 3A + B + C \qquad A_1 = B - C \qquad A_2 = B + C \tag{17.57}$$

These A_I are the scattering amplitudes for total isospin I; at the symmetry point $s = t = u = 4\mu^2/3$, we note that, according to (17.53), $A = B = C$ and

$$A_0 = \tfrac{5}{2}A_2 \qquad A_1 = 0 \qquad s = t = u \tag{17.58}$$

Use of these general properties of S will be made in Sec. 18.12 in an application of dispersion relation techniques.

17.9 Rules for Graphs in Quantum Electrodynamics

At this stage we have established all the rules for graphs stated in Appendix B and extensively applied in the companion volume, with the exception of quantum electrodynamics, for which the rules apparently differ. We have found two kinds of graphs, the first involving exchange of a transverse photon, with vertices $-ie_0\gamma_\mu$ and a propagator, according to (17.33)

$$iD_F^{\text{tr}}(q,\eta)_{\mu\nu} = \frac{i}{q^2 + i\epsilon}\left[-g_{\mu\nu} - \frac{q_\mu q_\nu}{(q\cdot\eta)^2 - q^2} + \frac{(q\cdot\eta)(q_\mu\eta_\nu + \eta_\mu q_\nu)}{(q\cdot\eta)^2 - q^2} - \frac{q^2\eta_\mu\eta_\nu}{(q\cdot\eta)^2 - q^2}\right] \tag{17.59}$$

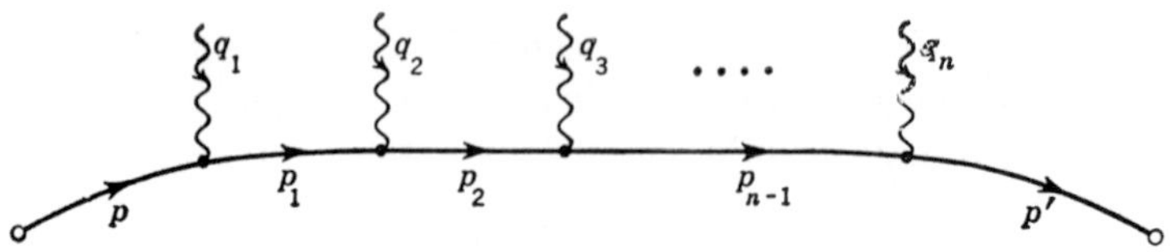

Fig. 17.13 n photon line ending on a charged fermion line.

where $\eta = (1,0,0,0)$ in the Lorentz frame in which quantization was carried out. The second kind of graph is due to the instantaneous Coulomb interaction in (15.28).

$$H_I = \frac{e_0^2}{2}\int \frac{d^3x\, d^3x'\ :\bar{\psi}(x)\not\eta\psi(x)::\bar{\psi}(x')\not\eta\psi(x'):}{4\pi|\mathbf{x}-\mathbf{x}'|} \tag{17.60}$$

We are now in a position to carry out our promise made in Sec. 14.6 that

1. The three middle terms proportional to q_μ or q_ν in (17.59) do not contribute to scattering amplitudes because of current conservation.[1]

2. The last term on the right in (17.59) cancels the Coulomb term (17.60).

As a consequence, we may use the covariant propagator

$$iD_F^{\mu\nu} = \frac{-ig^{\mu\nu}}{q^2 + i\epsilon}$$

in practical calculations, thereby guaranteeing the covariance of electrodynamic scattering amplitudes.

To establish assertion 1, we consider a fermion line to which n photons, real or virtual, have been attached. This graph, illustrated in Fig. 17.13, is given by the amplitude

$$g(p,p';q_1,a_1 \cdots q_n,a_n) = \frac{1}{\not p' - m}\not a_n \frac{1}{\not p_{n-1} - m}\not a_{n-1} \cdots \not a_2 \frac{1}{\not p_1 - m}\not a_1 \frac{1}{\not p - m} \tag{17.61}$$

We insert a photon line, real or virtual, which has momentum q and the interaction vertex $\not q$ as shown in Fig. 17.14. The effect of this insertion is to change g to $\tilde{g}_r$, given by

$$\tilde{g}_r(p,p',q;q_i,a_i) = \frac{1}{\not p' + \not q - m}\not a_n \frac{1}{\not p_{n-1} + \not q - m}\not a_{n-1} \times \cdots \not a_{r+1}\frac{1}{\not p_r + \not q - m}\not q \frac{1}{\not p_r - m}\not a_r \cdots \not a_1 \frac{1}{\not p - m} \tag{17.62}$$

[1] The validity of this assertion for real external photon lines was used in Chap. 7 in establishing the Feynman technique for polarization sums.

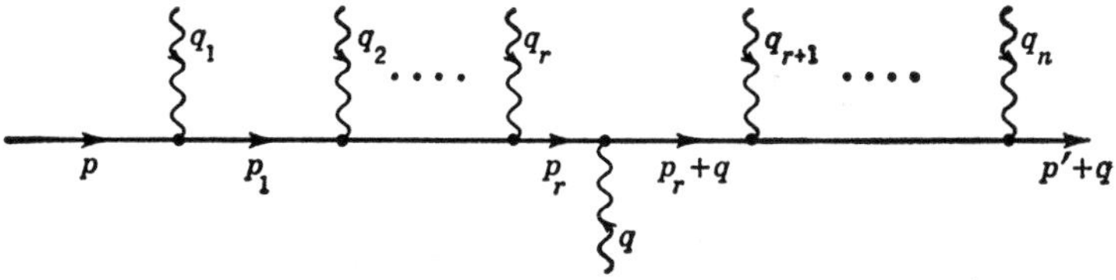

Fig. 17.14 Fermion line with n photon line endings plus an additional interaction vertex introducing momentum q.

The factors surrounding $\not q$ may be simplified, using the identity

$$\frac{1}{\not p_r + \not q - m} \not q \frac{1}{\not p_r - m} = \frac{1}{\not p_r - m} - \frac{1}{\not p_r + \not q - m}$$

and consequently[1]

$$\begin{aligned}\tilde{g}_r(p,p',q;q_i,a_i) &= \frac{1}{\not p' + \not q - m} \not a_n \cdots \frac{1}{\not p_{r+1} + \not q - m} \not a_{r+1} \\ &\times \frac{1}{\not p_r - m} \not a_r \cdots \not a_1 \frac{1}{\not p - m} - \frac{1}{\not p' + \not q - m} \not a_n \cdots \frac{1}{\not p_{r+1} + \not q - m} \\ &\times \not a_{r+1} \frac{1}{\not p_r + \not q - m} \not a_r \frac{1}{\not p_{r-1} - m} \not a_{r-1} \cdots \not a_1 \frac{1}{\not p - m} \qquad (17.63)\end{aligned}$$

Upon summation over all possible positions r where the photon can be inserted,[2] we see that the terms in (17.63) cancel pairwise; only two terms coming from the insertion of photon q to the right of q_n and the left of q_1 in Fig. 17.14 escape the cancellation:

$$\begin{aligned}\tilde{g}(p,p',q;q_i,a_i) &= \sum_r \tilde{g}_r = \frac{1}{\not p' - m} \not a_n \frac{1}{\not p_{n-1} - m} \not a_{n-1} \cdots \not a_1 \frac{1}{\not p - m} \\ &\quad - \frac{1}{\not p' + \not q - m} \not a_n \frac{1}{\not p_{n-1} + \not q - m} \not a_{n-1} \cdots \not a_1 \frac{1}{\not p + \not q - m} \\ &= g(p,p';q_1,a_1 \cdots q_n,a_n) \\ &\qquad - g(p+q,p'+q;q_1,a_1 \cdots q_n,a_n) \quad (17.64)\end{aligned}$$

If the fermion line is a line in an S-matrix element, p and $p' + q$ are momenta of real external particles and the contribution of the graph of Fig. 17.14 to the S matrix is obtained by removing the external

[1] If mass counterterms appear on the fermion line, the proof may still be carried through. However, the q_μ and q_ν terms do contribute to Z_2, owing to vanishing denominators on external lines.

[2] If q represents one end of a virtual photon line, *both* ends of which land on the same electron line in Fig. 17.14, the sum of diagrams for all permutations in the positions of both ends of this photon line gives twice the sum of Feynman diagrams for each fixed ordering of the remaining $n - 2$ vertices.

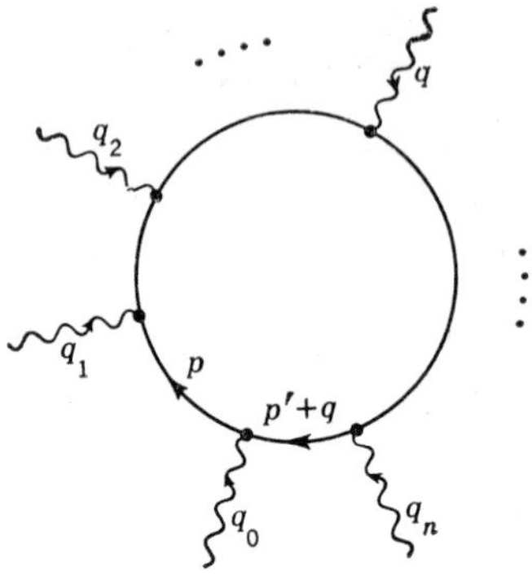

Fig. 17.15 Closed fermion loop with photon line endings.

propagator legs from $\tilde{g}$

$$S \propto \bar{u}(p' + q)[\lim_{\substack{p^2 \to m^2 \\ (p'+q)^2 \to m^2}} (\not{p}' + \not{q} - m)\tilde{g}(p,p',q;q_i,a_i)(\not{p} - m)]u(p) = 0 \tag{17.65}$$

On the other hand, if the fermion line is a closed loop, as illustrated in Fig. 17.15, the contribution is

$$\begin{aligned} \tilde{\mathfrak{B}} &\propto \int \frac{d^4p}{(2\pi)^4} \operatorname{Tr} \not{a}_0 \tilde{g}(p,p',q;q_i,a_i) \\ &= \int \frac{d^4p}{(2\pi)^4} \operatorname{Tr} \not{a}_0[g(p,p';q_i,a_i) - g(p + q, p' + q; q_i,a_i)] \end{aligned} \tag{17.66}$$

$\tilde{\mathfrak{B}}$ also vanishes provided the origin of integration in momentum space may be shifted, which is the case for a convergent integral. As in the case of the vacuum polarization integral discussed in Chap. 8, we assume that if the integral is divergent, it is regulated in such a way that this shift of origin may be carried out.

Thus assertion 1 has been established, and in computing S-matrix elements $D^{tr}_{F\mu\nu}$ may be replaced by

$$iD^{tr}_F(q)_{\mu\nu} \to +ig_{\mu\nu}D_F(q) - \frac{i\eta_\mu\eta_\nu}{(q\cdot\eta)^2 - q^2} \tag{17.67}$$

As remarked in (14.55), in coordinate space this means (in the preferred quantization frame)

$$iD^{tr}_F(x)_{\mu\nu} \to +ig_{\mu\nu}D_F(x) - \frac{i\delta(t)\eta_\mu\eta_\nu}{4\pi|\mathbf{x}|} \tag{17.68}$$

The last term has the same structure as the Coulomb term (17.60); we now establish assertion 2, which states that the last term in (17.68) cancels off the contributions of (17.60).

To establish 2 involves only a careful counting of graphs. The two ingredients in the Wick reduction are the interaction with the transverse photons:

$$\int H_I^{\mathrm{tr}}(t)\,dt = e_0 \int d^4y : \bar{\psi}_{\mathrm{in}}(y)\gamma_\mu \psi_{\mathrm{in}}(y) : A_{\mathrm{in}}^\mu(y) \equiv e_0 \int d^4y\, j_\mu^{\mathrm{in}}(y) A_{\mathrm{in}}^\mu(y) \quad (17.69)$$

and the instantaneous Coulomb interaction following from (17.60)

$$\int H_I^{\mathrm{coul}}\,dt = \frac{1}{2} e_0^2 \int d^4z\, d^4z'\, \delta(z_0 - z_0')\, j_\mu^{\mathrm{in}}(z) \frac{\eta^\mu \eta^\nu}{4\pi|\mathbf{z}' - \mathbf{z}|} j_\nu^{\mathrm{in}}(z') \quad (17.70)$$

We consider a term in the perturbation expansion (17.22) which is order $2n$ in H_I^{tr} and order m in H_I^{coul}. This will appear in the $(2n + m)$th $= N$th order in expanding $H_I = H_I^{\mathrm{tr}} + H_I^{\mathrm{coul}}$. However, the statistical weight is not $[N!]^{-1} = [(2n + m)!]^{-1}$ but $[(2n)!m!]^{-1}$, since

$$\frac{1}{N!}\left(\int H_I^{\mathrm{tr}}\,dt + \int H_I^{\mathrm{coul}}\,dt\right)^N = \frac{1}{N!}\sum_{m=0}^{N} \frac{N!}{m!(N - m)!}\left[\int H_I^{\mathrm{tr}}\,dt\right]^{N-m} \cdot \left[\int H_I^{\mathrm{coul}}\,dt\right]^m$$

These statistical factors cancel when one considers only topologically distinct graphs, since there are $(2n)!m!$ identical graphs obtained by permutation of y_i among themselves and z_i among themselves. In addition there are 2^m topologically equivalent graphs obtained by interchange of pairs of z_i and z_i', which cancels off all the factors $\frac{1}{2}$ in H_I^{coul}.

Thus the statistical weight associated with each topologically distinct graph is 1. For each graph involving exchange of a transverse photon (Fig. 17.16*a*) there is also a similar graph, with the same statistical weight, involving the instantaneous Coulomb interaction (Fig. 17.16*b*). The contribution to the τ function from these interactions

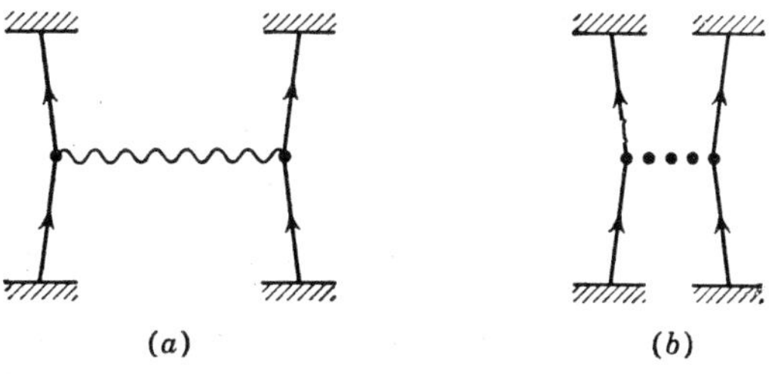

Fig. 17.16 Exchange of a transverse photon (*a*) and of an instantaneous Coulomb interaction (*b*) in a general graph.

is found from (17.68) and (17.70). For transverse photons, it is

$$\int d^4x\, d^4x'(-ie_0\gamma_\mu)\left[+ig^{\mu\nu}D_F(x-x') - \frac{i\delta(t-t')\eta^\mu\eta^\nu}{4\pi|\mathbf{x}-\mathbf{x}'|}\right](-ie_0\gamma_\nu) \tag{17.71}$$

and for the Coulomb force

$$-ie_0^2\int d^4x\, d^4x'\,\gamma_\mu \frac{\delta(t-t')\eta^\mu\eta^\nu}{4\pi|\mathbf{x}-\mathbf{x}'|}\gamma_\nu \tag{17.72}$$

Thus the Coulomb term removes the ugly right-hand term in (17.71) and leaves us with a covariant photon propagator.

17.10 Soft Photons Radiated from a Classical Current Distribution; the Infrared Catastrophe

We close this chapter by discussing the interaction of the quantized radiation field with a fixed classical current density. This physically interesting problem may be solved exactly; in particular, we are interested in calculating the number and frequency spectrum of photons radiated by the current. To this end we construct the S operator and then evaluate the matrix elements of interest.

In the radiation gauge the field equation to be solved is[1]

$$\square\mathbf{A}(x) = \mathbf{j}(x) \tag{17.73}$$

where $\mathbf{j}(x)$ is a prescribed c-number transverse current

$$\nabla\cdot\mathbf{j}(x) = 0 \tag{17.74}$$

The integral form of (17.73), which solves the equation, is, as in (16.146),

$$\begin{aligned}\mathbf{A}(x) &= \mathbf{A}_{\text{in}}(x) + \int d^4y\, D_{\text{ret}}(x-y)\mathbf{j}(y)\\ &= \mathbf{A}_{\text{out}}(x) + \int d^4y\, D_{\text{adv}}(x-y)\mathbf{j}(y)\end{aligned} \tag{17.75}$$

where we have set $Z_3 = 1$, since $\mathbf{A}$, $\mathbf{A}_{\text{in}}$, and $\mathbf{A}_{\text{out}}$ differ only by a c number and satisfy the same commutation relations. The S matrix is defined, according to (16.66), by

$$S^{-1}\mathbf{A}_{\text{in}}(x)S = \mathbf{A}_{\text{out}}(x) \tag{17.76}$$

and is a solution of

$$S^{-1}\mathbf{A}_{\text{in}}(x)S = \mathbf{A}_{\text{in}}(x) + \int d^4y\, D(x-y)\mathbf{j}(y) \tag{17.77}$$

[1] Notice that we now include the charge within the current $\mathbf{j}$.

where (see Appendix C)

$$-D(z) = D_{\text{adv}}(z) - D_{\text{ret}}(z) = -i \int \frac{d^4k}{(2\pi)^3} e^{-ik\cdot z}\delta(k^2)\epsilon(k_0) \quad (17.78)$$

It is convenient to work in k space and to expand $\mathbf{j}(y)$ in a Fourier series. With

$$\mathbf{j}(y) = \int_0^\infty \frac{dk_0}{2\pi} \int \frac{d^3k}{(2\pi)^{3/2}} \sum_{\lambda=1}^{2} \boldsymbol{\varepsilon}(k,\lambda)[j(k,\lambda)e^{-ik\cdot x} + j^*(k,\lambda)e^{ik\cdot x}] \quad (17.79)$$

(17.77) becomes

$$a_{\text{out}}(k,\lambda) = S^{-1}a_{\text{in}}(k,\lambda)S = a_{\text{in}}(k,\lambda) + \frac{ij(k,\lambda)}{\sqrt{2|\mathbf{k}|}}$$

$$S^{-1}a_{\text{in}}^\dagger(k,\lambda)S = a_{\text{in}}^\dagger(k,\lambda) - \frac{ij^*(k,\lambda)}{\sqrt{2|\mathbf{k}|}} \qquad k^2 = 0 \quad (17.80)$$

We may explicitly solve for S, since the field $a_{\text{in}}(k,\lambda)$ is displaced by a c number. The key for doing this is found in the identity, valid for $[A,B]$ a c number,

$$e^BAe^{-B} = A + [B,A] \quad (17.81)$$

Therefore, if B is chosen to be a linear combination of the a_{in} and $a_{\text{in}}^\dagger$ operators, we can hope to accomplish our goal. Unitarity of S restricts us to the form

$$S = \exp\left\{i \int d^3k \sum_{\lambda=1}^{2} [f(k,\lambda)a_{\text{in}}^\dagger(k,\lambda) + f^*(k,\lambda)a_{\text{in}}(k,\lambda)]\right\} \quad (17.82)$$

and upon inserting (17.82) into (17.80), we find consistency provided

$$f(k,\lambda) = \frac{j(k,\lambda)}{\sqrt{2|\mathbf{k}|}} \quad (17.83)$$

In order to evaluate matrix elements of S between in-states, it is convenient to normal-order S. For this purpose the following theorem is useful:

Theorem

If $[A,B] = C$ is a c number, then

$$e^{A+B} = e^Ae^Be^{-\frac{1}{2}[A,B]} \quad (17.84)$$

Proof. We define

$$F(\lambda) = e^{\lambda(A+B)}e^{-\lambda B}e^{-\lambda A}$$

Then

$$\frac{dF}{d\lambda} = e^{\lambda(A+B)}[A,e^{-\lambda B}]e^{-\lambda A}$$

It follows directly from (17.81) that

$$[A,e^{-\lambda B}] = -\lambda C e^{-\lambda B} \qquad \text{and} \qquad \frac{dF(\lambda)}{d\lambda} = -\lambda C F(\lambda)$$

Since $F(0) = 1$, it follows that $F(\lambda)$ is a c number; in particular,

$$F(1) = e^{-\frac{1}{2}C}$$

Applying this theorem to (17.82), we find

$$S = \exp\left[i\int \frac{d^3k}{\sqrt{2|\mathbf{k}|}} \sum_{\lambda=1}^{2} j(k,\lambda)a_{\text{in}}^{\dagger}(k,\lambda)\right]$$
$$\times \exp\left[i\int \frac{d^3k}{\sqrt{2|\mathbf{k}|}} \sum_{\lambda=1}^{2} j^*(k,\lambda)a_{\text{in}}(k,\lambda)\right]$$
$$\times \exp\left[-\frac{1}{2}\int \frac{d^3k}{2|\mathbf{k}|} \sum_{\lambda=1}^{2} |j(k,\lambda)|^2\right] \tag{17.85}$$

Knowing S, we may now calculate the probability P_n that n photons of specified polarizations and momenta are radiated by the current into some region $\mathcal{R}$ of the $3n$-dimensional phase space of the photons:

$$P_n(\mathcal{R},\lambda_1 \cdots \lambda_n) = \sum_{k_i \text{ in } \mathcal{R}} |\langle k_1\lambda_1 \cdots k_n\lambda_n \text{ out}|0 \text{ in}\rangle|^2$$
$$= \sum_{k_i \text{ in } \mathcal{R}} |\langle k_1\lambda_1 \cdots k_n\lambda_n \text{ in}|S|0 \text{ in}\rangle|^2 \tag{17.86}$$

In the expansion of S, as given by (17.85), in a power series, only the term with n creation operators and no destruction operators contributes to (17.86); therefore

$$P_n(\mathcal{R},\lambda_1 \cdots \lambda_n) = \left\{\exp\left[-\int \frac{d^3k}{2|\mathbf{k}|} \sum_{\lambda=1}^{2} |j(k,\lambda)|^2\right]\right\}$$
$$\times \sum_{k_i \text{ in } \mathcal{R}} \left|\langle k_1\lambda_1 \cdots k_n\lambda_n \text{ in}\right| \frac{i^n}{n!}$$
$$\times \left(\int \frac{d^3k}{\sqrt{2|\mathbf{k}|}} \sum_{\lambda=1}^{2} j(k,\lambda)a_{\text{in}}^{\dagger}(k,\lambda)\right)^n |0 \text{ in}\rangle|^2 \tag{17.87}$$

If we wish to calculate the radiation probability of photons into a given momentum interval R (that is, k_1 in R, . . . k_n in R) and with arbitrary polarizations, we may further simplify (17.87) by using closure techniques. We first write

$$P_n(R) = P_0 \sum_{\lambda_i=1}^{2} \sum_{k_i \text{ in } R} \left|\langle k_1\lambda_1 \cdots k_n\lambda_n \text{ in}\right| \frac{i^n}{n!} \times \left[\int_R \frac{d^3k}{\sqrt{2|\mathbf{k}|}} \sum_{\lambda=1}^{2} j(k,\lambda) a_{\text{in}}^\dagger(k,\lambda) \right]^n |0 \text{ in}\rangle|^2 \tag{17.88}$$

where
$$P_0 = \exp\left[-\int \frac{d^3k}{2|\mathbf{k}|} \sum_{\lambda=1}^{2} |j(k,\lambda)|^2 \right] \tag{17.89}$$

is the probability that no photons whatever are radiated.[1] Because all k_i are in the region R, the integral over $a_{\text{in}}^\dagger(k,\lambda)$ in (17.88) can be limited to this region R without changing $P_n(R)$. Then the sum $\sum_{\lambda_i=1}^{2} \sum_{k_i \text{ in } R}$ can be extended to all k_i not in R; furthermore, the sum can be extended to include all other in-states without changing $P_n(R)$:

$$\begin{aligned} P_n(R) &= P_0 \sum_\alpha \left|\langle \alpha \text{ in}\right| \frac{i^n}{n!} \left[\int_R \frac{d^3k}{\sqrt{2|\mathbf{k}|}} \sum_{\lambda=1}^{2} j(k,\lambda) a_{\text{in}}^\dagger(k,\lambda) \right]^n |0 \text{ in}\rangle|^2 \\ &= \frac{P_0}{(n!)^2} \langle 0 \text{ in}| \left[\int_R \frac{d^3k}{\sqrt{2|\mathbf{k}|}} \sum_{\lambda=1}^{2} j^*(k,\lambda) a_{\text{in}}(k,\lambda) \right]^n \\ &\quad \times \left[\int_R \frac{d^3k}{\sqrt{2|\mathbf{k}|}} \sum_{\lambda=1}^{2} j(k,\lambda) a_{\text{in}}^\dagger(k,\lambda) \right]^n |0 \text{ in}\rangle \\ &= \frac{P_0}{n!} \left[\int_R \frac{d^3k}{2|\mathbf{k}|} \sum_{\lambda=1}^{2} |j(k,\lambda)|^2 \right]^n \end{aligned} \tag{17.90}$$

If the region R is chosen to be all of momentum space, we see that the total probability P_n of radiating n photons follows a Poisson distribution:

$$P_n = \frac{e^{-\bar{n}}(\bar{n})^n}{n!} \tag{17.91}$$

[1] For a quantum-mechanical current source such as an electron scattering from a Coulomb potential (Chap. 7), we saw that graphs due to virtual intermediate photons had to be taken into account in order to obtain a finite result. These graphs take the place of the factor $P_0 = \langle 0 \text{ out}|0 \text{ in}\rangle$, since $|0 \text{ in}\rangle$ and $|0 \text{ out}\rangle$ "include" the electron state.

with the average number $\bar{n}$ of photons emitted given by

$$\bar{n} = \int \frac{d^3k}{2|\mathbf{k}|} \sum_{\lambda=1}^{2} |j(k,\lambda)|^2 = \sum_{n=0}^{\infty} nP_n \tag{17.92}$$

Evidently $\sum_{n=0}^{\infty} P_n = 1$ and probability is conserved.

The Poisson distribution indicates the statistical independence of successive emissions of the photons. This is a consequence of our assumption that the current source is fixed and is not altered by the emitted radiation; each photon is emitted under identical circumstances.

The total energy radiated may be computed from (17.86); it is

$$\begin{aligned}\bar{E} &= \sum_{n=0}^{\infty} \sum_{k_1\lambda_1,\ldots,k_n\lambda_n} |\langle k_1\lambda_1 \cdots k_n\lambda_n \text{ in}|S|0 \text{ in}\rangle|^2(k_1 + k_2 + \cdots + k_n) \\ &= \langle 0 \text{ in}|S^{-1}H_0(A_{\text{in}})S|0 \text{ in}\rangle \end{aligned} \tag{17.93}$$

where $$H_0(A_{\text{in}}) = \int d^3k\, k \sum_{\lambda=1}^{2} a_{\text{in}}^\dagger(k,\lambda)a_{\text{in}}(k,\lambda)$$

is the hamiltonian of the free radiation field. Referring back to (17.80), we find

$$\begin{aligned}\bar{E} &= \langle 0 \text{ in}| \int d^3k\, k \sum_{\lambda=1}^{2} \left[a_{\text{in}}^\dagger(k,\lambda) - \frac{ij^*(k,\lambda)}{\sqrt{2|\mathbf{k}|}}\right] \left[a_{\text{in}}(k,\lambda) + \frac{ij(k,\lambda)}{\sqrt{2|\mathbf{k}|}}\right] |0 \text{ in}\rangle \\ &= \frac{1}{2} \int d^3k \sum_{\lambda=1}^{2} |j(k,\lambda)|^2 \end{aligned} \tag{17.94}$$

in agreement with the classical result for the energy radiated from a current source.

If the current source radiates a finite amount of energy per unit frequency interval as $k \to 0$, $j(k,\lambda) \sim 1/k$ as $k \to 0$. For instance, for a point charge moving with the velocity $\boldsymbol{\beta}$ for $t < 0$, being kicked at $t = 0$, and moving with velocity $\boldsymbol{\beta}'$ thereafter,

$$\sum_{\lambda=1}^{2} \boldsymbol{\varepsilon}(k,\lambda)j(k,\lambda) \sim e\left(\frac{\boldsymbol{\beta}}{k - \mathbf{k}\cdot\boldsymbol{\beta}} - \frac{\boldsymbol{\beta}'}{k - \mathbf{k}\cdot\boldsymbol{\beta}'}\right)$$

For currents j which behave as k^{-1} as $k \to 0$, the average number of photons radiated is infinitely large, owing to the emission of an infinite number of photons of frequency $k \sim 0$, according to (17.92). This is the "infrared catastrophe," already encountered in Chaps. 7 and 8. Although the probability to emit any finite number of photons vanishes for such a current, (17.90) shows that the probability to emit any number of photons with $k \sim 0$ sums to a finite contribution. For instance, if we choose R to be the region $0 \leq |\mathbf{k}| \leq \Delta$,

$$\sum_{n=0}^{\infty} P_n(|\mathbf{k}| \leq \Delta) = \exp\left[-\int_{k>\Delta} \frac{d^3k}{2|\mathbf{k}|} \sum_{\lambda=1}^{2} |j(k,\lambda)|^2\right] \tag{17.95}$$

Since the ratio of the probability to emit $n + 1$ photons to that to emit n is given by

$$\frac{1}{n+1} \int d^3k \frac{|j(k,\lambda)|^2}{2|\mathbf{k}|} \sim \alpha \int \frac{dk}{k}$$

and diverges for $k \to 0$, a perturbation theory calculation based on an expansion in powers of α is not valid for long-wavelength radiation from an accelerated charge. The analysis and resolution of this divergence, in a manner similar to that presented above, was given by Bloch and Nordsieck[1] in 1937.

Problems

1. Prove (17.5) for the time development of the canonical momentum.

2. Solve for the U matrix of a Klein-Gordon particle coupled to a constant c-number current, that is

$$(\Box + m^2)\varphi = f$$

Discuss this in the light of Haag's theorem (footnote page 175) which states that $U(t)$ does not exist in such circumstances. Construct the S matrix.

3. Show that the vacuum bubbles comprising $\sum_k D_k$ in (17.38) add up to form the exponential $\left(\sum_i \frac{1}{\eta_i} B_i\right)$, where B_i is a connected bubble calculated by the rules of Appendix B and η_i is its order.

4. Give the general spin and isotopic spin structure of the π-nucleon scattering amplitude.

5. Show in detail that the S matrix for π-nucleon scattering is isospin independent in the limit of vanishing pion four-momenta (see Sec. 17.7).

[1] F. Bloch and A. Nordsieck, *Phys. Rev.*, **52,** 54 (1937); D. Yennie, S. Frautschi, and H. Suura, *Ann. Phys.* (*N.Y.*) **13,** 379 (1961).

6. Calculate the divergent part of (17.49) and show that a counterterm of the form (17.50) may be chosen to cancel it out.

7. Establish the crossing symmetry (17.53) for π-π scattering.

8. Repeat the argument (17.61) to (17.64) when mass counterterms are included, and show that the result is unchanged.

9. Starting with the lagrangian

$$\mathcal{L} = \mathcal{L}_D + \mathcal{L}_S + g\!:\!\bar{\Psi}\gamma_\mu\Psi \frac{\partial\varphi}{\partial x_\mu}\!:$$

with $\mathcal{L}_D$ and $\mathcal{L}_S$ the free lagrangians for a Dirac and scalar particle, respectively, solve for φ_{in} and φ_{out} and show that $S = 1$. Are the exact and free Feynman propagators of the Dirac particle equal; that is, is $S'_F = S_F$?

18

Dispersion Relations

18.1 Causality and the Kramers-Krönig Relation

The techniques discussed so far for computing transition amplitudes have been limited to a weak-coupling perturbation expansion in terms of the electric charge $\alpha = 1/137$, for instance. Using the apparatus of field theory, we have derived in a systematic (if formal) way the rules for drawing Feynman graphs and writing down the corresponding amplitudes.

However, these methods are of little value when we turn to the realm of strong interactions. The need to develop techniques more appropriate for this realm was illustrated earlier in the discussion of the meson-nucleon and nucleon-nucleon interactions in Chap. 10. At present the most successful approach to the strong-interaction problem appears to come from the method of "dispersion relations," which is based upon the exploration of the complex plane. The local structure of the field theory commutation relations and the field equations pose certain restrictions on the behavior of scattering amplitudes, studied as functions of energy and momentum transfer, when these variables are analytically continued from the physical domain of values which can be achieved in the laboratory to an unphysical domain in the complex plane. From these restrictions useful relations for evaluating these amplitudes or expressing them in terms of other measurable quantities are then constructed.[1]

This technique was inspired by the Kramers-Krönig relation in optics,[2] which we describe briefly as an illustrative example of the basic idea as well as of the utility of the dispersion method. The Kramers-Krönig relation expresses the real part of the amplitude for forward scattering by atoms of light of a fixed frequency ω as an integral over the cross section for absorption by atoms of light of all frequencies. In macroscopic terms, the real part of the index of refraction of a medium of such atoms is given by an integral over all frequencies of the imaginary part. This relation is derived by establishing that the forward-scattering amplitude is analytic in the upper half of the complex ω plane, a mathematical property based upon the physical limitation that electromagnetic signals cannot travel with a speed greater than that of light.

To be specific, we consider a monochromatic light wave

$$a_{\text{inc}}(\omega)e^{-i\omega(t-x)}$$

[1] M. Gell-Mann, M. L. Goldberger, and W. Thirring, *Phys. Rev.*, **95**; M. L. Goldberger, *Phys. Rev.*, **99**, 979 (1955). See also references in the preface.

[2] R. Krönig, *J. Op. Soc. Am.*, **12**, 547 (1926); H. A. Kramers, *Atti Congr. Intern. Fisici Como* (1927).

propagating up the x axis and impinging upon a scattering center. The wave scattered forward along the x axis will be linearly related to the incident wave by the forward-scattering amplitude

$$a_{\text{scatt}}(\omega) = f(\omega)a_{\text{inc}}(\omega)$$

and becomes asymptotically

$$a_{\text{scatt}}(x,t) \underset{x\to\infty}{\longrightarrow} a_{\text{scatt}}(\omega)\,\frac{e^{-i\omega(t-x)}}{x}$$

Superposing waves of different frequency to form a general packet, we write for the incident and forward scattered waves

$$A_{\text{inc}}(x,t) = \int_{-\infty}^{\infty} d\omega'\; a_{\text{inc}}(\omega')e^{-i\omega'(t-x)} \tag{18.1}$$

$$A_{\text{scatt}}(x,t) \underset{x\to\infty}{\longrightarrow} \frac{1}{x}\int_{-\infty}^{\infty} d\omega'\, f(\omega')a_{\text{inc}}(\omega')e^{-i\omega'(t-x)} \tag{18.2}$$

Let us suppose that the incident packet (18.1) represents a signal which vanishes for $x > t$. This physical condition implies a mathematical one on the Fourier amplitudes

$$a_{\text{inc}}(\omega) = \frac{1}{2\pi}\int_{-\infty}^{0} dx\; A_{\text{inc}}(x,0)e^{-i\omega x} \tag{18.3}$$

where the upper limit comes from the condition that $A_{\text{inc}}(x,0)$ vanishes for $x > 0$. We observe that $a_{\text{inc}}(\omega)$ may be analytically continued into the upper half of the complex ω plane, since for $\omega \to \omega + i|\gamma|$

$$a_{\text{inc}}(\omega + i|\gamma|) = \frac{1}{2\pi}\int_{-\infty}^{0} dx\; A_{\text{inc}}(x,0)e^{-i\omega x-|\gamma||x|} \tag{18.4}$$

and the integral is absolutely convergent. The physical demand of causality is

$$A_{\text{scatt}}\;(x,t) = 0 \qquad x > t$$

that is, no signal propagates to $x > t$, ahead of the incident wave front. By the same argument as above, we conclude from (18.2) that $a_{\text{inc}}(\omega)f(\omega)$ may also be analytically continued into the upper half plane. Consequently,[1] $f(\omega)$ is analytic in the upper half plane and we may write a Cauchy relation

$$f(z) = \frac{1}{2\pi i}\int_C \frac{d\omega'\, f(\omega')}{\omega' - z} \tag{18.5}$$

[1] Possible poles in f associated with zeros in $a_{\text{inc}}(\omega)$ are ruled out because the form of $a_{\text{inc}}(\omega)$, the Fourier transform of the incident packet, is quite arbitrary.

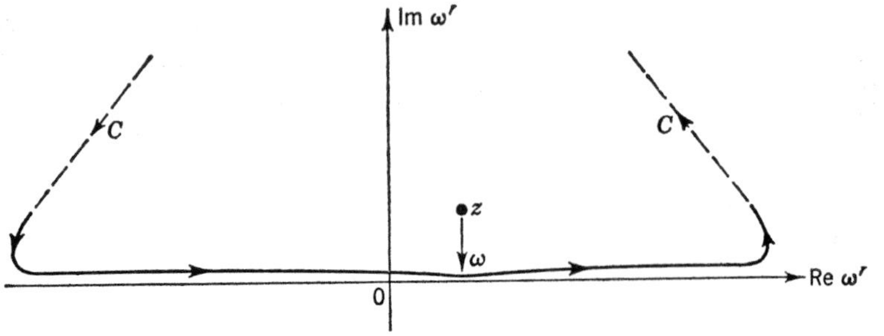

Fig. 18.1 Contour in upper half ω' plane for Cauchy relation (18.5).

for any $z = \omega + i|\gamma|$ in the upper half plane and for C the contour drawn in Fig. 18.1. Letting z approach real values ω from the upper half plane we find

$$f(\omega) = \lim_{\epsilon\to 0^+} f(\omega + i\epsilon) = \frac{1}{2\pi i} P \int_{-\infty}^{\infty} \frac{d\omega'\, f(\omega')}{\omega' - \omega} + \frac{1}{2} f(\omega) + \frac{1}{2}\mathcal{C}_\infty \quad (18.6)$$

where $P\int$ denotes the principal value of the integral along the real axis from $-\infty$ to ∞ in Fig. 18.1. The half-circuit around the pole at $\omega' = \omega$ gives the second term, and the contribution from the infinite semicircle is the complex quantity $\mathcal{C}_\infty = C_\infty + iC'_\infty$. The real and imaginary parts of (18.6) are

$$\operatorname{Re} f(\omega) = \frac{1}{\pi} P \int_{-\infty}^{\infty} \frac{d\omega' \operatorname{Im} f(\omega')}{\omega' - \omega} + C_\infty \quad (18.7a)$$

$$\operatorname{Im} f(\omega) = -\frac{1}{\pi} P \int_{-\infty}^{\infty} \frac{d\omega' \operatorname{Re} f(\omega')}{\omega' - \omega} + C'_\infty \quad (18.7b)$$

Equation (18.7*a*) is the real part of the equation

$$f(\omega) = \lim_{\epsilon\to 0^+} f(\omega + i\epsilon) = \lim_{\epsilon\to 0^+} \frac{1}{\pi} \int_{-\infty}^{\infty} \frac{d\omega' \operatorname{Im} f(\omega')}{\omega' - \omega - i\epsilon} + C_\infty \quad (18.8)$$

and is the more generally useful form of a dispersion relation.

The contribution C_∞ from the big semicircle at ∞ will not vanish if $f(\omega)$ does not approach zero as $\omega \to \infty$. It may be suppressed by making "subtractions." We may consider the amplitude $f(\omega)/\omega$ instead of $f(\omega)$ in constructing the Cauchy relation (18.5); the only difference is that $f(\omega)$ has an extra pole at $\omega = 0$ and better behavior at ∞. If $f(\omega)$ is bounded by a constant as $\omega \to \infty$, we find in place of (18.7*a*)

$$\frac{\operatorname{Re} f(\omega)}{\omega} = \frac{\operatorname{Re} f(0)}{\omega} + \frac{1}{\pi} P \int_{-\infty}^{\infty} \frac{d\omega' \operatorname{Im} f(\omega')}{\omega'(\omega' - \omega)} \quad (18.9)$$

which we call a dispersion relation with one subtraction. With the stronger convergence assumption that $f(\omega) \to 0$ as $\omega \to \infty$ we have the no subtraction version of (18.7*a*) with $C_\infty = 0$.

In either the form (18.7) or (18.9), the dispersion relation permits computation of the complete scattering amplitude from a knowledge of the imaginary part, plus its value at $\omega = 0$ if a subtraction is necessary [and derivatives of $f(\omega)$ at $\omega = 0$, if more subtractions are required]. As compensation for this advantage one needs the imaginary part for all frequencies in order to calculate the complete scattering amplitude at any one frequency.

Actually, in the present example this is scarcely a disadvantage because the imaginary part of the forward-scattering amplitude for positive frequencies ω is related by the optical theorem to the total cross section for absorption of light of that frequency

$$\operatorname{Im} f(\omega) = \frac{\omega}{4\pi} \sigma_{\text{tot}}(\omega) \qquad \omega > 0 \tag{18.10}$$

Furthermore, we can remove the range of negative ω from the dispersion integral (18.7) by observing from (18.1) and (18.2) that reality of the incident and scattered waves requires

$$a_{\text{inc}}(-\omega) = a_{\text{inc}}^*(\omega) \qquad f(-\omega) = f^*(\omega)$$

Therefore

$$\operatorname{Im} f(-\omega) = -\operatorname{Im} f(\omega)$$

and the dispersion integrals need be taken over the positive-frequency spectrum only:

$$\operatorname{Re} f(\omega) = \frac{2}{\pi} P \int_0^\infty \frac{\omega' \, d\omega'}{\omega'^2 - \omega^2} \operatorname{Im} f(\omega') \tag{18.11a}$$

or

$$\operatorname{Re} f(\omega) = \operatorname{Re} f(0) + \frac{2\omega^2}{\pi} P \int_0^\infty \frac{d\omega' \operatorname{Im} f(\omega')}{\omega'(\omega'^2 - \omega^2)} \tag{18.11b}$$

Thus with the use of the optical theorem (18.10) we arrive at a definite prediction based very generally only upon causality in the propagation of light signals and upon conservation of probability (unitarity) in the scattering process via the optical theorem. The real part of the coherent forward scattering of light in a medium of atoms, that is, the real part of the index of refraction, can be computed by the dispersion relation by either measuring or calculating the simpler quantity describing the absorption of light in the medium. This relation

$$\operatorname{Re} f(\omega) = \operatorname{Re} f(0) + \frac{\omega^2}{2\pi^2} P \int_0^\infty \frac{d\omega' \, \sigma_{\text{tot}}(\omega')}{\omega'^2 - \omega^2} \tag{18.12}$$

is the original Kramers-Krönig relation.

18.2 Application to High-energy Physics

The general program of dispersion relations follows closely the example of the Kramers-Krönig relation. The basic physical ingredient is a causality condition leading again to statements of analyticity properties of transition amplitudes of interest. The domain of analyticity having been established, these amplitudes are expressed via Cauchy's theorem in terms of residues at any poles and of discontinuities across branch cuts, or "absorptive parts," such as in (18.8):[1]

$$\lim_{\epsilon \to 0^+} [f(\omega + i\epsilon) - f(\omega - i\epsilon)] = 2i \operatorname{Im} f(\omega) \tag{18.13}$$

Then to obtain, hopefully, relations between physically observable quantities, one applies the analogue of (18.10) expressing unitarity of the S matrix in order to relate the absorptive parts to observable or computable amplitudes. Although the complications encountered in carrying out each of these four steps become considerably more intricate than in the Kramers-Krönig example, the basic ideas remain the same.

The common starting point for rigorous derivations of dispersion relations in field theory are the axioms pronounced in Chap. 16, including the crucial condition that the commutator of Bose-Einstein fields and the anticommutator of Fermi-Dirac fields must vanish for space-like separations, that is

$$\begin{aligned} \{\psi_i(x),\bar{\psi}_j(y)\} &= \{\psi_i(x),\psi_j(y)\} = 0 \\ [\varphi_i(x),\varphi_j(y)] &= 0 \\ [\varphi_i(x),\psi_j(y)] &= 0 \qquad (x-y)^2 < 0 \end{aligned} \tag{18.14}$$

These relations of local commutativity replace the causality condition used in our discussion of forward light scattering. From (18.14) it follows that measurements of amplitudes of Bose-Einstein fields and of bilinear forms representing local densities of physical operators (charge or energy, for example) of Fermi-Dirac fields do not interfere with each other if separated by a space-like interval; that is, their commutators vanish. These commutator conditions therefore assure the microscopic causality of the field theory.

As a consequence of the microscopic causality condition, it is possible to explore the complex plane and establish analyticity domains of Green's functions and S-matrix elements when one or more of their arguments is continued away from physical values on the real axis.

[1] We use the Schwartz reflection principle $f(\omega^*) = f^*(\omega)$ to define f in the lower half plane, separated by a branch cut from the upper half plane.

This is a major part of the general program of dispersion theory, and rigorous derivations of the analytic properties are much more complex and difficult than in the above example of forward scattering of light. One of the simple examples of the rigorous derivations is the Källén-Lehmann representation for propagators discussed in Chap. 16. For a boson field $\varphi(x)$, for example, the axioms of field theory, in particular the commutator condition (18.14), lead to the spectral decomposition (16.33)

$$\begin{aligned} \Delta'_F(x-y) &= -i\langle 0|T(\varphi(x)\varphi(y))|0\rangle \\ &= \int_0^\infty d\sigma^2\, \rho(\sigma^2)\Delta_F(x-y,\sigma) \end{aligned} \tag{18.15}$$

The Fourier transform

$$\Delta'_F(q) = \int_0^\infty \frac{d\sigma^2\, \rho(\sigma^2)}{q^2-\sigma^2+i\epsilon}$$

is seen to be analytic throughout the q^2 plane except for a cut along the positive real axis. Therefore $\Delta'_F(q)$ is determined completely in terms of the discontinuity across this cut as given by the spectral weight function $\rho(\sigma^2)$, which in practice may be evaluated approximately from the sum (16.27):

$$\rho(\sigma^2) = (2\pi)^3 \sum_n \delta^4(p_n-\sigma)|\langle 0|\varphi(0)|n\rangle|^2 \tag{18.16}$$

Equation (18.16) is the analogue of the optical theorem in the Kramers-Krönig example. It expresses the discontinuity $2\pi\rho(\sigma^2)$ in terms of the total probability for a virtual meson state $\varphi(0)|0\rangle$ with mass σ to decay into energetically available physical states $|n\rangle$, and it has a form typical of formulas for discontinuities arising in dispersion theory problems.

In generalizing this method from propagators to vertex functions and scattering amplitudes, one runs into formidable mathematical difficulties. There are more variables to handle, and the number of singularities and the complexity of their structure increase rapidly. Not only does the task of exploring the complex plane become arduous, but the singularities that are found are often uncivilized, located in unpleasant unphysical regions inaccessible to experiment. Then the unitarity conditions used to evaluate discontinuities must also be continued in order to domesticate these contributions. Despite these obstacles, much has been accomplished within the rigorous approach, in particular the fixed momentum transfer dispersion relations for π-nucleon scattering.[1]

[1] N. Bogoliubov, B. Medvedev, and M. Polivanov, *Fortschr. Physik,* **6,** 169 (1958); H. Bremermann, R. Oehme, and J. G. Taylor *Phys. Rev.,* **109,** 2178, (1958).

Here we shall confine ourselves to the much more modest program of investigating in a systematic way the singularities of individual Feynman graphs. The domain of analyticity common to all Feynman graphs for a given process certainly includes that resulting from the rigorous axiomatic approach, and at times exceeds it.[1] Then combining these analyticity properties with unitarity of the S matrix, we shall illustrate the utility of the dispersion method with some applications to the study of vertex functions and scattering amplitudes.

18.3 Analytic Properties of Vertex Graphs in Perturbation Theory

To illustrate techniques for determining analytic properties of general Feynman graphs, we consider first the lowest-order contribution to the electromagnetic structure of a π meson due to its coupling with neutron-antiproton pairs. This is the graph shown in Fig. 18.2, and the corresponding amplitude up to irrelevant constant factors is

$$(q_2 - q_3)_\mu F_\pi(q^2) \sim \int d^4p_1 \, \mathrm{Tr}\, \gamma_5 \frac{1}{\not p_1 - M} \gamma_5 \frac{1}{\not p_3 - M} \gamma_\mu \frac{1}{\not p_2 - M} \tag{18.17}$$

The momentum labels are illustrated in the figure, with the constraints

$$p_2 = p_1 + q_3 \qquad p_3 = p_1 - q_2 \qquad q = -q_2 - q_3 \tag{18.18}$$

and the restriction to real external pions

$$q_2^2 = q_3^2 = \mu^2$$

The nucleon mass M is understood to have an infinitesimal negative imaginary part.

[1] G. Källén and A. S. Wightman, *Kgl. Danske Videnskab. Selskab. Mat.-Fys. Skrifter*, **1**(6) (1958); R. Jost, *Helv. Phys. Acta*, **31**, 263 (1958).

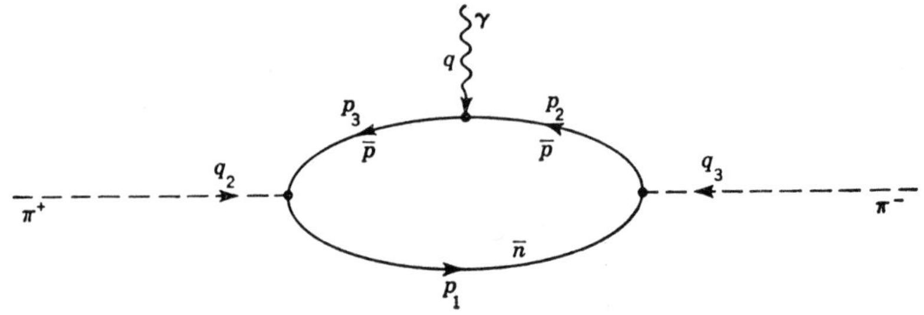

Fig. 18.2 Nucleon loop contribution to charged pion electromagnetic structure.

The vector form of (18.17) and the dependence of F_π on q^2 alone are determined by Lorentz invariance. The lack of a term proportional to $q_\mu = -(q_2 + q_3)_\mu$ is due to the requirement of current conservation, as discussed in Chap. 10. Here we are interested only in the analytic structure of $F_\pi(q^2)$, and to investigate this further, we introduce Feynman parameters α_1, α_2, and α_3 in order to combine denominators in (18.17) and do the momentum integration. Using the formula[1]

$$\frac{1}{ABC} = 2! \int_0^1 d\alpha_1 \int_0^1 d\alpha_2 \int_0^1 d\alpha_3 \frac{\delta(1 - \alpha_1 - \alpha_2 - \alpha_3)}{(A\alpha_1 + B\alpha_2 + C\alpha_3)^3} \tag{18.19}$$

we write

$$(q_2 - q_3)_\mu F_\pi(q^2) \sim \int d^4p_1 \int_0^1 d\alpha_1\, d\alpha_2\, d\alpha_3\, \delta\left(1 - \sum_{i=1}^{3} \alpha_i\right) \times \frac{\operatorname{Tr} \gamma_5(\not{p}_1 + M)\gamma_5(\not{p}_3 + M)\gamma_\mu(\not{p}_2 + M)}{\left[\sum_{i=1}^{3} (p_i^2 - M^2)\alpha_i\right]^3} \tag{18.20}$$

The momentum integral in (18.17) or (18.20) has an ultraviolet divergence corresponding to an infinite vertex renormalization, as studied in Chap. 8. It is defined with a regulator or ultraviolet cutoff, as discussed there, but this is irrelevant in our present discussion of the analytic properties of $F_\pi(q^2)$ as a function of q^2.

We interchange orders of integration[2] and first perform the momentum integral by shifting the origin in p_1 space:

$$p_1 = k_1 + l \qquad p_2 = k_2 + l \qquad p_3 = k_3 + l \tag{18.21}$$

where l is the new integration variable and the k_i are related by the same constraints of momentum conservation as the p_i in (18.18)

$$k_1 = k_2 - q_3 = k_3 + q_2 \tag{18.22}$$

The denominator of the integrand in (18.20) is the cube of

$$D = \sum_i (k_i^2 - M^2)\alpha_i + l^2 \sum_i \alpha_i + 2l \cdot \sum_i k_i\alpha_i \tag{18.23}$$

In order to remove the cross term in D, we set

$$\sum_{i=1}^{3} k_i\alpha_i = 0 \tag{18.24}$$

[1] Recall (8.59). Notice also that the formula is valid provided only that the imaginary parts of A, B, C all have the same sign.

[2] The momentum integral is absolutely convergent upon regulating, so that the interchange is legitimate.

These four equations (18.24) plus the eight momentum constraints (18.22) uniquely determine the k_i to be

$$k_1 = \frac{q_2\alpha_3 - q_3\alpha_2}{\Delta} \qquad k_3 = \frac{q\alpha_2 - q_2\alpha_1}{\Delta}$$
$$k_2 = \frac{q_3\alpha_1 - q\alpha_3}{\Delta} \qquad \Delta = \sum_{i=1}^{3} \alpha_i = 1 \tag{18.25}$$

The result of carrying out the l integrations[1] is

$$F_\pi(q^2) \sim \int_0^1 d\alpha_1 \int_0^1 d\alpha_2 \int_0^1 d\alpha_3\, \delta\left(1 - \sum_{i=1}^{3} \alpha_i\right) \times \left[\frac{N_1(q^2,\alpha_i)}{J(q^2,\alpha_i)} + N_2(q^2,\alpha_i) \ln \frac{\Lambda^2}{-J(q^2,\alpha_i)}\right] \tag{18.26}$$

N_1 and N_2 come from the numerator in (18.20) and are linear polynomials in q^2 with functions of α_i as coefficients. Λ^2 is an ultraviolet cutoff on the integral. The analytic properties are determined by the function $J(q^2,\alpha_i)$, which is what is left of the denominator D after the l integration:

$$\begin{aligned} J(q^2,\alpha_i) &= \sum_{i=1}^{3} (k_i^2 - M^2)\alpha_i \\ &= q^2\alpha_2\alpha_3 + q_3^2\alpha_2\alpha_1 + q_2^2\alpha_1\alpha_3 - M^2 \\ &= q^2\alpha_2\alpha_3 + \mu^2\alpha_1(1 - \alpha_1) - M^2 \end{aligned} \tag{18.27}$$

The last two forms of (18.27) follow from the conditions (18.25) and the mass shell conditions $q_2^2 = q_3^2 = \mu^2$ for the π meson.

We see from (18.26) that there can be trouble in either term of (18.26) if and only if J vanishes in the integration domain of the α's. Since the zeros of J control the analytic structure of $F_\pi(q^2)$, F_π has the same analytic properties as

$$I(q^2) = \int_0^1 d\alpha_1 \int_0^1 d\alpha_2 \int_0^1 d\alpha_3 \frac{\delta(1 - \alpha_1 - \alpha_2 - \alpha_3)}{q^2\alpha_2\alpha_3 + \mu^2\alpha_1(1 - \alpha_1) - M^2 + i\epsilon} \tag{18.28}$$

The inessential numerator factors have been dropped and the infinitesimal negative imaginary part of the mass in the Feynman propagator is now explicitly written.

$I(q^2)$, and thus $F_\pi(q^2)$, is analytic in the upper half of the complex q^2 plane since the integral (18.28) exists for all $q^2 = u + iv$ with $v > 0$. If there exist real values of q^2 such that $I(q^2)$ is real, we may continue

[1] As in Chap. 8; see footnote page 170.

$F_\pi(q^2)$ into the lower half plane by the Schwartz reflection principle, giving

$$F_\pi(u - iv) = F_\pi^*(u + iv)$$

We expect this to be so on the basis of a physical argument. When q^2 is space-like ($q^2 < 0$), $F_\pi(q^2)$ contributes to the amplitude for electron-pion scattering, which is real in first Born approximation in the electric charge. This may be checked explicitly. The denominator J is negative definite at $q^2 = 0$, being given by

$$J(0,\alpha_i) = \mu^2\alpha_1(1 - \alpha_1) - M^2 \leq \frac{\mu^2}{4} - M^2 < 0 \qquad 0 \leq \alpha_1 \leq 1$$

for $\mu^2 < 4M^2$, as in the physical case of pions and nucleons.[1] Thus there is no contribution to an imaginary part to (18.28) in the limit $\epsilon \to 0^+$, a result which evidently holds for $q^2 < 0$ as well.

Consequently, $F_\pi(q^2)$ can be extended into the lower half q^2 plane and is analytic throughout the entire complex plane with the possible exception of a cut along the real positive q^2 axis, due to the vanishing of J in this region. The onset of this cut occurs at the lowest value of q^2 such that $J = 0$. We may determine this branch point by maximizing J in (18.27) with respect to parameters α_i. Maximizing with respect to α_2 and α_3, we find

$$\alpha_2 = \alpha_3 = \frac{1 - \alpha_1}{2}$$

and

$$J \leq q^2 \left(\frac{1 - \alpha_1}{2}\right)^2 + \mu^2\alpha_1(1 - \alpha_1) - M^2 \qquad (18.29)$$

The right side of (18.29) attains its maximum either at $\alpha_1 = 0$ or somewhere between 0 and 1. However, by direct calculation one finds that for $q^2 > 2\mu^2$ the maximum value must occur at the end point $\alpha_1 = 0$. For physical values of the π and N masses, $\mu^2/M^2 \cong 0.022$, J is still negative definite at $q^2 = 2\mu^2$; therefore, the threshold value q_t^2 for the onset of the branch cut occurs[2] at $q_t^2 = 4M^2$, corresponding to the point in the α_i-parameter space

$$\alpha_1 = 0 \qquad \alpha_2 = \alpha_3 = \tfrac{1}{2} \qquad q_t^2 = 4M^2 \qquad (18.30)$$

when J vanishes. This threshold energy $q_t^2 = 4M^2$ is just that energy for which a virtual time-like photon, such as produced by a colliding

[1] Notice that $\mu^2 < 4M^2$ is just the condition that the physical π^- be stable against decay into $n\bar{p}$.

[2] For $\mu^2 > 2M^2$, the cut starts at $q_t^2 = 4\mu^2(1 - \mu^2/4M^2)$, known as an anomalous threshold. We discuss this case later.

electron-positron pair, can produce a real nucleon-antinucleon pair which then annihilates into a π^+-π^- pair via the graph of Fig. 18.2, turned on its side. Discussion of the analytic properties of the vertex function $F_\pi(q^2)$ is thus seen to require simultaneous study of the pair production as well as the scattering of π mesons—the momentum transfer being space-like when a pion of mass μ is scattered between positive-energy states and time-like when scattered from a negative-energy to a positive-energy state, that is, pair-produced.

The result (18.30) suggests that singularities of the Green's functions occur at momenta for which absorptive processes are physically allowed. This general connection between the singularities and probabilities for absorptive processes is the great virtue in the dispersion approach.

18.4 Generalization to Arbitrary Graphs and the Electrical Circuit Analogy

Proceeding from this special perturbation example to arbitrary graphs, it is useful for us to isolate the important factors from the inessential complications and to develop a standard notation and convention for the momentum variables and integration parameters. Several points should already be clear:

1. A general matrix element will take the form

$$\mathfrak{M} = \sum_i \mathcal{O}_i F_i(q_1, \ldots, q_m) \tag{18.31}$$

where the $\mathcal{O}_i$ are appropriate products of external momenta, spin matrices, and wave functions and the F_i are invariant functions of scalar products of the external momenta. Our interest is in the study of the analytic structure of the F_i.

2. Numerators of the integrands such as the trace in (18.20) are irrelevant, since they lead only to polynomial factors in the external momenta after the integrand has been parametrized and integrals over internal momenta have been carried out. These polynomial factors appear in the F_i only as polynomials of the scalar products $q_i \cdot q_j$.

3. The problem of ultraviolet divergences is not important here. If more powers of the momentum variables are needed in the denominators, as was the case in (18.20), they may be found by differentiating with respect to external momenta or with respect to masses of internal lines. This does not alter the analytic structure of the function the

integral represents. For example, an extra power of the denominator in (18.20) would converge the integral over p_1 without in any way altering the structure of singularities, which are determined by the zeros of J in (18.27).

We shall always denote the external momenta of the graphs we study by $q_1, \ldots, q_m$, defined with the sense of entering the graph. Connected graphs only are considered and for these

$$\sum_{s=0}^{m} q_s = 0$$

representing over-all momentum conservation. To each internal line we assign a momentum p_j with a specified direction and a mass m_j, as illustrated by Fig. 18.3. At each vertex, we have a law of momentum conservation of the form

$$\sum_{j=1}^{n} \epsilon_{ij} p_j + \sum_{s=1}^{m} \tilde{\epsilon}_{is} q_s = 0 \tag{18.32}$$

where
$$\epsilon_{ij} = \begin{cases} +1 & \text{if internal line } j \text{ enters vertex } i \\ -1 & \text{if internal line } j \text{ leaves vertex } i \\ 0 & \text{otherwise} \end{cases}$$

$\tilde{\epsilon}_{is}$ is similarly defined for the external lines, which by convention we take always to enter vertices.

Each graph has a definite number k of internal loops, or closed paths, over which momentum integrals over internal variables called l_r are to be carried out. Although the number of independent l_r is determined, there is an arbitrariness in how one chooses them. For example, two possible choices of the two loops for the diagram of Fig. 18.3 are shown in Fig. 18.4. Having chosen the loops in a particular way and assigned to them a direction, our task, after introducing

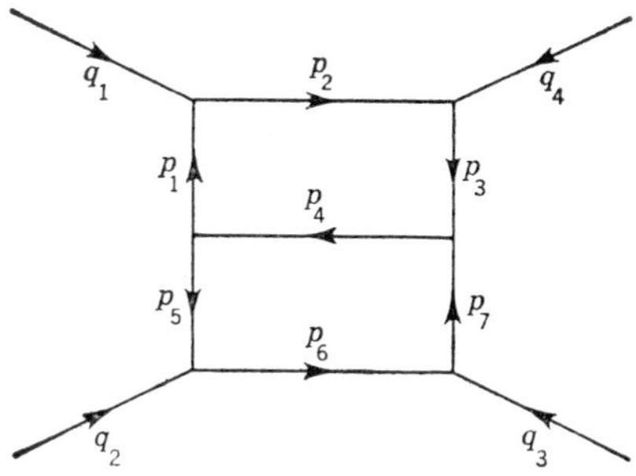

Fig. 18.3 General Feynman graph with internal lines labeled by momenta p_j and external ones by q_s entering the graph.

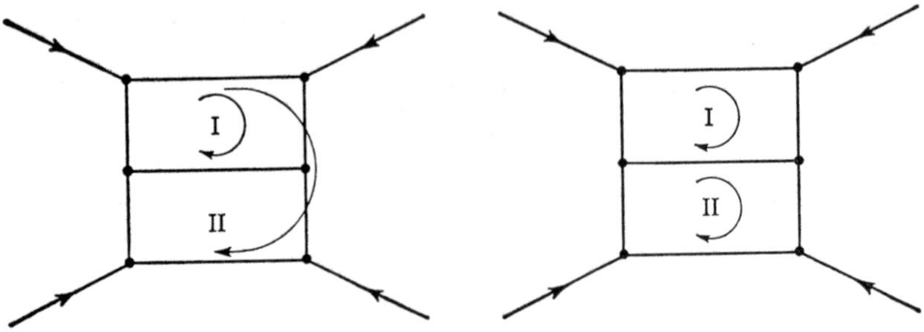

Fig. 18.4 Two possible choices of loops.

Feynman parameters, is to pick the l_r appropriately so that we can do the momentum integrals. In the perturbation example (18.20) with one loop we shifted the origin of integration to complete the square in the denominator D [Eqs. (18.21) to (18.24)]. The general procedure is to do the same thing for each loop. We write

$$p_j = k_j + \sum_{r=1}^{k} \eta_{jr} l_r \tag{18.33}$$

$$\text{where } \eta_{jr} = \begin{cases} +1 & \text{if the } j\text{th internal line lies on the } r\text{th loop and } p_j \text{ and } l_r \text{ are parallel} \\ -1 & \text{if the } j\text{th internal line lies on the } r\text{th loop and } p_j \text{ and } l_r \text{ are antiparallel} \\ 0 & \text{otherwise} \end{cases}$$

and the optimum choice of the k_j is made after introducing Feynman parameters α_j and combining denominators as follows.

To determine the analytic structure of a scalar "form factor" F_i in (18.31) for a given graph, it is sufficient to study the integral

$$I(q_1, \ldots, q_m) = \int \frac{d^4l_1 \cdots d^4l_k}{(p_1^2 - m_1^2) \cdots (p_n^2 - m_n^2)}$$

$$\sim \int d^4l_1 \cdots d^4l_k \int_0^1 \frac{d\alpha_1 \cdots d\alpha_n\, \delta\left(1 - \sum_{j=1}^{n} \alpha_j\right)}{\left[\sum_{j=1}^{n} (p_j^2 - m_j^2)\alpha_j\right]^n}$$

$$= \int d^4l_1 \cdots d^4l_k \int_0^1 d\alpha_1 \cdots d\alpha_n\, \delta\left(1 - \sum_j \alpha_j\right)$$

$$\times \frac{1}{\left[\sum_j (k_j^2 - m_j^2)\alpha_j + 2\sum_{j,r} k_j \alpha_j \eta_{jr} \cdot l_r + \sum_{j,r,r'} \alpha_j \eta_{jr} \eta_{jr'}\, l_r \cdot l_{r'}\right]^n} \tag{18.34}$$

In order to eliminate the cross term in the denominator of (18.34), we choose

$$\sum_{j=1}^{n} k_j \alpha_j \eta_{jr} = 0 \qquad \text{for each loop } r = 1, \ldots, k \tag{18.35}$$

The integrations over the l_r may then be carried out by diagonalizing the hermitian matrix

$$z_{rr'} = \sum_{j=1}^{n} \eta_{jr}\eta_{jr'}\alpha_j$$

with the result

$$I \propto \int_0^\infty \frac{d\alpha_1 \cdots d\alpha_n\, \delta\left(1 - \sum_{j=1}^{n} \alpha_j\right)}{\Delta^2 \left[\sum_{j=1}^{n} (k_j^2 - m_j^2)\alpha_j\right]^{n-2k}} \tag{18.36}$$

with

$$\Delta = \det ||z||$$

The momenta k_j are functions of the external momenta q_s and the Feynman parameters. They are determined by (18.35) and the momentum conservation laws at each vertex (18.32) with p_j replaced by k_j

$$\sum_{j=1}^{n} \epsilon_{ij} k_j + \sum_{s=1}^{m} \tilde{\epsilon}_{is} q_s = 0 \tag{18.37}$$

which follow from (18.32) and

$$\sum_j \epsilon_{ij}\eta_{jr} = 0$$

the latter a consequence of the definitions of ϵ_{ij} and η_{jr} given in (18.32) and (18.33).

The equations (18.35) and (18.37) determining the k_j suggest a precise analogue in electrical circuit theory when we put them in the more heuristic form (ignoring minus signs)

$$\sum_{k_j \text{ in loop } r} k_j \alpha_j = 0 \tag{18.38a}$$

$$\sum_{\substack{k_j, q_s \text{ entering} \\ \text{vertex } i}} (k_j + q_s) = 0 \tag{18.38b}$$

Think of the Feynman diagram as an electrical circuit and associate the momenta—as is natural—with the currents. The k_j are the internal currents flowing in the circuit and the q_s the external currents entering it. If we associate the parameters α_j with the resistance of the jth line, Eqs. (18.38) become simply Kirchhoff's laws in this circuit analogy.[1] Equation (18.38*a*) states that the sum of "voltage drops" around any closed loop is zero, and (18.38*b*) states that the sum of "currents" entering a vertex is zero.[2]

It is instructive to return to the calculations leading from (18.34) to (18.36) with this circuit analogy in mind. Aside from the constant term $-\sum_j m_j^2\alpha_j$, we recognize the denominator in (18.34)

$$\sum_j p_j^2\alpha_j$$

as the power burned up in the circuit with external current sources q_s and the internal sources l_r. By energy conservation this is just the sum of powers delivered by the external and internal sources

$$\sum_j p_j^2\alpha_j = \sum_j k_j^2\alpha_j + \sum_{r,r'} l_r\cdot l_{r'}z_{rr'}$$

where the k_j are the currents, given by Kirchhoff's laws, which run through the internal lines in the absence of internal sources l_r.

In our picture we have currents and resistances as the analogues of physical momenta and of Feynman parameters. To complete this picture, we may ask now what the physical analogue to the "voltage drop" means. Since "voltage" is a potential and evidently is a property of a vertex, it is natural to associate voltage with the coordinate x_μ of the vertex. The first of Kirchhoff's laws (18.38*a*) is then translated back to a simple statement that the sum of coordinate displacements around any closed loop in a Feynman graph vanishes.

We may continue to reconstruct a physical picture of the electrical circuit analogy by inquiring into the physical meaning of Ohm's law

$$V = IR$$

In physical language this becomes

$$\Delta x_\mu = k_\mu\alpha \tag{18.39}$$

where k_μ is the momentum in some line, α is the corresponding Feynman parameter, and Δx_μ is the coordinate displacement between the vertices which the line connects. Equation (18.39) is just the equation of

[1] The α_j are real; therefore, no capacitances or inductances appear in the circuit.

[2] This circuit analogy ensures that the k_j are determined uniquely.

motion of a free particle, as better seen in component form:

$$\mathbf{\Delta x} = \mathbf{k}\alpha \qquad \Delta t = k_0\alpha \qquad \frac{\mathbf{\Delta x}}{\Delta t} = \frac{\mathbf{k}}{k_0} \tag{18.40}$$

Since the parameter α is never negative, causal propagation of the particle is guaranteed:

$$\frac{\Delta t}{k_0} = \alpha > 0$$

As the particle progresses in the $\mathbf{k}$ direction according to (18.40), it moves either forward or backward in time depending on whether the sign of the energy k_0 is positive or negative. This is the same interpretation we used in the Feynman propagator theory development in Chap. 6. That α is ≥ 0 is in fact a direct consequence of the use of the Feynman propagator with its negative imaginary $i\epsilon$ added on to the mass. This is best seen by referring back to (8.12) and (8.18)

$$\prod_{j=1}^{n} \frac{1}{a_j + i\epsilon}$$

$$= i^{-n} \int_0^\infty d\alpha_1 \cdot\cdot\cdot d\alpha_n \left\{\exp\left[i \sum_j \alpha_j a_j - \epsilon\left(\sum_j \alpha_j\right)\right]\right\} \int_0^\infty \frac{d\lambda}{\lambda}\, \delta\left(1 - \frac{\sum_j \alpha_j}{\lambda}\right)$$

$$= (n-1)! \int_0^\infty d\alpha_1 \cdot\cdot\cdot d\alpha_n \frac{\delta(1 - \sum_j \alpha_j)}{\left(\sum_j a_j\alpha_j + i\epsilon\right)^n} \tag{18.41}$$

We have, therefore, a complete physical picture of the circuit analogy with the following correspondences:

Coordinate $\leftrightarrow$ voltage
Momentum $\leftrightarrow$ current
Feynman parameter $\alpha = \dfrac{\text{proper time}}{\text{mass}} \leftrightarrow$ resistance > 0
Free-particle equation of motion $\leftrightarrow$ Ohm's law

and with the requirement of positive "resistance" α related to causality of propagation of the particles.

This circuit analogy will be very valuable to us, because we can use intuitive understanding as well as the established lore and theorems of circuit theory in analyzing a Feynman diagram. For example, consider once again the vertex function with three external lines and arbitrary internal complications. Such a black box with only resistors

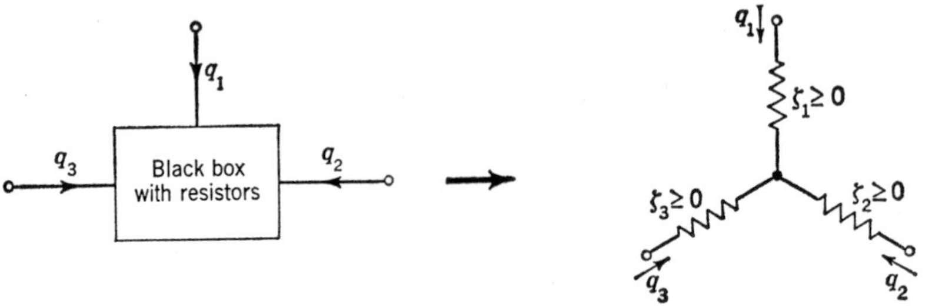

Fig. 18.5 Equivalent lumped circuit diagram for a vertex with three external lines.

inside can always be reduced to an equivalent lumped circuit[1] of Fig. 18.5. The power P burned up in such a circuit is

$$P = \zeta_1 q_1^2 + \zeta_2 q_2^2 + \zeta_3 q_3^2 \qquad \zeta_i \geq 0 \tag{18.42}$$

where the equivalent resistances ζ_i of the lumped circuit are nonnegative functions of the original network resistors α_j. This result is of immediate use, since from (18.36) we may write

$$I(q_1,q_2,q_3) \sim \int_0^1 \frac{d\alpha_1 \cdots d\alpha_n \delta(1 - \Sigma\alpha_j)}{\Delta^2(\zeta_1 q_1^2 + \zeta_2 q_2^2 + \zeta_3 q_3^2 - \Sigma m_j^2 \alpha_j + i\epsilon)^{n-2k}} \tag{18.43}$$

where we again exhibit explicitly the $i\epsilon$ behavior. Equation (18.43) shows the generalization to an arbitrary order of perturbation theory of the result found in the lowest order for Fig. 18.2. Since the ζ_i are nonnegative, $I(q_1^2,q_2^2,q_3^2)$, as a function of any one of the three external momenta when the other two are held constant and real, is analytic throughout the entire upper half complex q_i^2 plane. In fact, $I(q_1^2,q_2^2,q_3^2)$ is an analytic function of all three variables as these are simultaneously extended into the upper half plane.

Our method for arriving at these analyticity properties is a long way from the causality requirement used in constructing the Kramers-Krönig relation. Nevertheless, it was causality in the equation of free-particle motion (18.40), expressed by the statement $\alpha \geq 0$, that ensured the conditions in (18.42) leading to the derived analyticity properties.

18.5 Threshold Singularities for the Propagator

We now attack the problem of determining necessary conditions for a singularity of a general graph, starting with the propagator. We shall

[1] E. Guillemin, *Introductory Circuit Theory*, John Wiley & Sons, Inc., New York, 1953.

continue to make use of the electrical circuit analogy to guide us through the tangle of propagator lines in a general graph and to find singularities of form factors F_i from the structure of denominators

$$J = \sum_j (k_j^2 - m_j^2)\alpha_j \tag{18.44}$$

appearing in (18.36) or (18.43).

The analytic properties of the propagator of a spin-0 particle are summarized in the Källén-Lehmann representation (18.15), with similar results for nonzero spins, as discussed in Chap. 16. It is instructive, however, to rederive these analytic properties by using the electrical circuit analogy, because the techniques we develop can then be applied to vertex functions and scattering amplitudes. In the case of a self-energy diagram there is only one external momentum q, and the analytic properties of the form factors F_i are the same as those of the integral (18.36)

$$I(q^2) = \int_0^\infty \frac{d\alpha_1 \cdots d\alpha_n \, \delta\left(1 - \sum_j \alpha_j\right)}{\Delta^2 \left(\zeta q^2 - \sum_j m_j^2\alpha_j + i\epsilon\right)^{n-2k}} \tag{18.45}$$

The quantity ζq^2 in the denominator of (18.45) is the power burned up in the two-terminal network corresponding to the self-energy graph, and the equivalent lumped resistance ζ must therefore be nonnegative. Using the same arguments as we did for the perturbation example of the vertex, we conclude[1] from (18.45) that (1) I is analytic in the upper half q^2 plane, (2) I is real for q^2 real and negative, and (3) by the Schwartz reflection principle I is analytic in the lower half q^2 plane, possessing at most a branch cut from $q^2 = 0$ to $q^2 = \infty$.

To find the threshold value of the branch cut q_t^2, we look for the lowest value of q^2 for which the denominator J in (18.45) can vanish. Starting from zero, we increase q^2 until we reach q_t^2; at this value there will be at least one point $(\alpha_1^0, \ldots, \alpha_n^0)$ in the n-dimensional α space at which J vanishes

$$J(q_t^2; \alpha_1^0, \alpha_2^0, \ldots, \alpha_n^0) = 0 \tag{18.46}$$

This point can be chosen to satisfy the constraint $\sum_j \alpha_j = 1$ since J satisfies a scaling law:

$$J(\lambda\alpha_j) = \lambda J(\alpha_j) \tag{18.47}$$

[1] In addition, $\Delta > 0$ for $\alpha_j \geq 0$, except when all $\alpha_j = 0$ (Prob. 1). Δ will play no role in discussion of analytic properties.

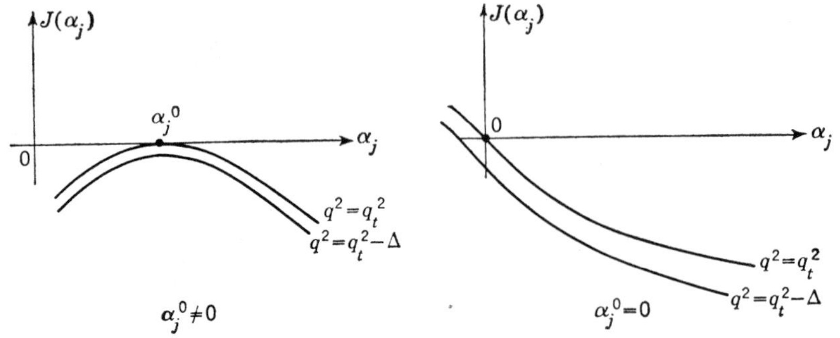

Fig. 18.6 Vanishing of "action" $J(\alpha)$ at a threshold singularity $q^2 = q_t^2$, with a maximum at α_j^0 or at an end point $\alpha_j^0 = 0$.

This scaling law follows from the form (18.44) and the observation that if all resistances in an electrical circuit are doubled for fixed external currents, the power doubles. Therefore the δ-function constraint on the n parameters α_j may be ignored and we can study J as function of n independent nonnegative parameters α_j.

Although J vanishes when $q^2 = q_t^2$ and the α_j are set equal to the α_j^0, by continuity in the variable q^2, J cannot be greater than zero within the integration domain:

$$J(q_t^2;\alpha_1^0, \ldots ,\alpha_j, \ldots ,\alpha_n^0) \le 0 \qquad \text{for all } \alpha_j \ge 0 \quad (18.48)$$

We conclude that either J is a maximum with respect to the individual α_j at $\alpha_j = \alpha_j^0$ or that α_j^0 lies at an end point. That is, for each α_j either

$$\left.\frac{\partial J(q_t^2;\alpha_1^0, \ldots ,\alpha_j, \ldots ,\alpha_n^0)}{\partial \alpha_j}\right|_{\alpha_j=\alpha_j^0} = 0 \qquad \text{if } \alpha_j^0 \ne 0 \quad (18.49a)$$

or

$$\alpha_j^0 = 0 \qquad (18.49b)$$

The two cases are illustrated in Fig. 18.6. The condition $\alpha_j^0 = 0$ corresponds to a short circuit in the electrical circuit analogy. The vertices connected by such branches with zero resistance therefore have the same voltage—or coordinate. Contracting together all such vertices which are short-circuited by condition (18.49*b*), we arrive at a "reduced graph." In the remaining lines of the reduced graph, the α_j^0 do not vanish. For example, in the perturbation calculation of the vertex discussed in Sec. 18.3, the α_j^0 corresponding to conditions (18.49) were given by (18.30); upon short-circuiting the vertices connected by internal line 1 ($\alpha_1 = 0$), we find the reduced graph illustrated in Fig. 18.7.

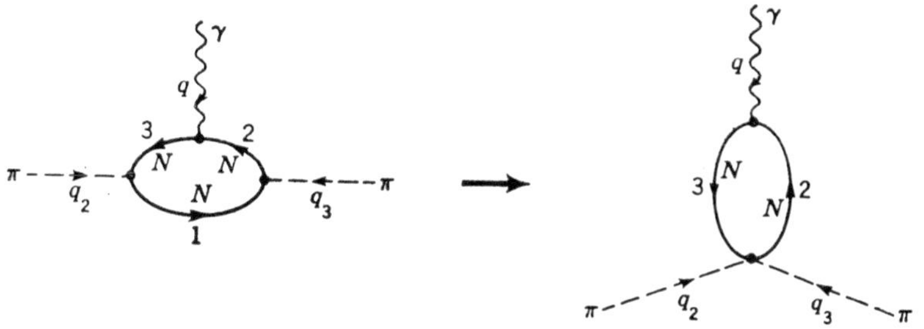

Fig. 18.7 Reduced graph with line 1 "short-circuited."

In all lines remaining in a reduced graph, as in Fig. 18.7, the α_j^0 do not vanish. Therefore, J is stationary with respect to variations of these nonvanishing α_j^0 according to (18.49a)

$$\delta J = \delta \left[\sum_j (k_j^2 - m_j^2)\alpha_j\right]$$
$$= \sum_j (k_j^2 - m_j^2)\,\delta\alpha_j + 2\sum_j k_j\alpha_j^0 \cdot \delta k_j \qquad \text{at } \alpha_j = \alpha_j^0 \tag{18.50}$$

The second term, present because the k_j are functions of the α_j through Kirchhoff's laws, vanishes because the process of "completing the square" has already made J stationary[1] with respect to variation of internal momenta. To see this, we observe that because the external currents are not being varied ($\delta q_s = 0$) and because the δk_j satisfy momentum conservation at the vertices, it is possible to find loop variables $\delta\hat{l}_r$ in terms of which the δk_j can be expressed

$$\delta k_j = \sum_r \eta_{jr}\delta\hat{l}_r \tag{18.51}$$

The sign factor η_{jr} is as defined in (18.33). Substituting (18.51) into (18.50) and using Kirchhoff's law (18.35), we obtain the desired cancellation, and

$$\delta J = \sum_j (k_j^2 - m_j^2)\,\delta\alpha_j = 0 \qquad \text{at } \alpha_j = \alpha_j^0 \tag{18.52}$$

Thus for each line of the reduced graph, $k_j^2 = m_j^2$. We see that the necessary conditions for the existence of the threshold singularity $q^2 = q_t^2$ of $I(q^2)$ are

[1] J is also the action function in circuit theory; Kirchhoff's laws can be derived from the variational principle $\delta J = 0$ subject to the momentum constraints (18.38b).

1. The graph must be reducible to short-circuited blobs, within which all α_j are zero, connected to one another by "real" lines on the mass shell for which $k_j^2 = m_j^2$.

2. Kirchhoff's laws (18.38) must be satisfied with positive resistances α_j in all the lines; that is, the correct causal behavior is maintained.

The physical meaning of condition 1 should be clear. A singularity in an amplitude occurs if the interaction is unbounded in space and/or time. Ordinarily the uncertainty principle limits the space-time interval over which a process extends. However, if it is possible for intermediate particles to be "real," that is, satisfy the kinematical relations for a free particle, then the process becomes unbounded in space-time and the corresponding amplitude singular. The restriction to positive resistances in condition 2 ensures causal propagation of the real intermediate particles, while Kirchhoff's laws express the geometrical and kinematical constraints imposed on the process.

These results are useful provided we can supply some graphical or intuitive criteria for determining the possible existence of real intermediate states. For the propagator, this is not a difficult task. For time-like q^2, we choose the coordinate frame for which

$$q_\mu = (q,0,0,0)$$

Since the current source has only a time-like component, so must all the internal currents k_j in order to satisfy Kirchhoff's laws. For a singularity of the propagator we must admit real intermediate states as in the diagrams of Fig. 18.8. Kirchhoff's laws allow us to assign a "time" to each vertex of the reduced graphs. The currents all flow in the same direction from t_1 to t_2 to t_3 by the "causality" in Ohm's law: the α_i are positive. The magnitudes of the currents are just

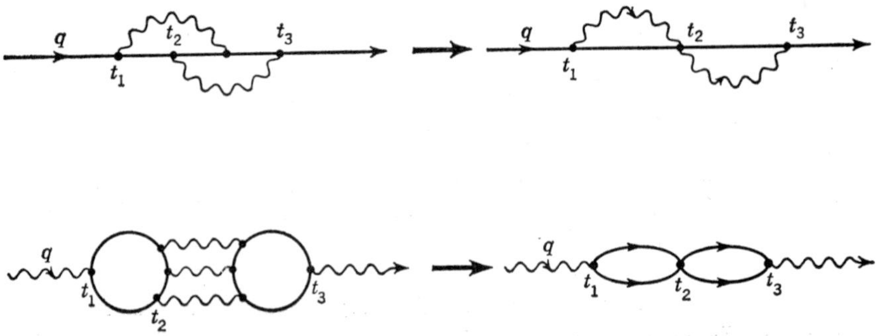

Fig. 18.8 Some propagator reduced graphs.

the masses of the intermediate particles, and the possible singularities associated with the reduced graphs are therefore those values of q for which

$$q = \Sigma m_i$$

where the sum goes over the masses of the intermediate state appearing in the reduced graph. The threshold value q_t is therefore given by the total mass of the lightest intermediate state coupled to the external particle in question. This result is the same[1] as that constructed from a more general starting point in Chap. 16.

18.6 Singularities of a General Graph and the Landau Conditions

Moving on from the propagator example to an arbitrary Feynman graph, we expect that the result of (18.52) for locating the threshold singularity will still apply. This is in fact the case, as we shall here prove. Not only do the conditions (18.52) have an intuitively understandable character, but they were derived by a very general method in the development of (18.46) to (18.52). Notice that the amplitude (18.36)

$$I(q_1, \ldots, q_m) = \int_0^1 \frac{d\alpha_1 \cdots d\alpha_n\, \delta(1 - \Sigma\alpha_j)}{\Delta^2 \left[\sum_j (k_j^2 - m_j^2)\alpha_j + i\epsilon\right]^{n-2k}} \tag{18.53}$$

provides a common starting point for an arbitrary Feynman graph. The Kirchhoff momenta k_j are linear functions of the external momenta q_s according to (18.37), and we may rewrite the denominator J in (18.53) as

$$J = \sum_j (k_j^2 - m_j^2)\alpha_j + i\epsilon = \sum_{i,j=1}^{m} \zeta_{ij} q_i \cdot q_j - \sum_j m_j^2 \alpha_j + i\epsilon \tag{18.54}$$

with the ζ_{ij} again coefficients multilinear in the α_j parameters. Equations (18.53) and (18.54) form what is known as the Nambu[2] representation; the propagator (18.45) and the vertex (18.43) are special cases of it. Since this Nambu representation provides a common form for studying arbitrary Feynman amplitudes as a function of complex variables $q_i \cdot q_j$, it is not at all surprising that we can prove

[1] It actually goes a little further, since it shows that if one ignores gauge terms in computing the electron propagator S_F' (a legitimate procedure if one is interested in computing S-matrix elements only), this propagator also satisfies a spectral representation.

[2] Y. Nambu, *Nuovo Cimento*, **9**, 610 (1958).

in general that the necessary conditions for *any* singularity of a given Feynman graph are

1. To each singularity there corresponds a "reduced graph" obtained by shrinking some subset of internal lines to a point.
2. The intermediate particles in the reduced graph must be real, that is

$$k_j^2 = m_j^2 \tag{18.55}$$

according to (18.52), where k_j is the Kirchhoff momentum assigned to each internal line along with a Feynman parameter α_j and satisfying Kirchhoff's laws (18.35) and (18.37) or, symbolically, (18.38).

The above requirements are known as the Landau conditions[1] and reduce to algebra the problem of finding singularities of Feynman graphs. For singularities which are $0(\epsilon)$ of real $q_i \cdot q_j$ on the "physical sheet" (that is, before any of these variables have been continued to complex values) there is the additional causality restriction

$$\alpha_j \geq 0 \qquad \text{Im}\ \alpha_j = 0 \qquad \text{for all } j \tag{18.56}$$

When we continue I to complex values of $q_i \cdot q_j$, ending up perhaps on the unphysical seventh sheet, the Landau conditions survive as necessary ones for determining singularities provided the α_j are allowed to become complex.

The Nambu representation (18.53) and (18.54) is of more limited power when applied to Feynman graphs such as scattering graphs with more than three external lines than it is for the propagator and the vertex. This is because the number of $q_i \cdot q_j$ then exceeds the number of independent invariants that can be formed. However, some statements can be made, and these will be discussed in Sec. 18.9 for the scattering amplitude. Nevertheless, (18.53) and (18.54) in the general case do allow a continuation for all real values of the invariants even if they are unphysical ones. From this it follows that all amplitudes obtainable from one another by the substitution rule—for example, Compton scattering, pair annihilation to two photons, and pair production by two photons—are described by the same analytic function for different values of the arguments.

Returning now to the main task of establishing necessary conditions for a singularity to occur in the integral I of (18.53), we remove the δ-function constraint on the α's by using the scaling property

[1] L. D. Landau, *Nucl. Phys.*, **13**, 181 (1959); J. D. Bjorken, doctoral dissertation, Stanford University, 1959.

(18.47) as before. We carry out in (18.34) and (18.36) the uniform scale change

$$\alpha_j \to \frac{\alpha_j}{\lambda} \qquad \lambda > 0$$

We can therefore rewrite (18.53) as

$$I = \int_0^\infty \frac{d\alpha_1 \cdots d\alpha_n \, \delta\left(1 - \sum_j \alpha_j/\lambda\right)}{\Delta^2(\alpha_j)[J(\alpha_j)]^{n-2k}}$$

I evidently is independent of λ in (18.34), and the δ-function constraint is removed by the trick

$$I = \int_0^\infty d\lambda \, e^{-\lambda} I = \int_0^\infty \frac{d\alpha_1 \cdots d\alpha_n \left(\sum_j \alpha_j\right) \exp\left(-\sum_j \alpha_j\right)}{\Delta^2(\alpha_j)[J(\alpha_j)]^{n-2k}} \tag{18.57}$$

We first consider the integral over α_1 in (18.57) for the $\alpha_2, \ldots, \alpha_n$ fixed and the invariants $q_i \cdot q_j$ held real and fixed:

$$\begin{aligned} \mathcal{I}_2(\alpha_2, \ldots, \alpha_n; q_i \cdot q_j) &= \int_0^\infty \frac{d\alpha_1 \left(\sum_j \alpha_j\right) \exp\left(-\sum_j \alpha_j\right)}{\Delta^2[J(\alpha_j)]^{n-2k}} \\ &= \int_0^\infty d\alpha_1 \mathcal{I}_1(\alpha_1 \cdots \alpha_n; q_i \cdot q_j) \end{aligned} \tag{18.58}$$

Clearly, $\mathcal{I}_2$ will exist unless $J = 0$ for some positive value of α_1, that is, unless $\mathcal{I}_1$ is singular. Even in this case the singularity can be avoided in general by distorting the integration contour into the complex α_1 plane, as shown in Fig. 18.9. $\mathcal{I}_2$ will then still be analytic, that is, a small variation in the q_s or $\alpha_2, \ldots, \alpha_n$ will cause a small variation in the positions and strengths of the singularities in the α_1 plane, with the consequence that the integral and its derivatives exist. The contour is trapped and the resultant $\mathcal{I}_2$ integral singular only if the singularity in $\mathcal{I}_1$ lies at the end point $\alpha_1 = 0$ or if two singularities

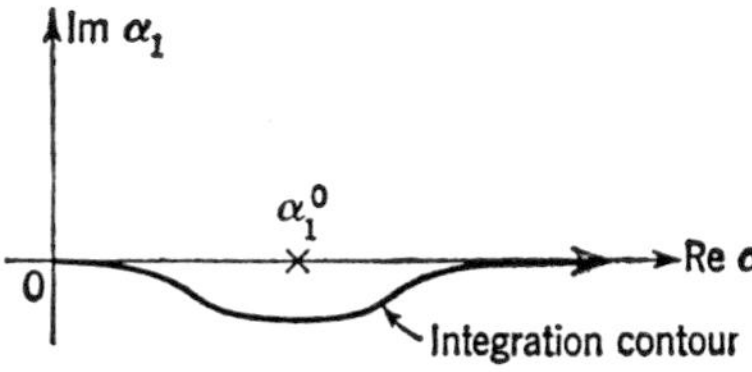

Fig. 18.9 Contour for α_1 integration distorted to avoid singularity at $\alpha_1{}^0$.

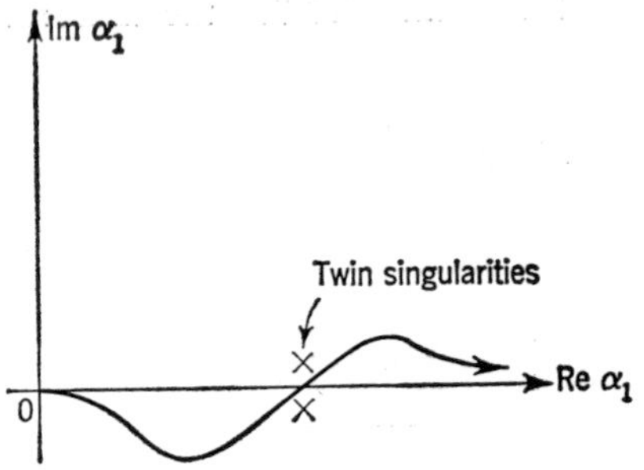

Fig. 18.10 Twin singularities pinching the contour.

coincide and pinch the contour at $\alpha_1 = \alpha_1{}^0$, as shown in Fig. 18.10. These two cases lead to conditions necessary for a singularity in $\mathcal{G}_2$:

$$J(\alpha_1{}^0,\alpha_2, \ . \ . \ . \ ,\alpha_n;q_i\cdot q_j) = 0 \tag{18.59a}$$

and either

$$\alpha_1{}^0 = 0 \qquad \text{or} \qquad \left.\frac{\partial J}{\partial \alpha_1}\right|_{\alpha_1=\alpha_1{}^0} = 0 \tag{18.59b}$$

since in the latter case there is a double zero at $\alpha_1 = \alpha_1{}^0$. We begin to see the Landau conditions emerge, since (18.59) has the same form as (18.49).

We proceed to the α_2 integration:

$$\mathcal{G}_3(\alpha_3, \ . \ . \ . \ ,\alpha_n;q_i\cdot q_j) = \int_0^\infty d\alpha_2 \mathcal{G}_2(\alpha_2, \ . \ . \ . \ ,\alpha_n;q_i\cdot q_j) \tag{18.60}$$

Again $\mathcal{G}_3$ can be singular only if $\mathcal{G}_2$ has twin singularities within $0(\epsilon)$ of each other, as in Fig. 18.10, or if there is a singularity at the end point $\alpha_2 = 0$. In the former case

$$J(\alpha_1{}^0,\alpha_2{}^0,\alpha_3, \ . \ . \ . \ ,\alpha_n;q_i\cdot q_j) = 0 \tag{18.61a}$$

and

$$\left.\frac{dJ}{d\alpha_2}\right|_{\alpha_2=\alpha_2{}^0} = 0 \tag{18.61b}$$

Equation (18.61*b*) includes an implicit variation of $\alpha_1{}^0$, because $\alpha_1{}^0$ was determined by (18.59*a*) in terms of α_2 and the other variables. Therefore,

$$\left.\frac{dJ}{d\alpha_2}\right|_{\alpha_2=\alpha_2{}^0} = 0 = \left(\frac{\partial J}{\partial \alpha_2} + \frac{\partial J}{\partial \alpha_1}\frac{\partial \alpha_1}{\partial \alpha_2}\right)_{\alpha_2=\alpha_2{}^0} = \left.\frac{\partial J}{\partial \alpha_2}\right|_{\alpha_2=\alpha_2{}^0}$$

by virtue of (18.59*b*). This procedure may be continued by induction until all α-integrations have been performed, leading to the conditions

$$\text{either} \qquad \left.\frac{\partial J}{\partial \alpha_j}\right|_{\alpha_j=\alpha_j{}^0} = (k_j{}^2 - m_j{}^2) = 0 \tag{18.62}$$

$$\text{or} \qquad \alpha_j{}^0 = 0$$

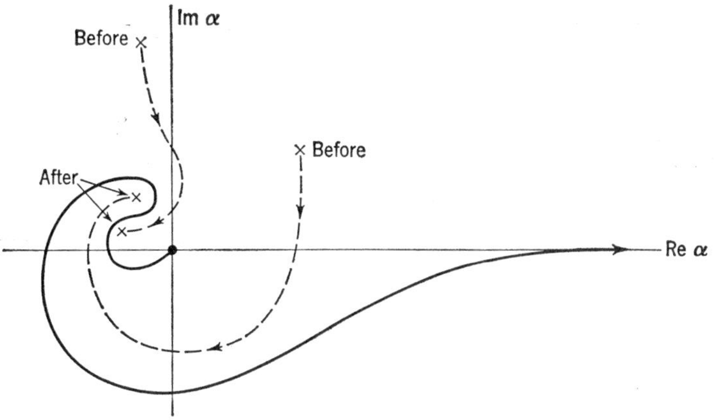

Fig. 18.11 Displacements of zeros of J from "before" for real $q_i \cdot q_j$ to "after" for possible complex values.

Referring back to (18.52) we see that these are just the Landau conditions. We notice that if the $q_i \cdot q_j$ are not allowed to wander off into the complex plane, the α contours will not be distorted, since the $i\epsilon$ prevents any zero of J coming closer than $0(\epsilon)$ to the real axis. Thus the additional causality constraint $\alpha_j \geq 0$, as given by (18.56), is applicable to these singularities, which include most of those of physical interest.

If the $q_i \cdot q_j$ are allowed to wander at will throughout the complex plane, zeros of J can approach the contours, which must be deformed to avoid them, such as in Fig. 18.11. Again, however, the only way the integral I can become singular is by means of pinches in the α contours or singularities at the end points $\alpha_j = 0$. Thus the Landau conditions survive this generalization provided the α's are allowed to become complex.[1]

18.7 Analytic Structure of Vertex Graphs; Anomalous Thresholds

We now apply the general Landau conditions to find the singularities of vertex graphs, which are of a richer variety than those of the propagator. Returning to our earlier example of the π-meson elec-

[1] There are only a finite number of solutions of the Landau conditions, since these are algebraic equations. Therefore, no natural boundaries occur when the $q_i \cdot q_j$ are continued into other sheets in the complex plane. Singularities associated with $\alpha = \infty$ are treated by Fairlee et al., *J. Math. Phys.*, **3**, 594 (1962).

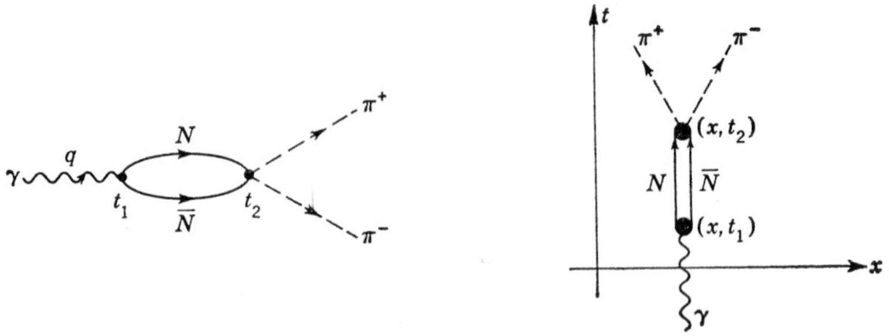

Fig. 18.12 Reduced graph and coordinate space picture for pion electromagnetic vertex.

tromagnetic form factor, we conclude from the representation (18.43) that to all orders in perturbation theory $F_\pi(q^2)$ is analytic in the upper half q^2 plane. Furthermore, on the same physical grounds as discussed in Sec. 18.3 for the perturbation example, we expect $F_\pi(q^2)$ to be real for $q^2 < 0$ to all orders in the strong interactions. We may establish this result by observing that for $q_\mu = 0$, the vertex graphs become essentially propagator graphs with regard to their kinematics. However as in Sec. 18.5 the denominator J in any such propagator graph is negative definite on the mass shell $q_2{}^2 = q_3{}^2 = \mu^2$, since $\mu < \sum_i m_i$, where $\sum_i m_i$ is the mass of any intermediate state coupled to the π meson. It follows that the denominator $J(q^2)$ appearing in any Feynman integral contributing to F_π is negative definite at $q^2 = 0$ and therefore negative definite for $q^2 < 0$ as well. The Schwartz reflection argument used in Sec. 18.3 may again be used to define $F_\pi(q^2)$ in the lower half q^2 plane, and the singularities of $F_\pi(q^2)$ are thus confined to at most a branch cut running from $q^2 = 0$ to $q^2 = \infty$.

To find the threshold value $q_t{}^2$ of this cut, as well as the branch points lying along the cut, we must investigate all reduced graphs which contribute to the singularities of $F_\pi(q^2)$. In the perturbation example the reduced graph of Fig. 18.12 contributed the singularity at $q^2 = 4M^2$. As in the case of reduced graphs for the propagator, this may be viewed as a literal picture in coordinate space. The timelike photon with $q_\mu = (q,0,0,0)$ decays at time t_1 into a nucleon-antinucleon pair at rest. The pair remain there an arbitrarily long time until they recombine at time t_2 into the two final escaping pions. This can happen only at the threshold $q_0 = 2M$.

As we proceed to investigate diagrams beyond the perturbation approximation, there occur additional reduced graphs, similar to Fig.

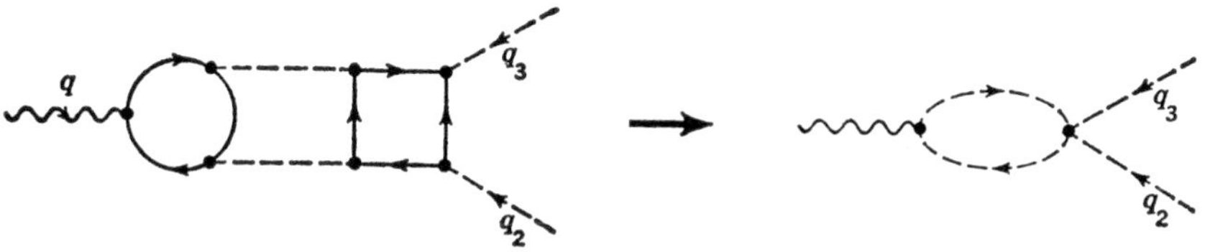

Fig. 18.13 Higher-order reduced graph with threshold singularity at $q_t^2 = 4\mu^2$.

18.12, which lower the threshold of the cut from $q_t^2 = 4M^2$ to $q_t^2 = 4\mu^2$. Such graphs involve the rescattering of the final two pions as illustrated in Fig. 18.13. $q^2 = 4\mu^2$ is also the threshold of the physical region for experimental study of $F_\pi(q^2)$ in reactions such as

$$e^- + e^+ \leftrightarrow \pi^+ + \pi^-$$

For space-like $q^2 < 0$ elastic electron-pion scattering measures F_π, but in between, for $0 \le q^2 \le 4\mu^2$, there lies an unphysical interval.

To study reduced graphs in the unphysical region, we must make the external pion momenta q_2 and q_3 complex and see what kinds of reduced graphs can crop up. For convenience we choose a coordinate system in which the momentum components are

$$q = (q,0,0,0) \qquad q_2 = \left(-\frac{q}{2}, i\sqrt{\mu^2 - \frac{q^2}{4}}, 0,0\right)$$
$$q_3 = \left(-\frac{q}{2}, -i\sqrt{\mu^2 - \frac{q^2}{4}}, 0,0\right) \tag{18.63}$$

Because of Kirchhoff's laws, all internal "currents" or momenta k must have the form

$$k = (k_0, ik_1, 0,0) \tag{18.64}$$

with k_0 and k_1 real. No internal currents flow in the direction of the second or third components of k in the absence of current sources in these directions. If, as in (18.64), the current sources in the direction of the one-component are all pure imaginary, so are the one-components of the internal currents by Kirchhoff's laws. Then, since all spatial momenta in (18.63) and (18.64) are pure imaginary, we may change from our metric tensor $g_{\mu\nu}$ to a metric $\delta_{\mu\nu}$ appropriate to a euclidean space if at the same time we drop the i's on the space components. The Landau conditions for singularities of the graphs for real intermediate states at $k_j^2 = m_j^2$ still apply, and again these conditions may be determined geometrically in a plane in a very simple way, since the

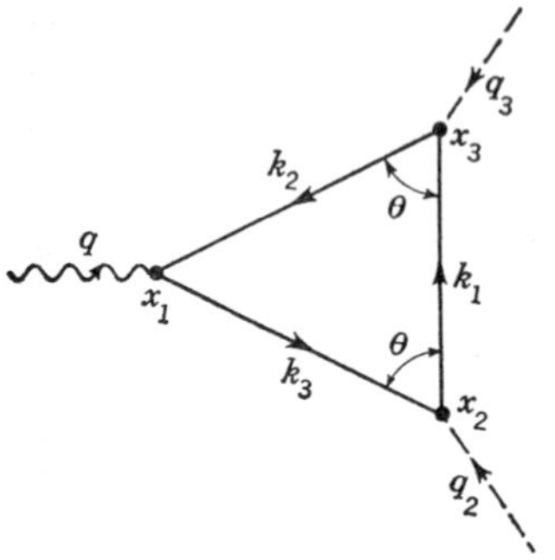

Fig. 18.14 Two-dimensional reduced graph.

geometry is euclidean. At each vertex of a reduced graph we can again assign two-dimensional real "voltages" or coordinates, and again a singularity can exist only if it is possible to draw such pictures with geometrical constraints appropriate to the internal particles on the mass shell. For example, consider again the perturbation example discussed in Sec. 18.3 and suppose a two-dimensional reduced graph were possible, as shown in Fig. 18.14. The angle θ is fixed by the requirement of momentum conservation and the condition that the intermediate particles be "real"

$$q_3^2 = \mu^2 = (k_1 - k_2)^2 = k_1^2 + k_2^2 - 2k_1k_2 \cos(\pi - \theta) = 2M^2(1 + \cos\theta)$$

However, in order to draw the figure at all, θ must be less than $\pi/2$, or $\mu^2 > 2M^2$, a condition[1] not satisfied by π mesons and nucleons. Therefore, this reduced graph does not contribute a singularity to $F_\pi(q^2)$. We emphasize that in drawing these closed figures with angles determined by momentum (or current) conservation at each vertex, the sides have a length given by the voltage drop (or current times resistance) between the vertices they connect.

To arbitrarily high orders of the interaction we can rule out singularities in $F_\pi(q^2)$ due to two-dimensional reduced graphs ($q^2 \leq 4\mu^2$) by observing that any candidate for a reduced graph can be enclosed by a triangle whose vertices x_1, x_2, and x_3 are the coordinates of the points at which the external particles first interact. The sides of the triangle cannot bulge, that is, no internal lines can wander through the sides of the triangle, as in Fig. 18.15. If this were to occur and a particle were to wander out through the sides, it would continue out because there is no external momentum available to push it back toward a vertex to be absorbed. Having constructed this triangle, we observe

[1] Compare with (18.29) and footnote on page 219.

that at at least one of the pion vertices the angle θ of the triangle is acute. At this vertex, the external pion dissociates into a spray of particles of momenta k_i and masses m_i, and

$$q^2 = \mu^2 = \Big(\sum_i k_i\Big)^2 = \sum_i k_i^2 + \sum_{i \neq j} k_i \cdot k_j > \sum_i m_i^2$$

Thus there must exist a state coupled to the π for which

$$\mu^2 > \sum_i m_i^2 \tag{18.65}$$

if there is to be a singularity. This does not exist, provided we ignore weak and additional electromagnetic interactions. It is clear, again considering strong interactions only, that no one-dimensional reduced graphs contribute to $F_\pi(q^2)$ in the region $0 \leq q^2 < 4\mu^2$, since the 2π state is the lightest state composed of strongly interacting particles that is coupled to the photon. Therefore we can conclude, to all orders of perturbation theory in the strong interactions, that $F_\pi(q^2)$ is analytic in the cut q^2 plane, the threshold value being $q_t^2 = 4\mu^2$, corresponding to reduced graphs containing 2π intermediate states.

Singularities associated with the two-dimensional reduced graphs are known as "anomalous thresholds." Although not present in the case we discussed, they are of considerable interest in themselves. They occur in form factors of "weakly bound" or composite particles, as well as unstable particles satisfying (18.65). For example, if weak interactions are included in $F_\pi(q^2)$, there is an anomalous threshold occurring in the *physical* region $q^2 < 0$, corresponding to the reduced graph in Fig. 18.16. In words, the physical process is as follows: The

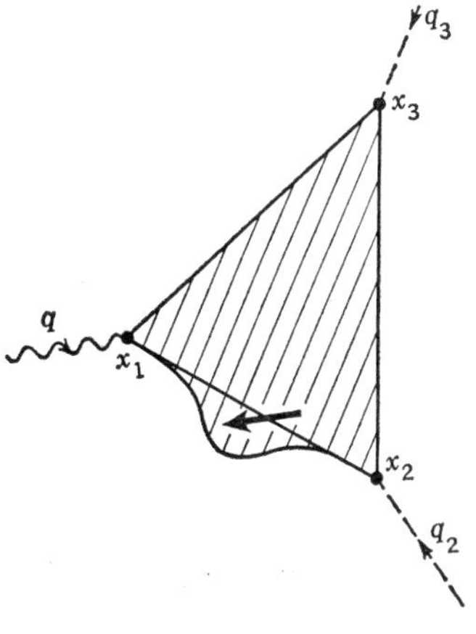

Fig. 18.15 Two-dimensional reduced graph with bulging side; an impossible case.

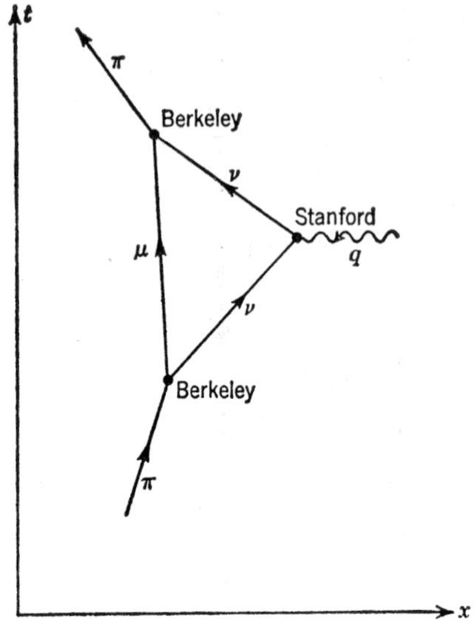

Fig. 18.16 Reduced graph, for example, of an anomalous threshold in the *physical* region corresponding to an unstable particle.

pion is produced in Berkeley with just the right momentum $\mathbf{q}/2$ that the decay μ meson remains at rest and the neutrino proceeds to Stanford with momentum $\mathbf{q}/2$. There it scatters off an electromagnetic potential and goes back to Berkeley with momentum $-\mathbf{q}/2$ and picks up the μ meson to make another π. The momentum transfer $\mathbf{q}$ for which F_π is singular is given by

$$\frac{|\mathbf{q}|^2}{4} = E_\pi{}^2 - \mu^2$$

with

$$m^2 + \mu^2 - 2mE_\pi = 0$$

where m is the μ-meson mass; therefore

$$|\mathbf{q}| = \frac{(\mu + m)(\mu - m)}{m} = 79 \text{ MeV}/c \tag{18.66}$$

The strength of this singularity is, to be sure, rather weak, but in principle it exists.

As one changes from unstable to stable particles, the anomalous thresholds move from space-like to time-like q^2, and for systems like a deuteron there will exist two-dimensional reduced graphs in the euclidean region. Two such graphs are shown in Fig. 18.17. For the graph in Fig. 18.17*a* the angle θ is very small, because of the small

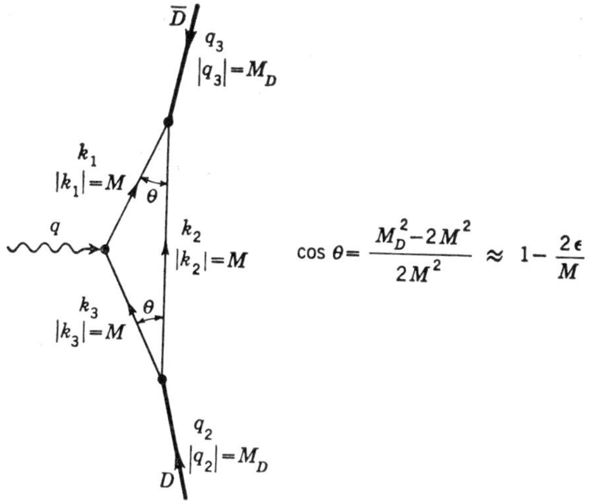

$$\cos\theta = \frac{M_D^2 - 2M^2}{2M^2} \approx 1 - \frac{2\epsilon}{M}$$

(a)

$$\cos\theta = \frac{M_D^2 - M^2 - (M+\mu)^2}{2M(M+\mu)} \approx 1 - \frac{2\mu}{M}$$

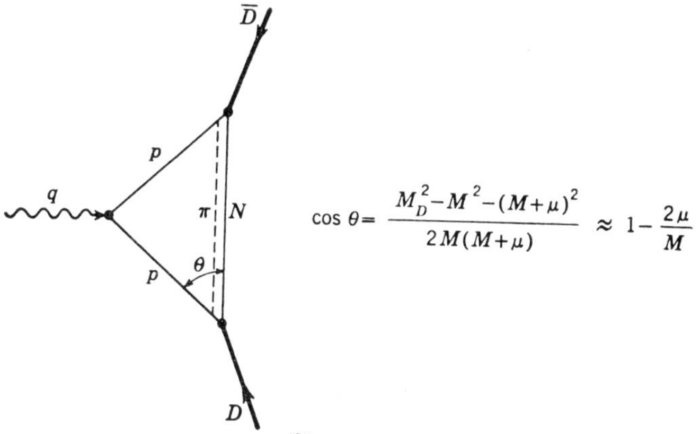

(b)

Fig. 18.17 Reduced graphs for anomalous threshold contributions to the deuteron form factor.

binding energy ϵ of the deuteron:

$$\cos\theta \cong 1 - \frac{\theta^2}{2} \cong \frac{(2M - \epsilon)^2 - 2M^2}{2M^2} \approx 1 - \frac{2\epsilon}{M}$$

From the figure, we see that the singularity occurs for

$$q = 2M\sin\theta \approx 2M\theta \approx 4\sqrt{M\epsilon} \tag{18.67}$$

and moves to $q = 0$ as the binding energy of the deuteron vanishes.

This result shows that the deuteron form factor becomes singular for time-like momentum transfers comparable to the mean momenta k in the deuteron wave function:

$$\frac{k^2}{2M} \sim \epsilon$$

As we discussed in connection with the pion form factor above (18.65), a necessary condition for anomalous thresholds is the existence of a state with particles of mass m_i such that if the external particle has mass μ, (18.65) is satisfied:

$$\mu^2 > \sum_i m_i^2$$

Thus for strong interactions, there appear no anomalous thresholds in the vertex functions of π mesons, K mesons, and nucleons, although they do exist for the Λ, Σ, and Ξ hyperons. In the physical region of time-like momentum transfers, no anomalous thresholds appear for stable particles, and only the one-dimensional reduced graphs contribute. We should emphasize the dependence of all these results upon physical mass parameters and selection rules for particle interactions at a vertex. Much more than the purely mathematical aspects of the theory are required to determine the region of singularities.

18.8 Dispersion Relations for a Vertex Function

Because the singularities of the pion electromagnetic form factor $F_\pi(q^2)$ to all orders of perturbation theory are confined to a cut along the real axis from $q^2 = 4\mu^2$ to $q^2 = \infty$, we can write a dispersion relation in the form

$$F_\pi(q^2) = \frac{1}{2\pi i} \int_C \frac{dq'^2\, F_\pi(q'^2)}{q'^2 - q^2} \tag{18.68}$$

with q^2 an arbitrary point in the complex plane contained within the contour C drawn in Fig. 18.18. Since $F_\pi(q^2)$ is real for real $q^2 < 4\mu^2$ and the Schwartz reflection principle holds, we may express the discontinuity across the branch cut in terms of the imaginary part of $F_\pi(q^2)$. As q^2 approaches the real axis from the upper half plane, we write

$$\lim_{\epsilon \to 0^+} F_\pi(q^2 + i\epsilon) = \operatorname{Re} F_\pi(q^2) + i \operatorname{Im} F_\pi(q^2) \qquad q^2 \text{ real}$$

and

$$\lim_{\epsilon \to 0^+} [F_\pi(q^2 + i\epsilon) - F_\pi(q^2 - i\epsilon)] = 2i \operatorname{Im} F_\pi(q^2) \tag{18.69}$$

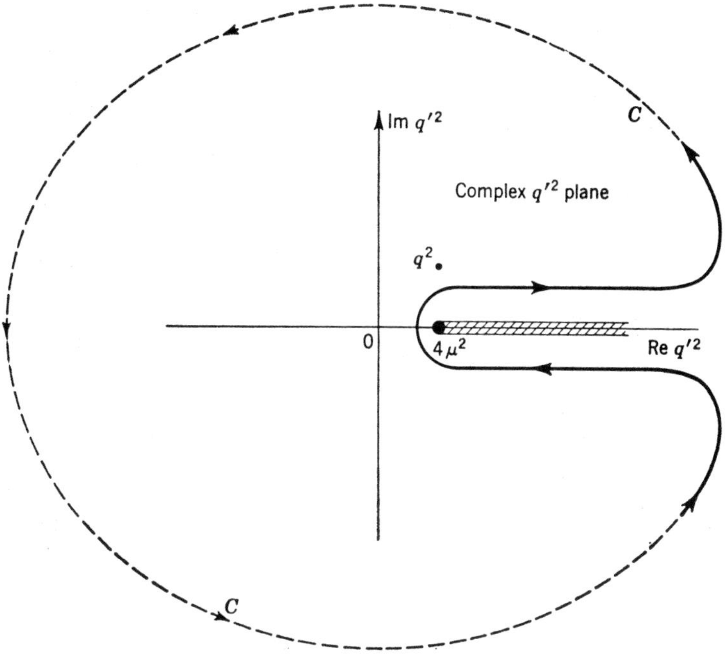

Fig. 18.18 Contour in complex q'^2 plane for dispersion integral for $F_\pi(q^2)$.

The dispersion integral (18.68) can then be separated into the contribution from the cut and the contribution C_∞ from the circle at infinity:

$$F_\pi(q^2) = \lim_{\epsilon\to 0^+} F_\pi(q^2 + i\epsilon) = \frac{1}{2\pi i} \lim_{\epsilon\to 0^+} \int_{4\mu^2}^{\infty} \frac{2i \operatorname{Im} F_\pi(q'^2)\, dq'^2}{q'^2 - q^2 - i\epsilon} + C_\infty \tag{18.70}$$

With the optimistic assumption that $F_\pi(q^2)$ vanishes at infinity, we have the unsubtracted form of the dispersion relation:

$$F_\pi(q^2) = \frac{1}{\pi} \int_{4\mu^2}^{\infty} \frac{\operatorname{Im} F_\pi(q'^2)\, dq'^2}{q'^2 - q^2 - i\epsilon} \tag{18.71}$$

If we are conservative and follow the indications of perturbation theory, we must make one subtraction and write a dispersion relation for $F_\pi(q^2)/q^2$ in order to provide enough convergence as $q^2 \to \infty$ that the contour at ∞ may be ignored. Proceeding as in the Kramers-Krönig example (18.9), we find

$$F_\pi(q^2) = F_\pi(0) + \frac{q^2}{\pi} \int_{4\mu^2}^{\infty} \frac{dq'^2 \operatorname{Im} F_\pi(q'^2)}{q'^2(q'^2 - q^2 - i\epsilon)} \tag{18.72}$$

Equations (18.71) and (18.72) are the analogues of the Kramers-Krönig relations (18.11). The Kramers-Krönig relations are useful in that they relate experimental quantities in (18.12) via a unitarity statement—the optical theorem. What we have achieved thus far for the form factor is a relation which reduces the task of computing $F_\pi(q^2)$ to that of computing its discontinuity for positive $q^2 > 4\mu^2$. Of what practical value is this? Our discussions of this section have shown that the necessary condition for the appearance of a singularity in F_π is the existence of reduced graphs in which the virtual time-like photon couples to real intermediate states. For the pion form factor in particular no anomalous thresholds appear and all reduced graphs are one-dimensional and in the physical region $q^2 > 4\mu^2$. Evaluation of the discontinuity $\text{Im } F_\pi(q^2)$ then involves evaluating physical amplitudes for a photon to dissociate into a real state $|n\rangle$ multiplied by the transition amplitudes for this state $|n\rangle$ to form a π^+-π^- pair. Such a discontinuity formula, which is similar to (18.16) expressing the discontinuity across the cut for a propagator, will be discussed in detail when we come to the practical side of dispersion theory calculations. However, we can already see that only a two-pion state can contribute to this sum in the interval $(2\mu)^2 \leq q^2 \leq (4\mu)^2$, since the three-pion state has the wrong quantum numbers to form the final pion pair.

It is on the basis of arguments such as this that approximation schemes have evolved for actually carrying out dispersion theory calculations. For example, a calculation of $F_\pi(q^2)$ for small values of q^2 with the once-subtracted relations (18.72) favors the threshold region of low q^2 by a factor of $1/q'^4$ in the integral over the absorptive part, and we may concentrate calculational efforts on the lowest mass state, the 2π state itself. In the extreme approximation of retaining only the contribution of the two-pion state to $\text{Im } F_\pi(q^2)$, (18.72) provides a connection between the pion electromagnetic form factor and the π-π scattering amplitude which can be checked against experiment. Later in the chapter we shall amplify how this is done. The point to be emphasized here is that the dispersion relations relate different physical amplitudes in a definite way and provide a basis for an approximation scheme free from an expansion in powers of large coupling constants.

Unfortunately, the relation with experimental parameters is often less direct than in the present example. In a study of proton electromagnetic form factors one encounters the reduced graph in Fig. 18.19, which gives a contribution proportional to the pion form factor multiplied by the amplitude for $\pi^+ + \pi^- \to \bar{p} + p$. Unfortunately, the reduced graph occurs in the euclidean unphysical region for $q^2 = 4\mu^2$; the physical annihilation process occurs only above the threshold $4M^2$.

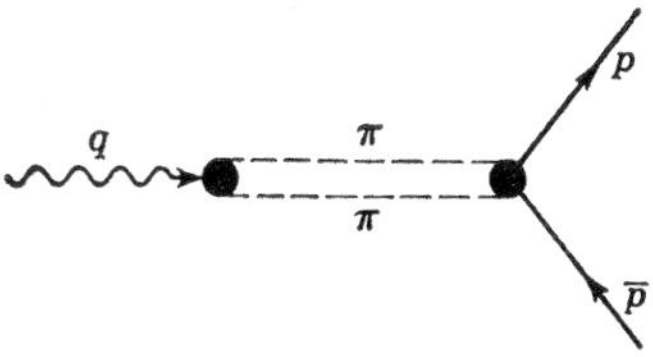

Fig. 18.19 Reduced graph for proton electromagnetic vertex.

Therefore, one must analytically continue the physical annihilation amplitude into this unphysical region, as well as any unitarity statements used in evaluating discontinuities when applying the dispersion approach. We now turn to these questions and study analytic properties of scattering amplitudes.

18.9 Singularities of Scattering Amplitudes

We may use the same methods to establish some of the analyticity properties of scattering amplitudes. As we noted below Eq. (18.56), the methods become less powerful in this application because of the constraints between the various invariants which can be formed from the external momentum variables q_i. We may choose these scalar variables to be, instead of inner products $q_i \cdot q_j$,

$$s = (q_1 + q_2)^2 \qquad t = (q_1 + q_3)^2 \qquad u = (q_1 + q_4)^2 \quad (18.73)$$

The kinematics is illustrated in Fig. 18.20. s is the square of the center-of-mass energy for the "s reaction" with incoming momenta q_1 and q_2 and outgoing momenta $-q_3$ and $-q_4$. t and u are the corresponding quantities for the "crossed" reactions. According to the substitution rule, the same Feynman diagram represents the set of three reactions with s, t, or u the center-of-mass energy. According to

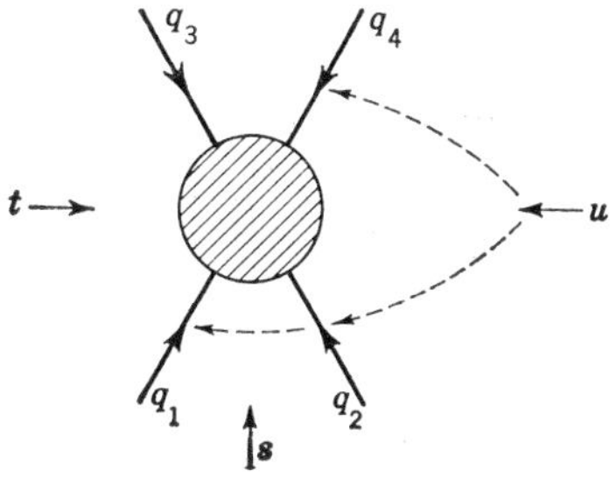

Fig. 18.20 Kinematic variables for scattering amplitude.

the Nambu representation (18.53) and (18.54), which we now write

$$I(s,t,u) = \int_0^\infty \frac{d\alpha_1 \cdots d\alpha_n \, \delta\left(1 - \sum_j \alpha_j\right)}{\Delta^2 \left(\zeta_1 s + \zeta_2 t + \zeta_3 u + \sum_{i=1}^{4} \zeta_i' m_i^2 - \sum_j m_j^2 \alpha_j + i\epsilon\right)^{n-2k}} \tag{18.74}$$

the same analytic function represents this amplitude for each of these reactions, the structure of (18.74)—that is, the $i\epsilon$—guaranteeing that the analytic continuation exists for all real s, t, and u. Because the quantities s, t, and u are not all independent, being related by the constraint

$$s + t + u = \sum_{i=1}^{4} m_i^2 \tag{18.75}$$

it is more difficult to infer analytic properties from the general structure of (18.74). For example, as s goes into the upper half plane for fixed t, u goes into the lower half plane owing to the constraint (18.75). Therefore, despite the fact that one can show ζ_1, ζ_2, and ζ_3 are positive,[1] the form of (18.74) itself does not guarantee the absence of singularities anywhere in the complex s plane.

The problem of determining the singularities of scattering amplitudes and from these constructing dispersion relations is therefore a much more difficult one than for the vertex and has not been completely solved. The nature of the known singularities is a sensitive function of the masses of the particles involved; and to keep these kinematical complications to a minimum, we shall consider π-π scattering.

The physical regions for the three s, t, and u reactions in this case are illustrated by the shaded areas in Fig. 18.21. The values of s, t, and u may be given simply in terms of the center-of-mass three-momentum $\mathbf{k}$ and the scattering angle θ of the s reaction, on which we shall concentrate

$$s = 4(\mathbf{k}^2 + \mu^2) \qquad t = -2\mathbf{k}^2(1 - \cos\theta) \qquad u = -2\mathbf{k}^2(1 + \cos\theta) \tag{18.76}$$

The line $t = 0$ defines forward π-π scattering, which we study first. We begin by recasting the representation (18.74) for the scattering amplitude $A(s,t)$ in the more compact form

$$A(s,t) = \int_0^\infty \frac{d\alpha_1 \cdots d\alpha_n F(\alpha_1, \ldots, \alpha_n)}{(\zeta s + \zeta' t - \sigma^2 + i\epsilon)^{n-2k}} \tag{18.77}$$

[1] See Prob. 2.

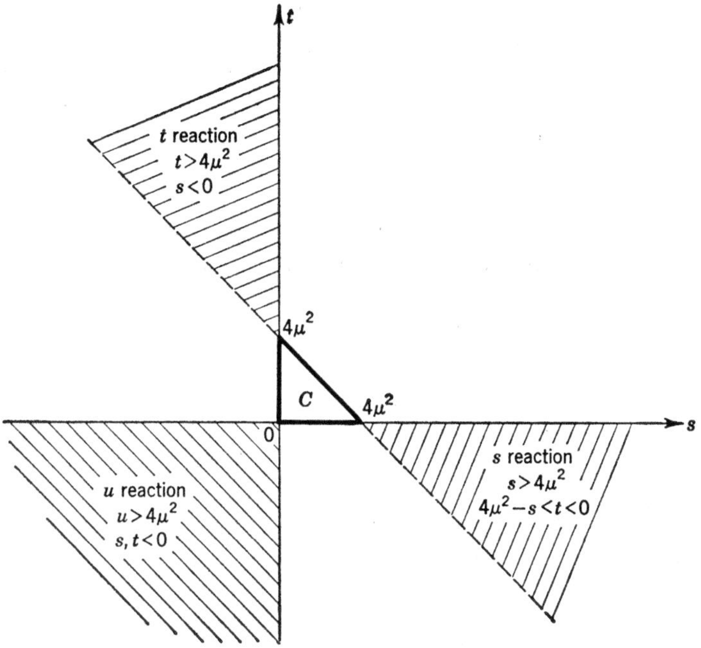

Fig. 18.21 Physical regions for s, t, and u reactions in π-π scattering.

Neither ζ, ζ', nor σ^2 is positive definite, and in order to establish analyticity properties of $A(s,0)$, we break it up into two terms:

$$A(s,0) = \int_0^\infty \frac{d\alpha_1 \cdots d\alpha_n F(\alpha_1, \ldots ,\alpha_n)}{(\zeta s - \sigma^2 + i\epsilon)^{n-2k}} [\theta(\zeta) + \theta(-\zeta)]$$
$$\equiv A_+(s) + A_-(s) \qquad (18.78)$$

By construction we have exhibited $A(s,0)$ as composed of two parts, one of which, $A_+(s)$, is analytic in the upper half complex s plane as $\epsilon \to 0^+$, and the other of which is analytic in the lower half plane. If there is some region of real s for which the denominator J in (18.78) does not vanish, then both $A_+(s)$ and $A_-(s)$ will be real in that region. By the Schwartz reflection principle we may then continue A_+ into the lower half s plane and A_- into the upper half plane. If this is the case we shall have proved that $A(s,0)$ is analytic throughout the entire s plane except for cuts along that part of the real axis for which the denominator can vanish. The desired dispersion relation may then be written.

The crux of the matter is the existence of a region of real s for which the denominator for $A(s,0)$ does not vanish and through which continuation of $A_+(s)$ and $A_-(s)$ can be made. In Fig. 18.21 we see that the interval $0 \leq s \leq 4\mu^2$ is the only suitable region through which to continue $A(s,0)$. For $s \geq 4\mu^2$ we are in the physical region for the s reaction. There we know that the scattering amplitude is complex, being the sum of terms of the form $e^{i\delta} \sin \delta$ for each partial wave. Similarly, when $s \leq 0$, we have $u \geq 4\mu^2$ and we are in the physical region for the u reaction, also described by the same analytic function $A(s,t)$ according to the substitution rule. The interval $0 \leq s \leq 4\mu^2$ for $t = 0$ is an unphysical region, as is the triangle $s \geq 0$, $t \geq 0$, $u \geq 0$, marked C in Fig. 18.21. This region resembles the euclidean region discussed for the vertex functions, and similar arguments can be used here.

We now prove that the denominator J in (18.77) has no singularities and is negative definite in this region C. The analytic continuation of A_+ and A_- can then be made. As in the discussion of the vertex, we change to a euclidean metric within the triangle C. Since s, t, and u are all positive, we can choose momenta in a special frame

$$\begin{aligned} q_1 &= \tfrac{1}{2}(\sqrt{s}, i\sqrt{t}, i\sqrt{u}, 0) \\ q_2 &= \tfrac{1}{2}(\sqrt{s}, -i\sqrt{t}, -i\sqrt{u}, 0) \\ q_3 &= \tfrac{1}{2}(-\sqrt{s}, i\sqrt{t}, -i\sqrt{u}, 0) \\ q_4 &= \tfrac{1}{2}(-\sqrt{s}, -i\sqrt{t}, i\sqrt{u}, 0) \end{aligned} \tag{18.79}$$

in accord with (18.73) and transform to a euclidean metric by removing all i's from the space components of the q_i and changing the metric from $g_{\mu\nu}$ to $\delta_{\mu\nu}$. Again Kirchhoff's laws assure that all internal momenta are real, and the analyticity properties are unaltered by this change of metric. The procedure for determining singularities in the triangular region C is similar to that in the study of the vertex. Now, however, the momenta and therefore coordinates ("voltages") at the vertices are three-dimensional, according to (18.79), and the reduced graphs will likewise be three-dimensional or less. We can immediately rule out one-dimensional reduced graphs, such as illustrated in Fig. 18.22, since the mass of the lightest state coupled to two pions is 2μ, which by momentum conservation cannot be constructed for s, t, and $u < 4\mu^2$ in (18.79). The argument is the same as for the pion electromagnetic

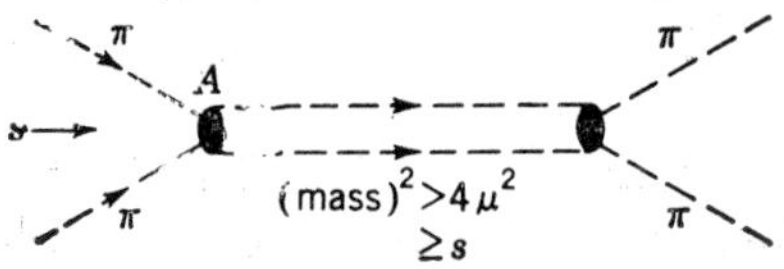

Fig. 18.22 One-dimensional reduced graph

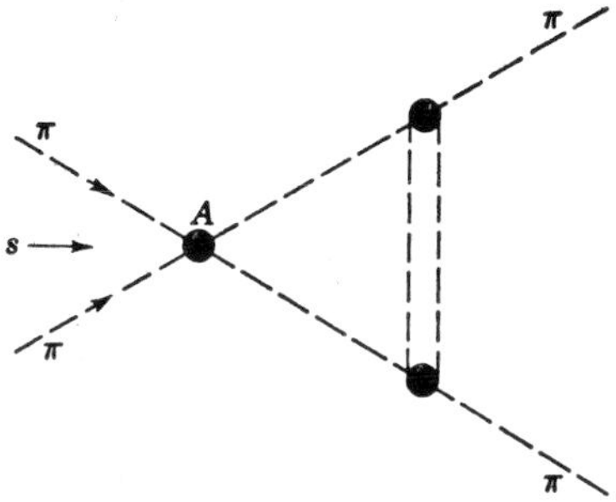

Fig. 18.23 Two-dimensional reduced graph.

form factor used earlier; the two external pions incident at A are kinematically equivalent to a photon of mass $\sqrt{s} < 2\mu$. In such a region the form factor F_π is free of singularities. Similarly, two-dimensional reduced graphs as in Fig. 18.23 cannot contribute anomalous thresholds in the euclidean triangle C for the same reason; none were found when computing the pion vertex.

Turning to the three-dimensional reduced graphs, we must confine all lines within a tetrahedron at the vertices of which the external particles are absorbed, as in Fig. 18.24. Bulging sides are ruled out for the reason given for the vertex function. It is a fact of euclidean solid geometry that at least one vertex is acute, that is, all three face angles are acute. Therefore, we can draw such a tetrahedron with real internal lines only if the masses are such that an acute angle exists. Suppose this to be vertex A in Fig. 18.24. Then any two particles with momenta k into which the pion at A dissociates must have momentum vectors differing in direction by an acute angle. The argument of

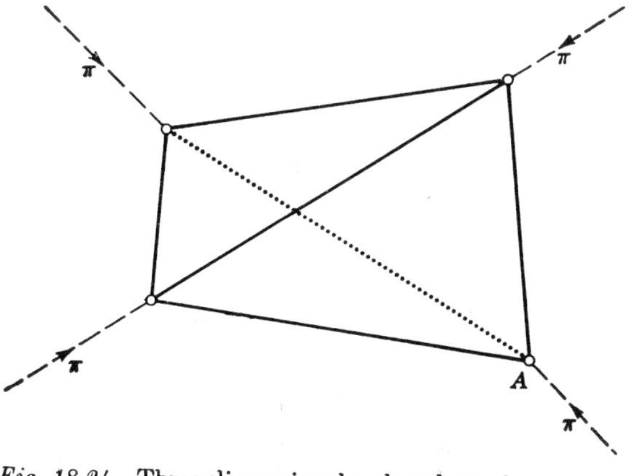

Fig. 18.24 Three-dimensional reduced graph.

(18.65) applies, but the condition

$$\mu^2 = \left(\sum_i k_i\right)^2 > \sum_i m_i^2 \tag{18.80}$$

is unattainable. Therefore, there are no singularities in the euclidean triangle C of Fig. 18.21 for π-π scattering. To complete the proof of the dispersion relation, we must still show that $A_\pm(s,0)$ are real in the interval $0 \leq s \leq 4\mu^2$. Then with the Schwartz reflection principle we may continue A_+ into the lower half s plane and A_- into the upper half plane as desired. To do this, we show that the denominator J is negative by turning off the external q_i^2 uniformly from some point in the Euclidean region, such as $s = t = u$; that is

$$q_i \rightarrow \lambda q_i \qquad \lambda \rightarrow 0$$

No singularities can be found as λ decreases because (18.80) becomes even more unattainable. In the limit $\lambda = 0$ the denominator $J = -\sum_j m_j^2\alpha_j < 0$. J remains negative definite as the external currents are turned back on, since if J became 0 on the way back, there would have to be a reduced graph corresponding to this condition. We have shown no such graph exists.

We now can write a dispersion relation for forward π-π scattering. The reality of $A(s,0)$ in the interval $0 \leq s \leq 4\mu^2$ allows us to continue A_+ and A_- through this gap; and we conclude that the only singularities of $A(s,0)$ are branch cuts extending from $s = 4\mu^2$ to $s = \infty$ and from $s = 0$ to $s = -\infty$, as illustrated in Fig. 18.25. Ignoring for the moment the question of subtraction terms, the dispersion relation has the form

$$A(s,0) = \frac{1}{\pi}\int_{4\mu^2}^{\infty} \frac{ds' \operatorname{Im} A(s',0)}{s' - s - i\epsilon} + \frac{1}{\pi}\int_{-\infty}^{0} \frac{ds' \operatorname{Im} A(s',0)}{s' - s - i\epsilon} + C_\infty \tag{18.81}$$

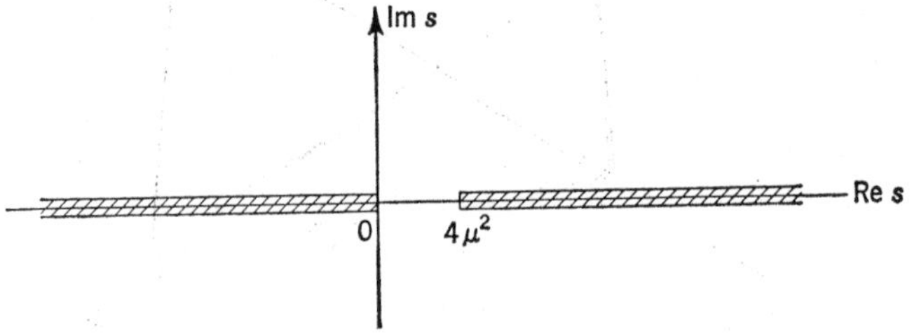

Fig. 18.25 Branch cuts in s for forward π-π scattering amplitude.

Equation (18.81) is very similar in form and content to the original Kramers-Krönig relation (18.8) which started off the whole dispersion relation study. Again we have expressed a forward-scattering amplitude $A(s,0)$ in terms of a dispersion integral over its imaginary part, which has all the charm attributed to the absorptive part in the Kramers-Krönig relation and the dispersion relation for the pion form factor $F_\pi(q^2)$ in that it is nonvanishing only in physical regions for scattering processes. Therefore, the optical theorem can be used to relate Im $A(s,0)$ to the total scattering cross sections in the s and u reactions, and we obtain a relation between (in principle) measurable quantities. The explicit connection between Im $A(s,0)$ and the total cross section comes via the unitarity condition $S^\dagger S = 1$ on the S matrix and requires only careful attention to normalization factors. It will be discussed in Sec. 18.12, where we use unitarity and crossing symmetry to apply analyticity properties derived here to the physics of the π-π system.

Thus far we have constructed forward-scattering dispersion relations at $t = 0$ by using the reality of $A(s,0)$ along the boundary of the euclidean triangle and using the representation (18.77). With the same representation we can establish dispersion relations for individual partial waves of $A(s,t)$ as well, an extremely useful result. The partial wave amplitudes $A_l(s)$ are defined by the expansion

$$A(s,t) = \sum_{l=0}^{\infty} (2l+1)A_l(s)P_l(\cos\theta)$$

$$A_l(s) = \frac{1}{2}\int_{-1}^{1} d(\cos\theta)P_l(\cos\theta)\hat{A}(s,\cos\theta) \tag{18.82}$$

where $\hat{A}(s, \cos\theta) \equiv A(s,t)$ and θ is the center-of-mass scattering angle. To find analytic properties of $A_l(s)$, we are led by (18.82) to study analytic properties of $\hat{A}(s, \cos\theta)$ at fixed $\cos\theta$. Solving for t in terms of $\cos\theta$ and s by using (18.76), we find

$$t = -\tfrac{1}{2}(s - 4\mu^2)(1 - \cos\theta)$$

Substituting this result into the integral representation (18.77) for $A(s,t)$, we see that the denominator J there can be written as

$$J = \hat{\xi}s - \hat{\sigma}^2 + i\epsilon \tag{18.83}$$

with the $\hat{\xi}$ and $\hat{\sigma}^2$ now depending upon $\cos\theta$ as well as the "resistances" α_j. Therefore $\hat{A}(s, \cos\theta)$ has the same representation (18.78) as the forward-scattering amplitude $A(s,0)$. The path in the *st* plane for fixed $\cos\theta$ is a straight line, shown in Fig. 18.26, which passes through the euclidean region C, within which we have shown that J cannot vanish. Therefore we may break up $\hat{A}(s, \cos\theta)$ into two parts as we did for $A(s,0)$ and, following the procedure leading from (18.78) to (18.81), conclude that for any $\cos\theta$, $\hat{A}(s, \cos\theta)$ is analytic in the s plane, with the possible exception of branch cuts running from $s = 4\mu^2$ to ∞ and $s = 0$ to $-\infty$. We may therefore write a Cauchy relation

$$\hat{A}(s, \cos\theta) = \frac{1}{2\pi i}\int_C \frac{ds'\,\hat{A}(s', \cos\theta)}{s' - s - i\epsilon} \tag{18.84}$$

for any contour C, as in Fig. 18.27, which avoids the branch cuts and encloses the point s. We then form the partial wave amplitudes $A_l(s)$ by multiplying by $P_l(\cos\theta)$ and integrating from -1 to 1. Because the integrals are absolutely convergent, orders of integration may be interchanged and $A_l(s)$ satisfies the same Cauchy relation

$$A_l(s) = \frac{1}{2\pi i}\int_C \frac{ds'\,A_l(s')}{s' - s - i\epsilon} \tag{18.85}$$

for any contour C enclosing s and avoiding the cuts; therefore, $A_l(s)$ has the same analytic structure as the forward-scattering amplitude and satisfies a dispersion relation analogous to (18.81). This result is of considerable practical importance.

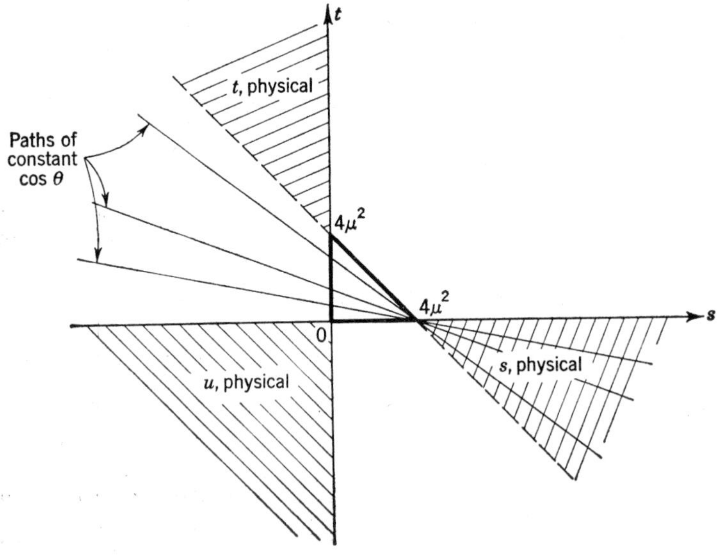

Fig. 18.26 Paths of constant cos θ for partial wave amplitude in π-π scattering.

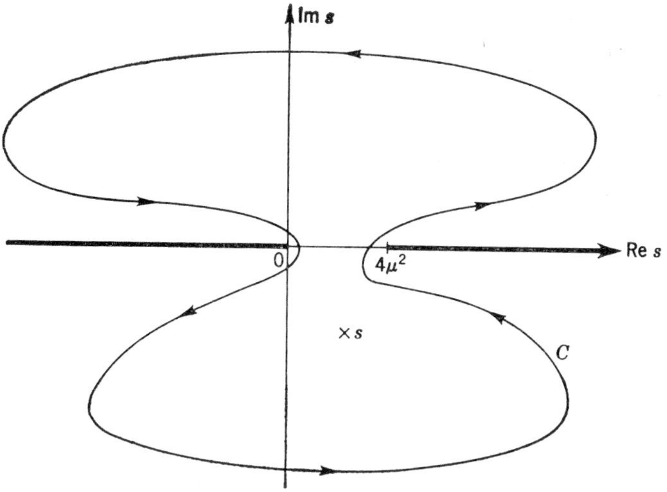

Fig. 18.27 Contour in s plane for dispersion relation for π-π partial wave amplitude.

The techniques used above determine similar analyticity domains and lead to dispersion relations for other scattering amplitudes such as forward π-nucleon and nucleon-nucleon scattering and for partial wave amplitudes in nucleon-nucleon scattering. When particles of more than one mass enter as external lines, the kinematical complications and troubles with anomalous threshold effects multiply rapidly.

18.10 Application to Forward Pion-Nucleon Scattering

The most important and successful application of dispersion relations has been the analysis of the pion-nucleon forward-scattering amplitude.[1] As in all applications of dispersion theory to actual calculation, there are three main steps to accomplish:

1. Separation of spin and isotopic factors to form functions F_i of scalar invariants, as in (18.31).

2. Derivation of the analytic properties of the form factors F_i by searching out the singularities coming from the denominator J in (18.54) by using reduced graphs or by other algebraic means.

3. Application of unitarity to calculation of the discontinuities across the branch cuts, so that physically interesting statements can

[1] M. L. Goldberger, *Phys. Rev.*, **97**, 508 (1955); **99**, 979 (1955).

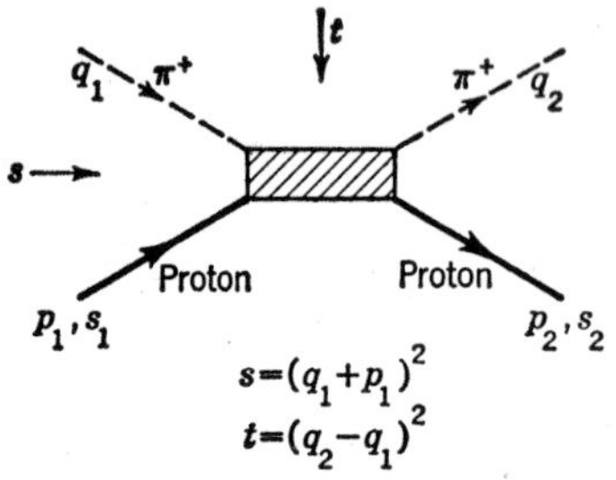

Fig. 18.28 Kinematics for π^+-proton scattering.

be made—as done in (18.11) and (18.12) for the Kramers-Krönig relation.

We shall specifically discuss π^+-proton scattering and therefore need not make a decomposition into isotopic spin channels. To obtain the structure of the invariant amplitude $\mathfrak{M}(q_2,p_2,s_2;q_1,p_1,s_1)$, corresponding to the kinematics of Fig. 18.28, we construct all quantities which become Lorentz scalars when sandwiched between Dirac spinors. These boil down to the structure

$$\mathfrak{M}(q_2,p_2,s_2;q_1,p_1,s_1) = \bar{u}(p_2,s_2)[A(s,t) + \tfrac{1}{2}(\not{q}_1 + \not{q}_2)B(s,t)]u(p_1,s_1) \tag{18.86}$$

because all other factors involving $\not{p}_1$ or $\not{p}_2$ are readily brought to this form by use of the Dirac equation, for example,

$$\not{p}_1 u(p_1,s_1) = Mu(p_1,s_1)$$

The invariants s and t are defined as in π-π scattering (see Fig. 18.28)

$$s = (p_1 + q_1)^2 \qquad t = (q_2 - q_1)^2 \tag{18.87}$$

In specializing to forward scattering, it is more convenient to express the amplitude in terms of the laboratory energy ω of the pion:

$$\omega = \frac{p_1 \cdot q_1}{M} = \frac{s - M^2 - \mu^2}{2M} \tag{18.88}$$

In this limit the two form factors A and B may be combined into a single form factor T; we find

$$\mathfrak{M}(q_1,p_1,s_1) = 4\pi\bar{u}(p_1,s_1)T(\omega)u(p_1,s_1) \tag{18.89}$$

where
$$4\pi T(\omega) \equiv A(s,0) + \omega B(s,0) \tag{18.90}$$

is the scalar quantity for which we wish to construct a dispersion relation.

The second step in the program is to determine the analytic properties of $T(\omega)$. This parallels the analysis of the π-π scattering amplitude. As in (18.78) the contribution of any graph to $T(\omega)$ possesses the structure

$$I(\omega) = \int_0^\infty \frac{d\alpha_1 \cdots d\alpha_n F(\alpha_1, \ldots, \alpha_n)}{(\zeta\omega - \sigma^2 + i\epsilon)^n} \tag{18.91}$$

We can establish the analyticity of $I(\omega)$ in the cut ω plane by the arguments used for the forward π-π amplitude and for the π-π partial wave amplitude if we find a region of real ω for which the denominator J does not vanish and for which $T(\omega)$ is therefore real.

$T(\omega)$ is certainly not real when $\omega > \mu$, or, by (18.88), $s > (M + \mu)^2$; for we are in the physical region for π^+-proton scattering. Similarly, $T(\omega)$ is not real for $u = (p_2 - q_1)^2 > (M + \mu)^2$, or $\omega < -\mu$, since we are then in the physical region for the crossed "u reaction" which is π^--proton scattering according to the substitution rule discussed below (17.45). Therefore, the only region along the real ω axis where $T(\omega)$ can be real is the interval $-\mu \le \omega \le +\mu$.

Pole contributions in this region were found in Chap. 10 in discussing the Born approximation contributions of Fig. 18.29 to π-nucleon scattering. For π^+-proton scattering only the crossed pole contributes. In searching for any additional contribution we choose the special coordinate system

$$\begin{aligned} p_1 &= p_2 = (M,0,0,0) \\ q_1 &= q_2 = (\mu \cos \varphi,\, i\mu \sin \varphi,\, 0,\, 0) \end{aligned} \tag{18.92}$$

and remove the i by converting to a euclidean metric. From the same kind of geometrical reasoning as discussed for the π-π system, no two-dimensional reduced graphs as in Fig. 18.30 can exist; there is only the "single-nucleon exchange," or Born contribution already mentioned above. Therefore, again using analytic continuation in the external currents $p_i,q_i \rightleftarrows \lambda p_i,\lambda q_i \rightleftarrows 0$ as for π-π scattering, we conclude that the denominator J is negative definite in the region $-\mu <$

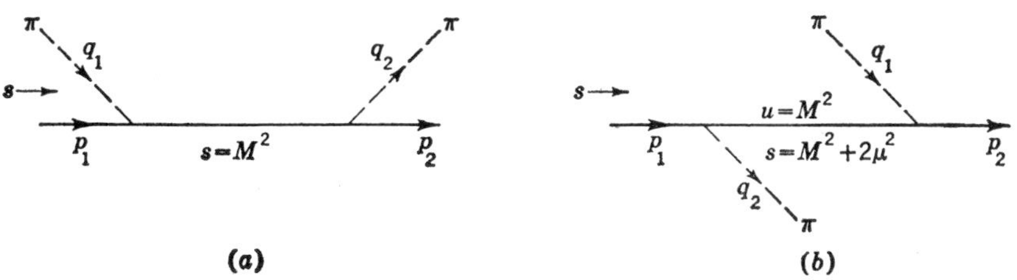

Fig. 18.29 Pole terms in forward π-nucleon scattering.

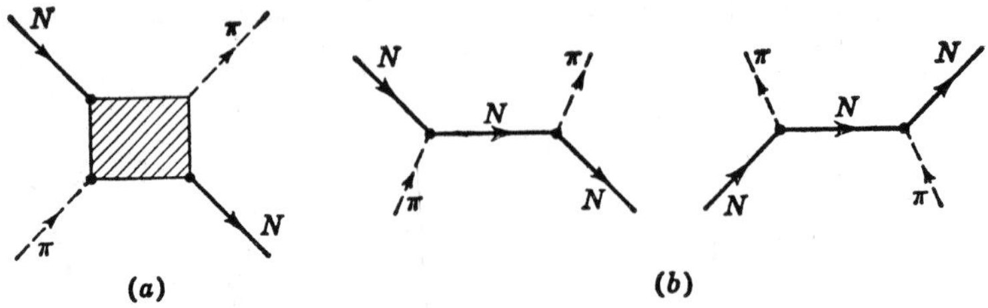

Fig. 18.30 Reduced graphs for π-nucleon scattering: (*a*) two-dimensional; (*b*) one-dimensional.

$\omega < \mu$ except for the pole contribution which occurs at

$$(p_1 - q_2)^2 = M^2 \qquad \text{or} \qquad \omega = +\frac{\mu^2}{2M} = +\omega_B \tag{18.93}$$

We have now established to any order of perturbation theory that the amplitude $T(\omega)$ is analytic in the ω plane except for cuts running from $+\mu$ to $+\infty$ and $-\mu$ to $-\infty$ and with one pole located at $\omega = +\mu^2/2M$. Using the contour C shown in Fig. 18.31, we may therefore write a dispersion relation for $T(\omega)$ relating the full amplitude to its absorptive part and to the contribution of the pole.

This brings us to the third part of the dispersion program—that of applying the unitarity condition in order to express the imaginary parts of the forward-scattering amplitudes in terms of total cross sections and thereby establish connections between measurable quantities. We have given a general proof in (16.69) that the S matrix is unitary, so that we may write

$$S^\dagger S = 1$$

or equivalently, for a transition from initial state i to final state f,

$$\sum_{\text{all } n} S_{nf}^* S_{ni} = \delta_{fi} \tag{18.94}$$

In terms of the transition amplitude $\mathfrak{I}$ defined by

$$S_{fi} = \delta_{fi} - i(2\pi)^4\,\delta^4(P_f - P_i)\mathfrak{I}_{fi} \tag{18.95}$$

the unitarity condition becomes, for the present case of forward scattering $f = i$:

$$\mathfrak{I}_{ii} - \mathfrak{I}_{ii}^* = -i \sum_n (2\pi)^4\,\delta^4(P_i - P_n)\mathfrak{I}_{ni}^*\mathfrak{I}_{ni} \tag{18.96}$$

The left-hand side of (18.96) is just twice the imaginary part of the forward-scattering amplitude, and the right-hand side is related to

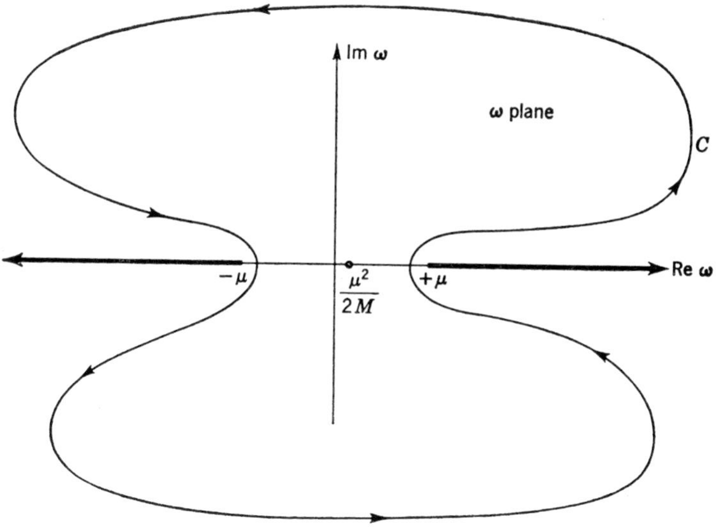

Fig. 18.31 Contour in ω plane for dispersion relation for forward π^+-proton scattering.

the total cross section. Recalling our normalization conventions in Chap. 7 [Eqs. (7.35) to (7.42) in particular], we see that we need only divide out an incident flux factor and some factors of (2π) in relating the right-hand side to the total meson-nucleon cross section for a given nucleon polarization

$$\sigma_{\text{tot}} = \frac{(2\pi)^6}{v_{\text{lab}}} \sum_n (2\pi)^4 \delta^4(P_i - P_n)|\mathfrak{I}_{ni}|^2 \tag{18.97}$$

The factor v_{lab} is the velocity of the pion in the laboratory frame, and the factor $(2\pi)^6$ removes the factors $(2\pi)^{-3/2}$ accompanying incident-particle wave functions according to the rules for constructing a cross section. By (18.96) we then have

$$\operatorname{Im} \mathfrak{I}_{ii}(\omega) = -\frac{1}{2} v_{\text{lab}} \frac{\sigma_{\text{tot}}^{\pi^+ p}(\omega)}{(2\pi)^6} \qquad \omega \geq \mu \tag{18.98}$$

We next relate the transition amplitude $\mathfrak{I}_{ii}$ to the invariant amplitude of interest by extracting the normalization factors of the external wave functions and the nucleon spin wave functions. Then in complete analogy with Eq. (10.54) we have from the definitions (18.86) and (18.89)

$$\mathfrak{I}_{ii} = \frac{1}{(2\pi)^6 2\omega} \mathfrak{M}(q_1,p_1,s_1) = \frac{1}{(2\pi)^6 2\omega} [4\pi \bar{u}(p_1,s_1) T(\omega) u(p_1,s_1)] \tag{18.99}$$

and therefore by (18.98)

$$\text{Im } T(\omega) = -\frac{k}{4\pi}\sigma_{\text{tot}}^{\pi^+p}(\omega) \qquad k = \omega v_{\text{lab}} = \sqrt{\omega^2 - \mu^2} \qquad \omega \geq \mu \tag{18.100}$$

With this normalization T is recognized from the optical theorem to be simply related to the forward differential elastic scattering amplitude by

$$\left(\frac{d\sigma(\omega)}{d\Omega}\right)_{\theta_{\text{lab}}=0}^{\pi^+p} = |T(\omega)|^2 \tag{18.101}$$

a result which can be directly checked by constructing the differential cross section from (10.54) and (10.68).

Equation (18.100) takes care of the discontinuity in the physical region $\omega \geq \mu$. There still remain the contributions from the left-hand cut $\omega \leq -\mu$ and the poles. Since the left-hand cut lies in the physical region for the crossed, or u, reaction which is π^--proton scattering, we make use of crossing symmetry to relate the discontinuity across the left-hand cut to the total π^--p cross section. As discussed in Sec. 17.7, the exact π-nucleon scattering amplitude is invariant under the interchange of pion isotopic indices (particle $\leftrightarrow$ antiparticle) and of the ingoing pion four-momentum q_1 with $-q_2$, the negative of the outgoing pion four-momentum. In the forward direction, this symmetry means changing $\omega \leftrightarrow -\omega$, along with the change $\pi^+ \leftrightarrow \pi^-$.

$$T(\omega) \equiv T^{(\pi^+p)}(\omega) = T^{(\pi^-p)}(-\omega) \tag{18.102}$$

Whereas the physical π^+-proton scattering amplitude lies above the right-hand cut $\omega \geq \mu$, the physical π^--proton scattering amplitude lies below the left-hand cut $\omega \leq -\mu$. This is seen from the form of (18.91), which coincides with (18.78) for forward π-π scattering and may be broken into two parts as done there for $\zeta > 0$ and $\zeta < 0$. We exhibit this explicitly by rewriting (18.102) as

$$T^{(\pi^+p)}(-\omega - i\epsilon) = T^{(\pi^-p)}(\omega + i\epsilon)$$

therefore,

$$\text{Im } T^{(\pi^+p)}(-\omega) \equiv \frac{1}{2i}[T^{(\pi^+p)}(-\omega + i\epsilon) - T^{(\pi^+p)}(-\omega - i\epsilon)]$$

$$= +\frac{k}{4\pi}\sigma_{\text{tot}}^{\pi^-p}(\omega) \tag{18.103}$$

Having evaluated the discontinuity on the left-hand cut in terms of a physical cross section, we must only calculate the residue at the pole $\omega = \omega_B = \mu^2/2M$ before expressing the forward-scattering amplitude in terms of experimental numbers. The pole contributions are the same as the second-order calculation with Feynman diagrams in Chap. 10 except for the replacement of the bare coupling constant g_0 by a renormalized coupling constant g.

To establish this result requires evaluating all Feynman graphs that look like Fig. 18.29, that is, with only a nucleon propagator between the vertices. Moreover, we need evaluate these graphs only in the neighborhood of the pole associated with the nucleon propagator according to our discussion of reduced graphs above (18.93). Since only the crossed graph, Fig. 18.29*b*, contributes to π^+-proton scattering, we find the general structure for the invariant amplitude $\mathfrak{M}_{\text{pole}}$ corresponding to this graph to be

$$-i\mathfrak{M}_{\text{pole}}(q_1,p_1,s_1) = Z_2 Z(-ig_0\sqrt{2})^2 \bar{u}(p_1,s_1) i\Gamma_5(p_1,p) \times iS'_F(p) i\Gamma_5(p,p_1) u(p_1,s_1) \qquad p = p_1 - q_1 \quad (18.104)$$

The kinematics is as in the graph specialized to forward scattering, $p_2 = p_1$ and $q_2 = q_1$, and Z_2 and Z are the nucleon and pion wave function renormalization factors according to the reduction formula (16.81) or (17.43). We acquire the factors $\sqrt{Z}$ in (18.104) as we amputate each external propagator leg in the graph from the typical combination appearing in the reduction formula (17.43):

$$\int d^4x \, \frac{1}{\sqrt{Z_2}} \, e^{ip_1\cdot x} \bar{u}(p_1)(i\overleftarrow{\nabla}_x - M) S'_F(x - y) \cdots = \sqrt{Z_2}\, e^{ip_1\cdot y} \bar{u}(p_1) \cdots$$

According to the spectral representation (16.122), the exact propagator $S'_F(p_1)$ approaches $Z_2/(\not{p}_1 - M)$ as $\not{p}_1 \to M$ on the mass shell and the Dirac operator just cancels this $\not{p}_1 - M$ factor so that we end up with the $\sqrt{Z_2}$ in the numerator. A similar result obtains for the pion propagator and wave function. Also, the propagator $S'_F(p)$ for the intermediate nucleon in (18.104), by the Källén-Lehmann representation (16.122), approaches $Z_2/(\not{p} - M)$ at the pole.

$\Gamma_5(p_1,p)$ is the sum of all proper vertex graphs, that is, all vertex graphs which cannot be separated into disconnected parts by removing one meson or nucleon line. As $\not{p} \to M$ at the pole for the intermediate nucleon, $\Gamma_5(p_1,p)$ may be reduced to a multiple of γ_5 by a procedure similar to that given above (10.157). All factors of $\not{p}_1$ may be moved adjacent to $\bar{u}(p_1)$ or $u(p_1)$ and set equal to M. All factors $\not{p}$ may be moved to the middle adjacent to $S'_F(p)$ and replaced by M; corrections proportional to $\not{p} - M$ cancel out the pole in which we are interested. Scalar arguments of the invariant form factors in Γ_5 may be set equal to the numerical values $p^2 = p_1{}^2 = M^2$ and

$q_1^2 = (p - p_1)^2 = \mu^2$ at the pole. After this is done, the Γ_5 can be set equal to a numerical multiple of γ_5 in evaluating the pole contribution of (18.104). This multiple is so defined as to remove all Z factors in the pole term

$$g\gamma_5 \equiv Z_2 Z_3^{1/2} g_0 \Gamma_5(p_1,p) \Big|_{\substack{p^2 \to M^2 \\ p_1^2 \to M^2 \\ q_1^2 = \mu^2}} \tag{18.105}$$

Inserting (18.105) into (18.104), we find the perturbation theory result (10.56) with the bare coupling g_0 replaced by g:

$$\mathfrak{M}_{\text{pole}}(q_1,p_1,s_1) = 4\pi T(\omega) = -2g^2 \frac{\bar{u}(p_1,s_1)\gamma_5(\not p_1 - \not q_1 + M)\gamma_5 u(p_1,s_1)}{(p_1 - q_1)^2 - M^2}$$

$$= \frac{2g^2\omega\bar{u}(p_1,s_1)u(p_1,s_1)}{2M\omega - \mu^2} = \frac{8\pi f^2}{\omega - \omega_B} \qquad \text{as } (p_1 - q_1)^2 \to M^2 \tag{18.106}$$

where

$$\left(\frac{g^2}{4\pi}\right)\frac{\mu^2}{4M^2} \equiv f^2$$

is the rationalized and renormalized pion-nucleon coupling constant.

Using the Cauchy formula and the contour C of Fig. 18.31, we now write a dispersion relation analogous to (18.81) for forward π-π scattering, but with the addition of the pole term. Putting together (18.100), (18.103), and (18.106), we find

$$T(\omega) = \frac{1}{\pi}\int_{-\infty}^{-\mu} \frac{d\omega \operatorname{Im} T(\omega')}{\omega' - \omega - i\epsilon} + \frac{1}{\pi}\int_{\mu}^{\infty} \frac{d\omega \operatorname{Im} T(\omega')}{\omega' - \omega - i\epsilon} + \frac{2f^2}{\omega - \omega_B} + C_\infty$$

$$= \frac{1}{4\pi^2}\int_{-\infty}^{-\mu} \frac{d\omega' \sqrt{\omega'^2 - \mu^2}\,\sigma_{\text{tot}}^{\pi^- p}(-\omega')}{(\omega' - \omega - i\epsilon)}$$

$$- \frac{1}{4\pi^2}\int_{\mu}^{\infty} \frac{d\omega' \sqrt{\omega'^2 - \omega^2}\,\sigma_{\text{tot}}^{\pi^+ p}(\omega')}{\omega' - \omega - i\epsilon} + \frac{2f^2}{\omega - \omega_B} + C_\infty$$

Upon taking the real part and combining the first and second terms, we obtain the final form of the dispersion relations for π^+-p and π^--p scattering:

$$\operatorname{Re} T^{(\pi^+ p)}(\omega) = \frac{2f^2}{\omega - \omega_B} - \frac{1}{4\pi^2} P \int_{\mu}^{\infty} \frac{d\omega' \sqrt{\omega'^2 - \mu^2}}{\omega'^2 - \omega^2}$$
$$\times \{\omega'[\sigma_{\text{tot}}^{\pi^+ p}(\omega') + \sigma_{\text{tot}}^{\pi^- p}(\omega')] + \omega[\sigma_{\text{tot}}^{\pi^+ p}(\omega') - \sigma_{\text{tot}}^{\pi^- p}(\omega')]\} + C_\infty$$

$$\operatorname{Re} T^{(\pi^- p)}(\omega) = -\frac{2f^2}{\omega + \omega_B} - \frac{1}{4\pi^2} P \int_{\mu}^{\infty} \frac{d\omega' \sqrt{\omega'^2 - \mu^2}}{\omega'^2 - \omega^2}$$
$$\times \{\omega'[\sigma_{\text{tot}}^{\pi^+ p}(\omega') + \sigma_{\text{tot}}^{\pi^- p}(\omega')] - \omega[\sigma_{\text{tot}}^{\pi^+ p}(\omega') - \sigma_{\text{tot}}^{\pi^- p}(\omega')]\} + C_\infty \tag{18.107}$$

Introducing the sum and difference

$$T^{+} = \tfrac{1}{2}[T(\omega) + T(-\omega)] = \tfrac{1}{2}[T^{(\pi^+p)}(\omega) + T^{(\pi^-p)}(\omega)]$$
$$T^{-} = -\tfrac{1}{2}[T(\omega) - T(-\omega)] = \tfrac{1}{2}[T^{(\pi^-p)}(\omega) - T^{(\pi^+p)}(\omega)] \tag{18.108}$$

we find a more convenient form for the dispersion relations. In particular for the odd amplitude T^-, assuming no subtractions are necessary, we arrive at the important relation

$$\frac{\operatorname{Re} T^-(\omega)}{\omega} = \frac{-2f^2}{\omega^2 - \omega_B{}^2} + \frac{1}{4\pi^2} P \int_\mu^\infty \frac{d\omega' \sqrt{\omega'^2 - \mu^2}\,[\sigma_{\text{tot}}^{\pi^+p}(\omega') - \sigma_{\text{tot}}^{\pi^-p}(\omega')]}{\omega'^2 - \omega^2} \tag{18.109}$$

It is a one-parameter fit to data at all energies provided

$$\lim_{\omega\to\infty} [\sigma_{\text{tot}}^{\pi^+p}(\omega) - \sigma_{\text{tot}}^{\pi^-p}(\omega)] \to 0 \tag{18.110}$$

which is indicated approximately in the 10-BeV region, where the difference has decreased to

$$\frac{\Delta\sigma}{\sigma} = \frac{\sigma_{\text{tot}}^{\pi^-p}(\omega) - \sigma_{\text{tot}}^{\pi^+p}(\omega)}{\sigma_{\text{tot}}^{\pi^+p}(\omega)} \approx \frac{1.2 \times 10^{-27}\text{cm}^2}{25 \times 10^{-27}\text{cm}^2} \approx 0.05$$

The dispersion relation (18.109) is experimentally checked and f^2 is measured[1] to be

$$f^2 = 0.080 \pm 0.001 \tag{18.111}$$

in a major triumph for this exact consequence of the theory.

This result is only very slightly sensitive to the cross-section behavior as $\omega \to \infty$. If, however, (18.110) is not experimentally verified by future studies at higher energies, a subtraction will have to be made and a dispersion relation written for $T^-(\omega)/\omega(\omega^2 - \omega_0{}^2)$, where $\omega_0{}^2 \leq \mu^2$ is some convenient subtraction point. We then obtain in place of (18.109), choosing $\omega_0 = \mu$,

$$\frac{\operatorname{Re} T^-(\omega)}{\omega} = \frac{T^-(\mu)}{\mu} + \frac{2f^2(\omega^2 - \mu^2)}{(\omega^2 - \omega_B{}^2)(\mu^2 - \omega_B{}^2)}$$
$$- \frac{\omega^2 - \mu^2}{4\pi^2} P \int_\mu^\infty \frac{d\omega'[\sigma_{\text{tot}}^{\pi^+p}(\omega') - \sigma_{\text{tot}}^{\pi^-p}(\omega')]}{\sqrt{\omega'^2 - \mu^2}\,(\omega'^2 - \omega^2)} \tag{18.112}$$

The dispersion integral now converges better at infinity, but the extra price one pays is that the threshold amplitude $T^-(\mu)$ must be inserted from measurements.

[1] U. Haber-Schaim, *Phys. Rev.*, **104,** 1113 (1956); T. D. Spearman, *Nuovo Cimento*, **15,** 147 (1960).

A violation of the condition (18.110) that the π^+-p and π^--p total cross sections approach the same limit as $\omega \to \infty$ leads us to a paradox which, although not rigorously disproved, is at least bizarre from the point of view of physics. Let us assume, for example, that $\sigma^{\pi^+ p}_{\text{tot}}$ and $\sigma^{\pi^- p}_{\text{tot}}$ remain finite as $\omega \to \infty$, continuing the presently observed trend, and that moreover the difference $\Delta\sigma$ does likewise. It follows from the dispersion relation (18.112) that

$$\begin{aligned} \operatorname{Re} T^-(\omega) &\cong -\frac{\omega^3}{4\pi^2}\,\Delta\sigma\, P\int_{\bar{\omega}}^{\infty}\frac{d\omega'}{\omega'(\omega'^2-\omega^2)} \\ &\sim \frac{\omega}{4\pi^2}\,\Delta\sigma\,\log\frac{\omega}{\bar{\omega}} \end{aligned} \tag{18.113}$$

for $\bar{\omega} \gg \mu$ and $\omega/\bar{\omega} \gg 1$, with $\bar{\omega}$ so chosen that the cross sections may be approximated by the asymptotic limit for $\omega > \bar{\omega}$. However, the optical theorem tells us, for $\omega \gg \mu$, that both for π^+ and π^- scattering from protons

$$\operatorname{Im} T(\omega) \sim -\frac{\omega}{4\pi}\,\sigma^{\pi p}_{\text{tot}}(\omega)$$

and we are forced to conclude therefore from (18.113) that the real part of $T(\omega)$ dominates the imaginary part as $\omega \to \infty$ for elastic π^+-proton scattering. This conclusion clashes with physical intuition. The observed total cross sections[1] exceed the elastic cross sections by a factor of more than 3 in the multi-BeV region. We therefore expect that diffraction scattering of the incident pion wave, accompanying its absorption into the numerous inelastic channels that can be populated for large ω, will give the dominant—and imaginary—contribution to the forward elastic cross section. High-energy elastic scattering experiments[2] support this intuitive picture of what is going on.

From such physical arguments, due originally to Pomeranchuk,[3] one concludes that (18.110), and therefore the unsubtracted dispersion relation (18.109), should be valid. This conclusion, referred to as Pomeranchuk's theorem, has been proved on somewhat weaker assumptions by Weinberg and others.[4] It may also be derived in the

[1] S. J. Lindenbaum, *Rept. 1964 Intern. Conf. High-energy Physics* (*Dubna*).
[2] *Ibid.*
[3] I. Ya. Pomeranchuk, *J.E.T.P.* (*USSR*), **34,** 725 (1958).
[4] S. Weinberg, *Phys. Rev.*, **124,** 2049 (1961); D. Amati, M. Fierz, and V. Glaser, *Phys. Rev. Letters*, **4,** 89 (1960); N. N. Meiman, *J.E.T.P.* (*USSR*), **16,** 1609 (1963).

same way for other particle-antiparticle cross sections, for example,

$$\begin{aligned} \sigma_{\text{tot}}^{pp}(\omega) - \sigma_{\text{tot}}^{\bar{p}p}(\omega) &\to 0 \\ \sigma_{\text{tot}}^{K^+p}(\omega) - \sigma_{\text{tot}}^{K^-p}(\omega) &\to 0 \end{aligned} \qquad \omega \to \infty$$

and appears to be supported experimentally.

We may write similar dispersion relations for the even amplitude $T^+(\omega)$ by the same methods. In this case the sum of cross sections $\sigma_{\text{tot}}^{\pi^-p}(\omega) + \sigma_{\text{tot}}^{\pi^+p}(\omega)$ appears in the dispersion integral and a subtraction is indeed required. Adding together the two dispersion relations in (18.107) and making a subtraction at threshold ($\omega = \mu$), we find

$$\begin{aligned} \operatorname{Re} T^+(\omega) = T^+(\mu) &- \frac{f^2\mu^2(\omega^2 - \mu^2)}{M(\omega^2 - \omega_B{}^2)(\mu^2 - \omega_B{}^2)} \\ &- \frac{\omega^2 - \mu^2}{4\pi^2} P \int_\mu^\infty \frac{d\omega'\, \omega'[\sigma_{\text{tot}}^{\pi^+p}(\omega') + \sigma_{\text{tot}}^{\pi^-p}(\omega')]}{(\omega'^2 - \omega^2)\sqrt{\omega'^2 - \mu^2}} \end{aligned} \tag{18.114}$$

Experimental verification of this relation has also been given.[1] Similar results are also possible for spin flip amplitudes in the forward direction.[2]

18.11 Axiomatic Derivation of Forward Pi-Nucleon Dispersion Relations

Since the dispersion relations (18.109) and (18.114) are definite connections between observable quantities, we may ask what a disagreement with experiment would mean. To attempt an answer, we may first review the ingredients. We based our S-matrix discussion in quantum field theory on very general postulates in Chap. 16, in particular on translational invariance, which implies the existence of an energy-momentum four-vector P_μ, and on Lorentz invariance. Furthermore, we demanded the existence of a unique normalizable ground state, the vacuum, along with a well-defined, complete spectrum of in- or out-states. Finally, we assumed the theory to be local, with the fields obeying differential equations and with P_μ constructed from a local density. The elementary Bose field commutators and Fermi field anticommutators vanish for space-like separations, as do the commutators of all local densities of "observables" formed from the fields. We call such a theory microscopically causal.

[1] H. L. Anderson, W. C. Davidon, and U. W. Kruse, *Phys. Rev.*, **100**, 339 (1955); J. Hamilton, *Phys. Rev.*, **110**, 1134 (1958).

[2] W. C. Davidon and M. L. Goldberger, *Phys. Rev.*, **104**, 119 (1956).

All of these are very fundamental concepts which we would be reluctant to abandon except as a last resort. There is, however, one further aspect of our discussion to which we could turn to rescue the theory in case of need, and that is the convergence of the perturbation expansion in Feynman diagrams. Perhaps the infinite sum of diagrams has different analytic properties from the individual terms. However, this escape hatch can be closed, and a rigorous derivation of the forward π-nucleon dispersion relations has been given by Symanzik and by Bogoliubov.[1] We follow the latter's procedure here.

Our starting point is the S matrix (16.77) for forward scattering, which we reduce halfway to the form of a vacuum expectation value by removing the two mesons from the in- and out-states.

$$\begin{aligned}\langle q_2p_1s_1\ \text{out}|q_1p_1s_1\ \text{in}\rangle &= \delta_{q_2q_1} - \frac{1}{Z}\int \frac{d^4x\, d^4y\, e^{+(iq_2\cdot y - iq_1\cdot x)}}{(2\pi)^3\sqrt{4\omega_2\omega_1}} \\ &\quad\times (\Box_y + \mu^2)(\Box_x + \mu^2)\langle p_1s_1\ \text{out}|T(\varphi(y)\varphi^\dagger(x))|p_1s_1\ \text{in}\rangle \\ &= \delta_{q_2q_1} - \frac{i(2\pi)^4\delta^4(q_2 - q_1)}{(2\pi)^6 2\omega}\,\mathfrak{M}(q_1,p_1,s_1) \qquad (18.115)\end{aligned}$$

where $\mathfrak{M}$, defined as in the perturbation discussion (18.95) and (18.99), is given by

$$\frac{\mathfrak{M}(q,p,s)}{(2\pi)^3} = \frac{-i}{Z}\int d^4y\, e^{+iq\cdot y}(\Box_y + \mu^2)^2\langle ps|T(\varphi(y)\varphi^\dagger(0))|ps\rangle \qquad (18.116)$$

φ is the π^+ field; that is, $\varphi^\dagger(x)$ acting on a state increases the charge by 1. In constructing (18.116) we have used the translational invariance of the theory in displacing $\varphi(x)$ to the origin, allowing one of the four-dimensional space integrals to be carried out. This is next brought to a causal form suitable for a dispersion analysis by converting the T product to a commutator with the aid of the identity

$$\begin{aligned}T(a(t)b(0)) &= \theta(t)a(t)b(0) + \theta(-t)b(0)a(t) \\ &= \theta(t)[a(t),b(0)] + b(0)a(t) \qquad (18.117)\end{aligned}$$

so that

$$\begin{aligned}\frac{\mathfrak{M}(q,p,s)}{(2\pi)^3} = \frac{-i}{Z}\int d^4y\, e^{+iq\cdot y}(\Box_y + \mu^2)^2\{\theta(y_0)\langle ps|[\varphi(y),\varphi^\dagger(0)]|ps\rangle \\ + \langle ps|\varphi^\dagger(0)\varphi(y)|ps\rangle\} \qquad (18.118)\end{aligned}$$

The second term vanishes because of the stability of the proton, as is

[1] K. Symanzik, *Phys. Rev.*, **105**, 743 (1957); N. N. Bogoliubov and D. V. Shirkov, "Introduction to the Theory of Quantized Fields," Interscience Publishers, Inc., New York, 1959.

seen by inserting a complete set of states $|n\rangle$ between the field operators and carrying out the y integration.

$$\int d^4y\, e^{+iq\cdot y}(\Box_y + \mu^2)^2\langle ps|\varphi^\dagger(0)|n\rangle\langle n|\varphi(y)|ps\rangle$$
$$= \sum_n (q^2 - \mu^2)^2|\langle ps|\varphi^\dagger(0)|n\rangle|^2(2\pi)^4\delta^4(q + p_n - p) \quad (18.119)$$

Since q_0 is positive, there is no state $|n\rangle$ such that $p = q + p_n$.

Equations (18.118) and (18.116) coincide on the mass shell for q_μ real and $q^2 = \mu^2$. They have, however, different analytic continuations into the complex $q_0 = \omega$ plane, and the retarded commutator in (18.118) has the causal form we can exploit in constructing a dispersion relation. This form is reminiscent of the opening classical discussions of this chapter. There we established a dispersion relation by requiring that no scattered wave arrive at velocities faster than the velocity of light in vacuum. Here we impose a causality condition with the requirement

$$[\varphi(y),\varphi^\dagger(0)] = 0 \qquad y^2 < 0 \quad (18.120)$$

With this limitation the desired analytic continuation for a dispersion relation is now possible.

To see this, we enter the rest frame of the nucleon $p = (M,0,0,0)$ and use the fact that $\mathfrak{M}$ is a scalar and therefore, for forward scattering, independent of spin s. The matrix element in (18.116) is therefore a function of $|\mathbf{y}|^2$ and y_0 alone, and the angular integrals can be carried out. In terms of the amplitude $T(\omega)$ defined in (18.99) we find, from (18.118),

$$T(\omega) = \frac{1}{4\pi}\,\mathfrak{M}(q,p,s) = \int_0^\infty y\,dy\,Y(\omega,y) \quad (18.121)$$

where the function

$$Y(\omega,y) \equiv -\frac{(2\pi)^3 i}{Z}\frac{\sin\sqrt{\omega^2-\mu^2}\,y}{\sqrt{\omega^2-\mu^2}}\int_{-\infty}^{\infty} dy_0\, e^{i\omega y_0}$$
$$\times (\Box_y + \mu^2)^2\theta(y_0)\langle Ms|[\varphi(y),\varphi^\dagger(0)]|Ms\rangle \quad (18.122)$$

is analytic in the upper half ω plane. It has no branch point at $\omega = i\mu$ because $(1/\sqrt{z})\sin\sqrt{z}$ depends on z and not on $\sqrt{z}$. Also, in analogy with the Kramers-Krönig example, the step function $\theta(y_0)$ ensures that the y_0 integration becomes exponentially damped as ω moves into the upper half plane. This would be evident if $\theta(y_0)$ stood to the left of the Klein-Gordon operator in (18.122), since the integral

$$\int_{-\infty}^{\infty} dy_0\,\theta(y_0) = \int_0^\infty dy_0 \quad (18.123)$$

extends only over the nonnegative range $0 \leq y_0 \leq \infty$. The effect of the time derivatives $\partial/\partial y_0$ in the Klein-Gordon operator as $\theta(y_0)$ is moved to the left is to introduce delta functions and derivatives via $d/dy_0\, \theta(y_0) = \delta(y_0)$. These additional terms thus involve nothing more than the harmless c-number commutators of the fields at equal times, such as

$$[\varphi(\mathbf{y},0),\varphi^\dagger(0)] = 0 \qquad [\dot\varphi(\mathbf{y},0),\varphi^\dagger(0)] = -i\delta^3(\mathbf{y}) \tag{18.124}$$

and add nothing worse than real polynomials in ω to the expression for $T(\omega)$ in (18.121). They are safely ignored in making the analytic continuation for constructing dispersion relations. Furthermore, the causality condition (18.120) assures decent behavior of $Y(\omega,y)$ as $\omega \to \infty$ in the upper half plane because the exponential factors in (18.122) are harmless if we are inside the forward light cone $y_0 \geq y$. That is,

$$\exp\,[i(\omega y_0 \pm \sqrt{\omega^2 - \mu^2}\, y)] \approx \{\exp\,[i\omega(y_0 \pm y)]\} \left[\exp\left(\mp i \frac{\mu^2}{2\omega}\, y\right)\right] \qquad \text{for } |\omega| \to \infty \tag{18.125}$$

causes no trouble when $\operatorname{Im} \omega \geq 0$ and $y_0 \geq y$. Although the second factor may give rise to an exponentially growing behavior for very large y, $\sim \exp\,(+\mu^2 y/2|\omega|)$, it may be suppressed by introducing a convergence factor $e^{-\epsilon y^2}$ into the integral over dy in (18.121) with $\epsilon \to 0$ eventually.

With these welcome properties of $Y(\omega,y)$ established, we can carry out the y-integral for all ω in the upper half plane, establishing thereby the analyticity of $T(\omega)$. We may therefore write a dispersion relation for $T(\omega)$

$$\operatorname{Re} T(\omega) = \frac{1}{\pi} P \int_{-\infty}^{\infty} \frac{d\omega'}{\omega' - \omega} \int_0^\infty dy\, y e^{-\epsilon y^2} \operatorname{Im} Y(\omega',y) + C_\infty \tag{18.126}$$

where C_∞, the contribution from the semicircle at ∞, may be disposed of by making a finite number of subtractions as required.[1]

We must still take the limit $\epsilon \to 0$ inside the dispersion integral in (18.126) before we are finished. This presents no problem for $|\omega| \geq \mu$ since the factor $(\sin \sqrt{\omega^2 - \mu^2}\, y)/\sqrt{\omega^2 - \mu^2}$ safely oscillates and the integral in (18.121) for the transition amplitude $T(\omega)$ exists for real physical values of ω. In this region the protection given by $e^{-\epsilon y^2}$ against an exponential increase in the second factor of (18.125) for complex ω is no longer needed. However, in that *bête noire* of all disper-

[1] Anything worse would require an essential singularity in the commutator of (18.122) as $y \to 0$ in order to put an infinite number of powers of ω upstairs.

sion studies, the unphysical region $|\omega| < \mu$, we cannot let $\epsilon \to 0$ because it is needed there to suppress $\sinh \sqrt{\mu^2 - \omega^2}\, y$. We might expect, therefore, that there is a violent behavior for $\epsilon \to 0$ and $|\omega| < \mu$ in (18.126). On the other hand, we already know from the Feynman graph analysis of the preceding section that the absorptive amplitude has a simple behavior in the unphysical region: $\text{Im } T(\omega)$ vanishes except for a simple pole at $\omega = \mu^2/2M$. Within the present framework how do we arrive at this result?

Returning to (18.122), we can extract the imaginary part of $Y(\omega,y)$ in the unphysical region $|\omega| < \mu$. We limit ourselves to this region and let ω approach the real axis from the upper half plane as usual. After inserting a complete set of states between fields φ and $\varphi^\dagger$ in the commutator with the familiar device $1 = \sum_n |n\rangle\langle n|$, we displace the coordinate of field φ to the origin and perform the y_0 integration:[1]

$$\begin{aligned} \text{Im } Y(\omega,y) = \text{Im} \Bigg[& \frac{(2\pi)^3}{Z} \frac{\sinh \sqrt{\mu^2 - \omega^2}\, y}{\sqrt{\mu^2 - \omega^2}} \\ & \times \sum_n \frac{\sin p_n y}{p_n y} [(E_n - M)^2 - p_n^2 - \mu^2]^2 \\ & \times \left(\frac{|\langle n|\varphi^\dagger|Ms\rangle|^2}{\omega + M - E_n + i\epsilon} - \frac{|\langle n|\varphi|Ms\rangle|^2}{\omega - M + E_n + i\epsilon} \right) \Bigg] \end{aligned} \tag{18.127}$$

The denominators vanish and $Y(\omega,y)$ acquires an imaginary part only when

$$E_n = M \pm \omega \tag{18.128}$$

However, the states $|n\rangle$ in (18.127) must couple to the state $\varphi^\dagger|Ms\rangle$ which has nucleon number $+1$ and charge $+2$, or to $\varphi|Ms\rangle$ with nucleon number $+1$ and charge 0. The only states with these quantum numbers and with energy $E_n < M + \mu$, as required by (18.128) when $|\omega| < \mu$, are single neutron states, and these contribute to the second term of (18.127). There are no contributions in the unphysical region from the continuous spectrum of nucleons and pions with a threshold at $M + \mu$. The contribution from the one-neutron states is evidently related to the pole contribution to $T(\omega)$ found in the preceding section.

We are led into somewhat intricate computations[2] if we attempt

[1] Time derivatives on the $\theta(y_0)$ in (18.122) lead only to real polynomials in ω which therefore do not appear in (18.127). The factor $(\sin p_n y)/p_n$ is introduced after we average over orientations of the spin s.

[2] K. Symanzik, *op. cit.*, and H. Lehmann, *Nuovo Cimento*, **10**, 579 (1958) and Suppl. 1, **14**, 153 (1959). The troubles are still worse when one goes to dispersion relations for nonforward scattering.

to evaluate Im $Y(\omega,y)$ in the unphysical region directly from (18.127). We avoid this by a trick introduced by Bogoliubov.[1] The imaginary part of $Y(\omega,y)$, for $\omega < \mu$, is given by

$$\begin{aligned}\text{Im } Y(\omega,y) &= \int d^3k\, f(\mathbf{k},y)\delta(\omega - M + \sqrt{\mathbf{k}^2 + M^2}) \\ &= \int d^3k\, F(\mathbf{k},y)\delta\left(\omega - \frac{\omega^2 - |\mathbf{k}|^2}{2M}\right)\end{aligned} \tag{18.129}$$

where $\mathbf{k}$ denotes the momentum of the neutron, $f(\mathbf{k},y)$ and $F(\mathbf{k},y)$ contain all the inessential factors, and $\sum_n \to \int d^3k$ in (18.127). Bogoliubov observed that it is possible to avoid this term by constructing an auxiliary function $\tilde{T}(\omega)$ and a corresponding $\tilde{Y}(\omega,y)$ with all the welcome properties permitting construction of a dispersion relation (18.126) but with

$$F(\mathbf{k},y) \to \tilde{F}(\mathbf{k},y) = \left(\omega - \frac{\omega^2 - |\mathbf{k}|^2}{2M}\right) F(\mathbf{k},y) \tag{18.130}$$

in (18.129). Then, since $x\delta(x) = 0$, $\tilde{T}(\omega)$ has no contribution at all in the unphysical region to harrass us.

The desired auxiliary function removing the neutron pole is simply

$$\tilde{T}(\omega) = \left(\omega - \frac{\mu^2}{2M}\right) T(\omega) \tag{18.131}$$

as we verify by repeating the entire calculation with the additional factor included. In (18.116) we multiply $\mathfrak{M}(q,p,s)$ by the factor

$$\omega - \frac{\mu^2}{2M} = \omega - \frac{\omega^2 - |\mathbf{q}|^2}{2M}$$

which may be brought inside the integral on the right-hand side and replaced by the local differential operator

$$\omega - \frac{\omega^2 + \nabla_y^2}{2M} \tag{18.132}$$

acting on $\varphi(y)$. Therefore, all steps leading from (18.116) to the dispersion relation (18.126) may be repeated for $\tilde{T}(\omega)$, the only change being the replacement

$$\varphi(y) \to \left(\omega - \frac{\omega^2 + \nabla_y^2}{2M}\right) \varphi(y) \tag{18.133}$$

Upon evaluating the absorptive part Im $\tilde{Y}(\omega,y)$ from (18.127), the only change from (18.129) is the appearance of the additional factor

[1] Bogoliubov and Shirkov, *op. cit.*

$\omega - (\omega^2 - |\mathbf{k}|^2)/2M$ as indicated in (18.130). Consequently

$$\text{Im } \tilde{Y}(\omega,y) = 0 \qquad |\omega| \leq \mu \tag{18.134}$$

and without further ado, $\tilde{T}(\omega)$ satisfies a dispersion relation exhibiting its analyticity in the cut ω plane:

$$\text{Re } \tilde{T}(\omega) = \frac{1}{\pi} P \int_{-\infty}^{-\mu} \frac{d\omega' \text{ Im } \tilde{T}(\omega')}{\omega' - \omega} + \frac{1}{\pi} P \int_{\mu}^{\infty} \frac{d\omega' \text{ Im } \tilde{T}(\omega')}{\omega' - \omega} + \tilde{C}_\infty \tag{18.135}$$

Additional subtractions are presumably required to compensate the powers of ω in (18.132) in order to suppress $\tilde{C}_\infty$, but this presents no special difficulty.

The physical amplitude

$$T(\omega) \equiv \frac{\tilde{T}(\omega)}{\omega - \mu^2/2M}$$

therefore possesses the same analytic properties as $\tilde{T}(\omega)$ with the exception of a single pole at $\omega = \mu^2/2M$. The residue at this pole may be expressed in terms of the matrix element $\langle\sqrt{k^2 + M^2}\, s'|\varphi(0)|Ms\rangle$ related to the π-nucleon vertex function. This task we leave to the reader; the result coincides with our earlier reduced graph analysis.

At this point it is clear we have returned to our conclusions of the preceding section but are freed from the reliance on the existence and convergence of a perturbation expansion. Evidently a failure of the forward pion-nucleon dispersion relations would have most profound consequences.

18.12 Dynamical Calculations of Pi-Pi Scattering Using Dispersion Relations

In the success of the forward-scattering pion-nucleon dispersion relations we have the most dramatic example of one aspect of the dispersion theory approach: a relation between experimental observables testing the basic structure of the theory. In recent years much effort has gone into a second aspect which uses the dispersion method as the basis of calculating dynamics of interacting particles. In particular, in the realm of strong interactions where weak-coupling perturbation methods are of little value, the dispersion approach is at present the most successful one.[1]

[1] G. F. Chew, "S Matrix Theory of Strong Interactions," New York, W. A. Benjamin, Inc., New York, 1962; S. Mandelstam, *Rept. Progr. Phys.*, **25**, 99 (1962).

Progress in the area of dynamical calculations has been made at the expense of drastic approximations, and the methods of calculation are still undergoing change and criticism. Validity of the approximations can rarely, if ever, be defended on purely theoretical grounds. Instead, one turns to salient experimental features in order to motivate a string of approximations at the end of which comes a prediction whose merit is judged only by comparison once more with observation. Correction terms to the approximate analysis are often not even well defined and in general are calculationally intractable. We therefore shall not dwell on these questions in great detail. Our goal here is to outline some of the main ideas involved in order to give one a view of the dispersion picture, as seen through the rose-colored glasses of all the most optimistic assumptions.

We consider the problem with the simplest kinematics—namely, pion-pion scattering—and set up a dynamical calculation for the phase shift and for the pion electromagnetic vertex. As in the forward pion-nucleon scattering dispersion relation, unitarity of the scattering amplitude, or conservation of probability, and crossing symmetry play key roles along with the analyticity properties.

Extracting kinematic and isotopic factors as in Sec. 17.8, we write for the π-π scattering amplitude in the center-of-mass frame of the incident pions

$$S_{fi} = \delta_{fi} - \frac{i(2\pi)^4\delta^4(q_1 + q_2 - q_3 - q_4)}{(2\pi)^6(2\omega)^2} \sum_{I=0}^{2} A_I(s,t,u)P_I \qquad (18.136)$$

where the kinematics are shown in Fig. 18.32, with $\omega = \sqrt{\mathbf{q}^2 + \mu^2}$, and the isotopic projection operators P_I are given in (17.54) and (17.55). We shall concentrate on the s reaction for which the physical region extends from $4\mu^2 \leq s \leq \infty$, with t, $u \leq 0$. The crossing relations

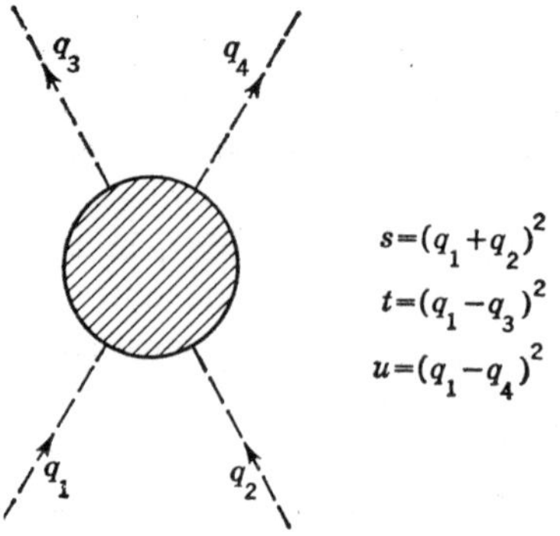

Fig. 18.32 Kinematic variables for π-π scattering.

(17.52) and (17.53) can be used for constructing physical amplitudes for the t and u reactions if desired.

Although we have no general results for the analytic behavior of $A(s,t,u)$ for fixed momentum transfer t, or u, we found the important result in Sec. 18.9, summarized in (18.84), that $\hat{A}_I(s, \cos\theta) \equiv A_I(s,t,u)$ for fixed $\cos\theta$ is analytic in the s plane with the exception of cuts from $-\infty \leq s \leq 0$ and $4\mu^2 \leq s \leq \infty$. Furthermore, it follows from this, according to (18.82) and (18.85), that the partial wave amplitudes

$$A_l^I(s) = \frac{1}{2}\int_{-1}^{1} d(\cos\theta)P_l(\cos\theta)\hat{A}_I(s, \cos\theta) \tag{18.137}$$

have the same very attractive analyticity properties which we shall exploit here along with unitarity and crossing symmetry.

To apply the unitarity condition, we need to extend (18.96), used in the forward π-N dispersion relations, for transitions to different possible final states and obtain thereby a relation for the discontinuity of $A_l^I(s)$ across the cut $s \geq 4\mu^2$ analogous to (18.100). We introduce temporarily the notation

$$S_{ni}^{(+)} \equiv \langle n \text{ out}|i \text{ in}\rangle = \delta_{ni} - (2\pi)^4 i\delta^4(P_n - P_i)\mathfrak{I}_{ni}^{(+)} \tag{18.138}$$

relating the familiar S-matrix elements and the transition amplitudes. If $|n\rangle$ as well as $|i\rangle$ represents a two-pion state, $\mathfrak{I}_{ni}^{(+)}$ is the transition amplitude for elastic π-π scattering and differs from the amplitude $\hat{A}(s, \cos\theta)$ by an over-all factor proportional to s^{-1}. In particular in the physical region $s \geq 4\mu^2$, $\mathfrak{I}_{ni}^{(+)}$ is given, according to the Nambu representation, by the limit as s approaches the cut along the real axis from the upper half plane.

$$\mathfrak{I}_{ni}^{(+)}(s, \cos\theta) \equiv \mathfrak{I}_{ni}(s + i\epsilon, \cos\theta)$$

Similarly, we may define

$$S_{fn}^{(-)} \equiv \langle f \text{ in}|n \text{ out}\rangle = S_{nf}^{(+)*} \tag{18.139}$$

and the unitarity condition of the S matrix may be written

$$\sum_n S_{fn}^{(-)}S_{ni}^{(+)} = \delta_{fi} \tag{18.140}$$

In analogy with (18.138), we introduce

$$S_{fn}^{(-)} = \delta_{fn} + i(2\pi)^4\delta^4(P_f - P_n)\mathfrak{I}_{fn}^{(-)} \tag{18.141}$$

and combining with (18.140) obtain as the unitarity condition

$$\mathfrak{I}_{fi}^{(+)} - \mathfrak{I}_{fi}^{(-)} = -i\sum_n (2\pi)^4\delta^4(P_n - P_i)\mathfrak{I}_{fn}^{(-)}\mathfrak{I}_{ni}^{(+)} \tag{18.142}$$

It follows from the definition given in (18.139) that $\mathfrak{J}^{(-)}$ is related to the physical transition amplitude $\mathfrak{J}^{(+)}$ according to

$$\mathfrak{J}_{fi}^{(-)} = \mathfrak{J}_{if}^{(+)*}$$

in terms of which (18.142) reduces to (18.96) for $i = f$.

A more useful form results upon showing that $\mathfrak{J}_{fi}^{(-)}$ is the analytic continuation of $\mathfrak{J}_{fi}^{(+)}$ to the point $s - i\epsilon$ below the physical cut, $s \geq 4\mu^2$, in the s plane. Then the left-hand side of (18.142) is just the discontinuity of $\mathfrak{J}(s)$ across the cut for physical energies. This result is proved by returning to the reduction procedure in Chap. 16 and repeating the steps in Sec. 16.7, Eqs. (16.72) to (16.81), with everywhere the role of in- and out-states interchanged. The main change is that the direction of the flow of time is reversed so that time-ordered products $T(AB \cdots)$ become anti-time-ordered products $\bar{T}(AB \cdots)$. For example, in place of (16.78) to (16.80), we have

$$\begin{aligned}\langle \gamma p \text{ in}|\varphi(x)|\alpha \text{ out}\rangle &= \langle \gamma \text{ in}|\varphi(x)|\alpha - p \text{ out}\rangle \\ &+ \langle \gamma \text{ in}|a_{\text{in}}(p)\varphi(x) - \varphi(x)a_{\text{out}}(p)|\alpha \text{ out}\rangle = \langle \gamma \text{ in}|\varphi(x)|\alpha - p \text{ out}\rangle \\ &- \frac{i}{\sqrt{Z}}\int d^4y\, \langle \gamma \text{ in}|\bar{T}(\varphi(x)\varphi(y))|\alpha \text{ out}\rangle \overleftarrow{(\Box_y + m^2)} f_p^*(y) \qquad (18.143)\end{aligned}$$

where

$$\bar{T}(\varphi(x)\varphi(y)) \equiv \varphi(x)\varphi(y)\theta(y_0 - x_0) + \varphi(y)\varphi(x)\theta(x_0 - y_0) \qquad (18.144)$$

defines the anti-time-ordered product. Upon contraction of all particles the result (16.81) is reproduced with only the change in $\rightarrow$ out, $T \rightarrow \bar{T}$, and $(i)^{m+n} \rightarrow (-i)^{m+n}$. To obtain the analogue of (17.21), giving the expansion of a τ function in terms of in-fields, we take the hermitian conjugate of (17.21). We see that this only changes time-ordering to anti-time-ordering, along with changing $-i$ to $+i$ in $\exp(-i\int H_I\, dt)$. It is then a straightforward matter to repeat the Wick reduction of Chap. 17, now to anti-time-ordered products of in-fields, viz., for scalar fields

$$i\bar{\Delta}_F(x - y) = \langle 0|\bar{T}(\varphi_{\text{in}}(x)\varphi_{\text{in}}(y))|0\rangle = -i\int \frac{d^4q}{(2\pi)^4}\frac{e^{-iq\cdot(x-y)}}{q^2 - \mu^2 - i\epsilon} \qquad (18.145)$$

This anticausal propagator $\bar{\Delta}_F(x - y)$ differs from the Feynman propagator $\Delta_F(x - y)$ by an over-all change of sign as well as the change in the denominator $i\epsilon \rightarrow -i\epsilon$. The same result is found for spinor and vector fields as well, and we may conclude that aside from questions of over-all signs, the only change in going from $\mathfrak{J}^{(+)}$ to $\mathfrak{J}^{(-)}$ is that of replacing the $i\epsilon$ by $-i\epsilon$ in propagator denominators.

Concerning signs, we have a relative factor -1 for each propagator by (18.145) and -1 for each vertex from the change

$$T \exp\left(-i\int H_I\,dt\right) \to \bar{T} \exp\left(+i\int H_I\,dt\right) \tag{18.146}$$

Finally, there is a factor -1 for each external field contracted, according to (18.143), which gives the operator difference $(a_{\text{in}}\varphi \cdots) - (\varphi a_{\text{out}} \cdots)$ in place of its negative[1] appearing in (16.79). This collection of minus signs cancels by the time we arrive at the Nambu representation (18.74). For trilinear couplings, vertices and fermion propagators are always added in pairs as one inserts a radiative correction.[2] Each additional boson propagator introduces an additional closed loop. Integration over each loop momentum d^4l changes sign with the $i\epsilon$ (because the contour is rotated in the opposite direction), and so altogether we have an even number of signs which may be ignored. Therefore, we need consider only the over-all sign of $\mathfrak{J}^{(-)}$ relative to $\mathfrak{J}^{(+)}$. The minus signs on the external propagator legs in the τ functions cancel those introduced with the wave functions as in (18.143). This leaves us with the simple task of comparing $\mathfrak{J}^{(-)}$ with $\mathfrak{J}^{(+)}$ in lowest-order perturbation theory and finding that they have the same sign, as follows from the definitions (18.138) and (18.141). Elaboration of all these assertions is left as an exercise for the reader.

Therefore, we[3] have shown that to obtain $\mathfrak{J}^{(-)}$ from $\mathfrak{J}^{(+)}$, one simply changes the sign of the $i\epsilon$ in the Nambu representation (18.74), that is, one goes from above the cut in the s plane to below the cut.[4] With this result established we return to (18.142) and apply this unitarity condition to the isotopic amplitudes $\hat{A}_I(s, \cos\theta)$ with the good analyticity properties. In the energy interval $4\mu^2 \le s \le 16\mu^2$ the unitarity equation is especially simple, since only the two-pion states contribute to the sum over states $\sum_n$ on the right-hand side of (18.142). This equation then reduces to a nonlinear condition on the elastic amplitude:

$$\frac{1}{(2\pi)^6 s}\left[\hat{A}_I(s+i\epsilon, \cos\theta_{fi}) - \hat{A}_I(s-i\epsilon, \cos\theta_{fi})\right]$$
$$= \frac{-i}{(2\pi)^{12}s^2}\int \frac{d^3k_1\,d^3k_2}{2}(2\pi)^4\delta^4(q_1+q_2-k_1-k_2)$$
$$\times \hat{A}_I(s-i\epsilon, \cos\theta_{fk})\hat{A}_I(s+i\epsilon, \cos\theta_{ki}) \tag{18.147}$$

[1] With $\bar{T} \to T$.

[2] For the $\lambda\varphi^4$ theory one finds a similar result, since this theory can be considered as the limit of one with trilinear coupling for purposes of counting minus signs.

[3] "We" includes the reader.

[4] Notice that time reversal invariance has *not* been used in deriving this result.

where we have used the closure property (17.56) of the isotopic projection operators. In (18.147) the factor ½ is necessary in converting the sum over states to integrals due to the identity of the two pions with momenta k_1 and k_2. In the center-of-mass frame

$$k_1 + k_2 = (\sqrt{s}, 0)$$

and

$$\cos\theta_{fi} = \frac{\mathbf{q}_3\cdot\mathbf{q}_1}{|\mathbf{q}_3|\,|\mathbf{q}_1|} \qquad \cos\theta_{fk} = \frac{\mathbf{q}_3\cdot\mathbf{k}_1}{|\mathbf{q}_3|\,|\mathbf{k}_1|} \qquad \cos\theta_{ki} = \frac{\mathbf{q}_1\cdot\mathbf{k}_1}{|\mathbf{q}_1|\,|\mathbf{k}_1|}$$

The right-hand side reduces further to an angular integral

$$\operatorname{Im}\hat{A}_I(s, \cos\theta_{fi}) = -\frac{1}{128\pi^2}\sqrt{\frac{s-4\mu^2}{s}}\int d\Omega_k\, \hat{A}_I^*(s, \cos\theta_{fk})\hat{A}_I(s, \cos\theta_{ki}) \tag{18.148}$$

where it is understood that s approaches the real axis from the upper half plane. In terms of the partial wave amplitudes (18.137), the unitarity condition is

$$\operatorname{Im} A_l^I(s) = -\frac{1}{32\pi}\sqrt{\frac{s-4\mu^2}{s}}\,|A_l^I(s)|^2 \tag{18.149}$$

so that in this elastic region $A_l^I(s)$ has the form

$$A_l^I(s) = -32\pi\sqrt{\frac{s}{s-4\mu^2}}\,\{\exp[i\delta_l^I(s)]\}\sin\delta_l^I(s) \tag{18.150}$$

with the phase shift δ_l^I real.

The unitarity equation is no longer so simple for energies above the four-pion threshold at $s = 16\mu^2$ owing to the presence of inelastic channels in $\sum_n$ in (18.142), and one writes generally

$$\operatorname{Im} A_l^I(s) = -\frac{1}{32\pi}\sqrt{\frac{s-4\mu^2}{s}}\,|A_l^I(s)|^2\,\mathcal{f}_l^I(s) \tag{18.151}$$

where $\mathcal{f}_l^I(s)$ is a real fudge factor which is identically 1 for $s \le 16\mu^2$ and at higher energies is the ratio of the contributions of all inelastic and elastic channels of angular momentum l to the elastic contribution alone; it may be written

$$\mathcal{f}_l^I(s) = \frac{\sigma_{\text{tot}}^{l,I}(s)}{\sigma_{\text{el}}^{l,I}(s)} = 1 + \frac{\sigma_{\text{inel}}^{l,I}(s)}{\sigma_{\text{el}}^{l,I}(s)} \tag{18.152}$$

The spirit of the dispersion approach to strong interactions is to parametrize the scattering amplitude in terms of functions which are analytic and slowly varying in the range of variables of interest. The

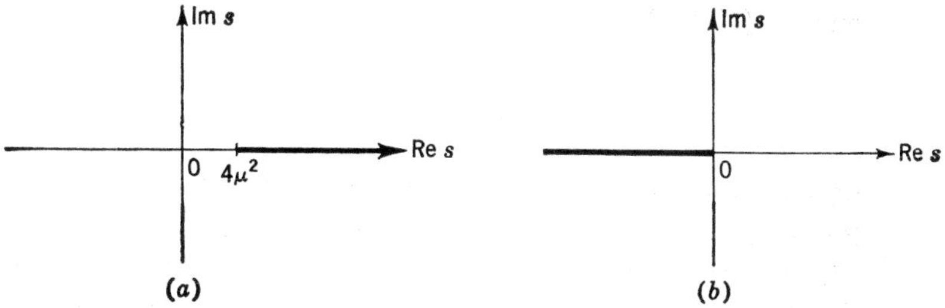

Fig. 18.33 Analytic properties of partial wave amplitude in π-π scattering; (*a*) shows right-hand cut of $D_l^I(s)$ and (*b*) the left-hand cut of $N_l^I(s)$.

aim is to construct in this way an effective range theory which honors as well as possible the analyticity, unitarity, and crossing symmetry properties of the full theory. To illustrate this program, we construct such a parametrization of the π-π scattering amplitude near threshold (Fig. 18.33). We know that $A_l^I(s)$ has a branch point at $s = 4\mu^2$ and a discontinuity across the cut $4\mu^2 \leq s \leq \infty$ determined by unitarity. If we can parametrize $A_l^I(s)$ in terms of a function without this branch point and cut, we may be able to predict the main energy dependence of $A_l^I(s)$ in the vicinity of $s = 4\mu^2$.

The Fredholm theory of potential scattering suggests such a parametrization in terms of a ratio of numerator and denominator function

$$\frac{1}{32\pi} A_l^I(s) = \frac{N_l^I(s)}{D_l^I(s)} \tag{18.153}$$

The numerator is analytic in the cut s plane except for the left-hand cut $-\infty \leq s \leq 0$ and is real for real $s > 0$, while $D_l^I(s)$, the analogue of the Fredholm determinant,[1] is analytic except for the right-hand cut $4\mu^2 \leq s \leq \infty$ and is real for real $s < 4\mu^2$. For a given or assumed $N_l^I(s)$, the structure of $D_l^I(s)$ is determined by the unitarity condition together with its known analytic properties. According to (18.153)

$$\frac{1}{32\pi} \operatorname{Im} A_l^I(s) = -\frac{N_l^I(s)}{|D_l^I(s)|^2} \operatorname{Im} D_l^I(s) \qquad s \geq 4\mu^2 \tag{18.154}$$

[1] R. Jost and A. Pais, *Phys. Rev.*, **82**, 840 (1951). D is known as the Jost function. The existence of this decomposition is assured by defining

$$D_l^I = \exp\left[-\frac{1}{\pi}\int \frac{\delta_l^I(s')\,ds'}{s' - s - i\epsilon}\right]$$

Thus

$$A_l^I D_l^I = N_l^I$$

is real for $0 < s < 16\mu^2$, that is, below inelastic thresholds.

and by (18.151)

$$\operatorname{Im} D_l^I(s) = \sqrt{\frac{s - 4\mu^2}{s}}\, N_l^I(s)\mathcal{J}_l^I(s) \qquad s \geq 4\mu^2 \tag{18.155}$$

The analytic properties of $D_l^I(s)$ together with the above expression for its discontinuity across the cut allow us to write a dispersion relation, with the number of subtractions determined by our assumptions as to the asymptotic behavior of $N_l^I(s)$ and $\mathcal{J}_l^I(s)$. In a nonrelativistic potential scattering problem, the analogue of $N_l^I(s)$ approaches Born approximation and $D_l^I(s)$ approaches 1 as the interaction approaches zero; also in this case as $s \to \infty$, $N_l^I(s) \to 0$ and $D_l^I \to 1$. If we, for simplicity, make the same assumption here, along with

$$\mathcal{J}_l^I(s) \underset{s\to\infty}{\longrightarrow} \text{const}$$

we can write

$$D_l^I(s) = 1 + \frac{1}{\pi}\int_{4\mu^2}^{\infty} \frac{ds'}{s' - s - i\epsilon}\sqrt{\frac{s' - 4\mu^2}{s'}}\, N_l^I(s')\mathcal{J}_l^I(s') \tag{18.156}$$

where the normalization $D_l^I(\infty) = 1$ is arbitrary, as is that of $N_l^I(s)$. Introducing (18.156) into (18.153), we have a parametrization of the partial wave amplitude

$$\frac{1}{32\pi} A_l^I(s) = \frac{N_l^I(s)}{1 + \dfrac{1}{\pi}\displaystyle\int_{4\mu^2}^{\infty} ds' \sqrt{\frac{s' - 4\mu^2}{s'}}\, \frac{N_l^I(s')\mathcal{J}_l^I(s')}{s' - s - i\epsilon}} \tag{18.157}$$

This form must be taken with due reservations in view of the optimistic assumptions made with regard to the asymptotic behavior of A_l^I as well as the simple form of (18.156). The advantage of writing A_l^I in the form (18.157) is that it is automatically unitary and has the correct analyticity properties. Furthermore, our input guesses and approximations lie in the numerator function N_l^I, with unblemished analytic behavior in the interesting neighborhood of $s = 4\mu^2$. As a first crude guess, one might estimate $N_l^I(s)$ by doing lowest-order perturbation theory on the interaction between the meson lines, that is, on the forces producing the scattering. Then as the coupling approaches zero, $A_l^I(s) \to 32\pi N_l^I(s)$ and is real. One hopes that the approximation is significantly improved for somewhat stronger coupling by including the correction in the denominator which ensures unitarity. Solving (18.157) in terms of the phase shift defined in

(18.150), we have

$$\sqrt{\frac{s-4\mu^2}{s}}\cot\delta_l{}^I(s)$$
$$= -N_l{}^I(s)^{-1}\left[1+\frac{1}{\pi}P\int_{4\mu^2}^{\infty}ds'\sqrt{\frac{s'-4\mu^2}{s'}}\,\frac{N_l{}^I(s')\,f_l{}^I(s')}{s'-s}\right] \tag{18.158}$$

Since $N_l{}^I(s)$ has only a left-hand cut, it is possible[1] to expand the right side of (18.158) in a power series about $s = 4\mu^2$ with radius of convergence $4\mu^2$, and thereby obtain a generalized effective range formula.

To do better, we must put more effort into the computation of $N_l{}^I(s)$ itself. There are vast numbers of singular points along the left-hand cut, corresponding to the various possible reduced graphs and Landau singularities. However, just as the low-energy part of the right-hand cut corresponds to the simple one-dimensional reduced graphs in Fig. 18.34*a*, so the low-energy part of the left-hand cut is due only to the existence of their crossed counterparts shown in Fig. 18.34*b*. It is possible to obtain the discontinuity corresponding

[1] See Prob. 10.

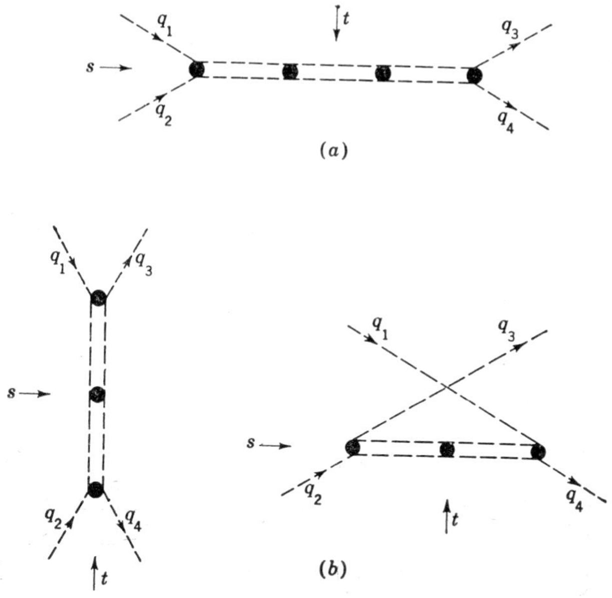

Fig. 18.34 One-dimensional reduced graphs in π-π scattering.

to these singularities by the use of elastic unitarity in the t and u channels, suitably analytically continued. This discontinuity is expressed again in terms of elastic π-π scattering; and if one believes that these singularities alone (Fig. 18.34) play the dominant role in determining the properties of $A_l^I(s)$, one obtains a closed set of integral equations to be solved for $A_l^I(s)$. Extensive studies of this closed set of equations have been carried out, but the approximation of retaining only the contribution of the singularities shown in Fig. 18.34 is not beyond doubt. The structure of the distant part of the left-hand cut and in particular its asymptotic behavior remain a mystery. This contribution comes from deep within the unphysical region, where intuition, as well as mathematics, is difficult to apply.

One additional consequence of the parametrization of $A_l^I(s)$ given in (18.157) is that the existence of bound states and/or resonances is exhibited by the behavior of the denominator function $D_l^I(s)$. If the sign and magnitude of the numerator function are such as to permit a zero in Re $D_l^I(s_R)$, at $s = s_R > 4\mu^2$, the phase shift δ_l^I passes through 90°, and a resonance occurs. Similarly, the vanishing of $D_l^I(s)$ for $s < 4\mu^2$ leads to a pole in $A_l^I(s)$, corresponding to a bound state. Evidently one wishes to avoid this in the π-π problem, while obtaining a p-wave resonance, associated with the ρ meson.

18.13 Pion Electromagnetic Structure

In the same spirit as used above, we can relate the calculation of the pion electromagnetic structure to pion scattering.[1] As we discussed in Sec. 18.8, only reduced graphs such as shown in Fig. 18.35 contribute singularities to $F_\pi(q^2)$ for $q^2 < 16\mu^2$. These are a product of the pion electromagnetic vertex itself multiplied by π-π scattering. We use the unitarity condition to translate this into a useful mathematical form for calculation.

[1] P. Federbush, M. Goldberger, and S. Treiman, *Phys. Rev.*, **112**, 642 (1958); W. Frazer and J. Fulco, *Phys. Rev.*, **117**, 1609 (1960).

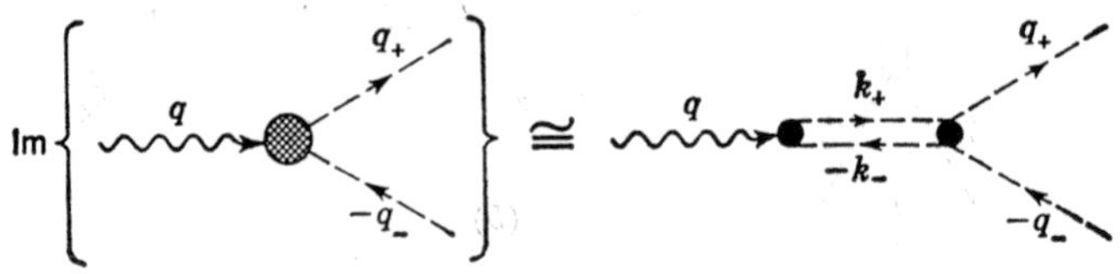

Fig. 18.35 Pion electromagnetic vertex.

Recall from Sec. 18.8 that the pion vertex $F_\pi(q^2)$ has the analytic properties that were essential in constructing the forward or partial wave scattering amplitudes. In fact, $F_\pi(q^2)$ lacks the left-hand cut, and its only singularity is the cut which extends from $4\mu^2$ to ∞. We may therefore repeat the unitarity discussion leading from (18.136) to (18.142), now applied to the vertex, with kinematics as shown in Fig. 18.35:

$$\begin{aligned} V_\mu^{(+)} &= \langle \pi^+(q_+)\pi^-(q_-) \text{ out}|j_\mu(q)|0\rangle \\ &= \frac{-i(2\pi)^4\delta^4(q - q_+ - q_-)}{(2\pi)^3\sqrt{4\omega_+\omega_-}} (q_+ - q_-)_\mu F_\pi(q^2 + i\epsilon) \qquad (18.159) \end{aligned}$$

where[1] $\Box A_\mu(x) = ej_\mu(x)$ and $j_\mu(q)$ is the Fourier transform of $j_\mu(x)$. The vertex evaluated below the cut is then given by

$$\begin{aligned} V_\mu^{(-)} &= \langle \pi^+(q_+)\pi^-(q_-) \text{ in}|j_\mu(q)|0\rangle \\ &= +\frac{i(2\pi)^4\delta^4(q - q_+ - q_-)}{(2\pi)^3\sqrt{4\omega_+\omega_-}} (q_+ - q_-)_\mu F_\pi(q^2 - i\epsilon) \qquad (18.160) \end{aligned}$$

and the unitarity condition becomes

$$\begin{aligned} &\langle f \text{ out}|j_\mu|0\rangle - \langle f \text{ in}|j_\mu|0\rangle \\ &\qquad = -\sum_n [\langle f \text{ in}|n \text{ out}\rangle - \delta_{fn}]\langle n \text{ out}|j_\mu|0\rangle \qquad (18.161) \end{aligned}$$

If we limit $\sum_n$ to π^+-π^- states only and go to the center-of-mass frame of the two pions, we find from the definitions (18.141) and (18.159)

$$\begin{aligned} &2(q_+ - q_-)^\mu \operatorname{Im} F_\pi(q^2) \\ &\quad = -i\int d^3k_+\, d^3k_-(2\pi)^4\delta^4(q - k_+ - k_-) \\ &\qquad\qquad \times (-i)\mathfrak{I}_{fn}^{(-)}(k_+ - k_-)^\mu F_\pi(q^2 + i\epsilon) \qquad (18.162) \end{aligned}$$

We observe that only the p-wave π^+-π^- scattering amplitude will survive the angular integration in (18.162). This amplitude is pure $I = 1$, as follows from the Bose statistics of the pions. We then pick out the $I = 1$ contribution to the expansion (18.136) of $\mathfrak{I}$, and if we do not forget that

$$\langle \pi^+\pi^-|P_1|\pi^+\pi^-\rangle = \tfrac{1}{2}$$

[1] To lowest order in the electromagnetic coupling, as is the case here, we can set $e = e_0$ and $Z_3 = 1$.

as follows from (17.54), we find after contracting V_μ with a space-like vector $\epsilon^\mu = (0,\boldsymbol{\varepsilon})$

$$\cos\theta \,\mathrm{Im}\, F_\pi(q^2) = -\sqrt{\frac{q^2-4\mu^2}{q^2}} \int \frac{d\Omega_n}{128\pi^2}\, \hat{A}_1(q^2 - i\epsilon, \cos\theta_{fn}) \times \cos\theta_n\, F_\pi(q^2+i\epsilon) \qquad (18.163)$$

where

$$\frac{\boldsymbol{\varepsilon}\cdot\mathbf{q}_+}{|\boldsymbol{\varepsilon}|\,|\mathbf{q}_+|} = \cos\theta \qquad \frac{\mathbf{q}_+\cdot\mathbf{k}_+}{|\mathbf{q}_+|\,|\mathbf{k}_+|} = \cos\theta_{fn} \qquad \frac{\boldsymbol{\varepsilon}\cdot\mathbf{k}_+}{|\boldsymbol{\varepsilon}|\,|\mathbf{k}_+|} = \cos\theta_n$$

Equation (18.163) simplifies after the angular integration to

$$\begin{aligned}\mathrm{Im}\, F_\pi(q^2) &= -\frac{1}{32\pi}\sqrt{\frac{q^2-4\mu^2}{q^2}}\, A_1^1(q^2-i\epsilon)F_\pi(q^2+i\epsilon)\\ &= \{\exp[-i\delta_1^1(q^2)]\}\sin\delta_1^1(q^2)F_\pi(q^2+i\epsilon) \qquad (18.164)\end{aligned}$$

This simple relation is the desired mathematical translation of Fig. 18.35. It follows from (18.164) that (because Im $F_\pi(q^2)$ is real) $F_\pi(q^2 + i\epsilon)$ has the same phase as the p-wave π-π scattering amplitude, a result known in potential scattering[1] as the final-state theorem.

The discontinuity formula (18.164), along with the dispersion relation (18.72), can now be used to determine $F_\pi(q^2)$. The solution to this problem[2] boils down essentially to constructing a function which has the following properties:

1. $F_\pi(q^2)$ is analytic in the cut q^2 plane with only the right-hand cut $\infty \geq q^2 \geq 4\mu^2$.
2. $F_\pi(q^2)$ has the phase $\exp(i\delta_1^1)$ for $q^2 \geq 4\mu^2$.
3. F_π is real for real $q^2 < 4\mu^2$.
4. $F_\pi(0) = 1$

We have already constructed in the preceding section a function which satisfies properties 1 to 3, namely, $1/D_1^1(q^2)$. Thus a solution for $F_\pi(q^2)$ is given in terms of $D_1^1(q^2)$ up to a polynomial $P(q^2)$ with real coefficients:

$$\begin{aligned}F_\pi(q^2) &= \frac{D_1^1(0)}{D_1^1(q^2)}\frac{P(q^2)}{P(0)}\\ &= \frac{P(q^2)}{P(0)}\left[\frac{1+\dfrac{1}{\pi}\displaystyle\int_{4\mu^2}^{\infty}\frac{ds'}{s'}\sqrt{\frac{s'-4\mu^2}{s'}}\,N_1^1(s')}{1+\dfrac{1}{\pi}\displaystyle\int_{4\mu^2}^{\infty}\frac{ds'}{s'-q^2}\sqrt{\frac{s'-4\mu^2}{s'}}\,N_1^1(s')}\right] \qquad (18.165)\end{aligned}$$

[1] See, for example, M. L. Goldberger and K. M. Watson, "Collision Theory," John Wiley & Sons, Inc., New York, 1964.

[2] R. Omnes, *Nuovo Cimento*, **8**, 316 (1958); N. I. Muskhelishvili, "Singular Integral Equations," Erven P. Noordhoff, NV, Groningen, Netherlands, 1953.

A more practical form of the solution which is liberated from the approximations associated with the calculation of $N_1^1(s)$ results from solving for F_π directly in terms of the phase shift $\delta_1^1(q^2)$. We observe that because

$$\frac{F_\pi(q^2 + i\epsilon)}{F_\pi(q^2 - i\epsilon)} = \exp\,[2i\delta_1^1(q^2)] \tag{18.166}$$

the discontinuity of $\log F_\pi(q^2)$ is just $\delta_1^1(q^2)$. This suggests defining the function

$$P(q^2) = F_\pi(q^2) \exp\left[-\frac{q^2}{\pi}\int_{4\mu^2}^{\infty}\frac{ds'\,\delta_1^1(s')}{s'(s'-q^2)}\right] \tag{18.167}$$

P has at worst the same analytic properties as $F_\pi(q^2)$. However, P is real for $q^2 > 4\mu^2$ and therefore is an entire function. Since we bar essential singularities at ∞ from consideration, $P(q^2)$ can only be a polynomial,[1] as in (18.165). Therefore,

$$F_\pi(q^2) = \frac{P(q^2)}{P(0)} \exp\left[\frac{q^2}{\pi}\int_{4\mu^2}^{\infty}\frac{ds'\,\delta_1^1(s')}{s'(s'-q^2)}\right] \tag{18.168}$$

provided $\qquad \dfrac{\delta_1^1(q^2)}{q^2} \to 0 \qquad \text{for } q^2 \to \infty$

We again emphasize the approximate nature of the whole calculation leading to (18.165) or, equivalently, (18.168). Higher mass contributions to the unitarity condition $q^2 \geq 16\mu^2$ were ignored. Again as in the π-π scattering calculations, the spirit here is that of an effective range theory. Equation (18.168) may be considered exact, provided $P(q^2)$ is allowed to have a branch cut from $16\mu^2 \leq q^2 \leq \infty$. $F_\pi(q^2)$ is thereby parametrized in terms of the phase shift[2] $\delta_1^1(q^2)$ and an unknown function $P(q^2)$ whose energy dependence is hoped to be small in the region of small $q^2 \leq 16\mu^2$. Although we cannot defend the validity of this assumption, it is still useful because of its simplicity and vulnerability to experimental test. In particular, (18.168), or more transparently (18.165), predicts a peak in $F_\pi(q^2)$ for energies $\sqrt{q^2}$ close to those of the observed p-wave π-π resonance or ρ meson. Experimental tests on this question are eagerly awaited.

Meanwhile, extensive studies have been made along the lines outlined above which provide insight into the behavior of the nucleon

[1] L. Castillejo, R. Dalitz, and F. Dyson, *Phys. Rev.*, **101**, 453 (1956).

[2] For $q^2 \geq 16\mu^2$ can define $\delta_1{}^1$ in various ways, for example,

$$A = \frac{e^{2i\delta} - 1}{2i} \qquad \text{or} \qquad A = \eta e^{i\delta}\sin\delta \qquad \eta \text{ real}$$

This is a matter of taste.

electromagnetic form factors. Having displayed here the basic method, we refer the reader to specialized references on this topic.[1]

Problems

1. Show that the network determinant Δ for a general Feynman graph may be written as

$$\Delta = \sum_S \prod_{j \epsilon S} \alpha_j$$

where the sum runs over all sets S of k internal lines j (k is the number of internal momenta d^4l to be integrated) with the property that removal of lines $j \epsilon S$ from the graph leaves a connected graph. This shows that $\Delta > 0$ unless all $\alpha_j = 0$.

2. Show that, in the Nambu representation (18.74) for a Feynman graph,

$$\zeta_1 = \sum_{S'} \frac{1}{\Delta} \prod_{i \epsilon S'} \alpha_i$$

where S' runs over all sets of $k + 1$ internal lines (Δ is composed of products of $k\ \alpha_i$) with the property that removal from the graph of the lines $i \epsilon S'$ leaves two and only two disconnected graphs, one containing external lines p_1 and p_2 and the other containing p_3 and p_4. This shows that $\zeta_1 > 0$, as asserted below (18.75).

3. Prove, to all orders of perturbation theory for π-π scattering, that $A(s,t)$ is real for $s < 4\mu^2$, $t < 4\mu^2$, and $u < 4\mu^2$ and thereby establish fixed-t dispersion relations in s for $0 < t < 4\mu^2$.

4. Elaborate the arguments between (18.91) and (18.93) establishing the analytic properties of the forward pion-nucleon scattering amplitude to all orders of perturbation theory.

5. Derive the forward spin-flip dispersion relations for π^+-p scattering. How is the absorptive part measured?

6. Derive, to all orders of perturbation theory, the K^+-p forward-scattering dispersion relations. What, besides physical scattering amplitudes and total cross sections, must be known to check the relation experimentally?

7. Consider the axial vector current $\mathcal{g}_\mu$ for nucleon β decay [Eq. (10.151)]. Assuming that its divergence satisfies an unsubtracted dispersion relation, give arguments for the validity of the Goldberger-Treiman relation (10.161).

8. Show without recourse to perturbation theory that the pion electromagnetic form factor $F_\pi(q^2)$ is real for $q^2 < 0$ when weak interactions are neglected.

9. Find the onset of the branch cut for the electromagnetic form factor of the Σ.

10. Using a dispersion relation for $N_l^I(s)$, state and prove the effective range formula discussed below (18.158). With N_0^I approximated by a pole, calculate $A_0^I(s)$ and the effective range parameters.

[1] For example, S. D. Drell and F. Zachariasen, "Electromagnetic Structure of Nucleons," Oxford University Press, Fair Lawn, N.J., 1961.

19

Renormalization

19.1 Introduction

In the low-order calculations in Chap. 8 we identified divergences appearing in self-energy and vertex parts of Feynman diagrams and gave a procedure for isolating them into renormalizations of the charge and mass of the electron. To order e^2, the differences between the observed charge and mass of the electron and the parameters introduced into the equations of motion were infinite. However, the observable predictions of the calculations of physical amplitudes, in terms of the renormalized charge and mass, were finite.

As we saw in the calculations of Chap. 8, renormalization is necessary in any physical theory of interacting fields. What is also necessary in any acceptable theory is that after the renormalization has been carried out, the S matrix is finite.

We shall assume that the perturbation expansion of the S matrix converges and therefore require S to be finite in each order of the expansion in powers of the interaction strength. It is another, and no less important, problem to verify convergence of the renormalized perturbation expansion. This question will not be discussed here, and we confine our attention to verifying that all cutoff-dependent terms in the expansion of τ functions and the S matrix may be collected into the renormalization constants. We shall work within the framework of perturbation theory, using the rules developed in Chap. 17 for Feynman graphs and using the general properties of Z_2 and Z_3 developed in Chap. 16. As a further specialization we confine our remarks to the quantum electrodynamics of electrons and photons.

The program is divided into three parts.[1] To begin, we discuss the topology of graphs and introduce the battery of nomenclature necessary to classify and analyze a general Feynman graph. Next we give the prescription for renormalizing a general nth-order graph. Finally, we show by induction that this prescription suffices to remove

[1] This program, its equations, and the criterion for renormalizability were first given by F. J. Dyson, *Phys. Rev.*, **75**, 486, 1736 (1949) and by A. Salam, *Phys. Rev.*, **82**, 217, **84**, 426 (1951).

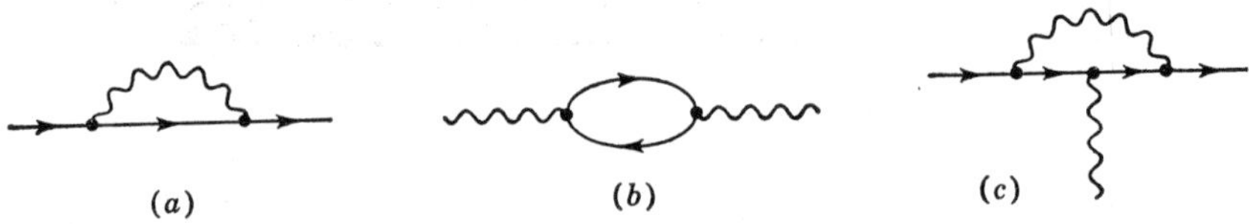

Fig. 19.1 Second-order graphs for (*a*) electron self-energy, (*b*) vacuum polarization, and (*c*) the vertex part.

Fig. 19.2 The complete electron propagator.

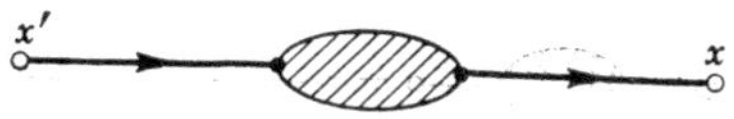

all divergent quantities[1] (that is, cutoff-dependent terms) from the S-matrix expansion.

19.2 Proper Self-energy and Vertex Parts, and the Electron-Positron Kernel

To begin, we shall extend to all orders of the interaction the connection with Feynman graphs of the renormalization constants Z_1, Z_2, Z_3, and δm, already encountered to order e^2 in the companion volume. Recall that Z_2 and δm were associated with the self-energy graph shown in Fig. 19.1a. Z_3 appeared in the computation of the vacuum polarization, Fig. 19.1b, and Z_1 in the vertex part, Fig. 19.1c. All of these numbers were logarithmically divergent to order e^2; their evaluation is in Chap. 8.

Starting with Fig. 19.1a, we recall from the perturbation theory discussion of Chap. 17 that it is one term in the expansion of the electron propagator

$$iS'_F(x - x') \equiv \langle 0|T(\psi(x)\bar{\psi}(x'))|0\rangle \tag{19.1}$$

which we represent by the general blob of Fig. 19.2. S'_F is the sum of all connected graphs for which there is one electron line entering at x'

[1] This is with the exception of infrared divergences which we consider understood. See D. Yennie, S. Frautschi, and H. Suura, *Ann. Phys. (N.Y.)*, **13**, 379 (1961).

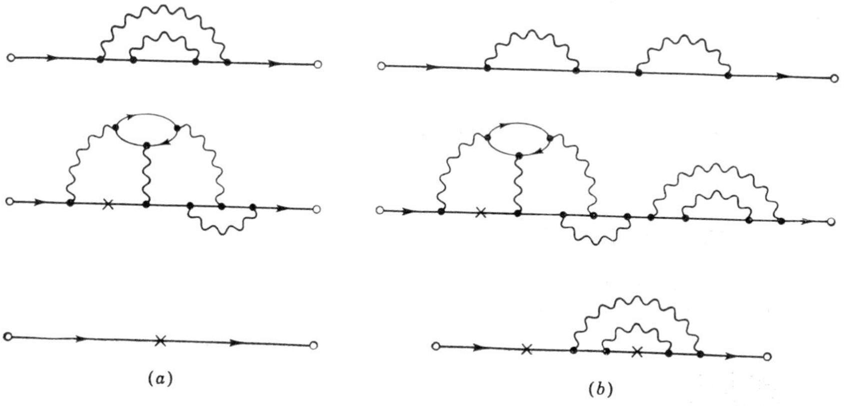

Fig. 19.3 Examples of (a) proper and (b) improper electron self-energy graphs.

$$iS'_F \quad = \quad iS_F \quad + \quad iS_F[-i\Sigma]iS_F \quad + \quad iS_F[-i\Sigma]iS_F[-i\Sigma]iS_F \quad + \cdots$$

Fig. 19.4 Pictorial representation of the series in Eq. (19.2) for the electron propagator as a sum of proper self-energy insertions.

and leaving at x and no external photon lines; it has the general structure discussed extensively in Chap. 16. The graphs composing S'_F may be divided into two distinct and unique classes, known as *proper* and *improper* graphs. The proper graphs cannot be divided into two disjoint parts by the removal of a single fermion line, whereas the improper graphs can be, as illustrated by Fig. 19.3a and b, respectively.

Going over to momentum space and removing the factors $iS_F(p) = i(\not p - m)^{-1}$ for the external legs, we denote by $-i\Sigma(p)$ the sum of all *proper* graphs[1] for an electron of momentum p. $S'_F(p)$ may then be written in terms of $\Sigma(p)$

$$\begin{aligned} iS'_F(p) = iS_F(p) &+ iS_F(p)[-i\Sigma(p)]iS_F(p) \\ &+ iS_F(p)[-i\Sigma(p)]iS_F(p)[-i\Sigma(p)]iS_F(p) + \cdots \end{aligned} \tag{19.2}$$

In terms of pictures, this equation is illustrated in Fig. 19.4. It may be formally summed to give the equation

$$S'_F(p) = \frac{1}{\not p - m - \Sigma(p)} \tag{19.3}$$

$\Sigma(p)$, the sum of all proper electron self-energy graphs with external lines removed, as illustrated in Fig. 19.5, is also known as the electron's mass operator.

We discuss the photon propagator in the same way. Figure 19.1b shows the term of order e^2 in the perturbation expansion of the complete photon propagator

$$iD'_F(x - x')^{\mu\nu} = \langle 0|T(A^\mu(x)A^\nu(x'))|0\rangle + \text{gauge and coulomb terms} \tag{19.4}$$

[1] The definition of $\Sigma(p)$ differs from the definition used in Chap. 8 by the inclusion of the mass counterterm δm. In the companion volume S'_F was expanded in terms of the *bare* propagator $(\not p - m_0)^{-1}$, whereas our starting approximation here is the propagator $(\not p - m)^{-1}$ for an in particle with physical mass m.

Fig. 19.5 The electron proper self-energy part, or mass operator Σ.

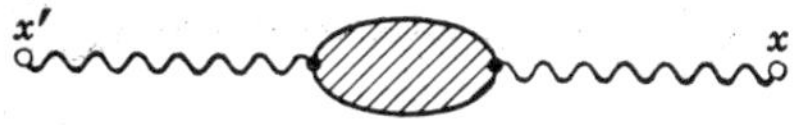

Fig. 19.6 The complete photon propagator.

which consists of the sum of all connected graphs with two external photon lines and no external electron lines. D'_F will be represented by the general blob illustrated in Fig. 19.6. We again single out proper graphs, defined as those graphs which cannot be made disjoint by removal of a single photon line. Some proper graphs are illustrated in Fig. 19.7*a*; improper graphs are shown in Fig. 19.7*b*. The sum over all proper graphs, with the external photon lines removed, is defined to be $ie_0{}^2\Pi_{\mu\nu}$, the proper photon self-energy part. $\Pi_{\mu\nu}$ is the analogue of $\Sigma(p)$; it is also known as the vacuum polarization tensor and was so referred to in Chap. 8. In analogy with (19.2) for the electron, we may express the complete photon propagator in terms of $\Pi_{\mu\nu}$, as illustrated in Fig. 19.8.

$$\begin{aligned} iD'_F(q)^{\mu\nu} &= iD_F(q)^{\mu\nu} + iD_F^{\mu\lambda}[+ie_0^2\,\Pi_{\lambda\sigma}]iD_F^{\sigma\nu} + \cdots \\ &= -\frac{ig^{\mu\nu}}{q^2} - \frac{ie_0^2}{q^2}[i\Pi^{\mu\nu}]\frac{(-i)}{q^2} - \frac{ie_0^4}{q^2}[i\Pi^{\mu\lambda}]\frac{(-i)}{q^2}[i\Pi^{\nu}{}_{\lambda}]\frac{(-i)}{q^2} + \cdots \end{aligned} \tag{19.5}$$

The general form of $D'_F(q)^{\mu\nu}$, studied in (16.172), contains terms involving η_μ, the time-like vector characterizing that frame in which quantization in transverse gauge was carried out. Here we take advantage of the discussion in Sec. 17.9 and ignore these terms proportional to $q_\mu\eta_\nu$, $\eta_\mu\eta_\nu$, $q_\mu q_\nu$ in writing the free photon propagator, since we know they do not contribute to S-matrix elements. Formally summing the series (19.5), we write

$$D'_F(q)_{\mu\nu} = -\frac{g_{\mu\nu}}{q^2} + \frac{e_0^2}{q^2}\Pi_{\mu\lambda}(q)D'_F(q)^{\lambda}{}_{\nu}$$

and analogously to (19.3),

$$[q^2 g_{\mu\lambda} - e_0^2\Pi_{\mu\lambda}(q)]D'_F(q)^{\lambda}{}_{\nu} = -g_{\mu\nu} \tag{19.6}$$

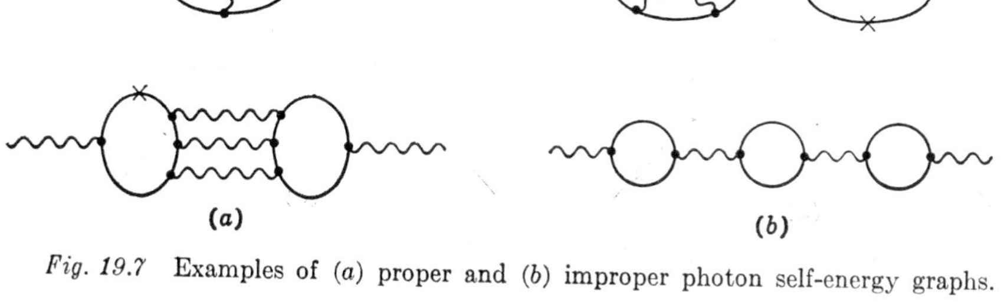

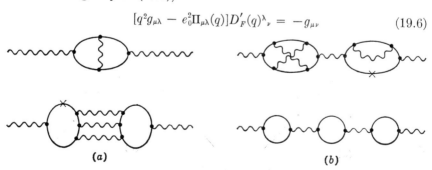

Fig. 19.7 Examples of (*a*) proper and (*b*) improper photon self-energy graphs.

$$iD_F'^{\mu\nu} = iD_F^{\mu\nu} + iD_F^{\mu\lambda}[ie_0^2\Pi_{\lambda\sigma}]iD_F^{\sigma\nu} + iD_F^{\mu\lambda}[ie_0^2\Pi_{\lambda\sigma}]iD_F^{\sigma\omega}[ie_0^2\Pi_{\omega\tau}]iD_F^{\tau\nu} + \cdots$$

Fig. 19.8 Series, Eq. (19.5), for the photon propagator as a sum of proper self-energy parts $\Pi_{\mu\nu}$, known also as the vacuum polarization tensor.

The final ingredient encountered in the second-order renormalization discussion of Chap. 8 was the vertex correction, Fig. 19.1*c*, associated with Z_1. We define the proper vertex part as the sum of all connected graphs which have two external fermion lines and one external photon line and which are proper, i.e., do not become disjoint upon removal of either a single electron line or photon line. Examples of proper and improper vertex graphs are shown in Fig. 19.9*a* and *b*, respectively. It is clear from these illustrations that electron and photon self-energy insertions in the external lines make proper into improper vertex parts. In momentum space we denote the sum of all proper vertex graphs by $\Gamma_\mu(p',p)$, where p' and p denote the four-

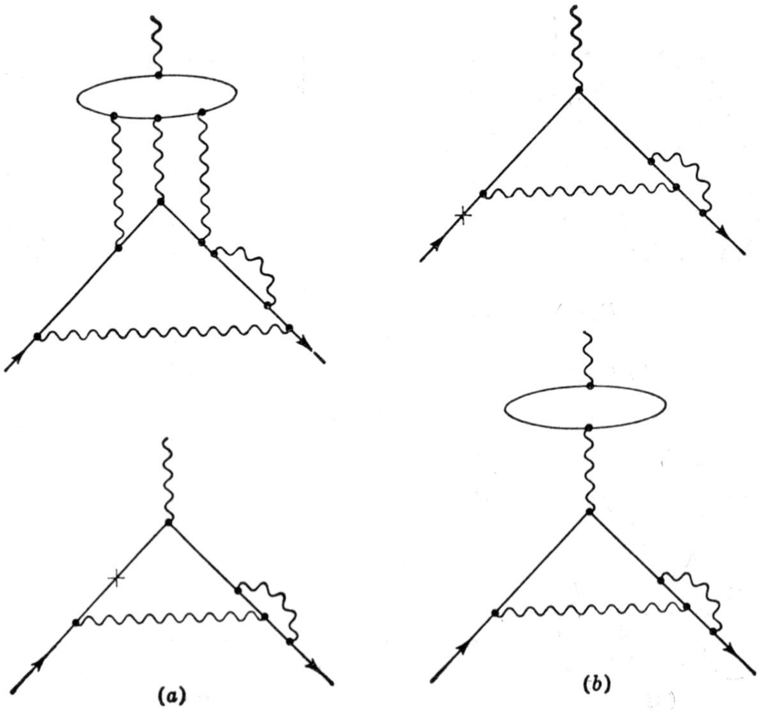

Fig. 19.9 Examples of (*a*) proper and (*b*) improper vertex graphs.

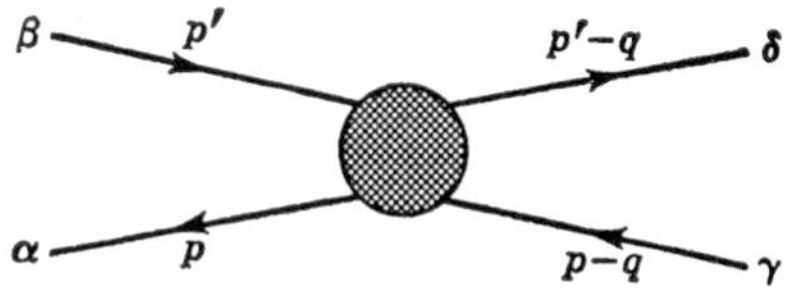

Fig. 19.10 The electron-positron scattering kernel K.

momenta of the emerging and entering electron lines, respectively. The propagators of the external fermion and photon lines are removed in defining $\Gamma_\mu(p',p)$ as well as the factor $-ie_0$ coming from our rules for the perturbation expansion as given in Chap. 17. Thus the proper vertex graphs are normalized such that $\Gamma_\mu(p',p) = \gamma_\mu$ in lowest order; in general we write

$$\Gamma_\mu(p',p) = \gamma_\mu + \Lambda_\mu(p',p) \tag{19.7}$$

In carrying out the renormalization program we shall need an additional quantity which did not explicitly appear in the e^2 calculations in Chap. 8. This quantity is the electron-positron scattering kernel, or, for short, the kernel. It is denoted in momentum space by

$$K(p,p',q)_{\alpha\beta,\gamma\delta}$$

and will be represented by the blob in Fig. 19.10. K consists of graphs with two external electron and two external positron lines; $-p$ and p' represent the initial positron and electron momenta and q the momentum transfer from the electron. α, β, γ, and δ are the corresponding spinor components as illustrated in Fig. 19.10. All graphs with the above properties are included in the definition of K *except* for two classes. We omit all graphs for which the external lines (p,α) and (p',β) join to one part A and the two external lines $(p-q,\ \gamma)$ and

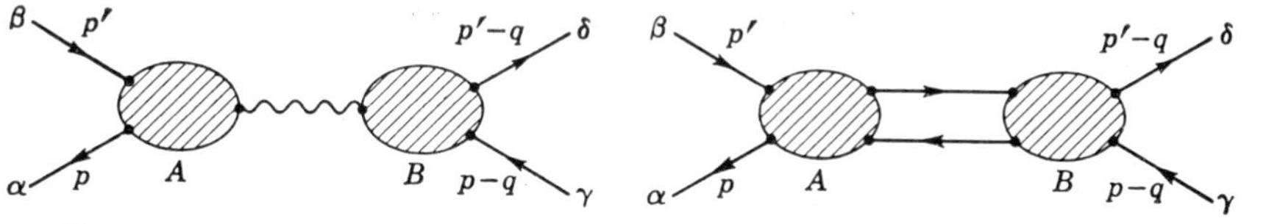

Fig. 19.11 Two classes of graphs excluded from the kernel K.

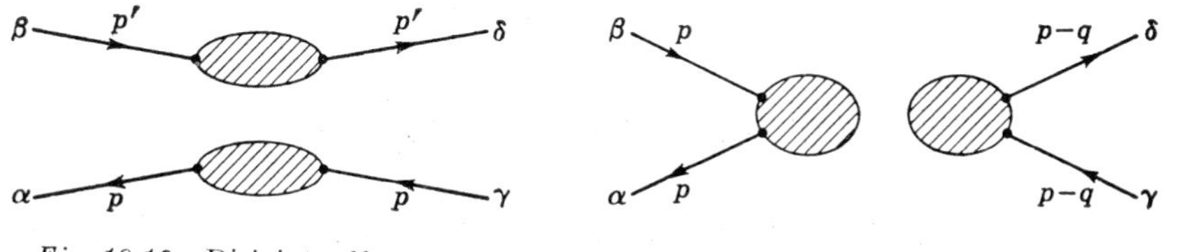

Fig. 19.12 Disjoint self-energy parts excluded from the kernel K.

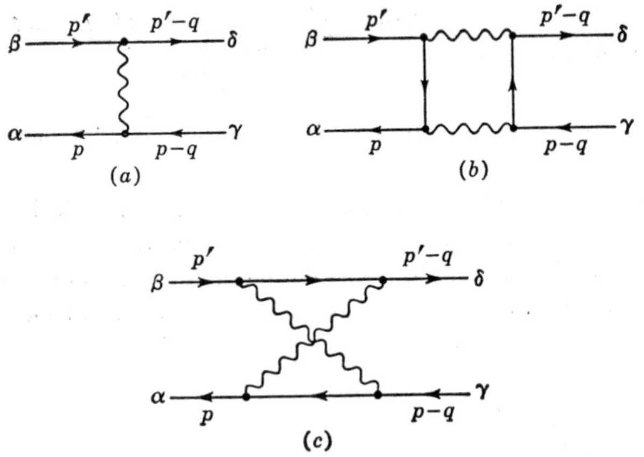

Fig. 19.13 Examples of graphs included in the kernel K.

$(p' - q, \delta)$ join to another part B such that A and B are connected by only one photon line or one electron-positron pair. Figure 19.11 shows the graphs which are excluded from K by these restrictions. Also excluded are graphs, such as shown in Fig. 19.12, which consist of two disjoint self-energy parts. Examples of graphs included in K are shown in Fig. 19.13. The lowest-order contribution to K is given by Fig. 19.13a; it is

$$K^{(0)}(p,p',q)_{\alpha\beta,\gamma\delta} = \frac{ie_0{}^2}{q^2}(\gamma_\mu)_{\alpha\gamma}(\gamma^\mu)_{\delta\beta} \tag{19.8}$$

In common with our definitions of Σ, $\Pi_{\mu\nu}$, and Γ_μ, the propagators of the external fermion lines, including all self-energy insertions on the external legs, are removed in defining K.

19.3 Integral Equations for the Self-energy and Vertex Parts

With the definitions of Σ, $\Pi_{\mu\nu}$, and Γ_μ given in the preceding section, we may proceed to write down integral equations relating these quantities to each other. For example, the electron proper self-energy part satis-

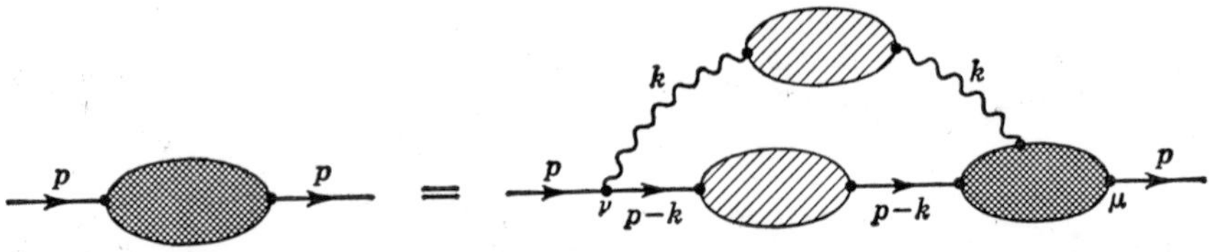

Fig. 19.14 Pictorial representation of integral Eq. (19.9) for the electron proper self-energy part $\Sigma(p)$.

Fig. 19.15 A contribution to $\Sigma(p)$.

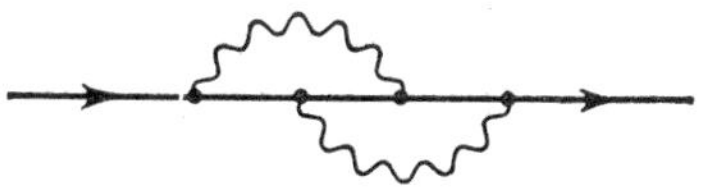

fies the integral equation illustrated in Fig. 19.14:

$$-i\Sigma(p) = (-ie_0)^2 \int \frac{d^4k}{(2\pi)^4} iD'_F(k)_{\mu\nu}\Gamma^\mu(p, p-k)iS'_F(p-k)\gamma^\nu \quad (19.9)$$

To be convinced that this equation is correct, we observe that after the first interaction at the vertex ν there are present an electron and photon which must interact in all possible ways consistent with the eventual absorption of the photon. After the full propagation of the electron and photon have been taken into account, this is just the definition of the proper vertex part. Notice that it is wrong to put the full vertex Γ_ν at ν; for instance, the graph of Fig. 19.15 would be counted twice.

The proper photon self-energy part $\Pi_{\mu\nu}(q)$ satisfies an equation similar to (19.9) which is illustrated in Fig. 19.16:

$$ie_0{}^2\Pi_{\mu\nu}(q) = (-ie_0)^2(-1) \int \frac{d^4k}{(2\pi)^4} \operatorname{Tr} \gamma_\mu iS'_F(k)\Gamma_\nu(k, k+q)iS'_F(k+q) \quad (19.10)$$

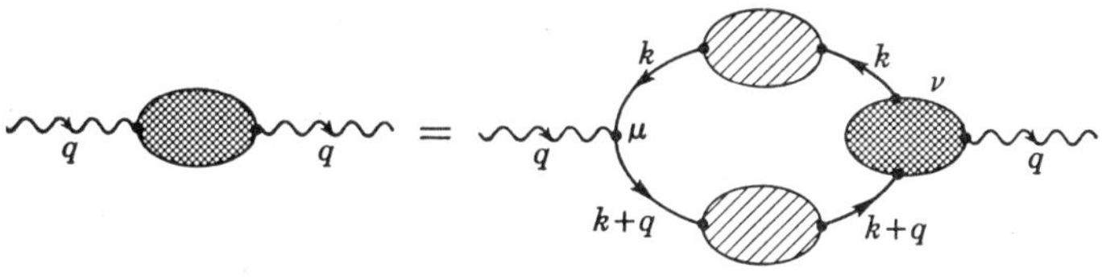

Fig. 19.16 Equation (19.10) for the photon proper self-energy part $\pi_{\mu\nu}(q)$.

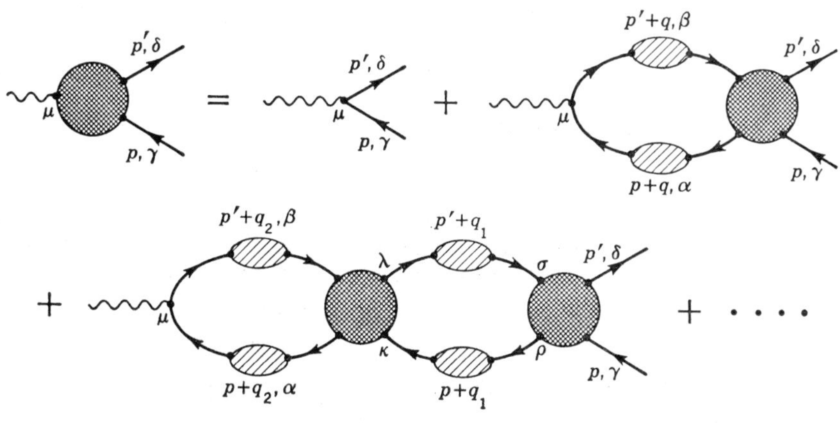

Fig. 19.17 Series in Eq. (19.11) for the vertex part Γ_μ in terms of the kernel K.

If Γ_μ were known, Eqs. (19.9) and (19.10), in addition to (19.3) and (19.6) for S'_F and D'_F, would give a closed set of nonlinear integral equations which could be solved, for example, by iteration in powers of e_0^2. We may express Γ_μ, in turn, in terms of the electron-positron kernel K by means of yet another integral equation, illustrated in Fig. 19.17:

$$\begin{aligned}\Gamma_\mu(p',p)_{\delta\gamma} = (\gamma_\mu)_{\delta\gamma} + \int \frac{d^4q}{(2\pi)^4} [iS'_F(p'+q)\gamma_\mu iS'_F(p+q)]_{\beta\alpha} \\ \times K_{\alpha\beta,\gamma\delta}(p+q, p'+q, q) + \int \frac{d^4q_1\, d^4q_2}{(2\pi)^8} \\ \times [iS'_F(p'+q_2)\gamma_\mu iS'_F(p+q_2)]_{\beta\alpha} K_{\alpha\beta,\kappa\lambda}(p+q_2, p'+q_2, q_2-q_1) \\ \times iS'_F(p'+q_1)_{\sigma\lambda} iS'_F(p+q_1)_{\kappa\rho} K_{\rho\sigma,\gamma\delta}(p+q_1, p'+q_1, q_1) \\ + \cdots \qquad (19.11)\end{aligned}$$

which we shall often abbreviate to

$$\Gamma = \gamma - \int \gamma S'_F S'_F K + \int\int \gamma S'_F S'_F K S'_F S'_F K + \cdots$$

To convince ourselves that (19.11) is correct, we observe that immediately after the external photon is absorbed there exists an electron-positron pair which must scatter in all possible ways in order to make up the full vertex Γ. This scattering can take place with either 0, 1, 2, . . . distinct virtual intermediate pair states, which gives rise to the series[1] (19.11). This series may be formally summed in the same manner as the Born series is summed into a closed integral equation in nonrelativistic potential scattering. In this way we obtain the integral equation (Fig. 19.18)

$$\begin{aligned}\Gamma_\mu(p',p)_{\delta\gamma} = (\gamma_\mu)_{\delta\gamma} + \int \frac{d^4q}{(2\pi)^4} [iS'_F(p'+q)\Gamma_\mu(p'+q, p+q) \\ \times iS'_F(p+q)]_{\beta\alpha} K_{\alpha\beta,\gamma\delta}(p+q, p'+q, q) \qquad (19.12)\end{aligned}$$

or in symbolic shorthand

$$\Gamma = \gamma - \int \Gamma S'_F S'_F K$$

[1] This series is analogous to the multiple scattering series developed in Chap. 6 for the Green's function. The corresponding equation relating the full electron-positron (off-shell) scattering amplitude T to the kernel K is

$$\begin{aligned}-iT = K + \int K\, iS'_F\, iS'_F\, K + \int K\, iS'_F\, iS'_F\, K iS'_F\, iS'_F\, K + \cdots \\ = K + \int K\, iS'_F\, iS'_F(-iT)\end{aligned}$$

which is known as the Bethe-Salpeter equation. It is the relativistic analogue of the integral form of the two-body Schrödinger equation, iK being the analogue of the potential V. E. Salpeter and H. Bethe, *Phys. Rev.*, **84**, 1232 (1951).

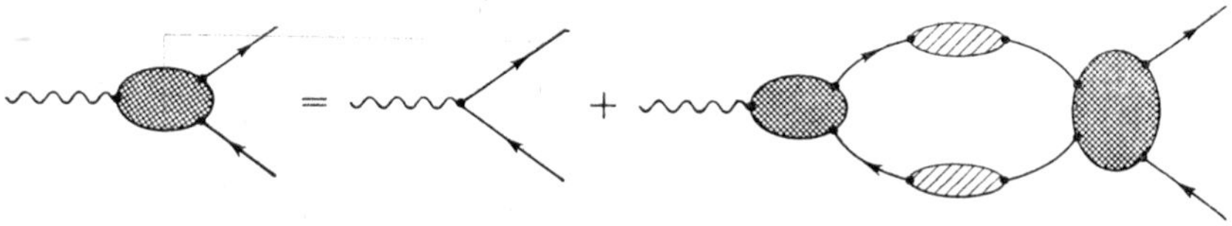

Fig. 19.18 Integral Eq. (19.12) for the vertex part.

The reader may wonder at this point what writing down all these integral equations has accomplished. From the point of view of making a practical calculation, we have accomplished little; the unknown quantities S'_F, D'_F, and Γ_μ have been expressed in terms of yet another unknown quantity—the kernel K. However, from the point of view of renormalization theory, expressing S'_F, D'_F, and Γ_μ in terms of K is of considerable use. The reason is that the divergences in K are due *only* to self-energy or vertex insertions in the internal lines. If there were no such insertions, K would be finite; we shall discuss this in detail later. Thus the renormalization problems involved in computing K are relatively mild, and the more difficult questions involving divergences in Σ, $\Pi_{\mu\nu}$, and Γ_μ may be discussed in the context of the integral equations relating them to each other and to K.

19.4 Integral Equations for Tau Functions and the Kernel K; Skeleton Graphs

The central ingredients of the renormalization program, introduced in Secs. 19.2 and 19.3, are Σ, $\Pi_{\mu\nu}$, and Γ_μ. All other quantities in the theory such as S-matrix elements, τ functions, and, in particular, the kernel K will be expressed in terms of these three, or more specifically in terms of S'_F, D'_F, and Γ_μ. However, we shall not be able to write down closed integral equations for τ and K as we did in the preceding section. Instead, we shall group together in the graphical expansion those terms which differ only by self-energy and vertex insertions; it is these insertions which are relevant to the problem of rendering the theory finite.

To begin, we consider the graphs which compose the kernel K and isolate the self-energy and vertex insertions in each graph. A simple way to do this is to draw a box around each such insertion; an example of this procedure is given in Fig. 19.19. We observe in this example that the boxes are either disjoint or nested (that is, one is contained entirely within the other) with one exception: within self-energy insertions the boxes may overlap. This topological property is actually

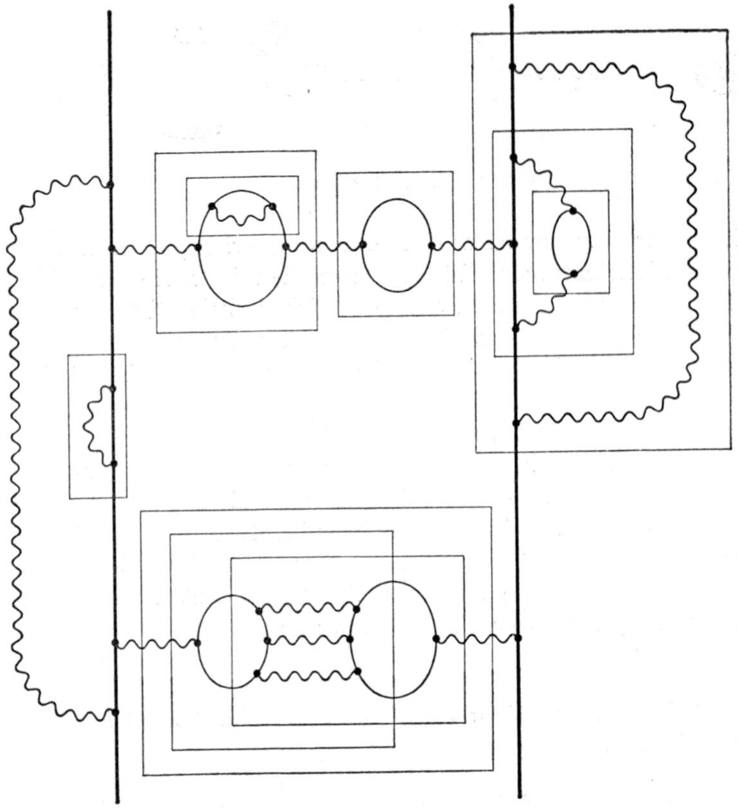

Fig. 19.19 A graph with boxes drawn around each self-energy and vertex insertion.

true for all graphs; the boxes enclosing vertex and self-energy insertions may always be drawn in such a way that they never overlap, except for vertex insertions within self-energy parts. We postpone proof of this statement to Sec. 19.5.

As a consequence of this nonoverlapping property, we may associate in a unique way with each graph $\mathcal{G}$ composing K another graph called the skeleton $\mathcal{S}$ of $\mathcal{G}$. The skeleton is obtained by shrinking the boxes we drew, along with all their contents, to points, that is, removing all vertex and self-energy insertions. Thus Fig. 19.20 is the skeleton

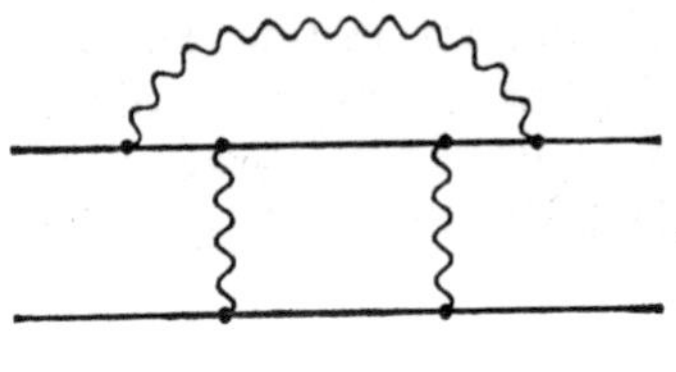

Fig. 19.20 Skeleton graph of Fig. 19.19.

graph of Fig. 19.19. Conversely, we can build up any graph composing K by starting with its skeleton and adding the meaty part, namely, the vertex and self-energy radiative corrections.

Denoting the contribution of an ordinary Feynman graph $\mathcal{G}$ which has the property of being its own skeleton ($\mathcal{G} = \mathcal{S}$) by $K^{\mathcal{S}}(p,p',q;S_F,D_F,\gamma_\mu,e_0)$, we may write the complete kernel K as a sum over only the skeleton graphs, with complete S'_F, D'_F, and Γ_μ, respectively, appearing at each electron line, photon line, and vertex:

$$K_{\alpha\beta,\gamma\delta}(p,p',q) = \sum_{\text{skeletons } \mathcal{S}} K^{\mathcal{S}}_{\alpha\beta,\gamma\delta}(p,p',q;S'_F,D'_F,\Gamma_\mu,e_0)$$

$$= (-ie_0)^2\Gamma^\mu_{\alpha\gamma}(p, p - q)iD'_F(q)_{\mu\nu}\Gamma^\nu_{\delta\beta}(p' - q, p') + \cdots \quad (19.13)$$

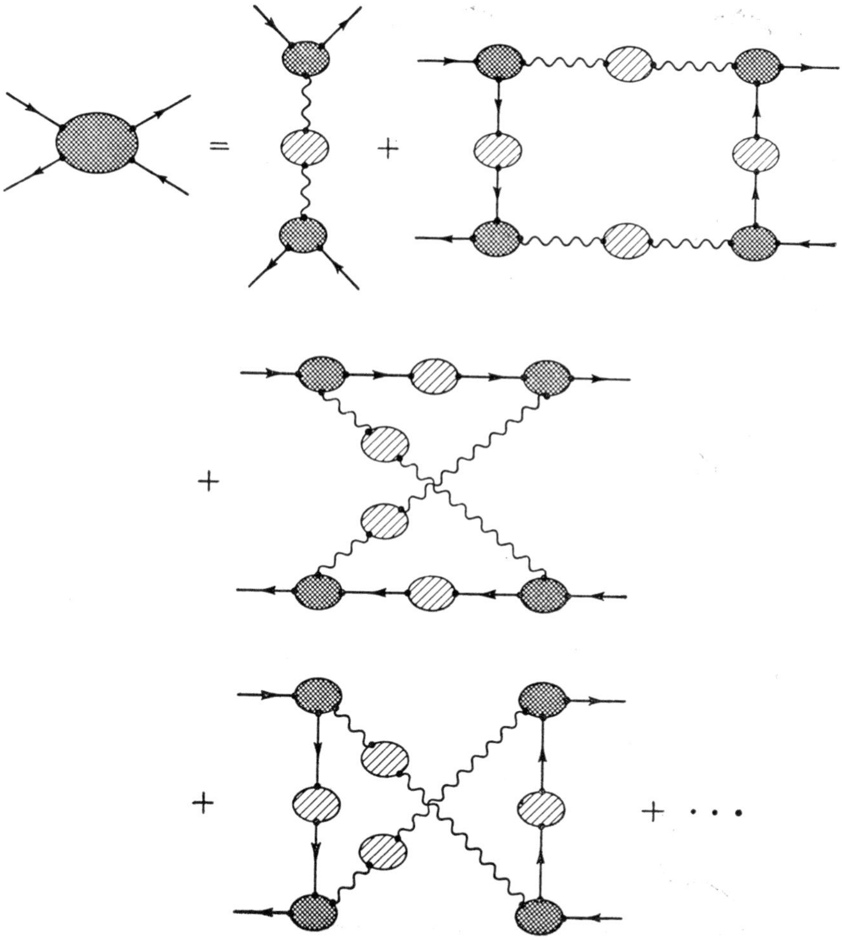

Fig. 19.21 Series, Eq. (19.13), for the kernel K.

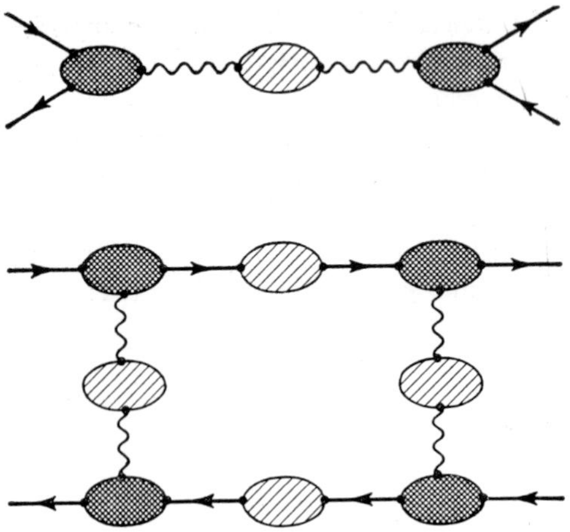

Fig. 19.22 Skeleton graphs not included in series for K in Fig. 19.21.

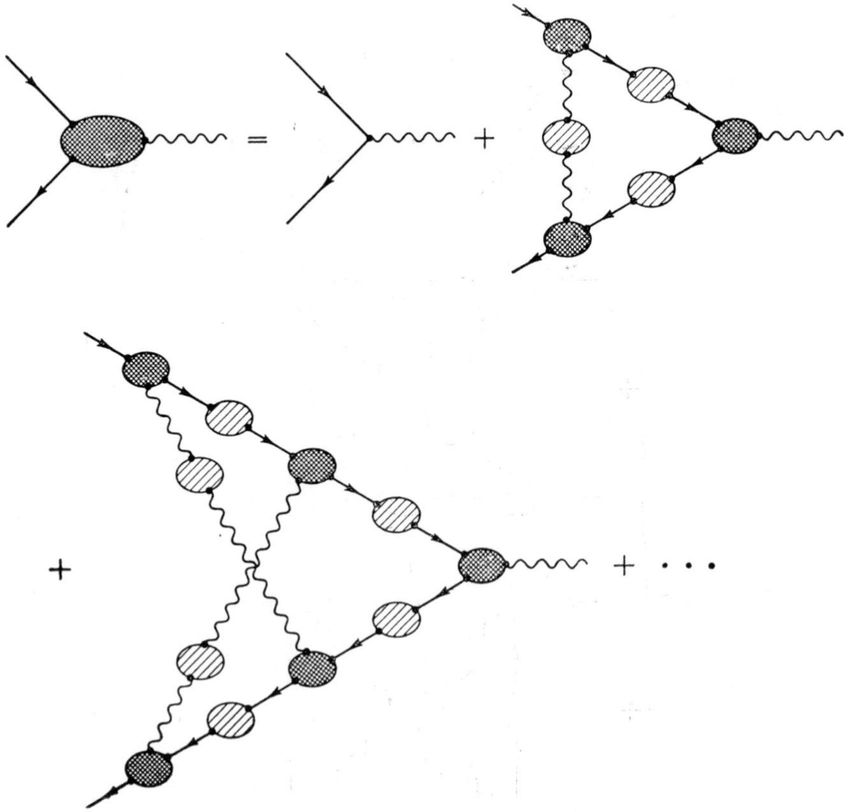

Fig. 19.23 Series, Eq. (19.15), for the vertex Γ_μ.

Equation (19.13) is illustrated in Fig. 19.21; notice that the skeleton graphs of Fig. 19.22 are not included in the expansion. In general, in accordance with the definition of K given in Sec. 19.2, those skeletons which can be separated into two parts joined only by a single photon line or an electron-positron pair are to be omitted from the expansion in (19.13).

In order to verify the correctness of (19.13), we observe that any Feynman graph $\mathcal{G}$ composing K appears once and only once in the skeleton expansion, since (1) the skeleton $\mathcal{S}$ of $\mathcal{G}$ is uniquely determined and (2) there is one and only one graph in the graphical expansion of $K^{\mathcal{S}}$ corresponding to $\mathcal{G}$. This latter statement is a consequence of the unique way of identifying vertex and self-energy insertions in a graph for K.

Skeleton expansions for general τ functions (excluding propagators and self-energy parts) may be made in the same way as for the kernel K. We write (in momentum space) for the τ function with m external fermion legs and n external photons:

$$\begin{aligned}\tau(p_1 \cdots p_m,q_1 \cdots q_n)_{\alpha_1\cdots\alpha_m,\mu_1\cdots\mu_n} \\ = \sum_{\text{skeletons } \mathcal{S}} \tau^{\mathcal{S}}(p_1 \cdots p_m,q_1 \cdots q_n;S'_F,D'_F,\Gamma,e_0)_{\alpha_1\cdots\alpha_m,\mu_1\cdots\mu_n} \qquad m+n>3\end{aligned} \tag{19.14}$$

where $\tau^{\mathcal{S}}(p_1 \cdots p_m,q_1 \cdots q_n;S_F,D_F,\Gamma,e_0)$ is the contribution to τ of the Feynman graph corresponding to skeleton $\mathcal{S}$.

For the vertex Γ_μ, a skeleton expansion may also be written[1]

$$\Gamma_\mu(p',p) = \gamma_\mu + \sum_{\text{skeletons } \mathcal{S}} \Lambda_\mu^{\mathcal{S}}(p',p;S'_F,D'_F,\Gamma_\mu,e_0) \tag{19.15}$$

Eq. (19.15) is illustrated in Fig. 19.23.

For self-energy parts, the skeleton expansion is not useful, owing to the overlap problem. For instance, for the graph of Fig. 19.24 the radiative correction may be considered to be either a vertex correction

[1] It is understood here that in the instruction to draw boxes around internal vertex parts, the box around the entire vertex is to be omitted.

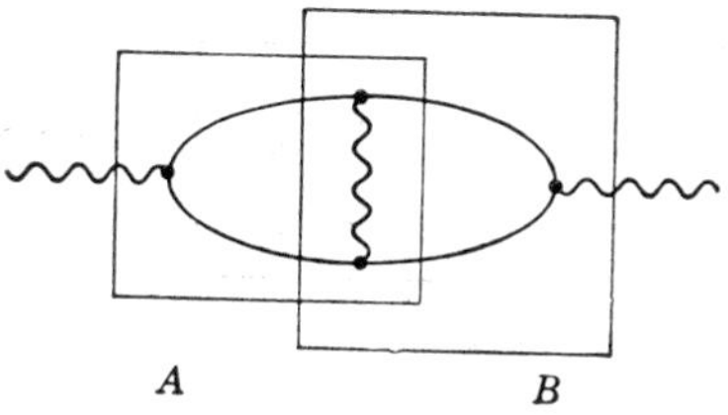

Fig. 19.24 Example of self-energy insertion with overlapping vertex correction.

at A or at B (but not both) and there is ambiguity in how to build up the full self-energy part from the skeleton.[1] Actually, the integral equation (19.10) for $\Pi_{\mu\nu}$ resolves this ambiguity, and we need not pursue this question further.

19.5 A Topological Theorem

In identifying vertex and self-energy insertions in the preceding section we stated that boxes may be drawn around them in such a way that their walls never overlap, except for vertex insertions within self-energy parts. We take it to be self-evident that boxes can be drawn around self-energy insertions such that they are either disjoint or nested, but never overlap. Before proving that vertex insertions need overlap only when within self-energy insertions, consider the example illustrated in Fig. 19.25. By definition only two electron lines and one photon line pass through the walls of each box presumed to be surrounding vertex insertions. Assuming that all lines passing through the nonoverlapping parts of the walls have been shown in the figure, we must locate the remaining three lines as in Fig. 19.26*a*. However, as redrawn in Fig. 19.26*b*, this reduces to disjoint self-energy and vertex insertions.

To prove the above-stated theorem that vertex boxes need overlap only when within self-energy insertions, we shall show by direct graphical construction that we can successively remove the overlap between pairs of boxes in all other cases. This is accomplished in Fig. 19.27 for all other cases besides that already illustrated in Fig. 19.25. Boxes with zero or one line emerging through the nonoverlapping parts of the walls are impossible and self-energy boxes with two such lines have already been dismissed and need not be included in this example.

[1] The second-order graph is the *only* skeleton graph of $\Pi_{\mu\nu}$ and Σ.

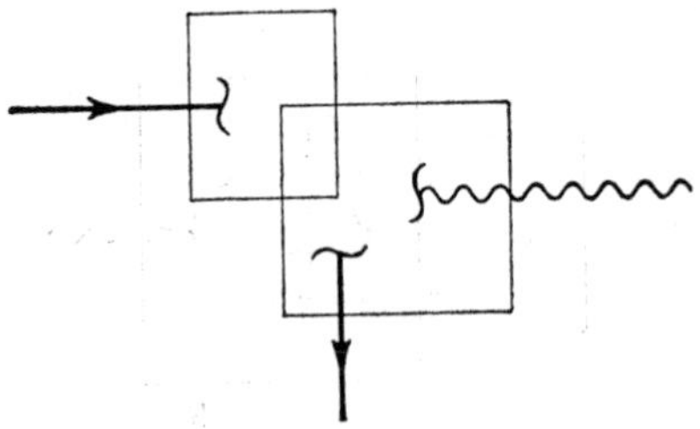

Fig. 19.25 Example of vertex insertion.

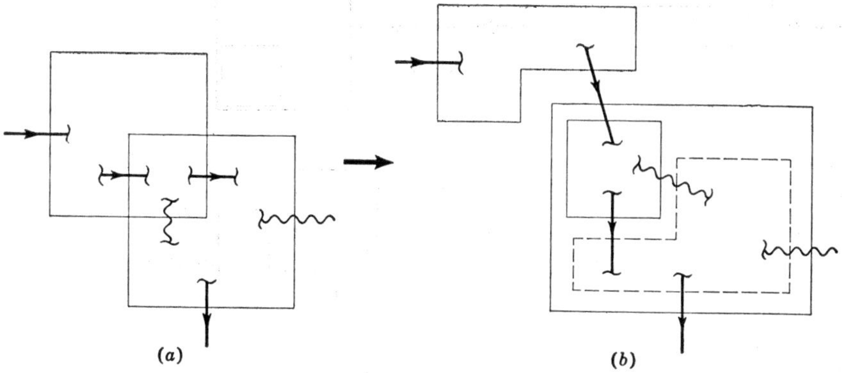

Fig. 19.26 Attempt to draw overlapping vertex insertions within a vertex part in (*a*) reduces to disjoint self-energy and vertex insertions in (*b*). Interchange of photon and incident electron line leads to a similar result.

19.6 The Ward Identity[1]

An important aid in the renormalization program for quantum electrodynamics is the generalized Ward identity, which allows us to compute S'_F directly from Γ_μ. This relation, a consequence of differential current conservation, states that

$$(p' - p)_\mu \Gamma^\mu(p',p) = [S_F'^{-1}(p') - S_F'^{-1}(p)] \tag{19.16}$$

It is obviously satisfied by the bare vertex γ_μ and the free propagator $S_F(p) = (\not{p} - m)^{-1}$ and can therefore be expressed in terms of the vertex corrections and proper self-energy parts with the aid of (19.3) and (19.7)

$$(p' - p)_\mu \Lambda^\mu(p',p) = -[\Sigma(p') - \Sigma(p)] \tag{19.17}$$

The crux of the identity is the very simple observation that $\Lambda_\mu(p',p)$ is obtained from $-\Sigma(p)$ by inserting a γ_μ bringing momentum $p' - p$ into any free fermion propagator occurring in $\Sigma(p)$ and summing over all possible insertions. Figure 19.28 illustrates this for the fourth-order self-energy graphs.

There is considerable simplification if we compute only the divergence of $\Lambda^\mu(p',p)$, that is, $(p' - p)_\mu \Lambda^\mu(p',p)$, as needed for the left-hand side of (19.17). In this case we may use the results of the discussion of the transverse photon propagator in Chap. 17, in particular (17.64) and (17.66). Equation (17.66) states that the sum of insertions onto closed loops vanishes, and (17.64) states that the sum of the insertions

[1] J. C. Ward, *Phys. Rev.*, **78,** 1824 (1950); Y. Takahashi, *Nuovo Cimento,* **6,** 370 (1957).

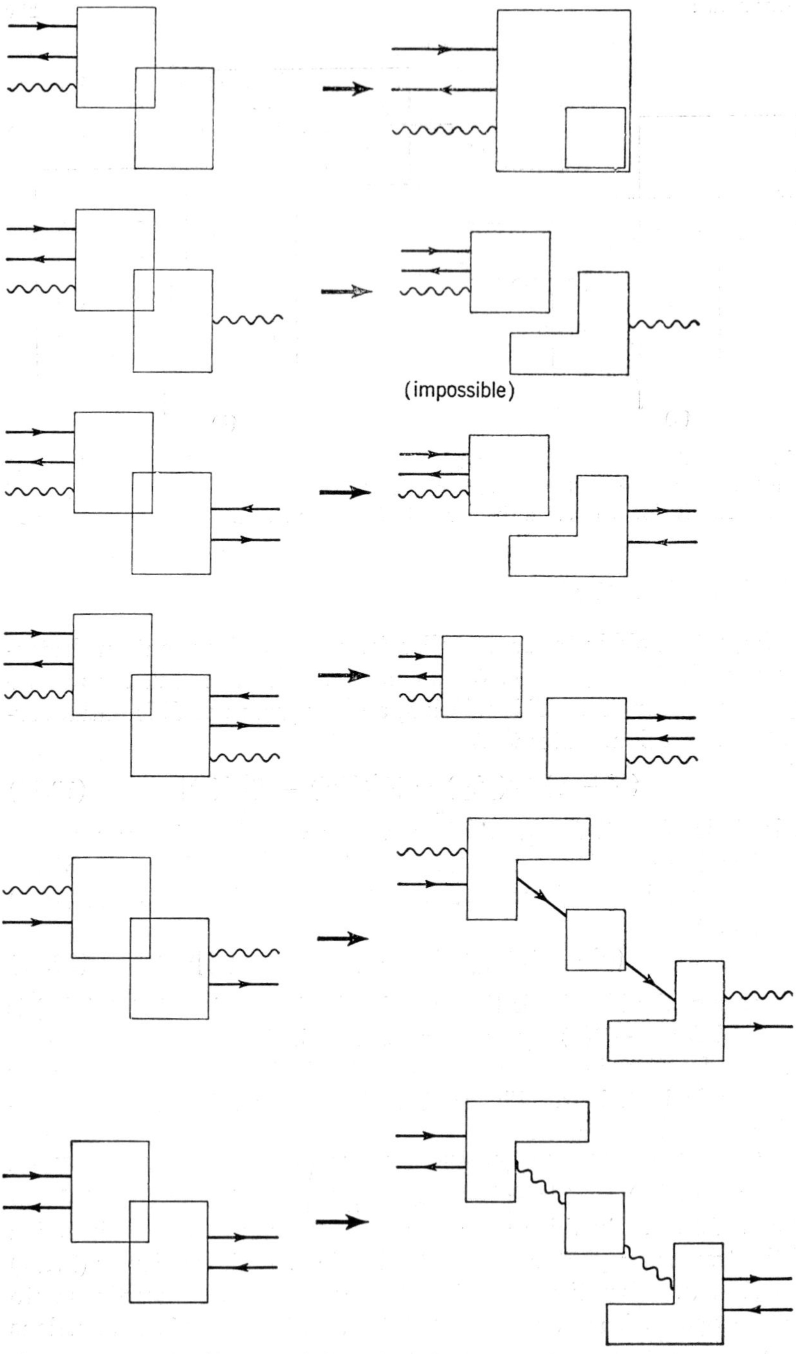

Fig. 19.27 Enumeration of diagrams showing removal of overlap in all cases except for vertex insertions within self-energy parts as in Fig. 19.24.

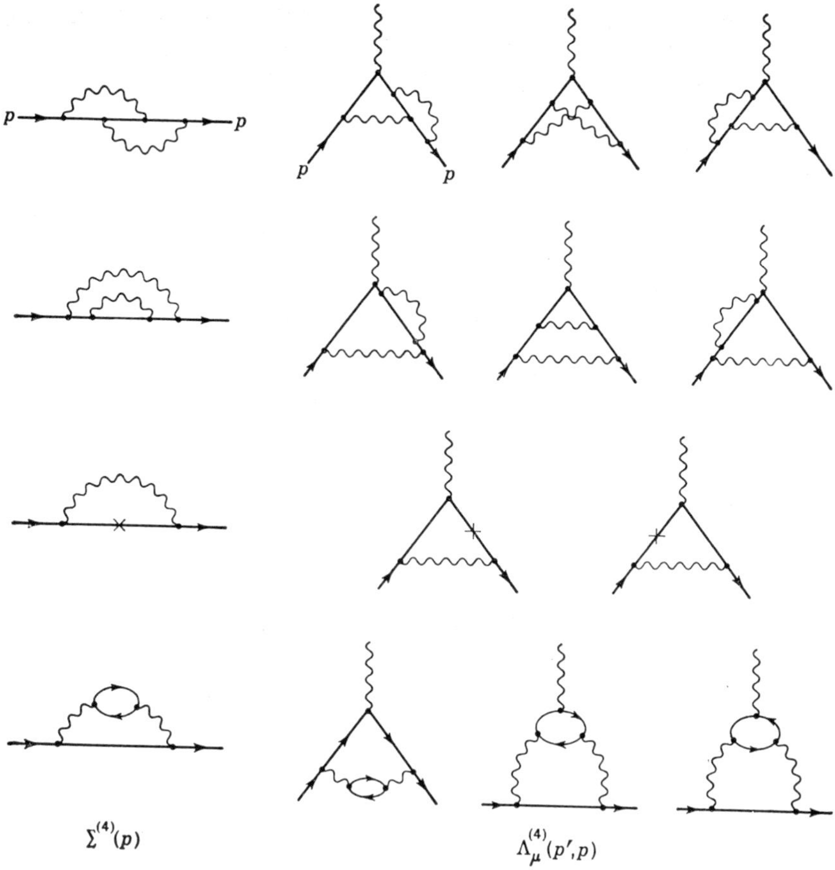

Fig. 19.28 Insertions of photon into fourth-order self-energy graph.

on the fermion line which enters and leaves the graph gives precisely the two terms on the right-hand side of (19.17) and establishes the Ward identity.

From (19.16) we can compute $S_F'^{-1}(p)$ and therefore $\Sigma(p)$ itself by taking the known limit of the self-energy as $\not p$ approaches the mass shell. From the perturbation discussion of Chap. 8 and the spectral representation discussed in Chap. 16, we recall that[1]

$$S'_F(p) \to \frac{Z_2}{\not p - m} \tag{19.18}$$

[1] The infrared divergence problem in Z_2 is ignored here. It may, in fact, be avoided altogether by the intermediate renormalization introduced in Sec. 19.9 or by introducing a photon mass. We also ignore the gauge problems associated with Z_2. See in particular the footnote on p. 231.

as $\not p$ approaches m, the physical mass of the electron. Therefore, we may derive an expression for $S_F'^{-1}(p')$ for arbitrary p' by putting $\not p$ on the mass shell in (19.16) and right-multiplying by a free-particle spinor $u(p)$:

$$\begin{aligned}(p' - p)_\mu \Gamma^\mu(p',p)u(p) &= S_F'^{-1}(p')u(p) - Z_2^{-1}(\not p - m)u(p) \\ &= S_F'^{-1}(p')u(p) \end{aligned} \tag{19.19}$$

Since $S_F'^{-1}(p')$ has, on grounds of Lorentz invariance and parity conservation, the structure

$$S_F'^{-1}(p') = \not p' A(p'^2) + B(p'^2)$$

the two scalar functions $A(p'^2)$ and $B(p'^2)$ may be determined in terms of Γ_μ by extracting the coefficients of $\not p' u(p)$ and $u(p)$, respectively, on the left-hand side of (19.19).

The Ward identity not only allows computation of $S_F'(p)$ given $\Gamma_\mu(p',p)$ but also determines Γ_μ in terms of S_F' for the special case $p' \to p$, that is, as the momentum transfer to the electromagnetic field approaches zero. In this limit (19.16) becomes

$$\Gamma_\mu(p,p) = \frac{\partial S_F'^{-1}(p)}{\partial p^\mu} \qquad \text{or} \qquad \Lambda_\mu(p,p) = -\frac{\partial \Sigma(p)}{\partial p^\mu} \tag{19.20}$$

which is the original form given by Ward. The generalization to (19.16) was proved in 1957 by Takahashi directly from the field equations and without recourse to diagrammatic expansions.

One further fruit of the Ward identity worth mentioning here is that it helps place a condition on the structure of the vacuum polarization integral $\Pi_{\mu\nu}(q)$. Taking the divergence of (19.10) and using the Ward identity (19.16), we find

$$q^\nu \Pi_{\mu\nu}(q) = i \int \frac{d^4k}{(2\pi)^4} \operatorname{Tr} \gamma_\mu [S_F'(k) - S_F'(k+q)] \tag{19.21}$$

Applying again the assumption[1] that we cut off the integral in such a way that the origin of integration in (19.21) may be shifted, even for the formally quadratically divergent second-order term, we end up with

$$q_\nu \Pi^{\mu\nu}(q) = 0 \tag{19.22}$$

The transformation property of $\Pi^{\mu\nu}(q)$ as a second-rank tensor coupled with the condition (19.22) allows us to write, as we did in the second-

[1] See Eqs. (8.10) and (17.66). See also K. Johnson, *Nucl. Phys.*, **25**, 431 (1961).

order calculation (8.20),

$$\Pi_{\mu\nu}(q) = (q_\mu q_\nu - g_{\mu\nu}q^2)\Pi(q^2) \tag{19.23}$$

Furthermore, to any finite order of $e_0{}^2$, $\Pi(q^2)$ has no pole at $q^2 = 0$ according to the discussions of analytic properties[1] in Chap. 18.

Introducing (19.23) into (19.6), the equation relating $\Pi_{\mu\nu}$ to $D'_F(q)_{\mu\nu}$ gives

$$q^2[1 + e_0{}^2\Pi(q^2)]D'_F(q)_{\mu\nu} - e_0{}^2\Pi(q^2)q_\mu q^\lambda D'_F(q)_{\lambda\nu} = -g_{\mu\nu}$$

with the solution

$$D'_F(q)_{\mu\nu} = -\frac{g_{\mu\nu}}{q^2[1 + e_0{}^2\Pi(q^2)]} - \frac{q_\mu q_\nu e_0{}^2\Pi(q^2)}{q^4[1 + e_0{}^2\Pi(q^2)]} \tag{19.24}$$

In practical calculations the $q_\mu q_\nu$ term in (19.24) will not contribute, since D'_F couples to conserved currents. In any case the structure of D'_F is completely determined by the scalar $\Pi(q^2)$ in the expression (19.23) for $\Pi_{\mu\nu}(q)$. Since $\Pi(q^2)$ has no pole at $q^2 = 0$, (19.24) shows that the photon propagator describes a zero-mass quantum.

19.7 Definition of Renormalization Constants and the Renormalization Prescription

Finally, we are able to begin the renormalization program itself. Using the definitions of the renormalization constants given in Chap. 16, along with our experience in the second-order calculations of Chap. 8, we rescale the propagators S'_F and D'_F, the vertex Γ_μ, and the charge e_0. The renormalization constants defined by this rescaling may be related back to Σ, Π, and Γ_μ. Then the whole battery of integral equations buried within the preceding pages of this chapter and defining the theory will be resurrected and rewritten in terms of the renormalized quantities. This brings us up to the point of facing the music and showing that, to each order in the renormalized charge e, these quantities and equations are finite, all cutoff-dependent terms being absorbed in the rescaling laws, as was seen to occur previously in the second-order calculations.

The constant Z_2, introduced in (16.84), is the probability to find a "bare" one-electron state in the physical one-electron state, as shown in (16.118). As $\not p \to m$, we found from the general structure of S'_F

[1] Actually, $\Pi(q^2)$ has a branch cut beginning at $q^2 = 0$ and extending to $q^2 = +\infty$. For the possibility and interpretation of a pole in the exact $\Pi(q^2)$ summed to all powers of $e_0{}^2$ see J. Schwinger, *Phys. Rev.*, **125**, 397 (1962).

discussed in (16.122) that

$$S'_F \to \frac{Z_2}{\not{p} - m} \qquad \text{for } \not{p} \to m \tag{19.25}$$

From the definition of $\Sigma(p)$ given in (19.3), this implies

$$\Sigma(p) \to -(\not{p} - m)(Z_2^{-1} - 1) \qquad \text{for } \not{p} \to m \tag{19.26}$$

Thus $\Sigma(p)$ must vanish as $\not{p} \to m$. This is ensured by the existence of the mass counterterm $\delta m = m - m_0$, which is one of the interaction terms which must be included according to the rules derived in Chap. 17. To any order we adjust δm in such a way that (19.26) is satisfied.

$$\Sigma(\not{p} = m, \delta m) = 0 \tag{19.27}$$

Aside from the inclusion of the mass counterterm in the definition of $\Sigma(p)$ that we made in this chapter, (19.26) has the same structure as found in the second-order calculation (8.42).

Similarly, the renormalization constant Z_3 is associated with the contribution of the one-photon state to the photon propagator, and it is the residue of the pole at $q^2 = 0$, according to (16.172),

$$D'_F(q^2)_{\mu\nu} \to -\frac{Z_3 g_{\mu\nu}}{q^2} + \text{gauge terms} \qquad \text{as } q^2 \to 0 \tag{19.28}$$

As always, we ignore the gauge terms which do not contribute to the S matrix.[1] Then from the form of (19.24) relating D'_F to the vacuum polarization, we find

$$e_0^2\Pi(0) = Z_3^{-1} - 1 \tag{19.29}$$

Referring to the second-order calculation in (8.23), we find to that order

$$\Pi(0) \approx \frac{1}{12\pi^2} \log \frac{M^2}{m^2}$$

with M the cutoff mass.

Finally, the renormalization constant Z_1 associated with the vertex is defined as in the perturbation example (8.50) in terms of the zero momentum transfer limit of Γ_μ for the electron lines on the mass shell; that is

$$\bar{u}(p)\Gamma_\mu(p,p)u(p) = Z_1^{-1}\bar{u}(p)\gamma_\mu u(p) \tag{19.30}$$

[1] It should be recognized that these gauge terms *do* modify τ functions, as defined by (19.14). We regard τ functions here as objects from which S-matrix elements can be calculated. From this point of view, the modifications due to the gauge terms can be safely ignored.

The right-hand side of (19.30) is the most general form for zero momentum transfer, as shown in (10.88). With the Ward identity the very useful identification

$$Z_1 = Z_2 \tag{19.31}$$

established in perturbation theory in (8.54), may be proved in general. For $p' = p + q$ and $p^2 = m^2$ we write, as in (19.19),

$$\bar{u}(p)q^\mu \Gamma_\mu(p',p)u(p) = \bar{u}(p)S_F'^{-1}(p')u(p) \tag{19.32}$$

Taking the derivative $\partial/\partial q_\alpha$ of (19.32) and using (19.30) and (19.25) in the limit $q_\mu \to 0$, we deduce

$$Z_1^{-1}\bar{u}(p)\gamma_\mu u(p) = Z_2^{-1}\bar{u}(p)\gamma_\mu u(p)$$

establishing the equality of Z_1 and Z_2.

The idea of renormalization is to *rescale* the propagators and the vertex function so that near the mass shell, and in the case of the vertex for zero momentum transfer, these quantities approach the corresponding free-particle quantities. To this end we follow Dyson[1] and introduce renormalized propagators and vertex functions, $\tilde{S}'_F, \tilde{D}'_F$, and $\tilde{\Gamma}_\mu$, and a renormalized charge e by the equations

$$\begin{aligned} S'_F(p) &= Z_2 \tilde{S}'_F(p) \\ D'_F(q)_{\mu\nu} &= Z_3 \tilde{D}'_F(q)_{\mu\nu} \\ \Gamma_\mu(p',p) &= Z_1^{-1}\tilde{\Gamma}_\mu(p',p) \\ e_0 &= \frac{Z_1 e}{Z_2\sqrt{Z_3}} = Z_3^{-1/2} e \end{aligned} \tag{19.33}$$

thus

$$\begin{aligned} &\tilde{S}'_F(p) \to \frac{1}{\not p - m} \qquad \not p \to m \\ &\tilde{D}'_F(q)_{\mu\nu} \to \frac{-g_{\mu\nu}}{q^2} + \text{gauge terms} \qquad q^2 \to 0 \\ &\tilde{\Gamma}_\mu(p,p) \to \gamma_\mu \qquad \not p \to m \end{aligned} \tag{19.34}$$

The great problem of renormalization theory facing us is to show that $\tilde{S}'_F$, $\tilde{D}'_F$, and $\tilde{\Gamma}_\mu$ as defined by (19.33) are finite, that is, cutoff-independent functions when expressed in terms of the renormalized charge e. What makes this task a formidable one is that the Z's

[1] F. J. Dyson, *Phys. Rev.*, **75**, 486, 1736 (1949).

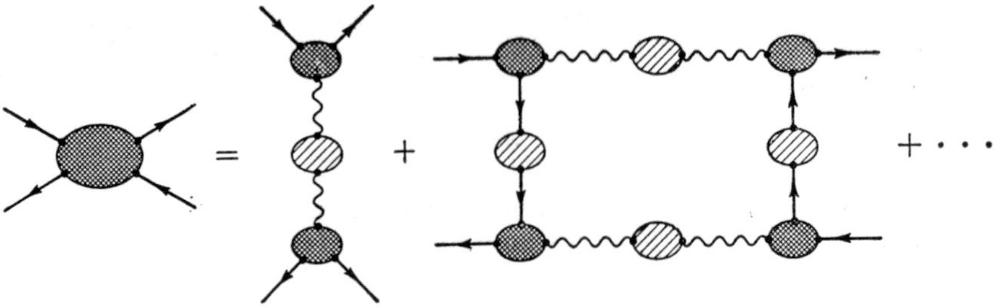

Fig. 19.29 First two terms in skeleton expansion for K.

diverge, that is, are cutoff-dependent quantities, as we discovered in Chap. 8.

To establish that $\tilde{S}'_F$, $\tilde{D}'_F$, and $\tilde{\Gamma}_\mu$, as well as the τ functions constructed from them, are finite, we shall make extensive use of the integral equations and skeleton expansions constructed in the first part of the chapter. It is of interest, therefore, to reexpress the elements of these equations in terms of renormalized quantities alone. Instead of an iteration in $e_0{}^2$, we shall be able to iterate the equations in a power series in e^2, the renormalized charge. It is fortunate for the renormalization program that the rescaling does not complicate the structure of the various integral equations.

To begin, we consider the kernel K, which is expressed in terms of S'_F, D'_F, Γ, and e_0 by the skeleton expansion (19.13), and write out the first two terms of the expansion (Fig. 19.29) in symbolic shorthand:

$$K = ie_0{}^2\Gamma D'_F\Gamma + e_0{}^4\int[\Gamma S'_F\Gamma]D'_FD'_F[\Gamma S'_F\Gamma] + \cdots \tag{19.35}$$

Applying the renormalization prescription (19.33), this may be rewritten in the form

$$\begin{aligned} K &= ie_0{}^2Z_1^{-2}Z_3\tilde{\Gamma}\tilde{D}'_F\tilde{\Gamma} + e_0{}^4Z_1^{-4}Z_2^2Z_3^2\int[\tilde{\Gamma}\tilde{S}'_F\tilde{\Gamma}]\tilde{D}'_F\tilde{D}'_F[\tilde{\Gamma}\tilde{S}'_F\tilde{\Gamma}] + \cdots \\ &= Z_2^{-2}\{ie^2\tilde{\Gamma}\tilde{D}'_F\tilde{\Gamma} + e^4\int[\tilde{\Gamma}\tilde{S}'_F\tilde{\Gamma}]\tilde{D}'_F\tilde{D}'_F[\tilde{\Gamma}\tilde{S}'_F\tilde{\Gamma}]\} + \cdots \end{aligned}$$

All that has happened is that unrenormalized quantities have been replaced by their renormalized counterparts and an over-all factor of Z_2^{-2} has appeared in front.

It is not difficult to show that this happens for each skeleton, so that the skeleton expansion (19.13) becomes

$$K(p,p',q) = Z_2^{-2} \sum_{\text{skeletons } \mathcal{S}} K^{\mathcal{S}}(p,p',q;\tilde{S}'_F,\tilde{D}'_F,\tilde{\Gamma},e) \tag{19.36}$$

At each vertex in the various skeletons there belong one-half of one photon line and half of each of two electron lines. The other halves of these lines join to other vertices or are external. Thus for each Z_1^{-1} appearing at a vertex due to renormalization of Γ_μ according to (19.33), there also appears a $(\sqrt{Z_2})^2$ from the adjacent electron lines and a $\sqrt{Z_3}$ from the photon line, provided these are internal lines. Together with the e_0 appearing at the vertex this gives an over-all factor for each vertex

$$e_0 Z_1^{-1} Z_2 \sqrt{Z_3} = e$$

At each of the vertices adjacent to the external electron and positron lines a factor $\sqrt{Z_2}$ per line is missing, since the propagators of the external lines are not included in the definition of K. Thus the renormalization prescription introduces an over-all factor of Z_2^{-2}, as given in (19.36). This result suggests defining a *renormalized kernel* $\tilde{K}$ as follows;

$$\tilde{K}(p,p',q) \equiv Z_2^2 K(p,p',q) = \sum_{\text{skeletons } \mathcal{S}} K^{\mathcal{S}}(p,p',q;\tilde{S}'_F,\tilde{D}'_F,\tilde{\Gamma},e) \qquad (19.37)$$

Similar scaling laws can be written for general Feynman amplitudes with arbitrary numbers of external lines. For instance, for a τ function with m external fermion legs and n external photons, the renormalization prescription following from its skeleton expansion (19.14) is

$$\begin{aligned}\tau(p_1 \cdots p_m, q_1 \cdots q_n)_{\alpha_1 \cdots \alpha_m, \mu_1 \cdots \mu_n} &= Z_2{}^{m/2} Z_3{}^{n/2} \\ &\times \sum_{\text{skeletons } \mathcal{S}} \tau^{\mathcal{S}}(p_1 \cdots p_m, q_1 \cdots q_n; \tilde{S}'_F, \tilde{D}'_F, \tilde{\Gamma}, e)_{\alpha_1 \cdots \mu_n} \\ &\equiv Z_2{}^{m/2} Z_3{}^{n/2}\, \tilde{\tau}(p_1 \cdots p_m, q_1 \cdots q_n)_{\alpha_1 \cdots \mu_n} \qquad (m + n > 3)\end{aligned} \qquad (19.38)$$

The proof of (19.38) follows that of (19.36) with only a small modification. In the τ functions are included the propagators associated with the external legs, while these were amputated in defining K. Therefore there are now sufficient $\sqrt{Z}$ factors available to renormalize all the vertices properly, and in fact a $\sqrt{Z}$ per external line is left over.

For S-matrix elements the situation is even simpler. As shown in the reduction formula derived in Chap. 16, for instance, in (16.139) and (16.148), one computes the invariant S-matrix element from the τ function by multiplying τ by $(\not{p} - m)/\sqrt{Z_2}$ for each electron and

$q^2/\sqrt{Z_3}$ for each photon, going onto the mass shell $\not p \to m$, $q^2 \to 0$, and inserting free-particle wave functions on the external lines, along with various factors of i. Since the renormalized propagators on the external legs approach free-particle propagators $(\not p - m)^{-1}$ and $-(q^2)^{-1}g_{\mu\nu}$, we are left with a remarkably simple prescription for computing S-matrix elements:

1. Draw all skeleton graphs.
2. Compute the amplitude according to the Feynman rules, with $e\tilde{\Gamma}_\mu$ inserted at the vertices and $\tilde{S}'_F$ and $\tilde{D}'_F$ in the lines.
3. Insert the free-particle wave functions

$$\frac{\epsilon_\mu}{\sqrt{2k(2\pi)^3}} \qquad \sqrt{\frac{m}{E(2\pi)^3}}\,u(p) \qquad \bar{u}(p)\sqrt{\frac{m}{E(2\pi)^3}} \qquad \text{etc.}$$

as appropriate for the external lines, *unmodified* by any Z factors or self-energy insertions. The $\sqrt{Z}$'s in (19.38) have canceled the $\sqrt{Z}$'s which appear in the reduction formula. Therefore, the question of the finiteness of the S-matrix elements has been decoupled from the Z factors and reduced to a study of the finiteness of $\tilde{S}'_F$, $\tilde{D}'_F$, and $\tilde{\Gamma}_\mu$ and of the convergence of the integrals over the internal momenta in the skeleton graphs.

The integral equations defining the vertex Γ_μ as well as the Ward identity, which determines the electron propagator S'_F in terms of the vertex Γ_μ, may also be renormalized. For the vertex, we insert (19.33) into (19.12); in symbolic shorthand we obtain

$$\tilde{\Gamma}_\mu = Z_1\gamma_\mu - \int \tilde{\Gamma}_\mu \tilde{S}'_F \tilde{S}'_F \tilde{K} \tag{19.39}$$

with the understanding that Z_1 is to be determined by the condition (19.34)

$$\tilde{\Gamma}_\mu(p,p)\Big|_{\not p = m} = \gamma_\mu \tag{19.40}$$

The Ward identity (19.16) preserves its form upon renormalization, since $Z_1 = Z_2$:

$$(p' - p)_\mu \tilde{\Gamma}^\mu(p',p) = \tilde{S}'^{-1}_F(p') - \tilde{S}'^{-1}_F(p) \tag{19.41}$$

In particular, the version given in (19.19) is

$$\tilde{S}'^{-1}_F(p') = (p' - p)_\mu \tilde{\Gamma}^\mu(p',p)\Big|_{\not p = m} \tag{19.42}$$

where the notation

$$\Big|_{\not p = m}$$

indicates that $p^2 = m^2$ and any factor $\not{p}$ at the right of Γ_μ is to be set equal to m.

Because the Ward identity determines S'_F completely in terms of Γ_μ, equations for the electron self-energy part need not be considered; we leave their renormalization as an exercise.

The equations for the photon propagator and vacuum polarization may also be renormalized. Applying the prescription (19.33) to (19.24), we find

$$\tilde{D}'_F(q)_{\mu\nu} = \frac{-g_{\mu\nu}}{q^2[Z_3 + Z_3 e_0{}^2\Pi(q^2)]} - \frac{q_\mu q_\nu e_0{}^2\Pi(q^2)}{q^4[Z_3 + Z_3 e_0{}^2\Pi(q^2)]} \tag{19.43}$$

Recalling the definition (19.29) of Z_3, and separating out the $q^2 = 0$ limit of $\Pi(q^2)$ by writing

$$\Pi(q^2) = \Pi(0) + \Pi_c(q^2)$$

with $$\Pi_c(0) = 0$$

and $$\Pi(0) = e_0^{-2}(Z_3^{-1} - 1) = \frac{1}{e^2} - \frac{1}{e_0^2} \tag{19.44}$$

we arrive at

$$\tilde{D}'_F(q)_{\mu\nu} = -\frac{g_{\mu\nu}}{q^2[1 + e^2\Pi_c(q^2)]} - \frac{q_\mu q_\nu}{q^2}\left[\frac{1}{Z_3 q^2} - \frac{1}{q^2[1 + e^2\Pi_c(q^2)]}\right] \tag{19.45}$$

Recall that terms proportional to $q_\mu q_\nu$ do not contribute to S-matrix elements because of current conservation, and only the first term of (19.45) survives in the calculation of S-matrix elements. Therefore, the calculation of $\tilde{D}'_F(q)_{\mu\nu}$ is reduced effectively to the calculation of $\Pi_c(q^2)$, which may be expressed in terms of renormalized quantities through (19.10), (19.23), (19.33), and (19.44)

$$\begin{aligned}\Pi_{\mu\nu}(q) &= (q_\mu q_\nu - g_{\mu\nu}q^2)[\Pi(0) + \Pi_c(q^2)] \\ &= iZ_1 \int \frac{d^4k}{(2\pi)^4} \operatorname{Tr} \gamma_\mu \tilde{S}'_F(k)\tilde{\Gamma}_\nu(k, k+q)\tilde{S}'_F(k+q)\end{aligned} \tag{19.46}$$

19.8 Summary: The Renormalized Integral Equations

Most of the formal machinery needed to carry out the renormalization program has now been constructed. Let us summarize here those equations necessary for computing S-matrix elements. Before renormalization, they have the form[1]

$$(a)\ \tau(p_1 \cdots p_m,q_1 \cdots q_n)_{\alpha_1 \cdots \mu_n} = \sum_{\text{skeletons } \mathbb{S}} \tau^{\mathbb{S}}(p_1 \cdots p_m,q_1 \cdots q_n;S'_F,D'_F,\Gamma_\mu,e_0)_{\alpha_1 \cdots \mu_n}$$

$$(b)\ K_{\alpha\beta,\gamma\delta}(p,p',q) = \sum_{\text{skeletons } \mathbb{S}} K^{\mathbb{S}}_{\alpha\beta,\gamma\delta}(p,p',q;S'_F,D'_F,\Gamma_\mu,e_0)$$

$$(c)\ \Gamma_\mu(p',p)_{\gamma\delta} = (\gamma_\mu)_{\gamma\delta} + \int \frac{d^4q}{(2\pi)^4} [iS'_F(p'+q) \times \Gamma_\mu(p'+q,\, p+q)iS'_F(p+q)]_{\beta\alpha}K_{\alpha\beta,\delta\gamma}(p+q,\, p'+q,\, q)$$
$$\equiv \gamma_\mu - \textstyle\int \Gamma_\mu S'_F S'_F K$$
$$\equiv \gamma_\mu + \Lambda_\mu(p',p)$$

$$(d)\ S_F'^{-1}(p') = (p'-p)_\mu \Gamma^\mu(p',p)\Big|_{\not p = m}$$

$$(e)\ D'_F(q)_{\mu\nu} = -\frac{g_{\mu\nu}}{q^2[1+e_0{}^2\Pi(q^2)]} + q_\mu q_\nu \text{ terms}$$

$$(f)\ \Pi_{\mu\nu}(q) = (q_\mu q_\nu - g_{\mu\nu}q^2)\Pi(q^2) = i\int \frac{d^4k}{(2\pi)^4} \operatorname{Tr} \gamma_\mu S'_F(k)\Gamma_\nu(k,\, k+q)S'_F(k+q) \quad (19.47)$$

These equations completely determine all τ functions as power series in $e_0{}^2$. To lowest nonvanishing order in $e_0{}^2$, the quantities S'_F, D'_F, and Γ_μ may be replaced by S_F, D_F, and γ_μ in computing the τ functions, K, and Π in (19.47*a*), (19.47*b*), and (19.47*f*). Γ_μ can then be computed to order $e_0{}^2$ and S'_F to the same order by using (19.47*c*) followed by (19.47*d*). Likewise, D'_F can be computed to order $e_0{}^2$ by inserting into (19.47*e*) the result of the zero-order calculation of Π in (19.47*f*). Indeed this was what we did in Chap. 8. These improved S'_F, D'_F, and Γ_μ, correct through order $e_0{}^2$, may then be inserted back into (19.47*a*), (19.47*b*), and (19.47*f*) and the whole procedure repeated until the desired order of accuracy is attained.

We note the absence from (19.47) of the equations involving the electron self-energy part $\Sigma(p)$:

$$S'_F = \frac{1}{\not p - m - \Sigma(p)}$$
$$\Sigma(p) = ie_0{}^2 \int \frac{d^4k}{(2\pi)^4} D'_F(k)_{\mu\nu}\Gamma^\mu(p,\, p-k)S'_F(p-k)\gamma^\nu \quad (19.48)$$

[1] Cf. Eqs. (19.14), (19.13), (19.12), (19.19), (19.24), and (19.10), for (*a*), (*b*), (*c*), (*d*), (*e*), and (*f*), respectively.

These equations are not needed, because the Ward identity (19.47*d*) determines S'_F completely from Γ_μ.

Introducing into (19.47) the renormalizations[1]

$$\begin{aligned} e_0 &= \sqrt{Z_3^{-1}}\, e \\ S'_F(p) &= Z_2 \tilde{S}'_F(p) \\ D'_F(q)_{\mu\nu} &= Z_3 \tilde{D}'_F(q)_{\mu\nu} \\ \Gamma_\mu(p',p) &= Z_1^{-1} \tilde{\Gamma}_\mu(p',p) \\ K(p,p',q) &= Z_2^{-2} \tilde{K}(p,p',q) \end{aligned} \tag{19.49}$$

$$\tau(p_1 \cdots p_m, q_1 \cdots q_n)_{\alpha_1 \cdots \mu_n} = Z_2^{m/2} Z_3^{n/2} \tilde{\tau}(p_1 \cdots p_m, q_1 \cdots q_n)_{\alpha_1 \cdots \mu_n}$$

along with

$$Z_1 = Z_2 \tag{19.50}$$

as a consequence of the Ward identity, we find instead the following set of equations:[2]

$$\tilde{\tau}(p_1 \cdots p_m, q_1 \cdots q_n)_{\alpha_1 \cdots \mu_n} = \sum_{\text{skeletons } \mathbb{S}} \tau^{\mathbb{S}}(p_1 \cdots p_m, q_1 \cdots q_n; \tilde{S}'_F, \tilde{D}'_F, \tilde{\Gamma}_\mu, e)_{\alpha_1 \cdots \mu_n} \tag{19.51a}$$

$$\tilde{K}_{\alpha\beta,\gamma\delta}(p,p',q) = \sum_{\text{skeletons } \mathbb{S}} K^{\mathbb{S}}_{\alpha\beta,\gamma\delta}(p,p',q; \tilde{S}'_F, \tilde{D}'_F, \tilde{\Gamma}_\mu, e) \tag{19.51b}$$

$$\begin{aligned} \tilde{\Gamma}_\mu(p',p)_{\gamma\delta} = Z_1(\gamma_\mu)_{\gamma\delta} &- \int \frac{d^4q}{(2\pi)^4} [\tilde{S}'_F(p'+q)\tilde{\Gamma}_\mu(p'+q,\, p+q) \\ &\times \tilde{S}'_F(p+q)]_{\beta\alpha} \tilde{K}_{\alpha\beta,\delta\gamma}(p+q,\, p'+q,\, q) \end{aligned} \tag{19.51c}$$

or

$$\begin{aligned} \tilde{\Gamma}_\mu &\equiv Z_1 \gamma_\mu - \int \tilde{\Gamma}_\mu \tilde{S}'_F \tilde{S}'_F \tilde{K} \\ &\equiv \gamma_\mu + \Lambda_\mu{}^c(p',p) \end{aligned}$$

$$\tilde{S}'^{-1}_F(p') = (p'-p)_\mu \tilde{\Gamma}^\mu(p',p)\Big|_{\not p = m} \tag{19.51d}$$

$$\tilde{D}'_F(q)_{\mu\nu} = -\frac{g_{\mu\nu}}{q^2[1 + e^2\{\Pi(q^2) - \Pi(0)\}]} + q_\mu q_\nu \text{ terms} \tag{19.51e}$$

$$\begin{aligned} \Pi_{\mu\nu}(q) = (q_\mu q_\nu - g_{\mu\nu} q^2)\Pi(q^2) &= iZ_1 \int \frac{d^4k}{(2\pi)^4} \\ &\times \operatorname{Tr} \gamma_\mu \tilde{S}'_F(k) \tilde{\Gamma}_\nu(k,\, k+q) \tilde{S}'_F(k+q) \end{aligned} \tag{19.51f}$$

with

$$\Pi(q^2) \equiv \Pi(0) + \Pi_c(q^2) \qquad \Pi_c(0) = 0$$

Along with these six equations goes the prescription to calculate S-matrix elements by removing the external propagators $\tilde{S}'_F$ and $\tilde{D}'_F$ in

[1] Cf. Eqs. (19.31), (19.33), (19.37), and (19.38).

[2] Cf. Eqs. (19.38), (19.37), (19.39), and Eqs. (19.42), (19.45), and (19.46).

the renormalized τ functions $\tilde{\tau}$ given by (19.51*a*) and replacing them by free-particle wave functions, unmodified by any Z factors.

The six equations (19.51) along with the boundary conditions[1]

$$\tilde{\Gamma}_\mu(p,p)\Big|_{\not p=m} = \gamma_\mu \qquad \frac{1}{e_0^2} = \frac{1}{e^2} - \Pi(0) = \frac{Z_3}{e^2} \tag{19.52}$$

which determine Z_1 and Z_3, comprise a complete set of equations determining τ functions and S-matrix elements in an expansion in the renormalized charge e. Exactly the same iteration scheme may be followed as we described for the expansion of the unrenormalized quantities. Beginning with (19.51*a*), (19.51*b*), and (19.51*f*) evaluated to lowest order in e, we insert $\tilde{K}$ and $\tilde{S}'_F \approx (\not p - m)^{-1}$ in (19.51*c*), and Γ_μ and Z_1 may be computed to order e^2, using the normalization condition (19.52). Then $\tilde{S}'_F$ is computed to order e^2 using the Ward identity (19.51*d*). $\tilde{D}'_F$ is found to order e^2 by inserting into (19.51*e*) the zero-order expression for $\Pi_{\mu\nu}$. With $\tilde{S}'_F$, $\tilde{D}'_F$, $\tilde{\Gamma}_\mu$, and Z_1 evaluated to order e^2, $\tilde{K}$ and the $\tilde{\tau}$ functions may be evaluated to one higher power of e^2. The process is then iterated.[2]

However, it is still not clear that the renormalized equations (19.51) are finite, that is, cutoff-independent. The appearance of a Z_1 in the inhomogeneous term in the vertex equation (19.51*c*) is relatively innocuous and appears to be able to be removed by a subtraction of the divergent part of the integral on the right-hand side of the equation. However, it certainly is not obvious how one disposes of the Z_1 factor in front of the integral defining $\Pi_{\mu\nu}(q)$. This problem is the most difficult we shall face. A similar problem occurs in the direct calculation of the electron self-energy via (19.48); this problem has been avoided by our use of the Ward identity. It is for this reason the identity was introduced and proves so valuable.

19.9 Analytic Continuation and Intermediate Renormalization

The algebraic and topological preliminaries are now completed. In Sec. 19.8 we have a full battery of equations for the $\tilde{S}'_F$, $\tilde{D}'_F$, $\tilde{\Gamma}_\mu$, $\tilde{K}_{\alpha\beta,\gamma\delta}$, and $\tilde{\tau}$ which may be solved by iteration to any finite order in the renor-

[1] Cf. Eqs. (19.40) and (19.44).

[2] The very presumptuous assumption we make in this section, which is not to be lost in the topological maze of the graphs or the jargon accessory to them, is this: In spite of cutoff dependences needed to *define* infinite integrals for all the Z_i and S_F, D_F, and Γ_μ, there exist uniformly convergent series in powers of both e^2 and e_0^2!!

malized charge e^2 by computing a finite sum of Feynman integrals. We have now arrived at the crux of the renormalization problem: we must look at these Feynman integrals and show that they converge and that, therefore, the renormalized perturbation expansion is finite—that is, cutoff-independent—term by term. The problem of convergence of the infinite series to all powers of e^2 will not be attacked here.

First let us dispose of a technical problem that arises when we attempt to put upper bounds on Feynman integrals. It has little to do with the real substance of the renormalization question. As discussed extensively in Chap. 18, the Feynman denominators $(p^2 - m^2)$ appearing in these integrals may vanish depending on the kinematics of the problem, and it is clear that it will be rather difficult to put limits on such integrals. For the renormalization problem, on the other hand, we are concerned only with questions of ultraviolet divergences as the $p^2 \to \infty$ and not with the question of the singularities giving rise to absorptive parts that were our main preoccupation in the dispersion theory discussions[1] of Chap. 18. Therefore we shall do away with this problem at the very beginning by performing an analytic continuation in the external momenta entering the Feynman amplitude to a region where it is certain that the denominators cannot vanish. Once we establish that the integral is convergent in this region, we can analytically continue back to physical values of the external momenta, with confidence that the integral must still exist almost everywhere except for a finite number of Landau singularities.

The representation (18.53) for Feynman integrals, with the action J in the denominator reexpressed in terms of external momenta via (18.35) and (18.37), allows us to make this continuation. The Feynman integral for a graph with m external momenta q_s and n internal momenta p_j is written, following (18.34) to (18.36),

$$\begin{aligned} I(q_1, \ldots, q_m) &= \int \frac{d^4l_1 \cdots d^4l_k}{(p_1^2 - m_1^2) \cdots (p_n^2 - m_n^2)} P(q_s, l_r) \\ &= \int_0^\infty \frac{d\alpha_1 \cdots d\alpha_n\, \delta\left(1 - \sum_j \alpha_j\right) Q(\alpha_j, q_s)}{[\Delta(\alpha_1 \cdots \alpha_n)]^2 \left[\sum_{j=1}^n \alpha_j(k_j^2 - m_j^2 + i\epsilon)\right]^p} \\ &= \int_0^\infty \frac{d\alpha_1 \cdots d\alpha_n\, \delta(1 - \Sigma\alpha_j) Q(\alpha_j, q_s)}{\Delta^2 \left[\sum_{i,j=1}^m \zeta_{ij} q_i \cdot q_j - \sum_{j=1}^n m_j^2 \alpha_j + i\epsilon\right]^p} \end{aligned} \tag{19.53}$$

[1] Also, the infrared question is of no concern. It is solved, moreover, by the method in this section, as we shall show.

where P and Q are polynomials in their arguments and may involve, as well, γ matrices and polarization vectors. The Kirchhoff momenta k_j are linear combinations of the q_s with coefficients depending on the α_j as defined in (18.35) and (18.37), and the ζ_{ij} are α-dependent coefficients constructed as below (18.35).

If the momenta in (19.53) could all be changed to a euclidean region by the replacement

$$k_j{}^0 \rightarrow ik_j{}^0$$
$$\mathbf{k}_j \rightarrow \mathbf{k}_j$$
$$k_j{}^2 \rightarrow -[|k_j{}^0|^2 + |\mathbf{k}_j|^2]$$

the denominators would all be negative definite and there would occur no singularities to plague us. This is our goal. To carry out this program, we observe from the linear relations among the p_j, k_j, and l_r in (18.32), (18.33), and (18.37) that whatever shift of this type we make on the external momenta q_s carries over to the k_j and l_r as well. Let us make the replacement for all external momenta

$$q_s{}^0 \rightarrow q_s{}^0 e^{i\varphi} \qquad 0 \leq \varphi \leq \pi/2$$
$$\mathbf{q}_s \rightarrow \mathbf{q}_s$$

We may verify that the integral I still exists, since the Kirchhoff momenta k_j also transform in the same way

$$k_j{}^0 \rightarrow k_j{}^0 e^{i\varphi} \qquad \mathbf{k}_j \rightarrow \mathbf{k}_j$$

and the denominator in (19.53) becomes

$$\begin{aligned} J &= \sum_{j=1}^{n} \alpha_j(k_j{}^2 - m_j{}^2 + i\epsilon) \\ &= \sum_{j=1}^{n} \{[(k_j{}^0)^2 \cos 2\varphi - \mathbf{k}_j{}^2 - m_j{}^2]\alpha_j + i[(k_j{}^0)^2 \sin 2\varphi + \epsilon]\alpha_j\} \end{aligned} \qquad (19.54)$$

Since

$$\operatorname{Im} J > 0$$

for all $\alpha_j \geq 0$ in the integral (19.53), I exists for all $0 < \varphi < \pi/2$, provided it is regulated as needed. We may now go to the limit $\varphi = \pi/2$ in (19.54) and, returning from the last to the first form of (19.53), we have

$$\bar{I}(\bar{q}_1, \ldots, \bar{q}_m) = \int \frac{d^4\bar{l}_1 \cdots d^4\bar{l}_k P(\bar{q}_s, \bar{l}_r)}{(\bar{p}_1{}^2 - m_1^2) \cdots (\bar{p}_n{}^2 - m_n^2)} \qquad (19.55)$$

with $\bar{p}_\mu = (ip_0, \mathbf{p})$ and $\bar{p}^2 = -(p_0{}^2 + \mathbf{p}^2) < 0$, and with the integrals extending over a euclidean four-space $d^4\bar{l}$.

With this analytic continuation procedure we have constructed in (19.55) a form for $\bar{I}$ with negative definite denominators which is more convenient for discussions of convergence. If we succeed in showing that $\bar{I}(\bar{q})$ exists, we shall also have succeeded in demonstrating existence of $I(q)$ for physical momenta q_s except for special values for which I has Landau singularities as discussed in Chap. 18. This is because we know we can analytically continue back to the physical q_s from the $\bar{q}_s$, just as we traversed the opposite route, by letting $\bar{q}_0 \to \bar{q}_0 e^{-i\varphi}$ and going to the limit $\varphi = \pi/2$. Since we start with a well-defined analytic function, we can be stopped in this analytic continuation procedure only by a natural boundary at the limit $\varphi = \pi/2$. However, the discussions in Sec. 18.4 indicate that only the Landau singularities of $I(q_s)$ are in the way of this continuation. We established this explicitly for a vertex with one external momentum variable q^2. We can extend the procedure to scattering graphs with more legs if we ignore the mass shell restrictions, e.g., (19.52), and continue all variables to the euclidean region as done here. Since the number of Landau singularities is less than or equal to the number of possible reduced graphs, which is finite, $I(q_s)$ exists almost everywhere for physical q_s if $\bar{I}(\bar{q}_s)$ exists.

Hereafter, we drop the distinction between q_s and $\bar{q}_s$ and assume the analytic continuation from physical to euclidean momenta is carried out whenever we discuss questions of convergence.

The Feynman integrals defining the renormalization constants Z_1 or Z_2 require a special discussion, since they are *evaluated* when external momenta p^2 are set equal to m^2. At these values we cannot use the convergence criteria, yet to be discussed, which will apply only to euclidean four-vectors[1] with $p^2 \leq 0$. To circumvent this additional nuisance, we shall first make our subtractions, designed to make the theory finite, at the point $p_\mu = 0$, which lies on the boundary of the euclidean region. We shall call this prescription *intermediate renormalization.* The propagator and vertex will consequently no longer be properly normalized on the mass shell $p^2 = m^2$. However, we shall have achieved the desired goal of demonstrating that $\tilde{S}'_F$ and $\tilde{D}'_F$ are finite if we can show:

1. The theory with the propagator and vertex intermediate-renormalized at $p_\mu = 0$ instead of $p^2 = m^2$ is convergent.
2. The renormalization constants relating propagators and vertex intermediate-renormalized at $p_\mu = 0$ to $\tilde{S}'_F$ and $\tilde{\Gamma}_\mu$ renormalized at $p^2 = m^2$ are finite (that is, cut-off independent).

[1] There is no problem with Z_3, since the photon propagator has already been renormalized at $q^2 = 0$.

Our procedure will therefore be to renormalize $S'_F(p)$ and $\Gamma_\mu(p,p)$ at $p_\mu = 0$; we may then use any theorems on convergence of integrals which are restricted to the euclidean region. However, we must investigate the effect of such an intermediate renormalization on the defining equations (19.51) and (19.52) and on the physical interpretation of the theory and then show that the additional renormalization necessary to compute $\tilde{S}'_F$ and $\tilde{D}'_F$ is finite.

In terms of the unrenormalized and the renormalized self-energy and vertex parts, we introduce the intermediate-renormalized quantities as follows:

$$\begin{aligned} \tilde{\tilde{S}}'_F(p) &= Z_2'^{-1} S'_F(p) = z_2^{-1}\tilde{S}'_F(p) \\ \tilde{\tilde{\Gamma}}_\mu(p',p) &= Z'_1\Gamma_\mu(p',p) = z_1\tilde{\Gamma}_\mu(p',p) \end{aligned} \tag{19.56}$$

so defined that

$$\begin{aligned} \tilde{\tilde{S}}'_F(p) &\to \frac{1}{\not p - \bar{m}} \qquad \text{as } p_\mu \to 0 \\ \tilde{\tilde{\Gamma}}_\mu(p,p) &\to \gamma_\mu \qquad \text{as } p_\mu \to 0 \end{aligned} \tag{19.57}$$

The defining equations (19.51) and (19.52) can therefore be written in this renormalized form by adding an extra tilde on the $\tilde{S}'_F$ and $\tilde{\Gamma}'_\mu$, changing Z_1 and Z_2 to Z'_1 and Z'_2, and replacing the boundary condition $\Big|_{\not p = m}$ by $\Big|_{p_\mu = 0}$. In particular, we retain Ward's identity and the equality

$$Z'_1 = Z'_2$$

since (19.20) and (19.57) in the limit $p_\mu \to 0$ give

$$(Z'_2)^{-1} \frac{\partial}{\partial p^\mu} (\not p - \bar{m}) = (Z'_1)^{-1}\gamma_\mu \qquad \text{as } p_\mu \to 0$$

or

$$Z'_1 = Z'_2$$

The discussion in the rest of the chapter will show that the theory with intermediate renormalization is finite. Then to show that $\tilde{S}'_F$ and $\tilde{\Gamma}_\mu$ are finite, we must, by (19.56) and (19.57), show that z_1, z_2, and $\bar{m}$ are finite. This we do by observing from the generalized Ward identity

$$\tilde{\tilde{S}}_F'^{-1}(p) - \tilde{\tilde{S}}_F'^{-1}(0) = p^\mu \tilde{\tilde{\Gamma}}_\mu(p,0)$$

that because $\tilde{\tilde{S}}_F'^{-1}(p)$ vanishes at $\not p = m$, the parameter $\bar{m}$ is given by

$$\bar{m} = p_\mu \tilde{\tilde{\Gamma}}^\mu(p,0)\Big|_{\not p = m} \tag{19.58}$$

Furthermore, from the normalization condition on $\tilde{\Gamma}_\mu$ and from (19.56) we deduce

$$z_1\gamma_\mu = z_1\tilde{\Gamma}_\mu(p,p)\Big|_{\not p = m} = \tilde{\tilde{\Gamma}}_\mu(p,p)\Big|_{\not p = m} \tag{19.59}$$

Thus if $\tilde{\tilde{\Gamma}}_\mu$ is finite, so are $z_1 = z_2$ and $\bar{m}$. Then the true renormalized theory (one tilde) will be finite provided the intermediate-renormalized (two tildes) is. Moreover, $\tilde{\tilde{\Gamma}}_\mu(p',p)$ is finite for physical p',p provided we prove it is finite for euclidean p' and p. This follows from the existence of the analytic continuation back to physical values of p' and p established in the first part of this section.

We shall hereafter ignore this "mathematical nicety" of intermediate renormalization, along with the analytic continuation to euclidean momentum four-vectors, in discussing convergence of integrals. It may always be assumed to be carried out when necessary. A more relevant point in quantum electrodynamics concerns the practical utility of the intermediate renormalization with regard to the infrared problem. Since the photon has zero mass, we are actually in the mathematically awkward situation of normalizing our functions at a branch point when we carry through the usual renormalization program at $p^2 = m^2$. Evidence of this was seen in the second-order self-energy part, (8.40), and in the evaluation of $Z_1 = Z_2$ in (8.42), both of which are infrared divergent as $p^2 \to m^2$. No such complications appear in the intermediate renormalization program at $p_\mu \to 0$.

19.10 Degree of Divergence; Criterion for Convergence

We can now concentrate on the problem of establishing criteria for the convergence of Feynman amplitudes and for identifying divergent graphs in quantum electrodynamics.[1] In the second-order calculations of Chap. 8 we encountered self-energy and vertex integrals for which the criterion for convergence was simply counting powers of internal momenta. The powers of momenta in the numerator, counting four for the internal-momentum integration d^4l, equaled or exceeded the powers in the denominator, and these integrals all had ultraviolet divergences for $l_\mu \to \infty$. After the subtractions of the renormalization program were carried out, the remaining parts had more powers of $l_\mu \to \infty$ in the denominator than in the numerator and were found to be finite.

As we go to higher orders of calculation with several or more internal-momentum integrations d^4l_r, it is by no means clear that it is only necessary to count total powers of the internal-momentum variables in the numerator and denominator in order to decide whether or not a Feynman integral converges. It is strongly suggestive, however,

[1] F. Dyson, *op. cit.*

that counting of powers has something to do with the question of determining how badly a Feynman integral diverges, and so in search of a convergence criterion we put the counting procedure on a systematic basis for arbitrary graphs.

First we introduce the *degree of divergence* of a graph $\mathcal{G}$ as the number of internal-momentum powers in the numerator, including a 4 for each d^4l_r, minus the number appearing in the denominator, and denote it by $D(\mathcal{G})$. The propagators of each internal photon, or boson, line contribute two powers to the denominator, and those of each internal electron, or fermion, line, one power. In quantum electrodynamics we may then write

$$D = 4k - 2b - f \tag{19.60}$$

where b = number of *internal* photon lines
f = number of *internal* electron lines
k = number of internal-momentum integrations

The remarkable feature of (19.60), and the key to the success of the renormalization program, is that D is in fact independent of all internal details of the Feynman graph and depends only on the number and kind (photon or electron) of the *external* lines of the graph. This is due essentially to the fact that the coupling strength e is dimensionless. The electron mass m is the only dimensional constant in the theory, but we ignore m in counting momentum powers as $l_\mu \to \infty$. As we go to higher and higher orders in the calculation of any given amplitude by inserting more and more internal photon lines its dimensionality cannot change. To verify this explicitly, we observe that each additional internal photon line inserted into a graph means creating two additional fermion propagators as well as creating one more $\int d^4l_r$. From (19.60) it follows then that the resulting change in the degree of divergence is

$$\Delta D = 4\,\Delta k - 2\,\Delta b - \Delta f = 4 - 2 - 2 = 0 \tag{19.61}$$

We find thus that D is determined by over-all dimensional considerations. For theories containing derivative couplings in the interaction and dimensional coupling constants, these arguments, as well as the traditional renormalization program, fail.

On the basis of (19.61) we can find D for any proper graph by removing internal lines until we come to an arbitrary simple one as in Fig. 19.30. If one now adds an additional *external* photon line to this graph, an additional internal fermion line is created and D decreases by 1, that is, $\partial D/\partial B = -1$, where B is the number of external photon or boson lines. Similarly, if one adds an additional *external* fermion, two additional fermion line endings, one extra fermion internal line, and one

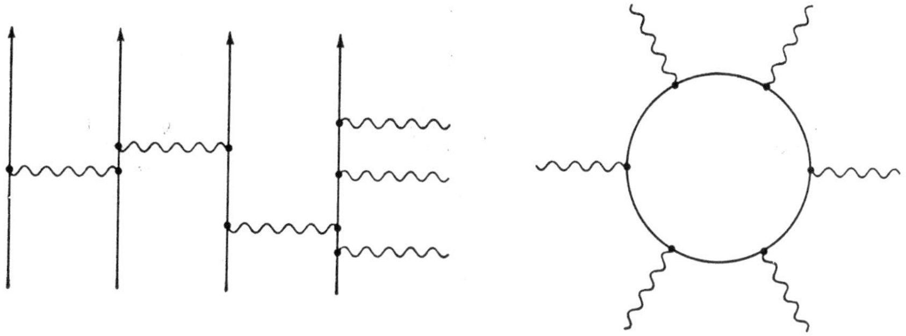

Fig. 19.30 Proper graphs simplified by removing extra internal lines for counting degree of divergence D.

extra photon line are created. Therefore D decreases by 3, or by $\frac{3}{2}$ per additional external fermion line F: $\partial D/\partial F = -\frac{3}{2}$. We then have an equation determining the degree of divergence in terms of B and F

$$D = 4 - \tfrac{3}{2}F - B \tag{19.62}$$

where the constant 4 is found by considering any particular graph. For example, the electromagnetic proper vertex, to lowest order in e, has $F = 2$, $B = 1$, and $D = 0$, since it is a constant with no momentum integrations.

In the renormalization program we shall also deal with Feynman integrals in which one or more "subtractions" have been made. This was encountered, as we have mentioned above, in the second-order calculations in Chap. 8, and it had the effect of adding an extra internal-momentum power in the denominator at the expense of introducing an external-momentum factor in the numerator. This is evidently true in general, and we include the effect of such subtractions on the definition of degree of divergence by adding the rule: For each subtraction with respect to some set of the external momenta q_s of a Feynman integral subtract 1 from the degree of divergence of this particular integral.

If $D < 0$, a Feynman graph has sufficient powers of momenta in the denominator to be finite, and it is said to be superficially convergent. It need not be finite, however, because a vertex insertion in this graph leaves D unchanged but would be sufficient to make it diverge. This illustrates that a condition on the convergence of *each* subintegration must be satisfied in addition to the requirement that the over-all degree of divergence is negative, $D < 0$, for a Feynman amplitude to converge.

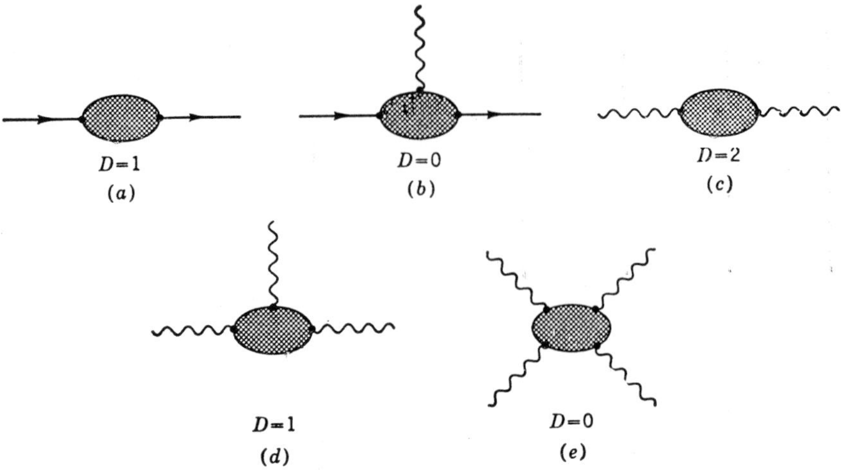

Fig. 19.31 Graphs with nonnegative degrees of divergence in quantum electrodynamics.

If, on the other hand, $D \geq 0$ for a graph, it has almost no chance to be finite. There is a very limited class of graphs with a nonnegative degree of divergence in quantum electrodynamics which we enumerate below and illustrate in Fig. 19.31.

(*a*)	Electron proper self-energy part $\Sigma(p)$	$D = 1$
(*b*)	Proper vertex part $\Gamma_\mu(p',p)$	$D = 0$
(*c*)	Photon proper self-energy part or vacuum polarization $\Pi_{\mu\nu}(q)$	$D = 2$
(*d*)	Proper vertex of three photons	$D = 1$
(*e*)	Photon-photon scattering amplitude	$D = 0$

Of these five, the only two of special concern are (*b*) and (*c*). The electron self-energy (*a*) can be determined from Γ_μ via the Ward identity and need not be discussed separately. The three-photon vertex (*d*) vanishes identically, since it is odd under charge conjugation. We recall from (15.96) and (15.97) that there exists a unitary operator $\mathcal{C}$ which is a constant of the motion $[\mathcal{C},H] = 0$ and under which the electromagnetic field is an odd operator,

$$\mathcal{C}A_\mu(x)\mathcal{C}^{-1} = -A_\mu(x)$$

Hence the vacuum state, if unique as assumed as usual in (16.2) for quantum electrodynamics, is an eigenstate of $\mathcal{C}$, $\mathcal{C}|0\rangle = |0\rangle$, and it follows that

$$\begin{aligned}\langle 0|T(A_\mu(x)A_\nu(y)A_\lambda(z))|0\rangle &= \langle 0|T(\mathcal{C}A_\mu(x)A_\nu(y)A_\lambda(z))|0\rangle \\ &= -\langle 0|T(A_\mu(x)A_\nu(y)A_\lambda(z))|0\rangle \\ &= 0\end{aligned}$$

and therefore the three-photon vertex vanishes. The generalization of this result to any graph with an odd number of external photon lines only, known as Furry's theorem,[1] was proved in 1937.

Although $D = 2$ for the photon self-energy part (*c*), a quadratically divergent term which would be associated with the photon mass has been legislated to zero with the aid of the Ward identity in (19.21) and (19.22). Thus two of the momentum powers in $\Pi_{\mu\nu}(q)$ are used up to form the $q_\mu q_\nu$ factors in (19.51), leaving $\Pi(q^2)$ with an effective $D_{\text{eff}} = 0$ and milder divergence problems.

With regard to the photon-photon scattering amplitude (*e*) the situation is even better. We lose four powers of internal momenta in forming electromagnetic field strengths $F_{\mu\nu}(q) \sim (q_\mu \epsilon_\nu - q_\nu \epsilon_\mu)$, one for each of the four external photons as required for a gauge-invariant amplitude. This gives an effective degree of divergence of $D_{\text{eff}} = -4$ instead of 0 and leads in perturbation theory to a finite amplitude as verified in explicit calculation by Karplus and Neumann.[2] The actual photon-photon scattering cross section is extremely small; at low energies

$$\sigma_{\gamma\gamma} \sim \frac{\alpha^4}{\pi^4}\frac{\omega^6}{m^8} \qquad \omega = \text{C.M. energy}$$

$$\omega \ll m$$

and for relativistic energies, $\omega \gg m$, decreases $\sim 1/\omega^2$.

We are then left with only two quantities, Γ_μ and $\Pi_{\mu\nu}$, to consider with $D \geq 0$, both of which have an effective degree of divergence $D_{\text{eff}} = 0$. This is a tremendous simplification to have the divergence troubles focused on just two quantities which in perturbation theory exhibit nothing worse than a logarithmic cutoff dependence.

From this it does not follow that all quantities with $D < 0$ are finite; indeed, any self-energy or vertex insertion leaves D unchanged but introduces infinities. Clearly, a more sophisticated convergence criterion than simply counting powers is needed, and something must be said about integrations over groups, or subsets, of internal momenta circulating through loops in the Feynman diagram. This criterion,

[1] W. H. Furry, *Phys. Rev.*, **51,** 125 (1937).

[2] R. Karplus and M. Neumann, *Phys. Rev.*, **80,** 380, **83,** 776 (1950).

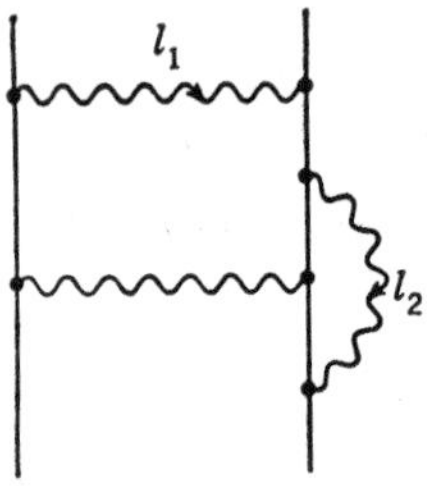

Fig. 19.32 A graph with two internal loop momenta l_1 and l_2.

stated by Dyson[1] in 1949 and completed by Salam[2] in 1951, was proved by Weinberg[3] in 1960. Weinberg's theorem, to which we now turn, provides the precise criterion for convergence of Feynman integrals and is basic to the success of the Dyson-Salam renormalization program.[4] Beyond its application to the renormalization program Weinberg's theorem does much more and determines, up to powers of logarithms, the asymptotic behavior of these integrals for any fixed set of external momenta tending to infinity. Here we discuss only that part of the theory relevant to the renormalization program, reserving to Sec. 19.14 the consideration of asymptotic behavior.

The notion of a subgraph and a subintegration must be introduced in order to state and explain the basic ideas of Weinberg's theorem. An integration over some subset S of internal loop momenta l_r is called a *subintegration*. Associated with a subintegration is a *subgraph*, obtained by removing all lines which do not depend on the momenta l_r in the subset S. For example, the graph in Fig. 19.32 has three possible subgraphs shown in Fig. 19.33. These correspond to subintegrations over l_1 with l_2 fixed as in (a), over l_2 with l_1 fixed as in (b), and over l_1 with $l_1 - l_2$ fixed as in (c). We may associate a degree of divergence $D(S)$ to each such subgraph and subintegration by counting powers in the same way as described for the over-all graph; thus $D = -3, 0$, and -5, respectively, for the three subgraphs (a), (b), and (c) in Fig. 19.33. It should be noted that an overall subtraction on a graph $\mathcal{G}$ does not in general reduce $D(S)$. The exceptions occur when the external momenta involved in the subtraction flow entirely within the subgraph.

However, there is an additional complication in assigning a degree of divergence to a subintegration. This arises because the Feynman

[1] F. J. Dyson, *op. cit.*

[2] A. Salam, *Phys. Rev.*, **82**, 217, **84**, 426 (1951).

[3] S. Weinberg, *Phys. Rev.*, **118**, 838 (1960).

[4] Meanwhile, Bogoliubov and collaborators have proved renormalizability by somewhat different techniques; see N. N. Bogoliubov and D. V. Shirkov, "Introduction to the Theory of Quantized Fields," Interscience Publishers, Inc., New York, 1959.

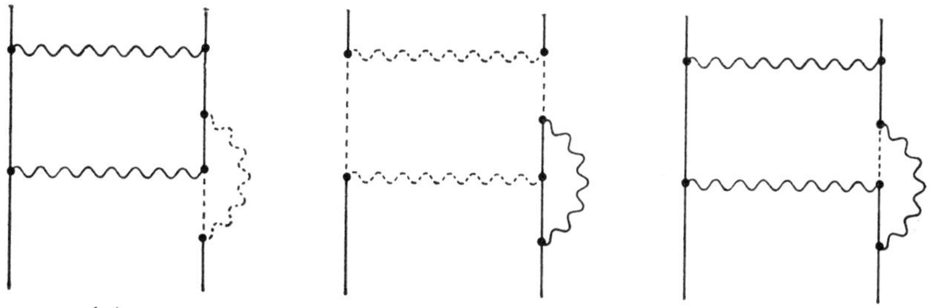

Fig. 19.33 Three subgraphs of Fig. 19.32 with (*a*) l_2 fixed, (*b*) l_1 fixed, and (*c*) $l_1 - l_2$ fixed. The internal-momentum variable passing through a dotted line in a subgraph is not integrated.

integrals appearing in the expansion of the *renormalized* equations (19.51) in powers of e also contain subtraction terms associated with the renormalization conditions (19.51*c*), (19.51*e*), and (19.52) defining $\tilde{\Gamma}_\mu$ and $\tilde{D}'_F$. To illustrate this, we return to our example in Fig. 19.33 and write out explicitly the vertex insertion

$$\tilde{\Gamma}_\mu{}^{(2)}(p, p + l_1) \sim \int d^4l_2\, \gamma_\nu S_F(p - l_2)\gamma_\mu S_F(p + l_1 - l_2)\gamma^\nu D_F(l_2) - \int d^4l_2\, \gamma_\nu S_F(\tilde{p} - l_2)\gamma_\mu S_F(\tilde{p} - l_2)\gamma^\nu D_F(l_2)\Big|_{\tilde{p}=m}$$

In the denominator of the first term we find one power of l_1 which contributed to our estimate $D(l_1) = -3$ for the l_1 subintegration. The second term, however, is independent of l_1, and therefore $D(l_1) = -2$ for a contribution involving this term. To handle such circumstances in general, we shall denote graphs corresponding to Feynman integrals in which there are such subtraction terms by drawing boxes with dotted lines around the part of the graph where the subtraction was performed. These are shown for our example in Fig. 19.34*a* and *b*. These two figures are the only two possible subgraphs associated with the vertex subtraction term; their degrees of divergence are $D = -2$ and 0, respectively.

According to our renormalization procedure we have straightforward rules for determining the degree of divergence of a subintegration associated with a Feynman graph in which there are subtraction terms:

1. Draw boxes with dotted lines about the vertex and self-energy insertions corresponding to the subtraction terms.

2. For subintegrations over momenta contained entirely within a box as in Fig. 19.34*b* count powers as usual.

3. In counting powers for integration momenta entering and leaving a box, as in Fig. 19.34*a*, ignore the box completely, that is, shrink

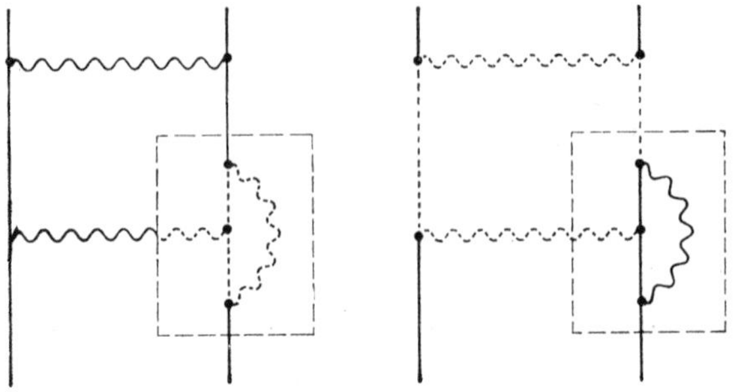

Fig. 19.34 Boxes with dashed lines surround portions of the two possible subgraphs of Fig. 19.32 where subtractions are performed.

the box to a point and forget it was ever there. This is equivalent to saying that everything inside the box is to be ignored, since the momenta leaving the box from within are fixed by the subtraction.

We must then assign a power 0 for a vertex box, +1 for an electron self-energy part, and +2 for a photon self-energy part, corresponding to the subtractions proportional to γ_μ, $\not{p}$, and $(q_\mu q_\nu - g_{\mu\nu}q^2)$ for these three insertions, respectively. However, these powers are just compensated by the additional fermion and boson propagators introduced by the corresponding self-energy insertions, and the result is the same as if we were to wipe out entirely the boxes with dotted lines corresponding to subtraction terms.

We may now state Weinberg's theorem: *A Feynman integral converges if the degree of divergence of the graph as well as the degree of divergence associated with each possible subgraph is negative.* Therefore the graph in Fig. 19.32 diverges owing to the vertex insertion Fig. 19.33*b* with $D(S) = 0$. However, when the renormalization prescription is applied to the vertex part and its divergent part with $D = 0$ is subtracted out (Fig. 19.34*b*) and absorbed into the rescaling laws, a finite part with $D(S) = -1$ remains behind and Weinberg's theorem assures us that the *entire* integral is thereby rendered convergent.

The idea behind the proof of Weinberg's theorem may be seen in our example of Fig. 19.32. The integral $d^4l_1\, d^4l_2$, considered as extending over the eight-dimensional l_1-l_2 space, is shown schematically in two dimensions in Fig. 19.35. The over-all degree of divergence according to (19.62) is $D = -2$ corresponding to the fact that, as we integrate radially outward from the origin toward ∞ in l_1-l_2 space, in most directions the integral behaves as

$$I \sim \int \frac{d^8l}{l^{10}}$$

the l^{10} denominator coming from four fermion and three photon propagators. However there are four-dimensional "tubes" where the integrand behaves differently. These correspond to the three regions

$$(a)\ l_2 \text{ small and } I \sim \left(\int_{\text{finite volume}} d^4l_2\right)\left(\int \frac{d^4l_1}{l_1{}^7}\right), \text{ with } D(a) = -3$$

$$(b)\ l_1 \text{ small and } I \sim \left(\int_{\text{finite volume}} d^4l_1\right)\left(\int \frac{d^4l_2}{l_2{}^4}\right), \text{ with } D(b) = 0$$

$$(c)\ l_1 - l_2 \text{ small and } I \sim \left(\int_{\text{finite volume}} d^4(l_1 - l_2)\right)\left(\int \frac{d^4l_1}{l_1{}^9}\right), \text{ with } D(c) = -5$$

Evidently these regions are just those covered by the subintegrations associated with the subgraphs (*a*), (*b*), and (*c*) in Fig. 19.33. Likewise, the subgraphs Fig. 19.34*a* and *b* associated with subtraction terms correspond to the two regions

$$(a)\ l_2 \text{ small and } I \sim \left(\int_{\text{finite volume}} d^4l_2\right)\left(\int \frac{d^4l_1}{l_1{}^6}\right), \text{ with } D(a) = -2$$

$$(b)\ l_1 \text{ small and } I \sim \left(\int_{\text{finite volume}} d^4l_1\right)\left(\int \frac{d^4l_2}{l_2{}^4}\right), \text{ with } D(b) = 0$$

The problem of slicing up the Feynman integral in an efficient way into pieces corresponding to these regions which can be bounded rigorously was solved by Weinberg. In particular his theorem justifies the estimates we made above based on the restriction "finite volume" in the integrals.

In higher orders the ideas shown in this example may be used in understanding the results of Weinberg's theorem. If there are k internal momenta l_r, the dimensionality of the integration volume is $4k$.

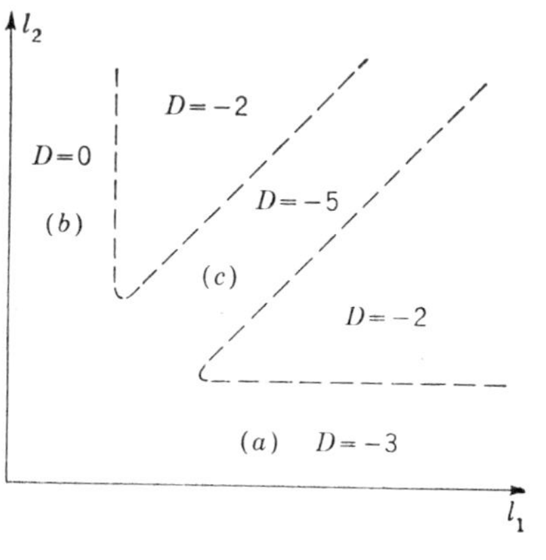

Fig. 19.35 Regions of l_1-l_2 space with corresponding degrees of divergence.

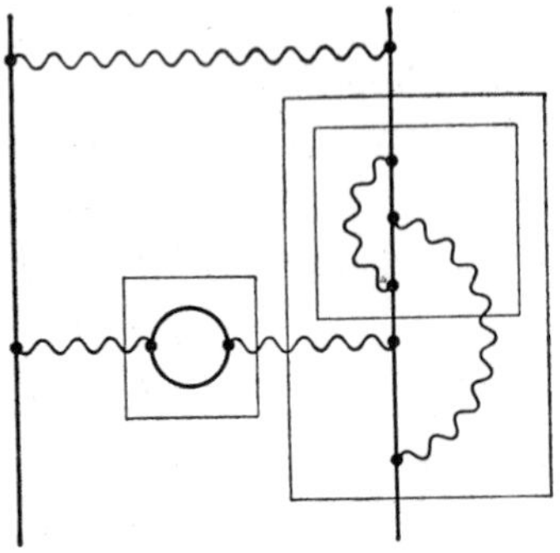

(a)

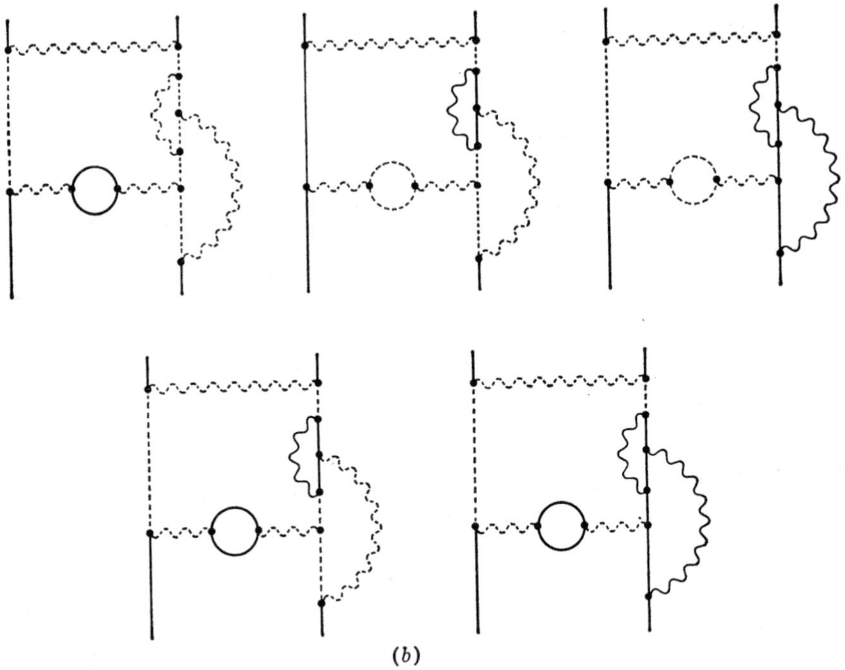

(b)

Fig. 19.36 Graph (a) and its subgraphs (b) corresponding to divergent subintegrations.

Again as one integrates radially outward toward ∞ in this $4k$-dimensional space the integrand will behave in almost all directions like l^{D-4k}, where D is the over-all degree of divergence of the graph. However, there will again be "tubes" of various dimensionality corresponding to subgraphs in each of which the propagator denominators not contained in the subgraph remain small and do not help to converge the integral. The contribution coming from these tubes is determined by ignoring lines not in the subgraph and counting powers on the lines within the subgraph to determine its degree of divergence $D(S)$. Confirmation of the validity of this prescription is given by Weinberg's theorem. The proof of this theorem involves a lengthy and intricate application of the Heine-Borel theorem, and the interested reader is referred to the original paper for the details.[1]

Armed with this theorem, we know that our task in establishing convergence of the terms in the renormalized perturbation theory is to search out all subgraphs S and subintegrations with $D(S) \geq 0$. Moreover, the result from (19.62) that only vertex and self-energy parts have nonnegative degrees of divergence allows us to determine directly those

[1] S. Weinberg, *op. cit.*

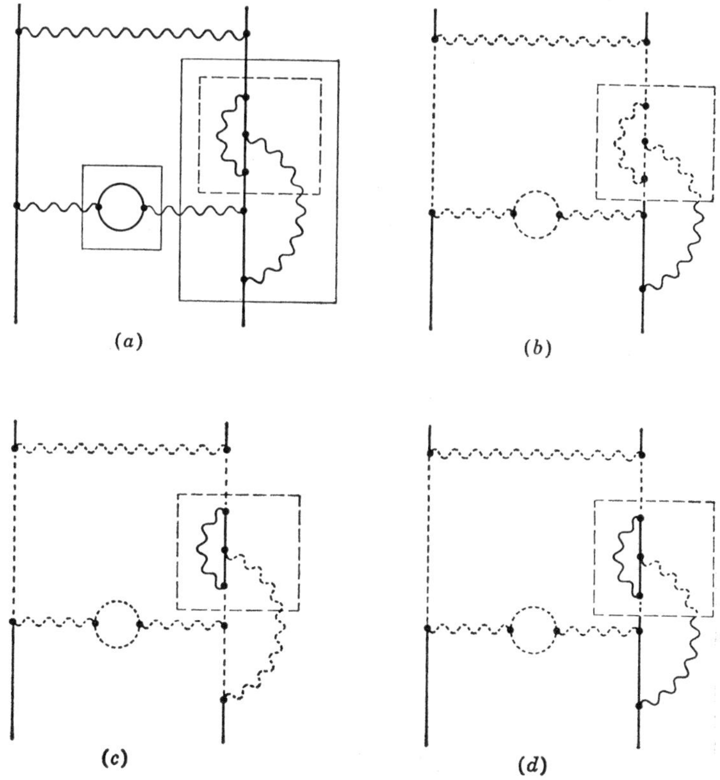

Fig. 19.37 Subtraction term of a vertex insertion (*a*) and associated divergent subintegrations (*b*), (*c*), and (*d*).

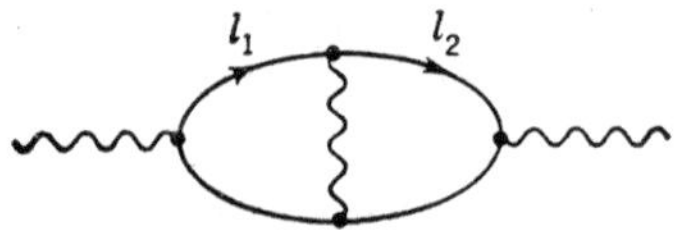

Fig. 19.38 Fourth-order photon self-energy graph.

subgraphs and subintegrations which are potential sources of divergence difficulties in Feynman integrals. In fact, they are just those portions of a graph enclosed within the boxes in Sec. 19.4 in identifying and isolating the vertex and self-energy insertions. As found there, these vertex and self-energy insertions may be made in an unambiguous and nonoverlapping manner, except within self-energy parts.

For example, we illustrate in Fig. 19.36*b* all subgraphs of Fig. 19.36*a* corresponding to divergent subintegrations. Some of these subgraphs are seen to be composed of disjoint parts. We need, however, consider explicitly only the connected subgraphs because, if the degree of divergence of all connected subgraphs is negative, the degree of divergence of all disconnected subgraphs will also be negative. In Fig. 19.37*a* we also show a subtraction term of the vertex insertion[1] enclosed by the dotted line, and in Fig. 19.37*b*, *c*, and *d* we show the associated divergent subintegrations. The subintegration of Fig. 19.37*b*, which was convergent in the unsubtracted term of Fig. 19.36, is now observed to be divergent. This divergence occurs within a vertex insertion, and it is evident from the discussion on page 323 that only within vertex and self-energy insertions will subtraction terms create divergent out of convergent subintegrations.

Special problems are posed by the overlapping vertex subgraphs within self-energy parts and will be overcome with the aid of the elec-

[1] Subgraphs associated with the vacuum polarization insertion present no new features or problems, and we ignore them.

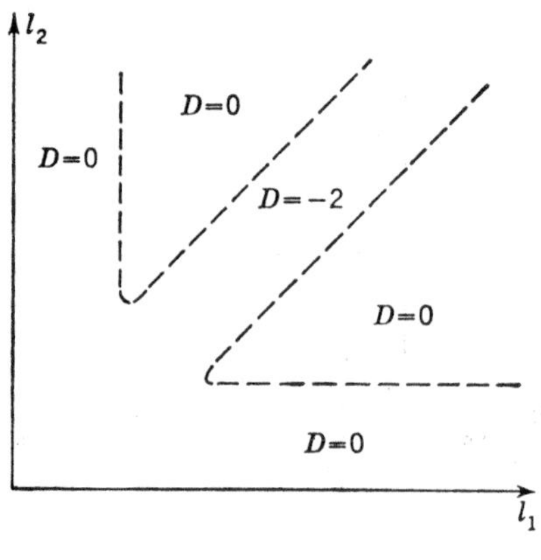

Fig. 19.39 Regions of l_1-l_2 space and associated degrees of divergence.

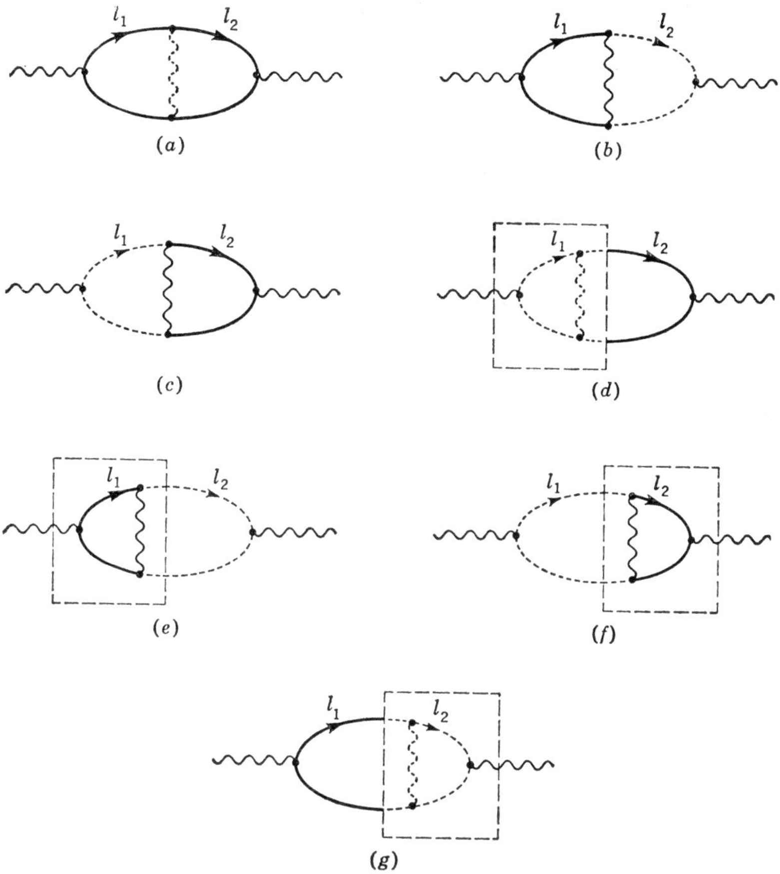

Fig. 19.40 Subgraphs of fourth-order photon self-energy graph, Fig. 19.38.

tron-positron kernel $\tilde{K}$. An example of this overlap problem is illustrated in Fig. 19.38 for the fourth-order photon self-energy graph. The over-all degree of divergence is $D = 0$ according to (19.62) after subtracting the two powers of momenta needed to form the $q_\mu q_\nu$ factor multiplying $\Pi(q^2)$ in (19.51). In addition, there are the three special tubes illustrated in Fig. 19.39 with associated degrees of divergence corresponding to the three subgraphs of Fig. 19.40*a*, *b*, and *c*:

(*a*) $l_1 - l_2$ small[1]	$D(a) = -2$
(*b*) l_2 small	$D(b) = 0$
(*c*) l_1 small	$D(c) = 0$

[1] Because the external momentum can be chosen to flow entirely within this subgraph, $D(a)$ is reduced by the over-all subtraction.

Also shown in d, e, f, and g of Fig. 19.40 are the subgraphs associated with subtraction terms which arise from renormalizing the vertex insertions. These correspond to the regions in l_1-l_2 space

(*d*) l_1 small $\quad D(d) = 0$

(*e*) l_2 small $\quad D(e) = 0$

(*f*) l_1 small $\quad D(f) = 0$

(*g*) l_2 small $\quad D(g) = 0$

In subtracting out the divergent parts, the degree of divergence of each subgraph must be reduced below zero by the vertex subtractions, as well as the D of the overall graph by the self-energy subtraction; the detailed prescription for this overlapping case was first spelled out by Salam.[1]

19.11 Proof That the Renormalized Theory Is Finite

We can now prove that the S matrix in quantum electrodynamics is finite to any finite order of expansion in the renormalized charge e. The iteration procedure in (19.51) allows us to compute the renormalized $\tilde{\Gamma}_\mu$, $\tilde{S}'_F$, $\tilde{D}'_F$, and $\tilde{\tau}$ functions and the S-matrix elements in an expansion in powers of e in terms of a finite sum of Feynman integrals. Also, Weinberg's theorem has provided us with a convergence criterion for these integrals: if the degree of divergence of all subgraphs as well as of the graph itself is negative, the integral converges.

The proof uses the method of mathematical induction. To lowest order in e^2, as verified by explicit calculation in Chap. 8, the degrees of divergence of $\tilde{\Gamma}_\mu$ and of $\Pi_c(q^2) = \Pi(q^2) - \Pi(0)$ are negative and they are given by convergent (that is, cutoff-independent) integrals. The same is true of $\tilde{S}'_F$ and $\tilde{D}'_F$, which are expressed in terms of $\tilde{\Gamma}_\mu$ and $\Pi_c(q^2)$. We *assume* this is also true to order $n - 2$ in e: the degree of divergence of all subgraphs of $\tilde{\Gamma}_\mu$ and $\Pi_c(q^2)$ is assumed to be negative and therefore, to this order in e, they as well as $\tilde{S}'_F$ and $\tilde{D}'_F$ are finite. We then use the iteration procedure of (19.51) and Weinberg's theorem to establish these assumptions to order n as well as to show that all S-matrix elements to order n are convergent.

We begin with the skeleton expansions (19.51*a*) and (19.51*b*) for the $\tilde{\tau}$ with four or more external legs and for the kernel $\tilde{K}$. A finite number of terms appear in the sum over skeletons if we evaluate $\tilde{\tau}$ and

[1] Salam, *op. cit.*

$\tilde{K}$ through any finite order e^n. Moreover, the kernel $\tilde{K}$ and the $\tilde{\tau}$ functions under consideration contain at least two interaction vertices and are at least of order e^2 as $e^2 \to 0$. Therefore, to order e^n, we need insert renormalized propagators and vertex functions to order e^{n-2} only. By the induction hypothesis, however, all such insertions S have $D(S) < 0$, since we are instructed in the skeleton sums (19.51a) and (19.51b) to insert renormalized vertex and self-energy parts. Therefore, there are no divergent subgraphs or subintegrations, since subgraphs S with $D(S) \geq 0$ can arise only from collections of vertex and self-energy parts.[1] This also applies to contributions from terms in which there are subtraction constants. As noted earlier, the degree of divergence of a subintegration for such a term can be nonnegative only if the subgraph lies within some vertex or self-energy insertion. Moreover, the over-all degree of divergence of these $\tilde{\tau}$ and $\tilde{K}$ functions is negative since the dubious distinction of having $D \geq 0$ is reserved for vertex and self-energy parts, as discussed in Sec. 19.10. We can conclude, then, that the kernel $\tilde{K}$ and four or more legged $\tilde{\tau}$ functions, or S-matrix elements, are finite to order e^n, since they are built up of a finite number of graphs $\mathcal{G}$ of which all subgraphs S have $D(S) < 0$ and for which $D(\mathcal{G}) < 0$ for the over-all integration as well.

We are not yet finished, because we must establish the induction hypothesis to order e^n for the $\tilde{\Gamma}_\mu$ and $\Pi_c(q^2)$, or $\tilde{D}'_F$, as well, using (19.51c) and (19.51f). As a result of the Ward identity, the convergence of $\tilde{S}'_F$ is assured by (19.51d) once it is proved for $\tilde{\Gamma}_\mu$.

To compute the vertex $\tilde{\Gamma}_\mu$ to order e^n from (19.51c), we can again insert $\tilde{S}'_F$, $\tilde{\Gamma}_\mu$, and $\tilde{K}$ to order e^{n-2}:

$$\int \tilde{S}'_F{}^{(n-2)} \tilde{\Gamma}_\mu{}^{(n-2)} \tilde{S}'_F{}^{(n-2)} \tilde{K}^{(n-2)} = Z_1{}^{(n)} \gamma_\mu - \tilde{\Gamma}_\mu{}^{(n)} = -\Lambda_\mu{}^{(n)} \qquad (19.63)$$

where the superscript (n) indicates the order of calculation in e. Our task is to show that after one subtraction, as required by the renormalization condition for $\tilde{\Gamma}_\mu$, $\Lambda_\mu{}^{(n)}$ is rendered finite, that is,

$$\Lambda_\mu{}^{c(n)}(p',p) = \Lambda_\mu{}^{(n)}(p',p) - \Lambda_\mu{}^{(n)}(p,p) \Big|_{\not{p}=m}$$

is given by a sum of unique, cutoff-independent integrals. In order to apply Weinberg's theorem, we first determine whether any subgraphs of (19.63) have nonnegative degrees of divergence. No subgraph S

[1] Again because of gauge considerations, photon-photon scattering parts may be ignored. They can be uniquely identified and isolated by putting boxes around them, and the boxes can be shown not to overlap by considerations as in Sec. 19.5. When graphs with all permutations of where the four external photon lines land in charged lines are added together, the effective divergence drops from $D = 0$ to $D = -4$.

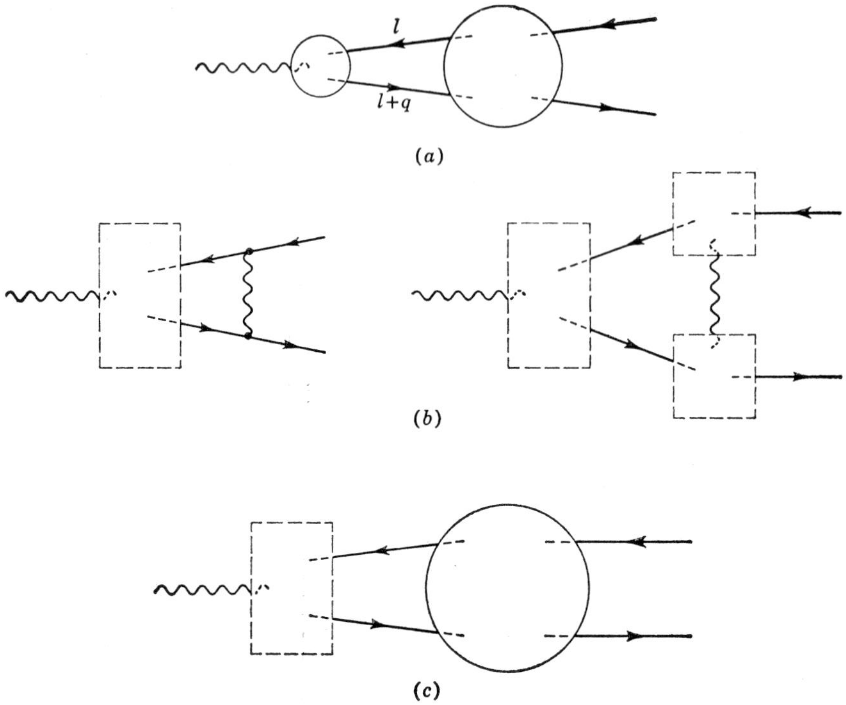

Fig. 19.41 Vertex subintegrations in which momenta l of propagators S_F participate.

with $D(S) \geq 0$ could be buried within the vertex $\tilde{\Gamma}_\mu$, propagators $\tilde{S}'_F$, or kernel $\tilde{K}$ because, by the induction hypothesis, the renormalization program[1] has already done away with this problem to the indicated order e^{n-2}. This leaves for consideration only those subintegrations $\mathfrak{L}$ in which the momenta l of the propagators $\tilde{S}'_F$ participate, as illustrated by Fig. 19.41.

We acquire at least two powers of momenta in the denominator from these two electron propagators, as drawn for the subintegration in Fig. 19.41*a*. Therefore, to obtain a divergent subgraph $\mathfrak{L}$, the portion within $\tilde{K}$ must have degree of divergence $D(\tilde{K}) \geq -2$. In our search for such subgraphs, we find one class of candidates to be the lowest-order perturbation contribution along with a class of subtraction terms illustrated in (19.41*b*); these have $D(\tilde{K}) = -2$. There will be similar higher-order terms in which *all* lines in $\tilde{K}$ (aside possibly from some within the dotted boxes associated with subtractions)

[1] Also, (19.63) is built up of a *finite* number of Feynman graphs each with a *finite* number of vertex and self-energy insertions.

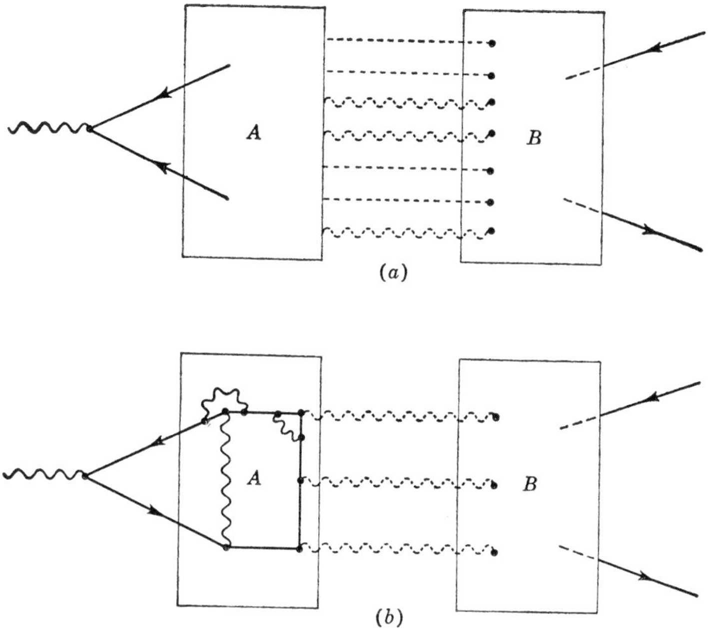

Fig. 19.42 Potentially dangerous subgraphs.

belong to the subgraph; all such graphs by the counting rule (19.62) have $D(\tilde{K}) = -2$. We next search for *subgraphs* in $\tilde{K}$ with $D(\tilde{K}) \geq -2$ and not involving the innocuous subtraction terms; these cannot be tamed by an over-all subtraction in (19.63). We describe such subgraphs in Fig. 19.42*a* by splitting up the kernel into two parts A and B. Through A run all the lines in the subgraph[1] $\mathfrak{L}$, while all those lines[2] not in the subgraph comprise B and the connecting links between A and B. Evidently only if the part A can lead to a contribution with $D(\tilde{K}) \geq -2$ does the subgraph $\mathfrak{L}$ diverge. However, the counting rule shows this to be impossible. The only possibilities are for two fermion and zero photon lines or zero fermion and one, two, or three photon lines to connect A to B. However, the two fermion and the one photon examples are ruled out by our definition of $\tilde{K}$ according to Fig. 19.11. The two-photon possibility is zero by charge conjugation, since joined to the vertex it corresponds to a closed loop with three external photon lines. The final possibility of three photon lines shown in Fig. 19.42*b* is rescued by the four very welcome powers

[1] Cases for which all external lines are attached to A are easily eliminated from consideration.

[2] Only connected subgraphs need be considered; see page 328.

by which the degree of divergence is reduced because of current conservation as we discussed earlier.[1]

Thus we have shown that the contribution of $\tilde{K}$ to the degree of divergence $D(\mathfrak{L})$ of any subintegration $\mathfrak{L}$ under consideration is ≤ -2; the only contributions with $D(\tilde{K}) = -2$ are the set of all lines within $\tilde{K}$ (aside from those within subtraction terms). Therefore, if a subintegration $\mathfrak{L}$ is to diverge, the contribution of the part of the subgraph within the vertex $\tilde{\Gamma}_\mu{}^{(n-2)}$ to the degree of divergence must be zero. By the induction hypothesis, however, all subintegrations within $\tilde{\Gamma}_\mu{}^{(n-2)}$ have $D < 0$, and this leaves us only one possibility: *no* lines within $\tilde{\Gamma}_\mu{}^{(n-2)}$ appear in the subgraph $\mathfrak{L}$. Such a subgraph can occur only for the over-all subtraction term illustrated in Fig. 19.41*c*. Its degree of divergence, however, since all lines within $\tilde{K}$ participate, is reduced to -1 upon making the over-all subtraction on (19.63) in computing $\Lambda_\mu{}^c$, in accordance with the renormalization program. This subtraction also reduces the over-all degree of divergence of the graph to -1, and Weinberg's theorem assures us that[2]

$$\Lambda_\mu{}^{c(n)}(p',p) = \Lambda_\mu{}^{(n)}(p',p) - \Lambda_\mu{}^{(n)}(p,p)\Big|_{\not p = m} \tag{19.64}$$

is given by a unique, convergent (that is, cutoff-independent) integral. Since

$$\Lambda_\mu{}^{c(n)}(p,p)\Big|_{\not p = m} = 0 \qquad \text{and} \qquad \Lambda_\mu{}^{(n)}(p,p)\Big|_{\not p = m} = L^{(n)}\gamma_\mu \tag{19.65}$$

by construction, where $L^{(n)}$ is a cutoff-dependent constant, we have from (19.52) and (19.63)

$$Z_1{}^{(n)} = 1 - L^{(n)} \tag{19.66}$$

to order e^n. To this order, then,

$$\tilde{\Gamma}_\mu{}^{(n)}(p',p) = \gamma_\mu + \Lambda_\mu{}^{c(n)}(p',p)$$

is finite, as is the electron propagator $\tilde{S}'_F$ according to (19.51*d*).

All that remains now is the more difficult problem of the vacuum polarization $\Pi^c(q^2)$ with the overlapping divergence difficulties. As found in Sec. 19.5, the boxes surrounding the vertex parts overlap for vertex insertions within self-energy parts, and this problem comes home to haunt us now. A typical graph for the vacuum polarization is shown in Fig. 19.43. Subgraphs that might diverge in this example are those which are vertex insertions at either end of the graph, and

[1] We, of course, need to include additional photon-photon subgraphs required to render the amplitude gauge invariant.

[2] Recall that $|_{\not p = m}$ means that $\not p = m$ when standing to the right.

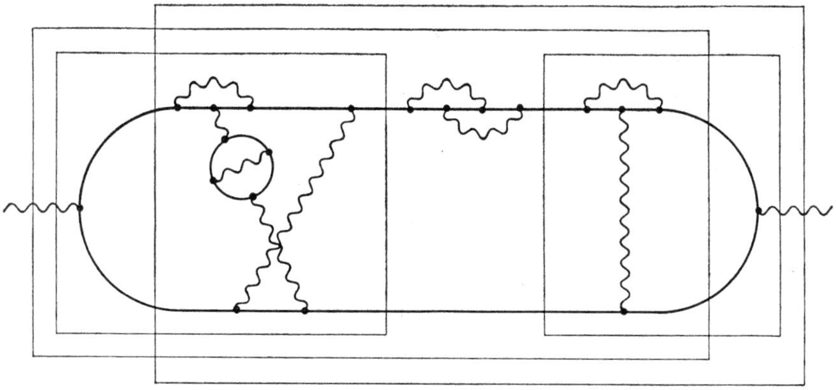

Fig. 19.43 A vacuum polarization graph.

these may overlap as shown by the boxes drawn in the figure. They cannot, therefore, be subtracted out separately, one by one, in series as for disjoint or nested insertions. In addition, many convergent subgraphs become divergent upon introducing the subtraction terms—in graphical terms by replacing one or more of the boxes in Fig. 19.43 by boxes with dotted lines.[1] We must also analyze such subgraphs and demonstrate that they have negative degree of divergence. For the $\tilde{S}_F'$ we sidestepped all these difficulties with the Ward identity. Here we must do battle with them—and this was foremost in our minds when we introduced the kernel K.

The effect of these overlapping divergences is to remove the factor Z_1 appearing in (19.51*f*). To see how this occurs, we return to (19.51*c*), solve for $Z_1\gamma^\mu$, and insert into (19.51*f*), obtaining in the symbolic notation[2]

$$\Pi_{\mu\nu}(q) = i\int\tilde{\Gamma}_\mu\tilde{S}_F'\tilde{S}_F'\tilde{\Gamma}_\nu + i\int\tilde{\Gamma}_\mu\tilde{S}_F'\tilde{S}_F'\tilde{K}\tilde{S}_F'\tilde{S}_F'\tilde{\Gamma}_\nu \qquad (19.67)$$

We want to show that the overlapping divergence is removed in this combination of the two terms on the right-hand side of (19.67) [which would appear more naturally as a difference of terms if the propagators were introduced always in the combination (iS_F') as in our rules for Feynman graphs]. To this end it is useful to construct an iteration for the vertex insertions in powers of the kernel $\tilde{K}$ instead of e^2, starting with $\tilde{\Gamma}_{(0)}{}^\mu \approx \gamma^\mu$. The first iteration in (19.51*c*) gives

$$\tilde{\Gamma}^\mu_{(1)} = Z_{1(1)}\gamma^\mu - \int\gamma^\mu\tilde{S}_F'\tilde{S}_F'\tilde{K} \qquad (19.68)$$

[1] It does not make sense to have such boxes *overlap*, however.

[2] The ordering of indices in the trace and the momentum arguments are clear from (19.51*c*) and (19.51*f*) and will be restored when needed.

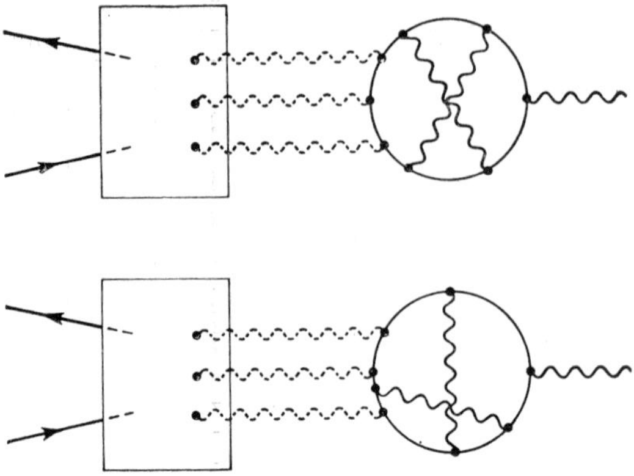

Fig. 19.44 Examples of photon-photon scattering subgraphs corresponding to different orders of iteration in $\tilde{K}$.

Aside from the subgraphs corresponding to photon-photon scattering insertions the subintegrations in (19.68) converge by the identical argument applied earlier in the discussion of (19.63). The subgraphs for photon-photon scattering insertions converge only when terms corresponding to different orders of iteration in the kernel $\tilde{K}$ are added together. Figure 19.44 illustrates an example requiring both one and two iterations of $\tilde{K}$ to be grouped together. However, we need not concern ourselves here with the photon-photon subgraphs. We know that among different orders of the iteration of (19.51*c*) in $\tilde{K}$ there exists a unique set of subgraphs, canceling off these divergent terms and lowering their effective degree of divergence to $D_{\text{eff}} = -4$. This applies to the photon-photon scattering subgraphs in vacuum polarization as well as in vertex graphs.

For the rest of the argument we shall forget about photon-photon scattering subintegrations and concentrate on the more intricate problem of vertex subintegrations. Hereafter we use "convergent" to mean "convergent with respect to vertex subintegrations." The individual terms really diverge because of the photon-photon scattering insertions and converge only after all pieces to a given order in e but to different orders of iteration in $\tilde{K}$ are added together. This is of no concern to us because our procedure is to consider each subintegration *in turn* according to Weinberg's theorem. The behavior of the integrand with regard to other subintegrations than the ones under consideration can be ignored.[1]

[1] The reader should by now be all too well aware of the wealth of qualifications and side issues with which a discussion of renormalization is burdened. This is characteristic of a proof by "enumeration of all possibilities." It is all too easy—as the authors will readily testify—to omit special cases from consideration. A similar situation holds in regard to the "what else can it be?" arguments used in

We now return to (19.68); the induction hypothesis assures us that there are no divergent subintegrations within $\tilde{K}$ to the required order e^n, and for momenta running through the $\tilde{S}'_F$ the analysis of page 332 applies unaltered, aside from the qualifications discussed above. Therefore one subtraction on (19.68) gives a convergent result

$$-\int \gamma^\mu \tilde{S}'_F \tilde{S}'_F \tilde{K} = L_1 \gamma^\mu + \Lambda_1{}^\mu(p',p) \tag{19.69}$$

where $\Lambda_1{}^\mu(p',p)$ is the "finite" part to first order in $\tilde{K}$. The subtraction is made as usual[1] at $\not{p}' = \not{p} = m$, so that

$$\Lambda_1{}^\mu(p,p)\Big|_{\not{p}=m} = 0 \tag{19.70}$$

We then have in (19.68) to first order in $\tilde{K}$

$$\tilde{\Gamma}^\mu_{(1)}(p',p) = (Z_{1(1)} + L_1)\gamma^\mu + \Lambda_1{}^\mu(p',p) \tag{19.71}$$

with

$$Z_{1(1)} = 1 - L_1 \tag{19.72}$$

according to the normalization condition (19.52). Inserting this improved $\tilde{\Gamma}^\mu$ back into (19.51*c*), we now obtain to second order in $\tilde{K}$

$$\begin{aligned}\tilde{\Gamma}^\mu_{(2)}(p',p) &= Z_{1(2)}\gamma^\mu - \int [\gamma^\mu + \Lambda_1{}^\mu(p',p)]\tilde{S}'_F \tilde{S}'_F \tilde{K} \\ &= Z_{1(2)}\gamma^\mu + L_1\gamma^\mu + \Lambda_1{}^\mu(p',p) - \int \Lambda_1{}^\mu(p',p)\tilde{S}'_F \tilde{S}'_F \tilde{K}\end{aligned}$$

In analogy with (19.69) we write

$$-\int \Lambda_1{}^\mu \tilde{S}'_F \tilde{S}'_F \tilde{K} = L_2\gamma^\mu + \Lambda_2{}^\mu(p',p)$$

with $\Lambda_2{}^\mu(p,p)\Big|_{\not{p}=m} = 0$

By arguments analogous to those given for $\Lambda_1{}^\mu(p',p)$, $\Lambda_2{}^\mu(p',p)$ is convergent and, to second order in $\tilde{K}$,

$$\tilde{\Gamma}^\mu_{(2)}(p',p) \approx \gamma^\mu + \Lambda_1{}^\mu(p',p) + \Lambda_2{}^\mu(p',p)$$

and $Z_{1(2)} \approx 1 - L_1 - L_2$.

Evidently after n iterations we find

$$\tilde{\Gamma}_{(n)}{}^\mu(p',p) = \gamma^\mu + \sum_{i=1}^{n} \Lambda_i{}^\mu(p',p) \tag{19.73}$$

with

$$Z_{1(n)} = 1 - \sum_{i=1}^{n} L_i \qquad \Lambda_i{}^\mu(p,p)\Big|_{\not{p}=m} = 0 \tag{19.74}$$

and

$$-\int \Lambda_i{}^\mu \tilde{S}'_F \tilde{S}'_F \tilde{K} = L_{i+1}\gamma^\mu + \Lambda^\mu_{i+1} \qquad \Lambda_0{}^\mu \equiv \gamma^\mu \tag{19.75}$$

---

constructing general forms for amplitudes consistent with prescribed symmetries. If you are not careful, someone will tell you what it can be.

[1] Cf. Sec. 19.7.

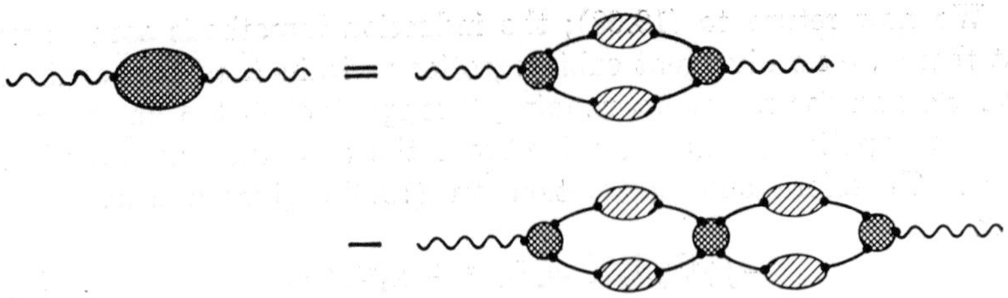

Fig. 19.45 Graphical expansion for $\Pi^{\mu\nu}(q)$.

What we have achieved here is a series for $\tilde{\Gamma}^\mu(p',p)$ as a sum of convergent integrals. The role of the subtraction constants L_i which determine Z_1 is to remove the divergence from the over-all vertex integral with $D = 0$ in (19.75) as well as the divergent sub-integrations associated with the subtraction terms buried within the integral. The series (19.73) is now introduced into (19.67) for $\Pi^{\mu\nu}(q)$, described by the graphical expansion illustrated in Fig. 19.45. Our task is to show that all subintegrations within a given graph converge. Once we have done this, we are assured by Weinberg's theorem that the over-all subtraction $\Pi_c(q^2) = \Pi(q^2) - \Pi(0)$ in (19.51*f*) renders finite $\Pi_c(q^2)$, with an effective $D = -1$.

Inserting (19.73) into (19.67) and collecting all terms of equal orders in the kernel $\tilde{K}$, we have

$$\begin{aligned}\Pi^{\mu\nu} &= i\int[\gamma^\mu + \Lambda_1{}^\mu + \Lambda_2{}^\mu + \cdots]\tilde{S}'_F\tilde{S}'_F[1 + \tilde{K}\tilde{S}'_F\tilde{S}'_F][\gamma^\nu + \Lambda_1^\nu + \cdots] \\ &= \Pi_0{}^{\mu\nu} + \Pi_1{}^{\mu\nu} + \Pi_2{}^{\mu\nu} + \cdots \end{aligned} \tag{19.76}$$

where

$$\begin{aligned}\Pi_0{}^{\mu\nu} &= i\int\gamma^\mu\tilde{S}'_F\tilde{S}'_F\gamma^\nu \\ \Pi_1{}^{\mu\nu} &= i\int\gamma^\mu\tilde{S}'_F\tilde{S}'_F\Lambda_1{}^\nu + i\int\Lambda_1{}^\mu\tilde{S}'_F\tilde{S}'_F\gamma^\nu + i\int\gamma^\mu\tilde{S}'_F\tilde{S}'_F\tilde{K}\tilde{S}'_F\tilde{S}'_F\gamma^\nu \\ \Pi_n{}^{\mu\nu} &= i\sum_{r+s=n}\int\Lambda_r{}^\mu\tilde{S}'_F\tilde{S}'_F\Lambda_s{}^\nu + i\sum_{r+s=n-1}\int\Lambda_r{}^\mu\tilde{S}'_F\tilde{S}'_F\tilde{K}\tilde{S}'_F\tilde{S}'_F\Lambda_s{}^\nu\end{aligned} \tag{19.77}$$

Each graph composing $\Pi_n{}^{\mu\nu}$ has n iterations of $\tilde{K}$ as shown in Fig. 19.46: by the construction (19.75), $\Lambda_r{}^\mu$ includes r iterations of $\tilde{K}$, and evidently $\tilde{K}$ occurs n times in each of the terms on the right-hand side of (19.77).

To show that $\Pi_c(q^2)$ is convergent, it is necessary to show that the degree of divergence of all subgraphs is negative. However, all subgraphs buried completely within a Λ^μ or $\tilde{K}$ are by the induction hypothesis convergent. Therefore, some subset of the momenta

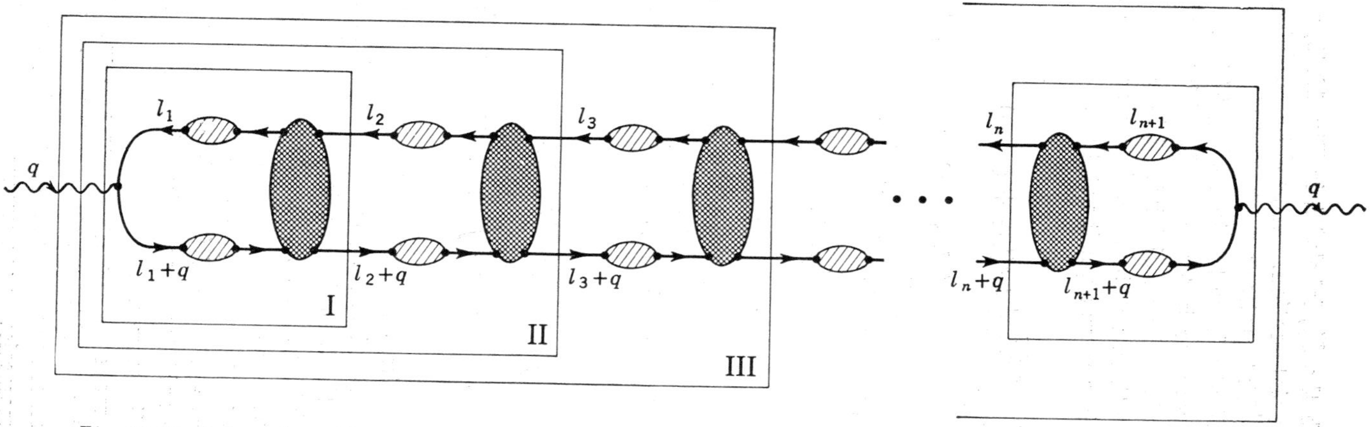

Fig. 19.46 Subgraphs with momenta l_i running through the electron propagators $S_F(l_i)$.

running through the $\tilde{S}'_F$, labeled l_i in Fig. 19.46, must contribute to any divergent subintegration. We call the subintegrations associated with box I l_1 subintegrations, with box II l_1l_2 subintegrations, and so on. We shall also have to discuss l_2l_3, $l_2l_3l_4$, . . . subintegrations as well, but ignore for instance $l_1l_2l_4l_5$ subintegrations, because the corresponding subgraphs are disjoint. Displaying the momenta l_i but suppressing all others within $\tilde{K}$ and Λ^μ along with spin indices we rewrite (19.77) in a more explicit notation:

$$
\begin{aligned}
\Pi_n{}^{\mu\nu}(q) = i \sum_{r+s=n} \int \frac{d^4l_{r+1}}{(2\pi)^4}\, \Lambda_r{}^\mu(l_{r+1}+q,\, l_{r+1}) \\
\times\, \tilde{S}'_F(l_{r+1})\tilde{S}'_F(l_{r+1}+q)\Lambda_s{}^\nu(l_{r+1},l_{r+1}+q) \\
+\, i \sum_{r+s=n-1} \int \frac{d^4l_{r+1}}{(2\pi)^4}\frac{d^4l_{r+2}}{(2\pi)^4}\, \Lambda_r^\mu(l_{r+1}+q,\, l_{r+1})\tilde{S}'_F(l_{r+1})\tilde{S}'_F(l_{r+1}+q) \\
\times\, \tilde{K}\tilde{S}'_F(l_{r+2})\tilde{S}'_F(l_{r+2}+q)\Lambda_s{}^\nu(l_{r+2},\, l_{r+2}+q) \qquad (19.78)
\end{aligned}
$$

Starting with the l_1 subintegrations, we need consider only the term with $r = 0$ and $\Lambda_0{}^\mu = \gamma^\mu$, since the l_1 subintegrations are buried *within* the $\Lambda_r{}^\mu(l_{r+1},l_{r+1}+q)$ for $r \geq 1$ and are evidently convergent by the construction in (19.73). The suspect term in (19.78) with $r = 0$ is, in shorthand,

$$
\begin{aligned}
i \int \gamma^\mu\tilde{S}'_F\tilde{S}'_F\Lambda_n^\nu + i \int \gamma^\mu\tilde{S}'_F\tilde{S}'_F\tilde{K}\tilde{S}'_F\tilde{S}'_F\Lambda_{n-1}^\nu \\
= -i \int \gamma^\mu\tilde{S}'_F\tilde{S}'_F\gamma^\nu L_n = -i \int \frac{d^4l_1}{(2\pi)^4}\, \gamma^\mu\tilde{S}'_F(l_1)\tilde{S}'_F(l_1+q)\gamma^\nu L_n \qquad (19.79)
\end{aligned}
$$

according to the iteration (19.75) for the Λ_n. The subtraction constant L_n, although divergent, is not relevant to the question of the convergence of the l_1 subintegration, since L_n is an integral over the remaining variables $l_2, \ldots, l_n$. As usual, the integral in (19.79) appears to have $D = +2$. Effectively, however, $D = 0$ after we extract the two powers of $q_\mu q_\nu$ to form the coefficient $\Pi(q^2)$. Constructing $\Pi_c(q^2) = \Pi(q^2) - \Pi(0)$ as in (19.51*f*) reduces D to -1 and renders the l_1 subintegration finite, since all subintegrations within the $\tilde{S}'_F$ in (19.79) converge.

We then move on the l_1l_2 subintegration. Analogously to our discussion of the l_1 subintegration, we need consider only terms with $r = 0$ and 1, since for $r \geq 2$ the l_1l_2 subintegrations are buried within $\Lambda_r{}^\mu$ and are automatically convergent.

The terms in (19.77) to be studied now are

$$i\int \gamma^\mu \tilde{S}'_F \tilde{S}'_F \Lambda_n{}^\nu + i\int \Lambda_1{}^\mu \tilde{S}'_F \tilde{S}'_F \Lambda^\nu_{n-1} + i\int \gamma^\mu \tilde{S}'_F \tilde{S}'_F \tilde{K} \tilde{S}'_F \tilde{S}'_F \Lambda^\nu_{n-1} + i\int \Lambda_1{}^\mu \tilde{S}'_F \tilde{S}'_F \tilde{K} \tilde{S}'_F \tilde{S}'_F \Lambda^\nu_{n-2} = -i\int \gamma^\mu \tilde{S}'_F \tilde{S}'_F \gamma^\nu L_n - i\int \Lambda_1{}^\mu \tilde{S}'_F \tilde{S}'_F \gamma^\nu L_{n-1}$$

where we again have made use of the iteration (19.75). Writing out the momentum arguments but still suppressing spinor indices, we have

$$i \int \frac{d^4 l_1}{(2\pi)^4} \frac{d^4 l_2}{(2\pi)^4} \gamma^\mu \tilde{S}'_F(l_1) \tilde{S}'_F(l_1 + q) \times [\tilde{K}(\tilde{p}, \tilde{p}, \tilde{p} - l_2) \tilde{S}'_F(l_2) \tilde{S}'_F(l_2) \Lambda^\nu_{n-1}(l_2, l_2)]_{\tilde{p}=m} - i \int \frac{d^4 l_2}{(2\pi)^4} \Lambda_1{}^\mu(l_2 + q, l_2) \tilde{S}'_F(l_2) \tilde{S}'_F(l_2 + q) \gamma^\nu L_{n-1} \tag{19.80}$$

where the divergent integral over l_2 has been written out explicitly. In counting powers on the l_1 integration in the first term we may subtract 2, which come from extracting the coefficient $\Pi(q^2)$ of $q_\mu q_\nu$, and another power when we make the over-all subtraction[1] to form $\Pi_c(q^2)$. In counting powers on the l_2 lines, which are all contained within the bracket, we cannot obtain a degree of divergence greater than zero under any circumstance, since this is a vertex insertion. Therefore, the degree of divergence of the $l_1 l_2$ subintegration for the first term is -1 or less. Likewise, in the second term an over-all subtraction reduces the degree of divergence of the $l_1 l_2$ subintegrations to -1 or less.[2] It is to be emphasized that neither the first nor the second term of (19.80) taken alone converges with respect to the l_2 subintegrations. This need not concern us, however, because the question we asked was whether the $l_1 l_2$ subintegrations, *not* the l_2 subintegrations, had negative degree of divergence: we answered our question in the affirmative. Each subintegration *in turn* must be brought to center stage and observed in the spotlight of Weinberg's theorem while the other subintegrations wait in the wings.

The way is now clear to generalize this result to the $l_1 \cdots l_r$ subintegration. Again all terms in (19.77) containing a factor $\Lambda_k{}^\mu(l_{k+1} + q, l_{k+1})$ with $k \geq r$ will automatically be convergent by induction,

[1] One must keep in mind that everything inside the bracket in (19.80) is a constant times γ.

[2] It must be realized that the over-all subtraction on Π does not reduce the D of *all* $l_1 l_2$ subgraphs. However, only the subgraph in which all lines in $\tilde{K}$ participate, and subgraphs such as in Fig. 19.40*a*, need the subtraction to make them convergent. For these the subtraction indeed decreases D, because the external momentum q flows only within the subgraph. The rest of the subgraphs already have $D \leq -1$.

because the $l_1 \cdots l_r$ subintegration lies completely within $\Lambda_k{}^\mu$. The remaining terms can be written

$$\sum_{k=0}^{r-1} \left(i \int \Lambda_k{}^\mu \tilde{S}'_F \tilde{S}'_F \Lambda^\nu_{n-k} + i \int \Lambda_k{}^\mu \tilde{S}'_F \tilde{S}'_F \tilde{K} \tilde{S}'_F \tilde{S}'_F \Lambda^\nu_{n-k-1} \right)$$

$$= -i \sum_{k=0}^{r-1} \int \Lambda_k{}^\mu \tilde{S}'_F \tilde{S}'_F \gamma^\nu L_{n-k}$$

$$= -i \sum_{k=0}^{r-1} \int \frac{d^4 l_{k+1}}{(2\pi)^4} \Lambda_k{}^\mu (l_{k+1} + q, l_{k+1}) \tilde{S}'_F(l_{k+1}) \tilde{S}'_F(l_{k+1} + q) \gamma^\nu L_{n-k}$$

The integrations over $l_1 \cdots l_k$ are buried within $\Lambda_k{}^\mu$ while the integrations from l_{k+2} to l_r, if any, are buried in the renormalization constant L_{n-k}. The degree of divergence of this latter subintegration is zero or less, because L_{n-k} is a vertex insertion. The degree of divergence of the $l_1 \cdots l_{k+1}$ subintegration, on the other hand, is lowered, as before, from $+2$ to -1 in forming $\Pi_c(q^2)$. Therefore, all subintegrations $l_1 \cdots l_r$ are convergent.

We must now investigate the l_2, $l_2 l_3$, . . . subintegrations in the same way. Our task is to bring these subintegrations to a form which permits us to make use of the over-all subtraction on the graph for $\Pi_{\mu\nu}$ to show that these subgraphs converge. Starting with the l_2 subintegrations, we spot the only suspect terms in (19.77) as those three in which the l_2 subintegrations do not lie harmlessly within the Λ_r or Λ_s. In writing these terms we underline the factors in which momentum l_2 appears:

$$\begin{aligned}
[\text{Suspect}]_{l_2} &= i\int \underline{\Lambda_1{}^\mu \tilde{S}'_F \tilde{S}'_F \Lambda^\nu_{n-1}} + i\int \underline{\gamma^\mu \tilde{S}'_F \tilde{S}'_F \tilde{K} \tilde{S}'_F \tilde{S}'_F \Lambda^\nu_{n-1}} \\
&\quad + i\int \underline{\Lambda_1{}^\mu \tilde{S}'_F \tilde{S}'_F \tilde{K}} \tilde{S}'_F \tilde{S}'_F \Lambda^\nu_{n-2} \\
&= -iL_1 \int \underline{\gamma^\mu \tilde{S}'_F \tilde{S}'_F \Lambda^\nu_{n-1}} + i\int \underline{\Lambda_1{}^\mu \tilde{S}'_F \tilde{S}'_F \tilde{K}} \tilde{S}'_F \tilde{S}'_F \Lambda^\nu_{n-2} \qquad (19.81)
\end{aligned}$$

Using the iteration (19.75) once again, we rewrite the last term of (19.81) as

$$\begin{aligned}
i\int \underline{\Lambda_1{}^\mu \tilde{S}'_F \tilde{S}'_F \tilde{K}} \tilde{S}'_F \tilde{S}'_F \Lambda^\nu_{n-2} &= -iL_1 \int \underline{\gamma^\mu \tilde{S}'_F \tilde{S}'_F \tilde{K} \tilde{S}'_F \tilde{S}'_F \Lambda^\nu_{n-2}} \\
&\quad - i\int \gamma^\mu \tilde{S}'_F \tilde{S}'_F \underline{\tilde{K} \tilde{S}'_F \tilde{S}'_F \tilde{K}} \tilde{S}'_F \tilde{S}'_F \Lambda^\nu_{n-2} \qquad (19.82)
\end{aligned}$$

The l_2 subintegration is clearly convergent in the last term of (19.82) since the four underlined factors involved correspond to a $D(l_2) \leq -2$ for any l_2 subintegration. Combining the first term of (19.82) with the first one of (19.81), we find for the remaining suspect terms

$$[\text{Suspect}]_{l_2} = -iL_1 \int \underline{\gamma^\mu \tilde{S}'_F \tilde{S}'_F \Lambda^\nu_{n-1}} - iL_1 \int \underline{\gamma^\mu \tilde{S}'_F \tilde{S}'_F \tilde{K}} \tilde{S}'_F \tilde{S}'_F \Lambda^\nu_{n-2}$$
$$= +iL_1 \int \underline{\gamma^\mu \tilde{S}'_F \tilde{S}'_F \gamma^\nu} L_{n-1}$$

which evidently converges with respect to the l_2 subintegrations after extracting $\Pi_c(q^2)$.

The same technique applies for study of general $l_2 \cdots l_s$ subintegrations. The only terms in the expansion (19.77) for $\Pi^{\mu\nu}$ which can have nonnegative degree of divergence for the $l_2 \cdots l_s$ subintegrations are, in shorthand, and with the aid of (19.75),

$$[\text{Suspect}]_{l_2 \cdots l_s} = \sum_{k=1}^{s-1} i \int \Lambda_k{}^\mu \tilde{S}'_F \tilde{S}'_F \Lambda^\nu_{n-k} + i \sum_{k=0}^{s-1} \int \Lambda_k{}^\mu \tilde{S}'_F \tilde{S}'_F \tilde{K} \tilde{S}'_F \tilde{S}'_F \Lambda^\nu_{n-k-1}$$
$$= -i \sum_{k=1}^{s-1} L_k \int \gamma^\mu \tilde{S}'_F \tilde{S}'_F \Lambda^\nu_{n-k} + i \int \Lambda^\mu_{s-1} \tilde{S}'_F \tilde{S}'_F \tilde{K} \tilde{S}'_F \tilde{S}'_F \Lambda^\nu_{n-s} \tag{19.83}$$

As we did for the l_2 subintegrations, we turn our attention to the last term in (19.83). Again using the iteration (19.75), we express Λ^μ_{s-1} in terms of Λ^μ_{s-2} and L_{s-1}; we then reexpress Λ^μ_{s-2} in terms of Λ^μ_{s-3} and L_{s-2}, and so on, until we obtain the series

$$i \int \underline{\Lambda^\mu_{s-1} \tilde{S}'_F \tilde{S}'_F \tilde{K} \tilde{S}'_F} \tilde{S}'_F \Lambda^\nu_{n-s}$$
$$= i \sum_{k=1}^{s-1} (-)^{s-k} \underline{L_k} \int \underline{\gamma_\mu \tilde{S}'_F \tilde{S}'_F \tilde{K} \tilde{S}'_F \tilde{S}'_F \cdots \tilde{K} \tilde{S}'_F} \tilde{S}'_F \Lambda^\nu_{n-s}$$
$$+ i(-)^{s-1} \int \gamma^\mu \tilde{S}'_F \underline{\tilde{S}'_F \tilde{K} \tilde{S}'_F \tilde{S}'_F \cdots \tilde{K} \tilde{S}'_F} \tilde{S}'_F \Lambda^\nu_{n-s} \tag{19.84}$$

where in (19.84) we have again underlined those terms in the expansion which depend upon[1] the momentum variables $l_2 \cdots l_s$.

The last term in (19.84) evidently has negative degree of divergence and can be dropped. The terms appearing within the summation sign can be simplified by reexpressing, via iteration (19.75), Λ^ν_{n-s} in terms of Λ^ν_{n-s+1}, Λ^ν_{n-s+2}, etc., until all factors of $\tilde{K}$ have been absorbed. All terms involving subtraction constants which will be obtained along the way will have the structure

$$\underline{L_k} \int \underline{\gamma^\mu \tilde{S}'_F \tilde{S}'_F \tilde{K} \cdots \tilde{S}'_F \tilde{S}'_F \gamma^\nu}\, \underline{L_l}$$

[1] In the first term, we understand that L_1 should not be underlined.

The extraction of $\Pi_c(q^2)$ from $\Pi^{\mu\nu}$ decreases the over-all degree of divergence of this integrand to -1, and since the subintegrations within the renormalization constants L_k all have $D \leq 0$, the $l_2 \cdots l_s$ subintegrations in all such terms have negative degree of divergence. Therefore, (19.84) simplifies to

$$i \int \Lambda^\mu_{s-1} \tilde{S}'_F \tilde{S}'_F \tilde{K} \tilde{S}'_F \tilde{S}'_F \Lambda^\nu_{n-s} = i \sum_{k=1}^{s-1} L_k \int \gamma^\mu \tilde{S}'_F \tilde{S}'_F \Lambda^\nu_{n-k}$$
$$+ \text{ terms with } D < 0 \quad (19.85)$$

The terms explicitly written in (19.85) cancel off the objectionable terms in (19.83), and the proof that the $l_2 \cdots l_s$ subintegration converges is thereby completed. The degree of divergence of a general $l_r \cdots l_s$ subintegration, with $1 < r < s < n$, is shown to be negative in just the same way, and we leave this task as an exercise for the reader.

Thus, vertex and self-energy subintegrations of $\Pi(q^2)$ converge and the over-all degree of divergence of $\Pi_c(q^2)$ is negative, $D = -1$. Furthermore, by arguments similar to those used for the vertex, apparently divergent photon-photon scattering subintegrations converge with $D \leq -4$ when gauge-invariant combinations of such subgraphs are considered together. Therefore it follows from Weinberg's theorem that $\Pi_c(q^2)$ is finite to any finite order of e^2. So is $\tilde{D}'_F(q)_{\mu\nu}$ by (19.51e), and our task is finished. By induction, we have shown that $\tilde{\Gamma}_\mu$, $\tilde{S}'_F$, $\tilde{D}'_{F\mu\nu}$, and elements of the S matrix are finite to order e^n provided they are finite to order e^{n-2}. To order e^2 we know they are finite by explicit calculation.

In concluding this section, we emphasize once more that this result implies nothing whatsoever about the convergence of the renormalized perturbation expansion. At high energies, for example, as discussed in Chap. 8 the expansion parameter may be $\alpha \log (E/m)$ and not α, indicating that the expansion may be asymptotic and at best converges at low energies only.

19.12 Example of Fourth-order Charge Renormalization[1]

The workings of the renormalization procedure were illustrated in the lowest-order calculations of Chap. 8. However, we did not encounter there the overlap problem just analyzed, which first crops up in e^4 calculations. As an illustration of how this analysis applies

[1] R. Jost and J. M. Luttinger, *Helv. Phys. Acta,* **23**, 201 (1950).

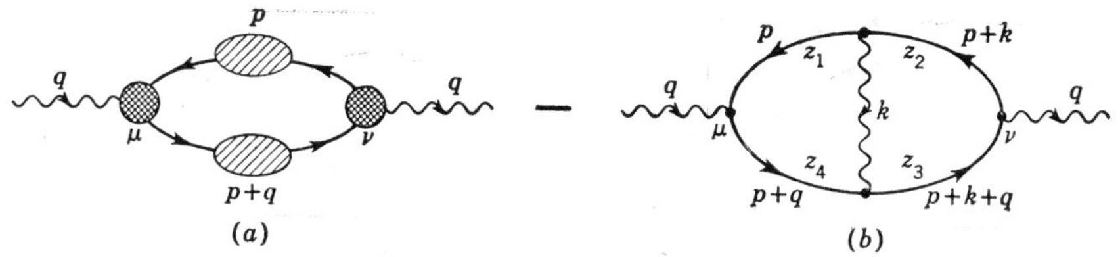

Fig. 19.47 Two terms for $\Pi^{\mu\nu}(q)$ through order e^4 in Eq. (19.86).

in detail, we calculate here explicitly the vacuum polarization and demonstrate that $\Pi_c(q^2)$ is finite through e^2 and therefore, by (19.51*e*), so is $\tilde{D}'_F$ through e^4. Starting from (19.67),

$$\Pi_{\mu\nu}(q) = i\int \tilde{\Gamma}_\mu \tilde{S}'_F \tilde{S}'_F \tilde{\Gamma}_\nu + i\int \tilde{\Gamma}_\mu \tilde{S}'_F \tilde{S}'_F \tilde{K} \tilde{S}'_F \tilde{S}'_F \tilde{\Gamma}_\nu \tag{19.86}$$

we introduce the iteration procedure (19.51) to collect all terms of order e^2. The first term on the right in (19.86) is illustrated in Fig. 19.47*a*, and through e^4 we must include, one at a time, each of the second-order self-energy and vertex insertions shown individually in Fig. 19.48. Actually, the two vertex insertions (a_1) and (a_2) are equal, as are the self-energy insertions (a_3) and (a_4), since the transformation $q \to -q$ and $\mu \leftrightarrow \nu$ leaves $\Pi_{\mu\nu}(q)$ unchanged but takes (a_1) into (a_2), (a_3) into (a_4), and vice versa. We can therefore simply make one vertex and one self-energy insertion and multiply by 2. The second term in (19.86) is already of order e^2 if we replace all $\tilde{\Gamma}_\mu$ and $\tilde{S}'_F$ by γ_μ and S_F, respectively, and approximate $\tilde{K}$ by its lowest-order skeleton as in Fig. 19.47*b*. Collecting and using the expansions (19.51), we have

$$\begin{aligned}\Pi^{(2)}_{\mu\nu}(q) &= 2i\int \gamma_\mu[\tilde{S}'^{(2)}_F - S_F]S_F\gamma_\nu + 2i\int \gamma_\mu S_F S_F[\tilde{\Gamma}_\nu^{(2)} - \gamma_\nu] \\ &\qquad + i\int \gamma_\mu S_F S_F K^{(0)} S_F S_F \gamma_\nu \\ &= 2i\int \gamma_\mu \delta\tilde{S}'^{(2)}_F S_F \gamma_\nu - 2i\int \gamma_\mu S_F S_F \gamma_\nu L^{(2)} \\ &\qquad - i\int \gamma_\mu S_F S_F K^{(0)} S_F S_F \gamma_\nu \end{aligned} \tag{19.87}$$

where $L^{(2)} \equiv 1 - Z_1$ in (19.65) and (19.66) is the second-order contribution to the vertex renormalization and $\delta\tilde{S}'^{(2)}_F \equiv \tilde{S}'^{(2)}_F - S_F$. Using (19.63), the last term of (19.87) is reexpressed in terms of the unrenormalized vertex function to order e^2

$$-\int K^{(0)} S_F S_F \gamma_\nu = \Lambda^{(2)}_\nu$$

and we can rewrite (19.87) in convenient form for calculating—

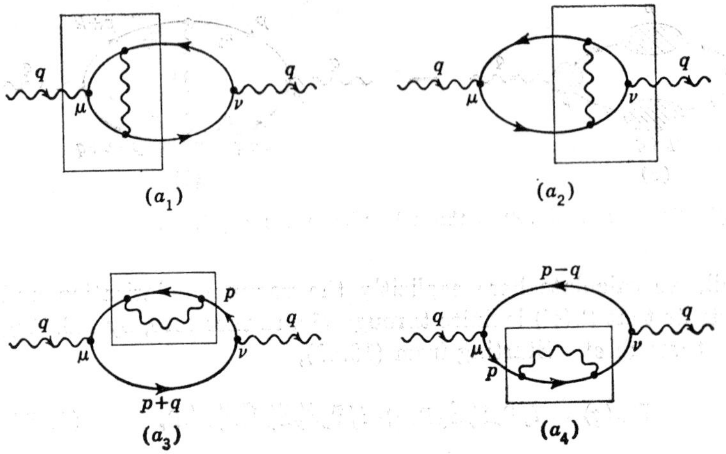

Fig. 19.48 Second-order self-energy and vertex insertions included through order e^4 in Fig. 19.47*a*.

inserting, finally, kinematics from Figs. 19.47 and 19.48

$$\Pi_{\mu\nu}{}^{(2)}(q) = 2i \int \frac{d^4p}{(2\pi)^4} \operatorname{Tr} \gamma_\mu \delta \tilde{S}_F'^{(2)}(p) \gamma_\nu S_F(p+q)$$
$$+ i \int \frac{d^4p}{(2\pi)^4} \operatorname{Tr} \gamma_\mu S_F(p) \Lambda_\nu^{(2)}(p, p+q) S_F(p+q)$$
$$- 2i \int \frac{d^4p}{(2\pi)^4} \operatorname{Tr} \gamma_\mu S_F(p) \gamma_\nu S_F(p+q) L^{(2)} \qquad (19.88)$$

The first term of (19.88) contains the effect of the self-energy insertions. It involves no overlapping divergence, and we expect, and shall verify, that the self-energy renormalization can be carried out and a convergent integral obtained for $\Pi_c(q^2)$ without much difficulty. The combination of the last two terms contains the second-order unrenormalized vertex insertion $\Lambda_\nu^{(2)}$ from which the two possible divergent vertex subintegrations at either vertex of $\Pi_{\mu\nu}$ must be subtracted in order to achieve a convergent contribution to $\Pi_c(q^2)$. Since the overlapping divergences lie in these two terms, the mechanics of doing the integrals is a bit more intricate, but in the end justice triumphs.

We begin with the easier task of calculating the self-energy insertions, Fig. 19.48 (a_3) and (a_4), given by

$$\Pi_{\mu\nu}^{(2a)}(q) = 2i \int \frac{d^4p}{(2\pi)^4} \operatorname{Tr} \gamma_\mu \delta \tilde{S}_F'^{(2)}(p) \gamma_\nu S_F(p+q) \qquad (19.89)$$

In calculating this, the spectral representations for $\delta S_F'^{(2)}(p)$ discussed in Chaps. 16 and 18 will be of use. From (16.122) we recall that the unrenormalized propagator has the spectral form in momentum space

$$S_F'(p) = \frac{Z_2}{\not p - m} + \frac{1}{\pi}\int_{(m+\lambda)^2}^{\infty} \frac{d\sigma^2[\not p\rho_1(\sigma^2) + \rho_2(\sigma^2)]}{p^2 - \sigma^2 + i\epsilon} \qquad (19.90)$$

Upon renormalization this becomes

$$\tilde S_F'(p) = \frac{1}{\not p - m} + \frac{1}{\pi}\int_{(m+\lambda)^2}^{\infty} \frac{d\sigma^2[\not p\tilde\rho_1(\sigma^2) + \tilde\rho_2(\sigma^2)]}{p^2 - \sigma^2 + i\epsilon} \qquad (19.91)$$

with $\tilde\rho \equiv Z_2^{-1}\rho$. The spectral functions are of order e^2, and they can be obtained to this order from the lowest-order calculations[1] in Chap. 8, in particular, Eq. (8.34):

$$\tilde S_F'^{(2)}(p) = \frac{1}{\not p - m} + \frac{1}{\not p - m}\left[-ie^2\int \frac{d^4k}{(2\pi)^4}\frac{\gamma_\alpha(\not p - \not k + m)\gamma^\alpha}{(k^2 - \lambda^2)[(p-k)^2 - m^2]} - \delta m + L^{(2)}(\not p - m)\right]\frac{1}{\not p - m} \qquad (19.92)$$

The mass counterterm in (19.92) was included in the definition of the propagator (19.1) because the physical mass[2] appears in the bare propagators $[\not p - m]^{-1}$. The wave function renormalization constant $L^{(2)}$ in (19.92) appears because the renormalization of $\tilde S_F'^{(2)}$ has been carried out. The momentum integral in (19.92) as carried out in (8.38) leads to the second-order correction

$$\delta\tilde S_F'^{(2)}(p) = \frac{1}{\not p - m}\left[\frac{\alpha}{4\pi}\int_0^1 dz\,\gamma_\mu[\not p(1-z) + m]\gamma^\mu \times \ln\frac{\Lambda^2(1-z)}{m^2z + \lambda^2(1-z) - p^2z(1-z) - i\epsilon} - \delta m + (\not p - m)L^{(2)}\right]\frac{1}{\not p - m} \qquad (19.93)$$

We need only extract the absorptive part of (19.93) to find the spectral functions $\tilde\rho_1$ and $\tilde\rho_2$ to order e^2 which comprise the absorptive part of (19.90):

$$-[\not p\tilde\rho_1(p^2) + \tilde\rho_2(p^2)] = \frac{\alpha}{4\pi}\frac{\not p + m}{p^2 - m^2}\int_0^1 dz\,\gamma_\mu[\not p(1-z) + m]\gamma^\mu \times \pi\theta[p^2z(1-z) - m^2z - \lambda^2(1-z)]\frac{\not p + m}{p^2 - m^2} \qquad (19.94)$$

[1] The photon mass λ is used to displace the continuous spectrum from the discrete one-particle pole contribution to S_F'. This is the treatment of the infrared problem already applied in Chap. 8.

[2] See footnote on p. 286.

with θ the step function

$$\theta(z) = \begin{cases} 1 & z > 0 \\ 0 & z < 0 \end{cases}$$

Inserting these spectral amplitudes into (19.91) and (19.89), we can evaluate the self-energy correction to the vacuum polarization:

$$\Pi^{(2a)}_{\mu\nu}(q) = 2i \int \frac{d^4p}{(2\pi)^4} \frac{1}{\pi} \int_{(m+\lambda)^2}^{\infty} \frac{d\sigma^2}{p^2 - \sigma^2 + i\epsilon} \times \mathrm{Tr}\left[\frac{\gamma_\mu[\not p\tilde\rho_1(\sigma^2) + \tilde\rho_2(\sigma^2)]\gamma_\nu(\not p + \not q + m)}{(p+q)^2 - m^2}\right] \tag{19.95}$$

We are interested in the coefficient $\Pi^{(2a)}(q^2)$ of $q_\mu q_\nu$ in (19.95), and we isolate this with the usual procedure of combining denominators with the aid of Feynman parameters and shifting the origin of the integral in momentum space:

$$\Pi^{(2a)}(q^2) = -\frac{16i}{\pi} \int \frac{d^4p'}{(2\pi)^4} \int_{(m+\lambda)^2}^{\infty} d\sigma^2\, \tilde\rho_1(\sigma^2) \int_0^1 dz\, z(1-z) \times \frac{1}{[p'^2 + q^2 z(1-z) - m^2 z - \sigma^2(1-z)]^2} \tag{19.96}$$

The d^4p' integral diverges logarithmically, reflecting the effective degree of divergence $D = 0$ for the vacuum polarization. However, upon renormalization of $\Pi^{\mu\nu}(q^2)$ we need compute only $\Pi_c(q^2) = \Pi(q^2) - \Pi(0)$ and with this subtraction the momentum integral converges to

$$\Pi_c^{(2a)}(q^2) = \frac{1}{\pi^3} \int d\sigma^2\, \tilde\rho_1(\sigma^2) \int_0^1 dz\, z(1-z) \times \ln\left[\frac{m^2 z + \sigma^2(1-z)}{m^2 z + \sigma^2(1-z) - q^2 z(1-z)}\right] \tag{19.97}$$

This expression for the self-energy insertion is now seen to be finite since, for $\sigma^2 \to \infty$, the logarithm behaves as $1/\sigma^2$. So does $\tilde\rho_1(\sigma^2)$ according to (19.94), and therefore the spectral integral converges. Equation (19.97) can now be integrated in terms of elementary functions in the limit $-q^2 \gg m^2$. Its contribution is found as

$$\Pi_c^{(2a)}(q^2) = -\frac{\alpha}{48\pi^3} \ln^2\left(\frac{-q^2}{m^2}\right) + \frac{\alpha}{6\pi^3} \ln\left(\frac{-q^2}{m^2}\right)\left[\ln\left(\frac{m}{\lambda}\right) - \frac{13}{24}\right] + \text{terms of order unity} \qquad \text{for } \left|\frac{q^2}{m^2}\right| \gg 1 \tag{19.98}$$

There was nothing very involved in the demonstration that the self-energy insertion leads to a convergent integral for $\Pi_c(q^2)$ because no overlapping divergences appeared. We could first carry out the self-energy renormalization and then insert the result into the vacuum polarization integral. This simplicity is not shared by the second and third terms of (19.88) arising from vertex insertions. The overlapping divergence in

$$\Pi_{\mu\nu}^{(2b)}(q) \equiv i \int \frac{d^4p}{(2\pi)^4} \operatorname{Tr} \gamma_\mu S_F(p)\Lambda_\nu^{(2)}(p,\, p+q)S_F(p+q) \tag{19.99}$$

is removed by the two possible divergent vertex subintegrations in

$$2\Pi_{\mu\nu}^{(2c)}(q) \equiv 2i \int \frac{d^4p}{(2\pi)^4} \operatorname{Tr} \gamma_\mu S_F(p)\gamma_\nu S_F(p+q)L^{(2)} \tag{19.100}$$

which must be subtracted according to (19.88). It is their difference, after the over-all renormalization is carried out by subtracting at $q^2 = 0$ to form $\Pi_c^{(2b)}(q^2) - 2\Pi_c^{(2c)}(q^2)$, which should be given by a convergent integral. In order to exhibit this cancellation explicitly, this calculation is given now in some detail.

According to the Feynman rules, $\Pi_{\mu\nu}^{(2b)}(q)$ is given by the expression

$$\Pi_{\mu\nu}^{(2b)}(q) = e^2 \int \frac{d^4k\, d^4p}{(2\pi)^8(k^2 - \lambda^2)} \times \operatorname{Tr} \frac{[\gamma_\mu(\not p + m)\gamma_\alpha(\not p + \not k + m)\gamma_\nu(\not p + \not k + \not q + m)\gamma^\alpha(\not p + \not q + m)]}{(p^2 - m^2)[(p+k)^2 - m^2][(p+k+q)^2 - m^2][(p+q)^2 - m^2]} \tag{19.101}$$

with the kinematics as illustrated in Fig. 19.47*b*. Our immediate task is to remove the ambiguous quadratic divergence in (19.101) by extracting the coefficient of the $q_\mu q_\nu$ term, $\Pi^{(2b)}(q^2)$. To do this, we combine the four denominators associated with the charged-particle loop, using the identity

$$\frac{1}{a_1a_2a_3a_4} = 3! \int_0^1 dz_1 \cdots dz_4 \frac{\delta\left(1 - \sum_{i=1}^4 z_i\right)}{\left(\sum_{i=1}^4 a_iz_i\right)} \tag{19.102}$$

where the a_i are identified with the four electron denominators in the order given and shown in Fig. 19.47*b*. The p integral is then carried out by shifting the origin of integration by[1]

$$p' = p + k(z_2 + z_3) + q(z_3 + z_4)$$

[1] For a well-defined mathematical expression which is finite one should regulate the momentum integrals. A gauge-invariant method for doing the momentum integrations was given in the companion volume [see Eq. (8.20)] in the second-order calculation and may be repeated here.

in order to complete the square and remove odd terms in p' from the denominator (19.102). Terms in the numerator trace can then be collected and classified by powers of p'. Those with odd powers of p' integrate to zero, and those with four powers of p' evidently lead to a contribution proportional to $g_{\mu\nu}$ and may be dropped; this is the only nonconvergent part of the p' integral. The terms quadratic in p' lead to a simple trace $\propto p'^2(k_\mu k_\nu - q_\mu q_\nu)$ plus inessential terms $\propto g_{\mu\nu}$, after considerable algebra[1] (!). Temporarily leaving the numerator terms of (19.101) that are independent of p' still in trace form and carrying out $\int d^4p'$, we find

$$\Pi_{\mu\nu}{}^{(2b)}(q) = \frac{i\alpha}{4\pi}\int\frac{d^4k}{(2\pi)^4}\int\frac{dz_1\cdots dz_4\delta(1-\sum_i z_i)}{k^2-\lambda^2} \times\left\{\frac{16(k_\mu k_\nu - q_\mu q_\nu)}{D_1}+\frac{N_{\mu\nu}}{D_1{}^2}\right\} + g_{\mu\nu}\text{ terms} \qquad (19.103)$$

where

$$D_1 = k^2(z_1+z_4)(z_2+z_3) + q^2(z_1+z_2)(z_3+z_4) + 2k\cdot q(z_1z_3 - z_2z_4) - m^2 + i\epsilon$$

and

$$N_{\mu\nu} = \operatorname{Tr}\gamma_\mu[-\not{k}(z_2+z_3) - \not{q}(z_3+z_4) + m]\gamma_\alpha \times[\not{k}(z_1+z_4) - \not{q}(z_3+z_4)+m]\gamma_\nu[\not{k}(z_1+z_4)+\not{q}(z_1+z_2)+m]\gamma^\alpha \times[-\not{k}(z_2+z_3)+\not{q}(z_1+z_2)+m] \qquad (19.104)$$

The next step is to carry out the k integration, which we expedite as follows:

1. Before the two remaining denominators are combined, the k^2 in the photon propagator is provided the same factor as in D_1 by writing

$$\frac{1}{k^2-\lambda^2} = \frac{(z_1+z_4)(z_2+z_3)}{(z_1+z_4)(z_2+z_3)(k^2-\lambda^2)}$$

This leads to a neater form when we parametrize analogously to (19.102), associating Feynman parameter x with D_1 and $y = 1 - x$ with the photon propagator.

2. The photon propagator is regulated by

$$\frac{1}{k^2-\lambda^2} \to \frac{1}{k^2-\lambda^2} - \frac{1}{k^2-\Lambda^2} \qquad (19.105)$$

[1] Terms antisymmetric under the interchange $(z_1, z_2) \leftrightarrow (z_3, z_4)$ may be dropped.

so as to render finite the logarithmically divergent contributions in (19.103) coming from terms proportional to k^2/D_1 and k^4/D_1^2. These cutoff-dependent terms drop out in the end when we make the over-all subtraction to form $\Pi_c(q^2)$.

3. The origin of integration in k space is shifted in order to complete the square in the denominator by writing

$$k' = k + \frac{qx(z_1z_3 - z_2z_4)}{(z_1 + z_4)(z_2 + z_3)}$$

After the k' integration, we are left with three distinct terms: $\Pi_{\mu\nu}{}^{(2b_1)}$ comes from the first term of (19.103) proportional to D_1^{-1}; $\Pi_{\mu\nu}{}^{(2b_2)}$ and $\Pi_{\mu\nu}{}^{(2b_3)}$ come from the D_1^{-2} term and correspond to contributions from terms in $N_{\mu\nu}$ independent of k' and proportional to $(k')^2$, respectively. The $(k')^4$ term in $N_{\mu\nu}$ gives rise to only an inessential $g_{\mu\nu}$ contribution.

The coefficient of $q_\mu q_\nu$ for the first term is given by

$$\Pi^{(2b_1)}(q^2) = \frac{\alpha}{4\pi^3}\int_0^1 dx \int_0^1 \frac{dz_1 \cdots dz_4\, \delta\left(1 - \sum_i z_i\right)}{(z_1 + z_4)(z_2 + z_3)} \times \left[1 - \frac{x^2(z_1z_3 - z_2z_4)^2}{(z_1 + z_4)^2(z_2 + z_3)^2}\right] \ln \frac{D_2(\Lambda^2,q^2)}{D_2(\lambda^2,q^2)} \tag{19.106}$$

where

$$D_2(\lambda^2,q^2) = \lambda^2(1 - x)(z_1 + z_4)(z_2 + z_3) + m^2x - q^2x\left[(z_1 + z_2)(z_3 + z_4) - \frac{x(z_1z_3 - z_2z_4)^2}{(z_1 + z_4)(z_2 + z_3)}\right] \tag{19.107}$$

The parameter integrals give no divergence trouble in (19.106). When $z_1 + z_4 \to 0$, $z_2 + z_3 \to 1$ and in the region of small z_1 and z_4 the integral

$$\int_0^\epsilon dz_1 \int_0^\epsilon dz_4 \frac{1}{z_1 + z_4} \sim 2\epsilon \ln 2 \tag{19.108}$$

converges. With more powers of the z_i in the denominator there arise troubles, as we shall see later. With one subtraction

$$\Pi_c{}^{(2b_1)}(q^2) = \Pi^{(2b_1)}(q^2) - \Pi^{(2b_1)}(0)$$

converges and may be evaluated; this is a challenge to the energetic student. Here we give the asymptotic limit for $-q^2 \gg m^2$,

$$\Pi_c{}^{(2b_1)}(q^2) = \frac{17\alpha}{72\pi^3} \ln\left(\frac{m^2}{-q^2}\right) + 0(1) \tag{19.109}$$

This brings us next to $\Pi^{(2b_2)}$, which arises from that part of $N_{\mu\nu}$ in (19.104) independent of k'. The d^4k' integral here behaves as $\sim d^4k'/k'^6$ for large k' and is manifestly convergent even without photon regularization. We find for the coefficient of $q_\mu q_\nu$

$$\Pi^{(2b_2)}(q^2) = \frac{\alpha}{2\pi^3}\int_0^1 x\,dx \int \frac{dz_1 \cdots dz_4\,\delta\left(1 - \sum_i z_i\right) m^2}{(z_1 + z_4)(z_2 + z_3)D_2(\lambda^2,q^2)} \times \left[(z_1 + z_2)^2 - \frac{x^2(z_1z_3 - z_2z_4)^2(1 + z_2 + z_3)}{(z_1 + z_4)(z_2 + z_3)^2} \right] \tag{19.110}$$

with $D_2(\lambda^2,q^2)$ given by (19.107). Evaluating the integrals in (19.110) as $-q^2 \to \infty$, we find

$$\Pi^{(2b_2)}(q^2) \sim \frac{m^2}{q^2} \ln \frac{-q^2}{m^2} \tag{19.111}$$

This leaves only the piece $\Pi_{\mu\nu}{}^{(2b_3)}$ coming from the $(k')^2$ terms in $N_{\mu\nu}$ to evaluate. Not only does this part of (19.103) and (19.104) lead to a divergent logarithm upon $\int d^4k'$ but, by default, it must harbor the overlapping divergence and converge only after the vertex renormalization counterterms (19.100) have been included. After taking the traces (!!) and carrying out the k' integral, with the regulator (19.105), we obtain for this contribution to the coefficient of $q_\mu q_\nu$ in (19.103)

$$\Pi^{(2b_3)}(q^2) = +\frac{\alpha}{4\pi^3}\int_0^1 x\,dx \int_0^1 \frac{dz_1 \cdots dz_4\,\delta\left(1 - \sum_i z_i\right)}{(z_1 + z_4)^2(z_2 + z_3)^2} \times (z_1 - z_3)(z_4 - z_2) \ln \frac{D_2(\Lambda^2,q^2)}{D_2(\lambda^2,q^2)} \tag{19.112}$$

in terms of D_2 in (19.107). A subtraction at $q^2 = 0$ removes the Λ^2 dependence in (19.112) but introduces in its place a divergence in the parameter subintegrations—this is precisely the overlap difficulty. Thus in (19.112) for $z_2,z_3 \to 0$ (and $z_1 + z_4 \approx 1$ by the δ-function constraint) we have the behavior

$$\int_0^1 dz_1\,dz_4\,\delta(1 - z_1 - z_4)z_1z_4 \int_0^\epsilon \frac{dz_2\,dz_3}{(z_2 + z_3)^2} \times \ln \frac{m^2x - q^2xz_1z_4 + \Lambda^2(1 - x)(z_2 + z_3)}{m^2x - q^2xz_1z_4 + \lambda^2(1 - x)(z_2 + z_3)} \tag{19.113}$$

which, for finite λ and Λ, is convergent as in (19.106). However, the subtraction at $q^2 = 0$ to form the cutoff-independent $\Pi_c{}^{(2b_3)}(q^2)$ changes the logarithm to $\ln [D_2(\lambda^2,0)/D_2(\lambda^2,q^2)]$ in (19.112). Now as $z_2,z_3 \to 0$

$$\ln \frac{D_2(\lambda^2,0)}{D_2(\lambda^2,q^2)}\bigg|_{z_2,z_3\to 0} \approx \ln \frac{m^2}{m^2 - q^2 z_1 z_4}$$

and the divergent factor $\int_0^\epsilon \frac{dz_2}{z_2}$ is introduced in (19.113).

This divergence trouble can be traced back to the existence of the divergent subintegrations in the vertex insertions. Referring back to Fig. 19.47*b* and recalling the electrical circuit analogy developed in Chap. 18, we see that for z_2 and $z_3 \to 0$ the resistances in the two lines with momenta $p + k$ and $p + k + q$ become vanishingly small. This means that a very large amount of current flows around the loop associated with the vertex insertion at ν. The logarithmic divergence in $\Pi_c^{(2b_3)}(q^2)$ arising from the region $(z_2,z_3) \approx 0$ is evidently related to the subintegration associated with this loop. A similar divergence at $(z_1,z_4) \approx 0$ is associated with the vertex subintegration at μ. These divergences are precisely removed by the two subtractions (19.100) required by the renormalization program.

To exhibit these cancellations, we recalculate the counterterms (19.100). $\Pi^{(2c)}(q^2)$ was actually evaluated in Chap. 8, but we repeat its calculation here in a manner which mimics that of $\Pi^{(2b_3)}(q^2)$ as closely as possible. The cancellations with (19.112) can then be shown before having to perform all the parameter integrals.

Into (19.100) we insert the definition of $L^{(2)}\gamma_\nu$ from (19.65), using Feynman rules,

$$\begin{aligned} L^{(2)}\gamma_\nu &= \Lambda_\nu^{(2)}(p,p)\bigg|_{\not p = m} \\ &= -ie^2 \int \frac{d^4k}{(2\pi)^4(k^2 - \lambda^2)} \left[\frac{\gamma_\alpha(\tilde{\not p} + \not k + m)\gamma_\nu(\tilde{\not p} + \not k + m)\gamma^\alpha}{[(\tilde p + k)^2 - m^2]^2}\right]\bigg|_{\tilde{\not p} = m} \end{aligned}$$

The kinematics is as illustrated for the ν vertex in Fig. 19.47*b*. The quantity $\tilde{\not p}$ is not an integration variable, but $\tilde p^2 = m^2$ and $\tilde{\not p} = m$ when standing to the right or left in the bracket [ ] because the vertex subtraction occurs at the mass shell. This gives[1]

$$\begin{aligned} \Pi_{\mu\nu}^{(2c)}(q) &= e^2 \int \frac{d^4k\, d^4p}{(2\pi)^8(k^2 - \lambda^2)} \operatorname{Tr} \gamma_\mu \frac{\not p + m}{p^2 - m^2} \\ &\times \left[\frac{\gamma_\alpha(\tilde{\not p} + \not k + m)\gamma_\nu(\tilde{\not p} + \not k + m)\gamma^\alpha}{[(\tilde p + k)^2 - m^2]^2}\right]\bigg|_{\tilde{\not p} = m} \frac{\not p + \not q + m}{(p + q)^2 - m^2} \end{aligned} \qquad (19.114)$$

[1] It is not wise to tamper with the part of the numerator within the square bracket until it is replaced by a constant times γ_ν.

Despite the fact that $\tilde{p}$ is not an integration variable, we combine the four charged particle denominators just as in (19.101) and (19.102) in order to stay as close as possible to the old result for $\Pi_{\mu\nu}^{(2b)}$. After a shift of origin

$$p' = p + \frac{qz_4}{z_1 + z_4}$$

to complete the square in the denominator, the p' integration can be done:

$$\Pi_{\mu\nu}^{(2c)}(q) = -\frac{i\alpha}{2\pi}\int \frac{d^4k}{(2\pi)^4}\int_0^1 \frac{dz_1 \cdots dz_4\, \delta\left(1 - \sum_i z_i\right) z_1 z_4}{(k^2 - \lambda^2)(z_1 + z_4)^4}$$
$$\times\, q_\mu \operatorname{Tr} \not{q} \left[\frac{\gamma_\alpha(\tilde{\not{p}} + \not{k} + m)\gamma_\nu(\tilde{\not{p}} + \not{k} + m)\gamma^\alpha}{[(\tilde{p} + k)^2(z_2 + z_3) + q^2 z_1 z_4/(z_1 + z_4) - m^2]^2}\right]_{\tilde{\not{p}} = m} + g_{\mu\nu} \text{ terms} \qquad (19.115)$$

To carry out the k integration, we rewrite the photon propagator as

$$\frac{1}{k^2 - \lambda^2} = \frac{z_2 + z_3}{(z_2 + z_3)(k^2 - \lambda^2)}$$

and combine with the denominator of (19.115) using the parameter x. After shifting the origin in k space

$$k' = k + \tilde{p}x$$

doing the integral d^4k', and reducing the trace, we are left with two terms, $\Pi^{(2c_1)}(q^2)$ and $\Pi^{(2c_2)}(q^2)$, as coefficients of $q_\mu q_\nu$ in $\Pi_{\mu\nu}^{(2c)}(q)$.

$\Pi^{(2c_1)}(q^2)$ comes from the part of the numerator of (19.115) independent of k' and is given by

$$\Pi^{(2c_1)}(q^2) = -\frac{\alpha}{4\pi^3}\int_0^1 x\,dx \int_0^1 \frac{dz_1 \cdots dz_4\, \delta(1 - \Sigma z_i) m^2 z_1 z_4}{(z_1 + z_4)^4(z_2 + z_3)}$$
$$\times \frac{2 - 2x - x^2}{m^2x - m^2x(1 - x)(z_2 + z_3) - q^2 x z_1 z_4/(z_1 + z_4) + \lambda^2(1 - x)(z_2 + z_3)} \qquad (19.116)$$

This integral diverges with respect to the z_1z_4 subintegration,

$$\int_0^\epsilon dz_1 \int_0^\epsilon dz_4 \frac{z_1 z_4}{(z_1 + z_4)^4} > \int_0^\epsilon \frac{dz_1}{z_1}\int_0^1 \frac{y\,dy}{(1 + y)^4}$$

and is rendered convergent by the subtraction at $q^2 = 0$. According to the circuit analogy and Fig. 19.47*b*, we associate this divergence with the vacuum polarization subintegration multiplying a finite part of the $L^{(2)}$ insertion at vertex ν. Although it is ultraviolet convergent, $\Pi_c^{(2c_1)}(q^2)$ is infrared divergent owing to the infrared divergence in the vertex renormalization constant $L^{(2)}$ itself. In the limit $-q^2 \gg m$ we evaluate (19.116) [multiplied by -2 as in (19.88)] to give

$$-2\Pi_c^{(2c_1)}(q^2) = +\frac{\alpha}{6\pi^3}\ln\left(\frac{m^2}{-q^2}\right)\left[\ln\left(\frac{m}{\lambda}\right) - \frac{5}{4}\right] + 0(1) \qquad (19.117)$$

$\Pi^{(2c_2)}(q^2)$ comes from the other part of the numerator in (19.115) proportional to $(k')^2$ and is logarithmically divergent. Using the same regulator prescription (19.105) as before, we find

$$\Pi^{(2c_2)}(q^2) = \frac{\alpha}{4\pi^3}\int_0^1 x\,dx\int_0^1 \frac{dz_1\cdots dz_4\,\delta\left(1-\sum_i z_i\right)z_1z_4}{(z_1+z_4)^4(z_2+z_3)^2}$$
$$\times \ln\frac{\Lambda^2(1-x)(z_2+z_3)+m^2x-m^2x(1-x)(z_2+z_3)-q^2\,xz_1z_4/(z_1+z_4)}{\lambda^2(1-x)(z_2+z_3)+m^2x-m^2x(1-x)(z_2+z_3)-q^2\,xz_1z_4/(z_1+z_4)} \qquad (19.118)$$

This diverges even after subtraction of $\Pi^{(2c_2)}(0)$ to form $\Pi_c^{(2c_2)}(q^2)$ as wanted for the finite part of the vacuum polarization. The divergence in (19.118) is seen in the z_2,z_3 subintegration as $z_2,z_3 \to 0$ and is traced in (19.115) to the divergence in the vertex renormalization constant $L^{(2)}$ multiplying the vacuum polarization integral, which the subtraction at $q^2 = 0$ has rendered finite.

Precisely the same divergence was found to occur in (19.112) after the subtraction at $q^2 = 0$ and was traced there via the electrical circuit analogy to the same origin. It cancels when we form the difference between (19.112) and (19.118) as required in (19.88) by the renormalization program. After making the subtraction at $q^2 = 0$, we have

$$\Pi_c^{(2b_3)}(q^2) - 2\Pi_c^{(2c_2)}(q^2) = \frac{\alpha}{4\pi^3}\int_0^1 x\,dx\int_0^1 \frac{dz_1\cdots dz_4\,\delta\left(1-\sum_i z_i\right)}{(z_1+z_4)^2(z_2+z_3)^2}$$
$$\left[(z_1-z_3)(z_4-z_2)\times\ln\frac{m^2}{m^2-q^2\left[(z_1+z_2)(z_3+z_4)-\dfrac{x(z_1z_3-z_2z_4)^2}{(z_1+z_4)(z_2+z_3)}\right]}\right.$$
$$-\frac{z_1z_4}{(z_1+z_4)^2}\ln\frac{m^2[1-(1-x)(z_2+z_3)]}{m^2[1-(1-x)(z_2+z_3)]-q^2\dfrac{z_1z_4}{z_1+z_4}}$$
$$\left.-\frac{z_2z_3}{(z_2+z_3)^2}\ln\frac{m^2[1-(1-x)(z_1+z_4)]}{m^2[1-(1-x)(z_1+z_4)]-q^2\dfrac{z_2z_3}{z_2+z_3}}\right] \qquad (19.119)$$

where the photon mass λ is neglected because there are no infrared difficulties and we have set $\Lambda \to \infty$ because this combination is independent of Λ. Also we have symmetrized the subtraction term $\Pi_c^{(2c_2)}(q^2)$ by the interchange $z_1,z_4 \leftrightarrow z_2,z_3$ corresponding to the vertex insertion being subtracted once at each of the two vertices of the vacuum polarization. Equation (19.119) is now finite and completely free of overlapping divergences; for $z_2,z_3 \to 0$ and $z_1 + z_4 \to 1$ the divergences in the first two terms cancel since

$$\int_0^\epsilon dz_2 \int_0^\epsilon dz_3 \frac{1}{(z_2+z_3)^2}\left[z_1z_4 \ln \frac{m^2}{m^2 - q^2z_1z_4} + 0(z_2,z_3) \right.$$
$$\left. - z_1z_4 \ln \frac{m^2}{m^2 - q^2z_1z_4} + 0(z_2,z_3)\right] \sim \epsilon \qquad \text{as} \qquad \epsilon \to 0$$

and the third term is finite. Similarly, for $z_1,z_4 \to 0$, $z_2 + z_3 \to 1$, the first and third terms cancel divergent parts and the second is finite. These two subtraction terms have therefore accomplished the desired cancellation of the divergent parts of the vertex insertions buried in the factor $\Pi_c^{(2b_3)}$. The over-all subtraction at $q^2 = 0$ has rendered the vacuum polarization finite. What we have seen here in some detail is an example of the inevitable success of the renormalization program as well as the apparently unavoidable toil in implementing it.

To actually evaluate (19.119), we introduce the very helpful variables

$$z_1 + z_4 = z \qquad z_1 = zu \qquad z_2 = (1-z)v \qquad 0 \le z,\, u,\, v \le 1$$

and find in the limit $-q^2 \gg m^2$

$$\Pi_c^{(2b_3)}(q^2) - 2\Pi_c^{(2c_2)}(q^2) = \frac{\alpha}{48\pi^3}\ln^2\left(\frac{-q^2}{m^2}\right)$$
$$+ \frac{\alpha}{18\pi^3}\ln\left(\frac{-q^2}{m^2}\right) + 0(1) \qquad (19.120)$$

Adding up all the pieces (19.98), (19.109), (19.111), (19.117), and (19.120) we have the complete fourth-order vacuum polarization in the limit $-q^2 \gg m^2$,

$$\Pi_c^{(2)}(q^2) = -\frac{\alpha}{16\pi^3}\ln\left(\frac{-q^2}{m^2}\right) + 0(1) \qquad (19.121)$$

The $\ln^2(-q^2/m^2)$ terms have canceled as well as the infrared ones between $\Pi^{(2a)}$ for the self-energy insertions and $\Pi^{(2b)}$ and $\Pi^{(2c)}$ for the vertex parts. Introducing (19.121) into the equation for the photon propagator (19.51*e*) along with the second-order vacuum polarization [see

Eq. (8.29)] we obtain through e^4 (and dropping gauge terms)

$$i\tilde{D}'_F(q^2)_{\mu\nu} = -\frac{ig_{\mu\nu}}{q^2}\left[\frac{1}{1-\frac{\alpha}{3\pi}\ln\left(\frac{-q^2}{m^2}\right)-\frac{\alpha^2}{4\pi^2}\ln\left(\frac{-q^2}{m^2}\right)+\cdots}\right] \qquad -q^2 \gg m^2 \tag{19.122}$$

This result was first derived by Jost and Luttinger[1] in 1950. Both correction terms have the same sign and tend to *increase* the force between particles at small distances. The same general conclusion was obtained to all orders in α from the spectral representation constructed in Chap. 16 for $\tilde{D}'_F(q^2)$. It follows from the positive definiteness of the weight function in (16.172) and (16.173).

19.13 Low-energy Theorem for Compton Scattering

The Ward identity and the renormalization program combine to yield a theorem on the low-energy behavior of the Compton scattering amplitude that is valid to all orders in e^2. As proved first by Thirring[2] in 1950, the amplitude for scattering of a photon from an electron becomes the classical Thomson limit as the photon energy approaches zero. In this limit its magnitude is α/m, where $\alpha = 1/137$ and m are without approximation, simply the renormalized charge and mass of the electron. With second-order perturbation theory we computed this result as an approximation in Chap. 7 [see Eq. (7.74)].

The proof of this theorem is based on the observation from Ward's identity that the complete vertex function for a zero-energy photon $\Gamma_\mu(p,p)$ can be made from the exact inverse propagator $S_F'^{-1}(p)$ by a differentiation:

$$\Gamma_\mu(p,p) = \frac{\partial}{\partial p^\mu} S_F'^{-1}(p) \tag{19.123}$$

Thus a zero-energy photon may be inserted into a charged line by differentiation with respect to a momentum flowing in that charged line.

On grounds of gauge invariance alone we can understand this result simply. The τ function for a single free charged electron is

$$\tau(p) \equiv iS'_F(p) \tag{19.124}$$

[1] Jost and Luttinger, *op. cit.*
[2] W. Thirring, *Phil. Mag.*, **41**, 1193 (1950).

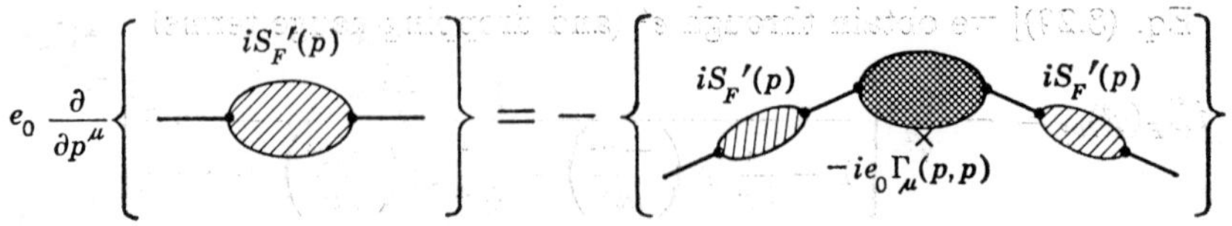

Fig. 19.49 Insertion of a zero-energy photon by differentiation illustrating the Ward identity.

We may compute from this the τ function for a charged particle with bare charge e_0 propagating in an external *constant* electromagnetic potential A_μ by making the gauge-invariant substitution

$$p_\mu \to p_\mu - e_0 A_\mu$$

This changes $\tau(p)$ to

$$\tau(p) \to \tau(p - e_0 A) = \tau(p) - e_0 A^\mu \frac{\partial}{\partial p^\mu} \tau(p) + \frac{1}{2} e_0^2 A^\mu A^\nu \frac{\partial^2}{\partial p^\mu \partial p^\nu} \tau(p) + \cdots \tag{19.125}$$

The coefficient of A^μ

$$-e_0 \frac{\partial}{\partial p^\mu} \tau(p) = ie_0 S_F'(p) \left[\frac{\partial}{\partial p^\mu} S_F'^{-1}(p) \right] S_F'(p) = iS_F'(p) e_0 \Gamma_\mu(p,p) S_F'(p) \tag{19.126}$$

is aside from a factor e_0 just the improper vertex function at zero energy, with self-energy parts standing on its electron legs but with the photon legs removed as in Fig. 19.49. The coefficient[1] of $\frac{1}{2}A_\mu A_\nu$ is the τ function for Compton scattering, representing a second-order interaction with two external lines for the two zero-energy photons before and after the scattering. The electron legs are included as usual in the definition of the τ function, but the photon legs are not, as illustrated in Fig. 19.50. We write then

$$\tau^{(c)}(p)^{\mu\nu} = ie_0^2 \frac{\partial^2}{\partial p_\mu \partial p_\nu} S_F'(p) \tag{19.127}$$

The graphical correspondence with the Compton amplitude is best seen by introducing the identity from (19.48),

$$\frac{\partial^2}{\partial p_\mu \partial p_\nu} S_F'^{-1}(p) = - \frac{\partial^2}{\partial p_\mu \partial p_\nu} \Sigma\,(p)$$

[1] The factor $\frac{1}{2}$ is included in this definition to compensate for the two possible ways of associating the two external photon lines with the factor $A_\mu A_\nu \to A_\mu^{(+)} A_\nu^{(-)} + A_\nu^{(+)} A_\mu^{(-)}$.

(19.123), and

$$\frac{\partial}{\partial p_\mu} S'_F(p) = -S'_F(p)\left[\frac{\partial}{\partial p_\mu} S'^{-1}_F(p)\right] S'_F(p)$$

into (19.127)

$$\tau^{(c)}(p)^{\mu\nu} = +[iS'_F(p)]\Big[(-ie_0)\Gamma^\mu(p,p)iS'_F(p)(-ie_0)\Gamma^\nu(p,p)$$

$$+ (-ie_0)\Gamma^\nu(p,p)iS'_F(p)(-ie_0)\Gamma^\mu(p,p) - ie_0{}^2\frac{\partial^2\Sigma}{\partial p_\mu \partial p_\nu}\Big][iS'_F(p)] \quad (19.128)$$

The three terms in (19.128) represent the sum of all graphs with two zero-energy photon insertions into the continuous propagator line bearing the charge and momentum p_μ of the external particle. Propagators along this line have the form $(\not p + \not k_i - m)^{-1}$, where k_i changes after each interaction with an internal photon within the

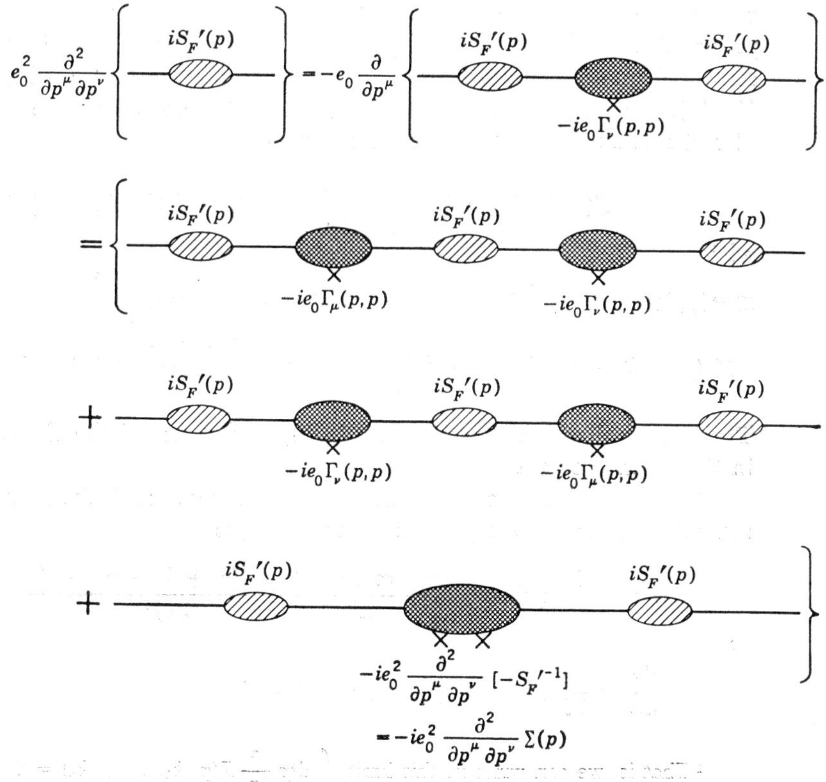

Fig. 19.50 A second differentiation to give the zero-energy Compton limit.

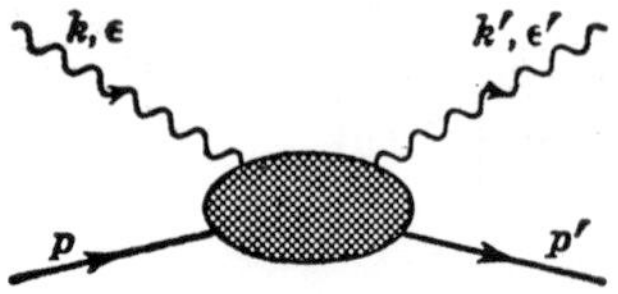

Fig. 19.51 Kinematics for Compton scattering.

self-energy blob. A differentiation $\partial/\partial p_\mu$ inserts a zero-energy photon everywhere along this line according to

$$e_0 \frac{\partial}{\partial p_\mu} \frac{i}{\not p + \not k_i - m} = - \frac{i}{\not p + \not k_i - m} (-ie_0\gamma^\mu) \frac{i}{\not p + \not k_i - m} \tag{19.129}$$

Insertions of a zero-energy photon into a closed loop vanish for reasons already discussed[1] in Sec. 19.10. Gauge invariance requires that the amplitude for a closed loop must be proportional to the field strength $\sim(q_\mu \epsilon_\nu - q_\nu \epsilon_\mu)$, and this vanishes at $q_\mu = 0$. Indeed then (19.127) contains the totality of Compton graphs at zero energy and we construct the S-matrix element by removing the external electron propagators and multiplying by wave functions for the incident and scattered photons and electrons according to our instructions in, say, (17.43). With the kinematics in Fig. 19.51 this gives for the electron at rest and as the photon momenta $k,k' \to 0$

$$S(p',k',\epsilon';p,k,\epsilon) = \frac{\epsilon_\mu \epsilon'_\nu}{(2\pi)^6 \sqrt{4kk'}} \mathfrak{M}^{\mu\nu}(p',k';p,k) \tag{19.130}$$

with

$$\mathfrak{M}^{\mu\nu}(p,0;p,0) = \frac{-ie_0^2}{Z_2 Z_3^{-1}} \lim_{k\to 0} \bar u(p)(\not p + \not k - m) \frac{\partial^2 S'_F(p+k)}{\partial p_\mu \partial p_\nu}$$
$$\times (\not p + \not k - m)u(p) = -ie^2 \lim_{k\to 0} \bar u(p)\, \not k \frac{\partial^2 \tilde S'_F(p+k)}{\partial p_\mu \partial p_\nu} \not k u(p) \tag{19.131}$$

The renormalization prescription (19.49) has absorbed the Z_2 and Z_3 in the second form above.

The differential cross section may be computed from this S matrix and, for a given photon polarization, is

$$d\sigma = \frac{1}{2k} |\epsilon_\mu \epsilon'_\nu \mathfrak{M}^{\mu\nu}|^2 \frac{d^3k'}{2k'} \frac{m\, d^3p'}{E'} \frac{(2\pi)^4 \delta^4(p+k-p'-k')}{(2\pi)^6}$$

or

$$\frac{d\sigma}{d\Omega} = \frac{1}{16\pi^2} |\epsilon_\mu \epsilon'_\nu \mathfrak{M}^{\mu\nu}|^2 \qquad \text{as } k,\, k' \to 0 \tag{19.132}$$

[1] That is, we can write in this limit $\int d^4p \frac{\partial}{\partial p_\mu} F(p, k_1, \ldots, k_n) = 0$, where F is the closed-loop contribution.

Evidently $d\sigma/d\Omega$ is finite at zero energy if $\mathfrak{M}^{\mu\nu}$ is, and in (19.131) we must evaluate terms of order k^{-2} in $\partial^2 \tilde{S}'_F(p+k)/\partial p_\mu \partial p_\nu$. We write the general form of the renormalized propagator as

$$\tilde{S}'_F(p') = \frac{p\!\!\!/' A(p'^2) + mB(p'^2)}{p'^2 - m^2} \qquad \text{with } A(m^2) = B(m^2) = 1 \quad (19.133)$$

according to the renormalization conditions. It simplifies the calculation of the second derivative of (19.133) greatly if we first choose the transverse gauge in the electron rest system. Then

$$\epsilon \cdot p = 0 \qquad \epsilon' \cdot p = 0 \tag{19.134}$$

and all terms proportional to p_μ or p_ν need not be computed. Carrying out the differentiation, we find the structure

$$\frac{\partial^2 \tilde{S}'_F(p+k)}{\partial p_\mu \, \partial p_\nu} = -2g^{\mu\nu}\left[\frac{p\!\!\!/ + m}{4(p\cdot k)^2} + 0\left(\frac{1}{k}\right)\right] + p^\mu \Phi^\nu + p^\nu \Phi^\mu + p^\mu p^\nu I \tag{19.135}$$

Φ^μ is the sum of many terms, none larger than $0(1/k^2)$ but all irrelevant owing to the gauge choice (19.134). The I may be of order $0(1/k^3)$ coming from two differentiations of the denominator

$$(p+k)^2 - m^2 = 2p\cdot k$$

It is a warning of the treacherous shoals which may wreck electromagnetic calculations if current is not accurately conserved at all times. Here we dispose of this singular contribution right away by a convenient gauge choice (19.134), with confidence that it must disappear for any gauge choice also.

In estimating the orders of the terms in (19.135) we have assumed that the derivatives of A and B exist at $p^2 = m^2$. This is assured by the spectral representation (16.112) for $\tilde{S}'_F(p)$ provided we insert a photon mass λ to separate the cut in the region $(m+\lambda)^2 \le p^2 \le \infty$ from the pole at $p^2 = m^2$. With this device $A(p^2)$ and $B(p^2)$ have a radius of convergence for a power series expansion about $p^2 = m^2$ of $\sim 2m\lambda$. There are problems in letting $\lambda \to 0$, and we return to discuss them shortly.

Combining (19.134) and (19.135) in (19.131), we find

$$\epsilon_\mu \epsilon_\nu \mathfrak{M}^{\mu\nu}(p,0;p,0) = 2ie^2 \epsilon\cdot\epsilon' \lim_{k\to 0} \frac{\bar{u}(p) k\!\!\!/ (p\!\!\!/ + m) k\!\!\!/ u(p)}{4(p\cdot k)^2} = \frac{ie^2 \epsilon\cdot\epsilon'}{m} \tag{19.136}$$

This is just the Thomson amplitude in terms of the physical charge

e and mass m. The cross section from (19.132) is

$$\frac{d\sigma}{d\Omega} = \frac{\alpha^2}{m^2}(\epsilon\cdot\epsilon')^2 \qquad \text{for } k \to 0 \tag{19.137}$$

which shows that the fine structure constant $\alpha = 1/137$ appearing in the renormalized expansion of the S matrix can be defined experimentally by the Thomson limit of Compton scattering at zero energy. An extension of this theorem to compute the term in $\mathfrak{M}_{\mu\nu}$ which is linear in k in terms of the static properties of the fermion, namely its charge and magnetic moment, has been given by Low and by Gell-Mann and Goldberger.[1] This gives the exact low-energy limit $\propto k^2$ for a neutral particle with a magnetic moment μ, such as a neutron:

$$\frac{d\sigma}{d\Omega} = 4k^2\mu^4\left(1 + \frac{1}{2}\sin^2\theta\right) \qquad \text{as } k \to 0$$

For charged particles the interference between Thomson and Rayleigh scattering, which cannot be computed exactly in terms of static properties of the particle, is also $\propto k^2$, and it cannot be separated from the magnetic moment contribution which appears first[2] to order k^2. A full analysis of the k^2 terms thus requires a study of dynamical details of the interaction as can be made, for example, with the aid of the Kramers-Krönig relation.

Before we are justified in discussing as above the experimental implications of the low-energy theorem (19.137), we should dispose of the infrared problem appearing for $\lambda \to 0$ in the above analysis. In the proof of (19.136) it was necessary to assume a smooth behavior for $(p'^2 - m^2)\tilde{S}'_F(p')$ in (19.133). However, this is true only for $p'^2 - m^2 \lesssim 2m\lambda$, with λ the fictitious photon mass. This condition limits the photon energy to

$$(p + k)^2 - m^2 = 2mk_0 + \lambda^2 \lesssim 2m\lambda \qquad \text{or} \qquad k_0 < \lambda$$

However, if the photon had a mass $\lambda > 0$, the limit (19.132) would not even make sense, since an observable physical cross section would vanish for $k_0 < \lambda$. Therefore we must reverse one of these inequalities and for a useful theorem that makes physical sense extend (19.137) to the energy region $m \gg k,k' \gg \lambda$.

[1] F. Low, *Phys. Rev.*, **96,** 1428 (1954); M. Gell-Mann and M. L. Goldberger, *Phys. Rev.*, **96,** 1433 (1954). For related discussions see N. M. Kroll and M. A. Ruderman, *Phys. Rev.*, **93,** 233 (1954), who first derived a low-energy theorem for photomeson production; A Klein, *Phys. Rev.*, **99,** 998 (1955); and E. Kazes, *Nuovo Cimento,* **13,** 1226 (1959).

[2] For an unpolarized target particle.

A heuristic argument based on the analysis of the infrared problem for the Bloch-Nordsieck model as presented in Sec. 17.10 will provide the basis for this extension. To order e^2 the same results were found in the calculations carried out earlier in Chaps. 7 and 8. According to (17.89) and (17.95) we expect the elastic Compton scattering amplitude to vanish as the photon mass approaches 0, behaving as

$$\mathfrak{M}_{\mu\nu} = \frac{ie^2}{m} g_{\mu\nu} \exp\left[- \frac{e^2}{2(2\pi)^3} \sum_{\epsilon} \int \frac{d^3q}{2q_0} \left(\frac{\boldsymbol{\varepsilon}\cdot\boldsymbol{\beta}}{q_0 - \mathbf{q}\cdot\boldsymbol{\beta}} - \frac{\boldsymbol{\varepsilon}\cdot\boldsymbol{\beta}'}{q_0 - \mathbf{q}\cdot\boldsymbol{\beta}'} \right)^2 \right]$$
$$\cong \left(\frac{ie^2}{m} g_{\mu\nu}\right) \exp\left[- \frac{\alpha}{8\pi^2} \sum_{\epsilon} \int \frac{d^3q}{q_0{}^3} (\boldsymbol{\varepsilon}\cdot\boldsymbol{\beta}')^2 \right] \tag{19.138}$$

where $\boldsymbol{\beta}' = \frac{1}{m}\mathbf{p}' = \frac{1}{m}(\mathbf{k} - \mathbf{k}')$ = velocity of recoiling electron

$\boldsymbol{\beta} = 0$

$q_0 \equiv \sqrt{|\mathbf{q}|^2 + \lambda^2}$

Equation (19.138) is obtained by treating the electron current distribution produced by scattering from the "hard" photon k classically and multiplying the Thomson amplitude by the amplitude to radiate no additional photons, $\langle 0 \text{ out}|0 \text{ in}\rangle$, as given in (17.89). This procedure agrees with the vertex correction in Chap. 8 [Eq. (8.62)] and also to second order with the radiative correction to Compton scattering found by Brown and Feynman.[1]

For forward scattering $\boldsymbol{\beta}' = 0$, the exponential factor is unity, and there is no infrared problem in (19.138). For all nonforward directions, however, $|\boldsymbol{\beta}'| \neq 0$ for any $k \neq 0$ and the infrared divergence in the exponential causes $\mathfrak{M}_{\mu\nu} \to 0$ for $\lambda \to 0$. The low-energy theorem (19.137) is no longer correct. It was proved for $k_0/\lambda \to 0$ in the $k_0 \to 0$ limit, but here we have $\lambda/k_0 \to 0$ as $k_0 \to 0$ instead.

We know from our previous experience with the infrared problem, however, that the vanishing of the elastic amplitude $\mathfrak{M}_{\mu\nu}$ is compensated by the inelastic amplitude for emitting any number of soft photons, $q < \Delta k$. Adding these back in the classical current approximation for the electron replaces the exponential (17.89) by (17.95) and leads in place of (19.138) to a differential cross section

$$\left(\frac{d\sigma}{d\Omega}\right)_{\text{rad}} = \frac{\alpha^2}{m^2} (\epsilon\cdot\epsilon') \exp\left[- \frac{\alpha}{4\pi^2} \int_{\sim\Delta k}^{\sim m} \frac{dq}{q} \int d\Omega_q \sum_{\epsilon} (\boldsymbol{\varepsilon}\cdot\boldsymbol{\beta}')^2 \right]$$
$$\approx \frac{\alpha^2}{m^2} (\epsilon\cdot\epsilon') \exp\left[- \frac{2\alpha}{3\pi} \beta'^2 \ln\left(\frac{m}{\Delta k}\right) \right] \tag{19.139}$$

[1] L. M. Brown and R. P. Feynman, *Phys. Rev.*, **85**, 231 (1952).

with $\beta' = (2k/m)\sin^2(\theta/2)$. The lower cutoff, $\sim\Delta k$, is the energy resolution of the detector, that is, the maximum energy that the soft photons can carry off which cannot be detected by the measuring apparatus. The upper cutoff of the q integration is put[1] at $1/\Delta t \sim m$, where Δt is the time interval during which the electron is accelerated from velocity zero to β'. It is necessary to introduce this cutoff only because of the classical current model for the electron used here for simplicity.

According to (19.139) it is possible in principle to design an experiment at any given photon energy k with an energy resolution Δk so tiny that the differential cross section is substantially modified from the Thomson limit by radiative corrections. However, for a fixed percentage energy resolution, $\Delta k/k$ constant, (19.139) does in fact approach the Thomson limit as $k \to 0$. As is the case for most electrodynamic processes, experimental conditions must be examined carefully before precise predictions can be made.

19.14 Asymptotic Behavior of Feynman Amplitudes

Weinberg's theorem,[2] discussed in Sec. 19.10 in connection with the convergence properties of Feynman integrals, actually states considerably more than we used there. It gives a simple rule for determining the asymptotic behavior of a Feynman amplitude when some subset, or all subsets, of the external momenta are allowed to approach infinity. In order to state this theorem, which is based on the ideas discussed in Sec. 19.10, we must generalize slightly the notion of subgraph introduced earlier. We consider some subset $\{Q_s\}$ of the external momenta q_s entering the graph; these momenta Q_s will later be allowed to approach ∞. Then, having chosen internal momentum variables $\{l_r\}$ in some way, we define the subgraph corresponding to a subset $\{L_r\}$ of internal momenta and $\{Q_s\}$ of external variables as that set of internal lines through which run at least one L_r or Q_s, provided the graph so obtained is connected. As an example of subgraphs, we illustrate in Fig. 19.52 a sixth-order graph for electron-positron scattering. We shall study the behavior of this amplitude as the momenta q_1 and q_3 of the electron approach infinity. The subgraphs corresponding to this situation are illustrated in Fig. 19.53.

[1] Recall that Compton scattering goes via negative-energy states with energy denominators $\sim 2m$ in the nonrelativistic limit.

[2] S. Weinberg, *Phys. Rev.*, **118,** 838 (1960).

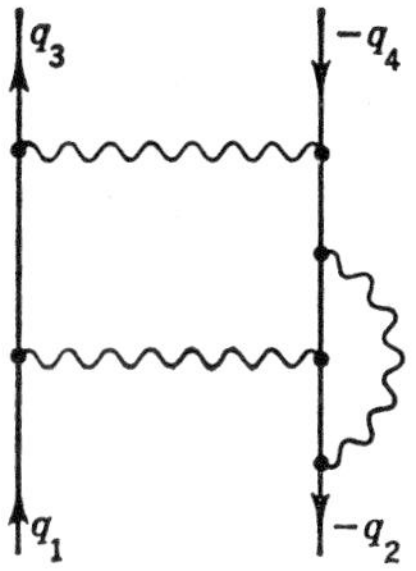

Fig. 19.52 Sixth-order electron-positron scattering graph.

To each subgraph as defined above, we assign as "asymptotic coefficient" α which is the analogue here of the degree of divergence D discussed in Sec. 19.10. This asymptotic coefficient α is obtained by counting powers for all lines in the subgraph, assigning -2 for each photon line, -1 for each electron line, and $+4$ for each internal momentum integral in the subgraph:

$$\alpha(g) = 4k - f - 2b$$

where b = number of photon lines in the subgraph g
f = number of electron lines in the subgraph g
k = number of internal-momentum integrals in the subgraph g

Thus, for example, considering Fig. 19.52 to be a *proper* graph (no external propagator legs), the asymptotic coefficients of the subgraphs illustrated in Fig. 19.53 are -1, -3, -5, -2, -6, -8, and -5, respectively. Weinberg's theorem states that if:

1. The external momenta q_s and internal momenta l_r are continued into the region with a euclidean metric as described in Sec. 19.9
2. The subset $\{Q_s\}$ of external momenta are allowed to approach ∞, according to

$$Q_s \to \chi Q_s \qquad \text{with} \qquad \chi \to \infty$$

then the asymptotic behavior of the amplitude $\mathfrak{M}(\chi Q_1 \cdots \chi Q_J, q_{J+1} \cdots q_m)$ as $\chi \to \infty$ is given by

$$\mathfrak{M} \sim \chi^\alpha (\log \chi)^\beta$$

where

$$\alpha = \max_{\text{subgraphs } g} \alpha(g)$$

with $\alpha(g)$ the asymptotic coefficient for the subgraph g. The coefficient β is not determined by Weinberg's analysis. Thus for the

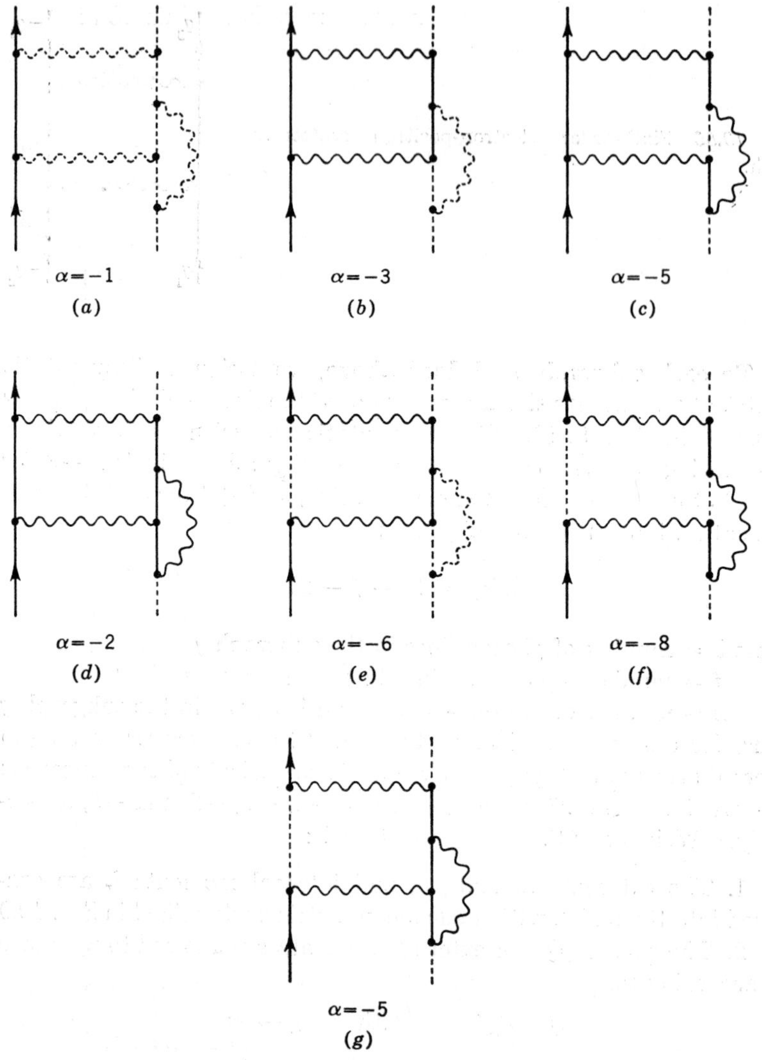

Fig. 19.53 Subgraphs of Fig. 19.52 and their asymptotic coefficients.

example in Fig. 19.52, the amplitude $\mathfrak{M}$ behaves asymptotically as

$$\mathfrak{M} \sim \frac{1}{\chi} (\log \chi)^{\beta}$$

with the main contribution coming from the region of integration where all momenta in the internal lines are small except for the heavy internal electron line shown in the subgraph (*a*) of Fig. 19.53.

The idea of the proof is the same as that of the convergence theorem discussed in Sec. 19.10. To each subgraph corresponds a subintegration over internal variables $\{L_r\}$, the remaining internal variables l_r remaining "small." The subintegration will be cut off at the lower limit by the large momenta χQ_s running through the subgraph, and the magnitude of the contribution will be obtained by over-all dimensional considerations, that is, by counting powers. Evidently the lines in the subgraph but *not* in any subintegration contribute -2 or -1 apiece to the asymptotic behavior. The asymptotic behavior is then obtained by inspecting all possible subgraphs, corresponding, as in Sec. 19.10, to the "tubes" in the $4k$-dimensional space of the internal momenta, and picking out the dominant contribution.

Renormalization does not alter the conclusions of this theorem; if renormalized vertex and self-energy parts are inserted (in order to define the integral!!), one may determine the asymptotic behavior by the rules we have given.[1] According to the theorem, the asymptotic behavior of vertex and self-energy parts is the same, within logarithmic factors, as that of their bare counterparts to any finite order of e:

$$\begin{aligned}
&\Sigma(p) \to \not{p}(\log(-p^2))^{\beta} \qquad \text{as } p^2 \to -\infty \\
&\Pi_{\mu\nu}(q) \to (q_\mu q_\nu - g_{\mu\nu}q^2)(\log(-q^2))^{\beta'} \qquad \text{as } q^2 \to -\infty \\
&\Gamma_\mu(p,p',q) \to \begin{cases} \gamma_\mu(\log(-p^2))^{\beta''} & \text{as } p^2 \to -\infty \\ \dfrac{\not{p}\gamma_\mu\not{p}}{p^2}(\log(-p^2))^{\beta'''} & \quad p'^2 \to -\infty \\ & \quad q \text{ fixed} \end{cases}
\end{aligned} \tag{19.140}$$

with similar statements for the asymptotic behavior as other combinations of p, p', and q approach ∞. We could have guessed this behavior for Σ, Π, and Γ_μ by the success of the renormalization program; had the asymptotic behavior been worse, the degree of divergence of a graph would increase upon adding more vertex and self-energy insertions. The asymptotic behavior predicted by the theorem is the same as obtained in the second-order calculations in Chap. 8 and the fourth-order calculation in Sec. 19.12.

The power of this theorem is limited by two factors. First of all, it does not touch at all the question of the asymptotic behavior of the exact functions summed to all orders of e. For instance, if we make the hypothetical expansion for $S_F'^{-1}(p)$, undoubtedly *wrong*

[1] The apparent circularity of this statement is removed by making an induction argument; see remarks in the following sentences of the text.

in detail beyond second order,

$$S_F'^{-1}(p) \cong \not{p}\left[1 - \frac{\alpha}{4\pi}\log\left(\frac{-p^2}{m^2}\right) + \frac{1}{2}\left(\frac{\alpha}{4\pi}\right)^2 \log^2\left(\frac{-p^2}{m^2}\right) - \frac{1}{6}\left(\frac{\alpha}{4\pi}\right)^3 \log^3\left(\frac{-p^2}{m^2}\right) + \cdots\right] = \not{p}\exp\left[-\frac{\alpha}{4\pi}\log\left(\frac{-p^2}{m^2}\right)\right]$$
$$= \not{p}\left(\frac{m^2}{-p^2}\right)^{\alpha/4\pi} \qquad \text{as } p^2 \to \infty$$

this would lead to an asymptotic behavior for S_F' entirely different from that obtained in an expansion to finite order of e.

A second constraint on the usefulness of Weinberg's theorem is its restriction to unphysical euclidean momenta, defined by the analytic continuation procedure described in Sec. 19.9. One would like to estimate, for example, the asymptotic behavior of physical scattering amplitudes when the external momenta q_s remain on the mass shell, $q_s^2 = m_s^2$, while energy s or momentum transfer t is allowed to approach infinity. This more difficult question has been attacked[1] but not yet completely solved.

19.15 The Renormalization Group

The development of the renormalization program was based on the possibility of rescaling propagators, vertex functions, and τ functions from bare quantities to renormalized ones according to the scaling laws (19.49). In addition to this (perhaps infinite) rescaling, there are finite rescalings that can be carried out. We have already encountered one such example in the intermediate renormalization of Sec. 19.9. These further rescalings reflect the arbitrariness as to where we choose to normalize our propagators and vertex functions. The physical S-matrix elements are necessarily unaffected by them.

As a by-product of the existence of these scaling transformations—known as the renormalization group—it is possible to obtain information about the rate of growth of higher-order graphs without actually doing detailed calculations. In this way one can improve on perturbation theory by turning an expansion in powers of e^2 into one whose individual terms exhibit a momentum dependence consistent with the scaling laws of the renormalization group. The renormalization group has also stimulated some interesting conjectures about the

[1] J. Polkinghorne, *J. Math. Phys.*, **4,** 503 (1963); P. Federbush and M. Grisaru, *Ann. Phys.* (*N.Y.*), **22,** 263 (1963); J. Bjorken and T. T. Wu, *Phys. Rev.*, **130,** 2566 (1963); G. Tiktopolous, *Phys. Rev.*, **131,** 480 (1963).

structure of the exact theory and the relation between bare and renormalized charges.[1]

Here we illustrate the renormalization group method by studying the photon propagator and associated charge renormalizations. The structure of the renormalized photon propagator in (19.51*e*) is

$$e^2\tilde{D}'_F(q)_{\mu\nu} = -\frac{g_{\mu\nu}}{q^2}\,d(q^2,0,e^2) + \text{gauge terms}$$

where the scalar function $d(q^2,0,e^2)$ is given by

$$d(q^2,0,e^2) = \frac{e^2}{1 + e^2\Pi_c(q^2,0,e^2)} \tag{19.141}$$

The first argument of d and Π is the momentum variable. The second argument, here set to zero, denotes the value of the momentum at which $\Pi_c(q^2)$ has been subtracted, that is

$$\Pi_c(q^2,0,e^2) \equiv \Pi(q^2,0,e^2) - \Pi(0,0,e^2) \qquad \Pi_c(0,0,e^2) = 0 \tag{19.142}$$

The third argument, e^2, is the expansion parameter of the renormalized perturbation scheme developed in this manner. It is the physical observable charge measured as equaling

$$e^2 = \frac{4\pi}{137}$$

and correctly predicting, say, the Coulomb cross section as $q^2 \to 0$ with

$$d(0,0,e^2) = e^2 \tag{19.143}$$

according to (19.141) and (19.142), and the Thomson limit of Compton scattering as discussed in Sec. 19.13.

Suppose now we were to change the entire renormalization scheme summarized in (19.51) by subtracting at some other point $q^2 = \lambda_1^2 \neq 0$. [To ensure that d remains real, we choose $\lambda_1^2 < 0$ and avoid the region of nonvanishing absorptive amplitudes in the spectral representation (16.172).] How would this affect what we have done?

The combination

$$e^2\tilde{D}'_F(q)_{\mu\nu} = e_0^2 D'_F(q)_{\mu\nu} \tag{19.144}$$

is an invariant under the renormalization theory scaling laws according to (19.49). Therefore, a renormalization scheme subtracting

[1] M. Gell-Mann and F. E. Low, *Phys. Rev.*, **95**, 1300 (1954), N. N. Bogoliubov and D. V. Shirkov, "Introduction to the Theory of Quantized Fields," Interscience Publishers, Inc., New York, 1959.

at an arbitrary point $q^2 = \lambda_1^2 < 0$ would lead to a similar form

$$e^2\tilde{D}'_F(q)_{\mu\nu} = -\frac{g_{\mu\nu}}{q^2}\,d(q^2,0,e^2) + \text{gauge terms}$$
$$= -\frac{g_{\mu\nu}}{q^2}\,d(q^2,\lambda_1^2,e_1^2) + \text{gauge terms} \qquad (19.145)$$

with

$$d(q^2,\lambda_1^2,e_1^2) = \frac{e_1^2}{1 + e_1^2[\Pi(q^2,\lambda_1^2,e_1^2) - \Pi(\lambda_1^2,\lambda_1^2,e_1^2)]} \qquad (19.146)$$

The new "charge" or expansion parameter e_1^2 is given by

$$d(\lambda_1^2,\lambda_1^2,e_1^2) = e_1^2 \qquad (19.147)$$

Removing gauge terms from (19.145), we obtain

$$d(q^2,0,e^2) = d(q^2,\lambda_1^2,e_1^2) \qquad (19.148)$$

which reduces to an equation relating e_1^2 to $e^2 = 4\pi/137$ at $q^2 = 0$:

$$e^2 = d(0,\lambda_1^2,e_1^2) \qquad (19.149)$$

In the same way we can renormalize at yet another subtraction point $\lambda_2^2 < 0$ to obtain relations analogous to (19.147) to (19.149) between propagator amplitudes for any two arbitrary points:

$$d(q^2,\lambda_1^2,e_1^2) = d(q^2,\lambda_2^2,e_2^2)$$
$$e_2^2 = d(\lambda_2^2,\lambda_2^2,e_2^2) = d(\lambda_2^2,\lambda_1^2,e_1^2)$$

These combine to form a single functional equation

$$d(q^2,\lambda_1^2,e_1^2) = d(q^2,\lambda_2^2,d(\lambda_2^2,\lambda_1^2,e_1^2)) \qquad (19.150)$$

a relation which must hold for all λ_2^2.

At this stage one may justifiably wonder what all this has accomplished. The functional equation (19.150) relates the propagator d back to itself, but only at the expense of introducing an extra parameter λ^2. This in itself hardly appears to be progress. In fact, before we can hope to achieve progress, we must provide some new input into the equation for d. This we do with the following *assumption:* If $-\lambda_1^2 \gg m^2$, $-\lambda_2^2 \gg m^2$, and $-q^2 \gg m^2$, it is a valid approximation to neglect the dependence of the propagator d on the mass m. Before backing up this assumption by a more detailed investigation of the structure of the theory, let us see what consequences derive from it.

One immediate consequence is based solely on dimensional arguments. If all m dependence is removed by fiat, d can be a function of only two variables, q^2/λ^2 and e^2, the same number with which we started this section. However, the functional equation (19.150) couples the dependence of d upon these two variables, and thus restricts the behavior of individual terms in the e^2 expansion of d at high momenta according to the renormalization group.

According to the above assumption, we can define

$$d(q^2,\lambda_1{}^2,e_1{}^2) \equiv d\left(\frac{q^2}{\lambda_1{}^2}, e_1{}^2\right) \qquad \text{for } -q^2, -\lambda_1{}^2 \gg m^2 \quad (19.151)$$

and rewrite (19.150) (dropping the subscript to write $\lambda_2 = \lambda$) as

$$d\left(\frac{q^2}{\lambda_1{}^2}, e_1{}^2\right) = d\left(\frac{q^2}{\lambda^2}, d\left(\frac{\lambda^2}{\lambda_1{}^2}, e_1{}^2\right)\right) \tag{19.152}$$

with the boundary condition for all $e_1{}^2$

$$d(1,e_1{}^2) = e_1{}^2 \tag{19.153}$$

It is remarkable that Eqs. (19.152) and (19.153) are easily solved. We differentiate first with respect to q^2 and then set $q^2 = \lambda^2$

$$\frac{\partial}{\partial\lambda^2} d\left(\frac{\lambda^2}{\lambda_1{}^2}, e_1{}^2\right) = \frac{1}{\lambda^2}\Phi\left(d\left(\frac{\lambda^2}{\lambda_1{}^2}, e_1{}^2\right)\right) \tag{19.154}$$

where, by (19.153),

$$\Phi(e^2) = \frac{\partial}{\partial x} d(x,e^2)\bigg|_{x=1} \tag{19.155}$$

Equation (19.154) can be integrated for fixed $e_1{}^2$, $\lambda_1{}^2$ by writing

$$\int_{\lambda_1{}^2}^{q^2} \frac{d\lambda^2}{\lambda^2} = \int_{e_1{}^2}^{d(q^2/\lambda_1{}^2,e_1{}^2)} \frac{d[d(\lambda^2/\lambda_1{}^2,e_1{}^2)]}{\Phi(d(\lambda^2/\lambda_1{}^2,e_1{}^2))}$$

and gives the functional form

$$\ln\frac{q^2}{\lambda_1{}^2} = \int_{e_1{}^2}^{d(q^2/\lambda_1{}^2,e_1{}^2)} \frac{du}{\Phi(u)} = F\left(d\left(\frac{q}{\lambda_1{}^2}, e_1{}^2\right)\right) - F(e_1{}^2) \tag{19.156}$$

Inverting to solve for d, we find the general solution of the functional equation (19.152)

$$d\left(\frac{q^2}{\lambda_1{}^2}, e_1{}^2\right) = F^{-1}\left[F(e_1{}^2) + \ln\left(\frac{q^2}{\lambda_1{}^2}\right)\right] \tag{19.157}$$

Assuming that F^{-1} exists, we have accomplished a form for d in (19.157) which exhibits it as a function of *one* variable only. As remarked earlier, the momentum dependence of the propagator has been coupled to its dependence on the charge, for large $-q^2 \gg m^2$.

We can use this result in general to solve for d, using a known estimate as our input at the start. The procedure is to:

1. Take an estimate of $d(x,e_1^2)$ and solve for $\Phi(e_1^2)$, using (19.155).
2. Integrate (19.156) and compute an improved $d(x,e_1^2)$ as in (19.157).

To illustrate the procedure, we suppose that all we know about d is that it may be expanded in a power series in e_1^2

$$d(x,e_1^2) = e_1^2 + e_1^4 f(x) + 0(e_1^6) \tag{19.158}$$

Then, using (19.155), we find

$$\Phi(e_1^2) = e_1^4 f'(1) + 0(e_1^6)$$

and insert into (19.156) to obtain

$$\ln \frac{q^2}{\lambda_1^2} = \int_{e_1^2}^{d(q^2/\lambda_1^2, e_1^2)} \frac{du}{u^2 f'(1)} + 0\left(\ln \frac{d}{e_1^2}\right)$$

or

$$\frac{1}{d(x,e_1^2)} = \frac{1}{e_1^2} - f'(1) \ln x + 0\left(\ln \frac{d}{e_1^2}\right)$$

and

$$d(x,e_1^2) = \frac{e_1^2}{1 - e_1^2 f'(1) \ln x + 0(e_1^4 \ln x)} \tag{19.159}$$

Therefore, without looking at a Feynman integral we have ascertained that the second-order vacuum polarization behaves as $\ln (q^2/\lambda^2)$ at large q^2 and that the fourth-order one behaves no worse than $\ln (q^2/\lambda^2)$ also. We conclude then without making further calculations that the $\ln^2 (q^2/m^2)$ terms which appear in Sec. 19.12 in the fourth-order calculation had to cancel, as indeed we found, after considerable effort, to occur in (19.121).

In a similar way we may start with a propagator d including all we learned from the fourth-order calculation, (19.122),

$$d(x,e_1^2) = \frac{e_1^2}{1 - \frac{e_1^2}{12\pi^2} \ln x - \frac{e_1^4}{64\pi^4} \ln x + 0(e_1^6)} \tag{19.160}$$

Then computing Φ from (19.155) and d from (19.156), we find

$$d(x,e_1^2) = \frac{e_1^2}{1 - \frac{e_1^2}{12\pi^2} \ln x + \frac{3e_1^2}{16\pi^2} \ln\left(1 - \frac{e_1^2}{12\pi^2} \ln x\right) + 0\left(e_1^2 \ln \frac{1+d}{1+e_1^2}\right)}$$

$$= \frac{e_1^2}{1 - \frac{e_1^2}{12\pi^2} \ln x - \frac{e_1^4}{64\pi^4} \ln x - \frac{e_1^6}{1536\pi^6} \ln^2 x + 0(e_1^6 \ln x)}$$

$$\text{for} \quad \frac{e_1^2}{12\pi^2} \ln x \ll 1 \tag{19.161}$$

Thus the leading term in the sixth-order calculation for large q^2 is completely determined, both as to form and numerical coefficient, by the equations of the renormalization group alone.[1]

Another remarkable property of (19.157) is that

$$d\left(\frac{q^2}{\lambda_1^2}, e_1^2\right) \to F^{-1}(\infty) \qquad \text{as } -q^2 \to \infty \tag{19.162}$$

showing that the propagator approaches a limit independent both of the renormalization point λ_1 and of the charge e_1 at large momenta. However, the spectral representation of the photon propagator suggests an interpretation of d in this limit as the bare charge e_0^2. According to (16.172) the complete unrenormalized propagator satisfies a spectral representation

$$D'_F(q)_{\mu\nu} = -g_{\mu\nu}\left[\frac{Z_3}{q^2} + \int_0^\infty \frac{dM^2 \Pi(M^2)}{q^2 - M^2}\right] + \text{gauge terms} \tag{19.163}$$

with $\Pi(M^2) \geq 0$ and a sum rule $1 = Z_3 + \int_0^\infty dM^2\, \Pi(M^2)$, as $q^2 \to \infty$, provided the spectral integral exists. Therefore the exact propagator should approach the bare one, up to gauge terms, that is,

$$D'_F(q)_{\mu\nu} \to \frac{-g_{\mu\nu}}{q^2} + \text{gauge terms} \qquad \text{as } q^2 \to \infty \tag{19.164}$$

Recalling that $e_0^2 D'_F(q)_{\mu\nu}$ is a renormalization invariant product, we combine (19.162) and (19.164) to find that

$$e_0^2 = F^{-1}(\infty) \tag{19.165}$$

We are presented with a dilemma. All our arguments in this section have been based on the renormalization program in perturbation theory, which permits us to expand propagators, vertex functions, etc., in power series in both the renormalized and bare charges e and e_0. However, in (19.165) we have come up with a result that puts a condition on the value of e_0 indicating that it cannot be chosen arbitrarily. This behavior is completely foreign to the perturbation development and forces us to conclude that at least one of our assumptions along the way has been wrong.[2] In particular, we suspect our

[1] With similar applications of this method, perturbation expansions of vertex functions and fermion propagators have been improved both in the ultraviolet and the infrared regions. See the book of Bogoliubov and Shirkov, *op. cit.*, for detailed discussions.

[2] See in this connection, in addition to the references on page 369, the work of Landau and coworkers [L. D. Landau in W. Pauli (ed.), "Niels Bohr and the Development of Physics," McGraw-Hill Book Company, New York, (1955)] and also P. J. Redmond, *Phys. Rev.*, **112**, 1404 (1958), and P. J. Redmond and J. L. Uretski, *Ann. Phys.* (*N.Y.*), **9**, 106 (1960).

use of a perturbation expansion for $q^2 \to \infty$, since (19.161) shows that the successive terms grow larger and larger when

$$\ln \frac{q^2}{\lambda_1^2} \gtrsim \frac{12\pi^2}{e_1^2} \tag{19.166}$$

With this sampling of applications of the renormalization group, let us return to the basic assumption which introduced physics into the group equation (19.150), namely, that the propagator becomes independent of m^2 for $-q^2$, $-\lambda^2 \gg m^2$. To support this assumption, we wish to show that the asymptotic solutions of the renormalized integral equations (19.51) are insensitive to mass m if the renormalization is made at a point $-\lambda^2 \gg m^2$. The argument to this end is based on Weinberg's theorem on the asymptotic behavior of Feynman amplitudes discussed in Sec. 19.14 and uses the induction method.

We assume that $\tilde{\Gamma}_\mu(p',p,\lambda)$, $\tilde{S}'_F(p,\lambda)$, and $\tilde{D}'_F(q,\lambda)$, all renormalized at some large mass $-\lambda^2 \gg m^2$, approach functions independent of m as p, p', $q \to \infty$ when computed to order e_λ^{n-2} in a perturbation expansion in the renormalized charge. We then argue from Weinberg's theorem that this is true also to order e_λ^n. For example, Weinberg's asymptotic theorem says that

$$\tilde{S}'^{-1}_F(p,\lambda) \to (\text{constant})\cdot \not{p} \cdot (\text{logarithms}) \qquad \text{as } \not{p} \to \infty \tag{19.167}$$

We assume to order e_λ^{n-2}, and wish to prove to order e_λ^n, that the asymptotic part of $\tilde{S}'^{-1}_F(p,\lambda)$ is independent of m^2, that is, that the constant and the logarithms in (19.167) are finite for $-\lambda^2 \gg m^2$ and $m^2 \to 0$. By checking back at Eq. (8.39) and making the subtraction at $-\lambda^2 \gg m^2$, we verify that this assumption is true to order e_λ^2. Similar inductive assumptions are made about the vertex and photon self-energy parts to order e_λ^{n-2}. By referring[1] back to Chap. 8, one can check directly that what we want to show to order e_λ^n is valid to order e_λ^2 for the asymptotic parts of Γ_μ and $\Pi_{\mu\nu}$ as any combination of virtual momenta $\to \infty$.

[1] With the normalization convention

$$\tilde{S}'_F(p,\lambda) = \frac{1}{\not{p} - \overline{m}(\lambda)}\bigg|_{\not{p}=\lambda\gg m}$$

and

$$\tilde{\Gamma}_\mu(p,p,\lambda)\bigg|_{\not{p}=\lambda} = \gamma_\mu$$

we still retain the Ward identity $Z_1(\lambda) = Z_2(\lambda)$ used there.

Let us begin with the vertex and use (19.51), understood to be subtracted at large λ instead of on the mass shell, to compute it to order $e_\lambda{}^n$. We are interested only in the asymptotic part which behaves according to Weinberg's theorem like (constant) $\times$ (logarithms) as shown in (19.140). According to Weinberg's theorem, this dominant part comes from those subintegrations corresponding to subgraphs with an asymptotic coefficient $\alpha = 0$. The only candidates for subgraphs with nonnegative asymptotic coefficients are evidently vertex and self-energy parts. If these appear as subgraphs within the vertex graph, however, at least two of their external propagator legs joining them to the rest of the graph will carry "large" momenta and therefore contribute to a negative asymptotic coefficient. This leaves only the entire vertex graph itself,[1] which has zero asymptotic coefficient because its own external lines are not included in defining the proper vertex part. The vertex graph also has zero degree of divergence, corresponding to $\alpha = 0$, and the associated integration region consists of those internal momenta such that all internal propagator denominators are large. Therefore, we may use the asymptotic forms for the propagator and vertex insertions into the graph. It appears, then, that the mass m does not appear in a critical way anywhere in the integral and the integral for Γ_μ exists after renormalization. Moreover, its asymptotic part will exist as $m \to 0$, since the lower limit cutting off the logarithms will be the parameter $-\lambda^2 \gg m^2$, or external momenta.

Continuing this argument, as presented here with a blatant disregard for rigor, we may turn to the Ward identity to compute $\tilde{S}'_F(p,\lambda)$ from $\tilde{\Gamma}_\mu$ and which therefore is also not dependent upon m. Finally, we may repeat a similar argument for the photon propagator as for the vertex. We search for a subgraph with asymptotic coefficient $+2$, according to (19.140), and we find the entire graph itself as the only candidate by the arguments applied to the vertex. This then establishes the induction assertion.

Fortified by the above arguments, we may feel justified in applying renormalization group arguments to improve perturbation theory estimates at high momenta. However, it is implicit in our arguments that an order-by-order sum of asymptotic parts in a perturbation series yields the asymptotic behavior of their sum. It may very well be that terms dropped along the way as relatively small at high momenta in each individual order of calculation actually become dominant relative to those retained when the perturbation series is

[1] Aside from subgraphs within subtraction terms; the situation here is just as in the renormalization discussion.

summed. Therefore, conclusions based on the renormalization group arguments concerning the behavior of the theory summed to all orders are dangerous and must be viewed with due caution.

So is it with all conclusions from local relativistic field theories.

Problems

1. Prove the generalized Ward identity (19.16) by forming the three-field vacuum expectation value

$$\langle 0|T(\psi(x)\bar{\psi}(y)j_\mu(z))|0\rangle$$

and using current conservation and the field equations.

2. Give the general structure of $\Gamma^\mu(p',p)$ and determine the restrictions imposed owing to the Ward identity and T, C, and P invariance.

3. Complete the proof of renormalizability given in Sec. 19.11 by showing that general $l_r \ldots l_s$ subintegrations converge.

4. Derive the results in Eqs. (19.109) and (19.111), (19.117), (19.118).

5. Verify Eqs. (19.120) and (19.122) and find the analogous results in the limit $-q^2/m^2 \ll 1$.

6. Verify Eq. (19.131).

7. Verify the structure of (19.135), evaluating Φ and I.

8. Use the Källén-Lehmann representation to show that F^{-1} in Eq. (19.157) exists.

9. Verify Eq. (19.161).

10. Write the general expression for the electromagnetic current for the emerging electron and photon on their mass shells ($p^2 = m^2$ and $l^2 = 0$). Using the generalized Ward identity show that it can be written in terms of four scalar functions $F_i(W^2)$ of the variable $W^2 = (p + l)^2$. Construct appropriate dispersion relations for the $F_i(W^2)$ and relate them to the irreducible vertex functions appearing in Prob. 2. In particular discuss and explain in physical terms the appearance of e as a necessary subtraction constant in the dispersion relations for this current.

11. Prove the renormalizability of the theory of pseudoscalar mesons interacting with nucleons with nonderivative coupling. No longer can gauge invariance be used to dodge the problem of π-π scattering subgraphs.

Appendix A

Notation

Coordinates and Momenta

The space-time coordinates $(t,x,y,z) \equiv (t,\mathbf{x})$ are denoted by the contravariant four-vector (c and $\hbar$ are set equal to 1):

$$x^\mu \equiv (x^0,x^1,x^2,x^3) \equiv (t,x,y,z)$$

The covariant four-vector x_μ is obtained by changing the sign of the space components:

$$x_\mu \equiv (x_0,x_1,x_2,x_3) \equiv (t,-x,-y,-z) = g_{\mu\nu}x^\nu$$

with

$$g_{\mu\nu} = \begin{bmatrix} 1 & 0 & 0 & 0 \\ 0 & -1 & 0 & 0 \\ 0 & 0 & -1 & 0 \\ 0 & 0 & 0 & -1 \end{bmatrix}$$

The summation convention, according to which repeated indices are summed, is used unless otherwise specified. It is likely that if two identical indices (to be summed) are both in the lower or the upper position, one has erred. The inner product is $x^2 = x_\mu x^\mu = t^2 - \mathbf{x}^2$.

Momentum vectors are similarly defined

$$p^\mu = (E,p_x,p_y,p_z)$$

and the inner product is

$$p_1 \cdot p_2 = p_1{}^\mu p_{2\mu} = E_1E_2 - \mathbf{p}_1 \cdot \mathbf{p}_2$$

Likewise

$$x \cdot p = tE - \mathbf{x} \cdot \mathbf{p}$$

Four-vectors p are always written in lightface type, while three-vectors $\mathbf{p}$ are in boldface.

The momentum operator in coordinate representation is written

$$p^\mu = i\frac{\partial}{\partial x_\mu} = \left(i\frac{\partial}{\partial t}, \frac{1}{i}\nabla\right) \equiv i\nabla^\mu$$

and transforms as a contravariant four-vector:

$$p^\mu p_\mu = -\frac{\partial}{\partial x_\mu}\frac{\partial}{\partial x^\mu} \equiv -\Box$$

In these units the Compton wavelength is $1/m$ ($\cong 3.86 \times 10^{-11}$ cm for the electron) and the rest energy is m($\cong 0.511$ MeV for the electron).

The four-vector potential of the electromagnetic field is defined by

$$\begin{aligned} A^\mu &= (\Phi, \mathbf{A}) \\ &= g^{\mu\nu}A_\nu \end{aligned}$$

The field strengths are defined by

$$F^{\mu\nu} = \frac{\partial}{\partial x_\nu}A^\mu - \frac{\partial}{\partial x_\mu}A^\nu$$

and the electric and magnetic fields in a noncovariant notation are given by

$$\begin{aligned} \mathbf{E} &= (F^{01}, F^{02}, F^{03}) \\ \mathbf{B} &= (F^{23}, F^{31}, F^{12}) \end{aligned}$$

Dirac Matrices and Spinors

A Dirac spinor for a particle of physical momentum p and polarization s is denoted by $u_\alpha(p,s)$, while for the antiparticle it is called $v_\alpha(p,s)$. In each case the energy $p_0 \equiv E_p = +\sqrt{\mathbf{p}^2 + m^2}$ is positive. In each case the vector s^μ, which in the rest frame has the form

$$s^\mu = (0, \hat{\mathbf{s}}) \qquad \hat{\mathbf{s}} \cdot \hat{\mathbf{s}} = 1$$

represents the direction of spin of the physical particle in the rest frame.

The γ matrices in the Dirac equation satisfy the anticommutation relations

$$\gamma^\mu\gamma^\nu + \gamma^\nu\gamma^\mu = 2g^{\mu\nu}$$

and are related to the α and β matrices by $\boldsymbol{\gamma} = \beta\boldsymbol{\alpha}$; $\gamma_0 = \beta$. A familiar representation is

$$\gamma^0 = \begin{bmatrix} 1 & 0 \\ 0 & -1 \end{bmatrix}$$

$$\{\gamma^i\} = \boldsymbol{\gamma} = \begin{bmatrix} 0 & \boldsymbol{\sigma} \\ -\boldsymbol{\sigma} & 0 \end{bmatrix}$$

where $\sigma^1 = \begin{bmatrix} 0 & 1 \\ 1 & 0 \end{bmatrix} \quad \sigma^2 = \begin{bmatrix} 0 & -i \\ i & 0 \end{bmatrix} \quad \sigma^3 = \begin{bmatrix} 1 & 0 \\ 0 & -1 \end{bmatrix}$

are the familiar 2×2 Pauli matrices and $1 = \begin{bmatrix} 1 & 0 \\ 0 & 1 \end{bmatrix}$ is the 2×2 unit matrix.

Frequently appearing combinations are

$$\sigma^{\mu\nu} = \frac{i}{2}[\gamma^\mu, \gamma^\nu] \qquad \text{and} \qquad \gamma^5 = i\gamma^0\gamma^1\gamma^2\gamma^3 = \gamma_5$$

σ^i and 1 form complete (i.e. indep't set of 2x2 matrices
" " " 4x4 matrices
$1, \gamma^\mu, \gamma^5, \gamma^\mu\gamma^5, \sigma^{\mu\nu}$ form (16) complete" for nxn matrices need (2^n) to form complete basis

In this representation the components of $\sigma^{\mu\nu}$ are

$$\sigma^{ij} = \begin{bmatrix} \sigma^k & 0 \\ 0 & \sigma^k \end{bmatrix}$$

with $i, j, k = 1, 2, 3$ in cyclic order and

$$\sigma^{0i} = i\alpha^i = i\begin{bmatrix} 0 & \sigma^i \\ \sigma^i & 0 \end{bmatrix} \qquad \gamma_5 = \gamma^5 = \begin{bmatrix} 0 & 1 \\ 1 & 0 \end{bmatrix}$$

The inner product of a γ matrix with an ordinary four-vector is often encountered and denoted by

$$\gamma_\mu A^\mu \equiv A\!\!\!/ = \gamma^0 A^0 - \boldsymbol{\gamma}\cdot\mathbf{A}$$
$$p_\mu\gamma^\mu \equiv p\!\!\!/ = E\gamma^0 - \mathbf{p}\cdot\boldsymbol{\gamma}$$
$$\mathsf{p}_\mu\gamma^\mu \equiv i\nabla\!\!\!\!/ = i\gamma_0\frac{\partial}{\partial t} + i\boldsymbol{\gamma}\cdot\boldsymbol{\nabla} = i\gamma^\mu\frac{\partial}{\partial x^\mu}$$

The spinors u and v satisfy the Dirac equation

$$(p\!\!\!/ - m)u(p,s) = 0$$
$$(p\!\!\!/ + m)v(p,s) = 0$$

and are given explicitly by Eq. (3.7), but for most applications the following projection operators suffice. In terms of the adjoint spinors

$$\bar{u} = u^\dagger\gamma^0$$
$$\bar{v} = v^\dagger\gamma^0$$

satisfying

$$\bar{u}(p,s)(p\!\!\!/ - m) = 0$$
$$\bar{v}(p,s)(p\!\!\!/ + m) = 0$$

the projection operators are

$$u_\alpha(p,s)\bar{u}_\beta(p,s) = \left[\frac{p\!\!\!/ + m}{2m}\cdot\frac{1 + \gamma_5 s\!\!\!/}{2}\right]_{\alpha\beta}$$
$$v_\alpha(p,s)\bar{v}_\beta(p,s) = -\left[\frac{m - p\!\!\!/}{2m}\cdot\frac{1 + \gamma_5 s\!\!\!/}{2}\right]_{\alpha\beta} \tag{A.1}$$

These lead to the normalization conditions

$$\bar{u}(p,s)u(p,s) = 1$$
$$\bar{v}(p,s)v(p,s) = -1 \tag{A.2}$$

and the completeness relation

$$\sum_s [u_\alpha(p,s)\bar{u}_\beta(p,s) - v_\alpha(p,s)\bar{v}_\beta(p,s)] = \boldsymbol{\delta}_{\alpha\beta}$$

In taking traces we form hermitian conjugates of matrix elements for which

$$[\bar{u}(p',s')\Gamma u(p,s)]^\dagger = \bar{u}(p,s)\,\bar{\Gamma}\,u(p',s')$$

with

$$\bar{\Gamma} \equiv \gamma^0 \Gamma^\dagger \gamma^0$$

For example

$$\bar{\gamma}^\mu = \gamma^0 \gamma^{\mu\dagger} \gamma^0 = \gamma^\mu$$

$$\bar{\sigma}^{\mu\nu} = \gamma^0 \sigma^{\mu\nu\dagger} \gamma^0 = \sigma^{\mu\nu}$$

$$\overline{i\gamma^5} = \gamma^0 (i\gamma^5)^\dagger \gamma^0 = i\gamma^5$$

Summing the projection operators (A.1) over spin leads to the energy projection operators

$$[\Lambda_+(p)]_{\alpha\beta} \equiv \sum_{\pm s} u_\alpha(p,s)\bar{u}_\beta(p,s) = \left(\frac{\not{p} + m}{2m}\right)_{\alpha\beta}$$

$$[\Lambda_-(p)]_{\alpha\beta} \equiv -\sum_{\pm s} v_\alpha(p,s)\bar{v}_\beta(p,s) = \left(\frac{-\not{p} + m}{2m}\right)_{\alpha\beta} \tag{A.3}$$

The Gordon decomposition of the current is a frequent and useful identity:

$$\bar{u}(p')\gamma^\mu u(p) = \bar{u}(p')\left[\frac{(p + p')^\mu}{2m} + \frac{i\sigma^{\mu\nu}(p' - p)_\nu}{2m}\right] u(p)$$

Trace Theorems and γ Identities

$$\not{a}\not{b} = a\cdot b - i\sigma_{\mu\nu}a^\mu b^\nu$$

Trace of odd number γ_μ's vanishes

$$\mathrm{Tr}\ \gamma_5 = 0$$

$$\mathrm{Tr}\ 1 = 4$$

$$\mathrm{Tr}\ \not{a}\not{b} = 4a\cdot b$$

$$\mathrm{Tr}\ \not{a}_1\not{a}_2\not{a}_3\not{a}_4 = 4[a_1\cdot a_2\ a_3\cdot a_4 - a_1\cdot a_3\ a_2\cdot a_4 + a_1\cdot a_4\ a_2\cdot a_3]$$

$$\mathrm{Tr}\ \gamma_5\not{a}\not{b} = 0$$

$$\mathrm{Tr}\ \gamma_5\not{a}\not{b}\not{c}\not{d} = 4i\epsilon_{\alpha\beta\gamma\delta}a^\alpha b^\beta c^\gamma d^\delta$$

$$\gamma_\mu\not{a}\gamma^\mu = -2\not{a}$$

$$\gamma_\mu\not{a}\not{b}\gamma^\mu = 4a\cdot b$$

$$\gamma_\mu\not{a}\not{b}\not{c}\gamma^\mu = -2\not{c}\not{b}\not{a}$$

For further rules see Sec. 7.2.

Appendix B

Rules for Feynman Graphs

Expressions for cross sections are divided into two parts: first the invariant amplitude $\mathfrak{M}$, which is a Lorentz scalar and in which physics lies, and second, the phase space and kinematical factors. In terms of $\mathfrak{M}$, the expression for a differential cross section $d\sigma$ is, *for spinless particles and for photons only,*

$$d\sigma = \frac{1}{|\mathbf{v}_1 - \mathbf{v}_2|}\left(\frac{1}{2\omega_{p_1}}\right)\left(\frac{1}{2\omega_{p}}\right)|\mathfrak{M}|^2 \frac{d^3k_1}{2\omega_1(2\pi)^3} \cdots \frac{d^3k_n}{2\omega_n(2\pi)^3} \times (2\pi)^4\delta^4\left(p_1 + p_2 - \sum_{i=1}^{n} k_i\right) S \quad \text{(B.1)}$$

where $\omega_p = \sqrt{|\mathbf{p}|^2 + m^2}$ as usual and $\mathbf{v}_1$ and $\mathbf{v}_2$ are velocities of the incident collinear particles. This expression is then integrated over all undetected momenta $k_1 \cdots k_n$ of the final particles. The statistical factor S is obtained by including a factor $1/m!$ if there are m identical particles in the final state:

$$S = \prod_i \frac{1}{m_i!}$$

For Dirac particles,[1] the factor $1/2\omega_p$ is replaced by m/E_p, and the statistical factor S is again included; all other factors remain the same.

A differential decay rate of a particle of mass M is given in its rest frame by

$$d\omega = d\left(\frac{1}{\tau}\right) = \frac{1}{2M}|\mathfrak{M}|^2 \frac{d^3k_1}{2\omega_1(2\pi)^3} \cdots \frac{d^3k_n}{2\omega_n(2\pi)^3}(2\pi)^4\delta^4\left(p - \sum_{i=1}^{n} k_i\right) S$$

[1] If one adopts the convention that Dirac spinors be normalized to $2m$ instead of to unity as in Eq. (A.2), Eq. (B.1) applies as well to fermions. The energy projection operators are then simply $(m \pm \not{p})$ in place of (A.3).

with factors defined as before. For any fermions in the final state, again $1/2\omega_i \to m/E_i$; the factor $1/2M$ is dropped if the initial particle is a fermion.

If desired, polarizations are *summed* over final and *averaged* over initial states.

The invariant amplitude $\mathfrak{M}$ is computed by drawing all Feynman graphs for the process in question except for graphs with *disconnected* bubbles and with self-energy insertions on *external* lines, which are specifically excluded. The amplitude $\mathfrak{M}(G)$ corresponding to graph G is built up by associating factors with the elements of the graph. Those factors independent of specific details of the interaction are:

1. For each spin-zero boson entering the graph a factor $\sqrt{Z}$. $\sqrt{Z}$ is found by computing the exact meson propagator $\Delta'_F(p)$ in the limit $p^2 \to \mu^2$; $\Delta'_F(p) \to Z\Delta_F(p)$ as $p^2 \to \mu^2$.

2. For each external fermion line entering the graph a factor $\sqrt{Z_2}\, u(p,s)$ or $\sqrt{Z_2}\, v(p,s)$ depending on whether the line is in the initial or final state; likewise, for each fermion line leaving the graph a factor $\sqrt{Z_2}\, \bar{u}(p,s)$ or $\sqrt{Z_2}\, \bar{v}(p,s)$. Z_2 is defined by the limit

$$\lim_{\not{p} \to m} S'_F(p) = Z_2 S_F(p)$$

3. For each external photon line a factor $\epsilon_\mu \sqrt{Z_3}$, where

$$D'_F(q)_{\mu\nu} \to \frac{-Z_3 g_{\mu\nu}}{q^2} + \text{gauge terms}$$

as $q^2 \to 0$.

In lowest order perturbation calculations these Z factors may be set equal to unity. In higher orders, together with self-energy and vertex insertions, they renormalize the charges from their bare to physical values.

4. For each internal fermion line with momentum p a factor

$$iS_F(p) = \frac{i}{\not{p} - m + i\epsilon} = \frac{i(\not{p} + m)}{p^2 - m^2 + i\epsilon}$$

5. For each internal meson line of spin zero with momentum q a factor

$$i\Delta_F(q) = \frac{i}{q^2 - \mu^2 + i\epsilon}$$

6. For each internal photon line with momentum q a factor

$$iD_F(q)_{\mu\nu} = -\frac{ig_{\mu\nu}}{q^2 + i\epsilon}$$

Gauge terms proportional to $q_\mu q_\nu$, $q_\mu \eta_\nu$, etc., may be ignored in a theory with conserved currents.

For meson-nucleon physics, an isotopic factor δ_{ij} appears on each internal meson line and for external lines there are factors:

7. χ or $\chi^\dagger$ for initial and final nucleon spinors; $\chi = \begin{bmatrix}1\\0\end{bmatrix}$ for a proton and $\chi = \begin{bmatrix}0\\1\end{bmatrix}$ for a neutron. (Similar factors appear for K and Ξ mesons.)

8. $\hat{\boldsymbol{\phi}}$ or $\hat{\boldsymbol{\phi}}^*$ for the isotopic wave function of a π in the initial or final state, respectively, with

$$\hat{\boldsymbol{\phi}}_{\pi\pm} = \frac{1}{\sqrt{2}}(1, \pm i, 0) \qquad \hat{\boldsymbol{\phi}}_{\pi_0} = (0,0,1)$$

(Similar factors appear for Σ particles.)

9. For each internal momentum l not fixed by momentum conservation constraints at vertices, a factor

$$\int \frac{d^4l}{(2\pi)^4}$$

10. For each closed fermion loop a factor -1.

11. A factor -1 between graphs which differ only by an interchange of two external identical fermion lines. This includes not only exchange of identical particles in the final state, but also interchange, for example, of initial particle and final antiparticle. The overall sign of a graph containing fermions is best obtained by returning to the Dyson-Wick rules.

The interactions determine the structure and type of the vertices. We present here the rules for four typical theories:

Spinor Electrodynamics

There are two kinds of vertices, shown in Fig. B.1, corresponding to the

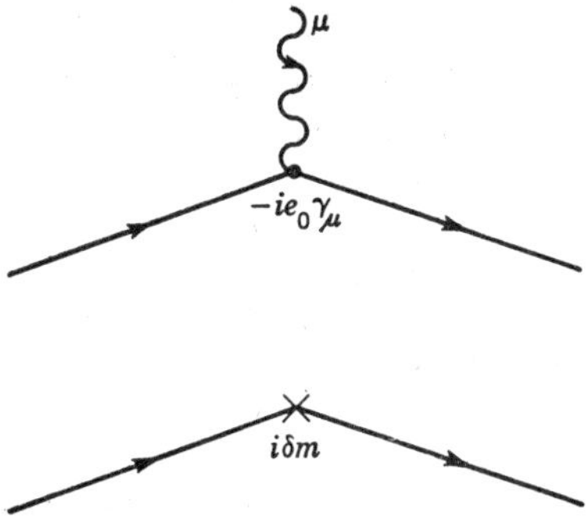

Fig. B.1

normal ordered interaction hamiltonian density

$$\mathcal{H}_I = -\mathcal{L}_I = :e_0\bar{\psi}\gamma_\mu\psi A^\mu: - \delta m: \bar{\psi}\psi:$$

The rules for these are:

1. A factor $-ie_0\gamma_\mu$ at each vertex.
2. A factor $i\delta m$ for each mass counterterm.
3. Renormalize the charge with $e = Z_2Z_1^{-1}\sqrt{Z_3}\,e_0 = \sqrt{Z_3}\,e_0$ where the exact vertex $\Gamma_\mu(p',p) \rightarrow Z_1^{-1}\gamma_\mu$ for $\not{p}' = \not{p} = m$ and $Z_1 = Z_2$ by Ward's identity.

Electrodynamics of a Spin-zero Boson

Here there are three vertices, shown in Fig. B.2, corresponding to the inter-

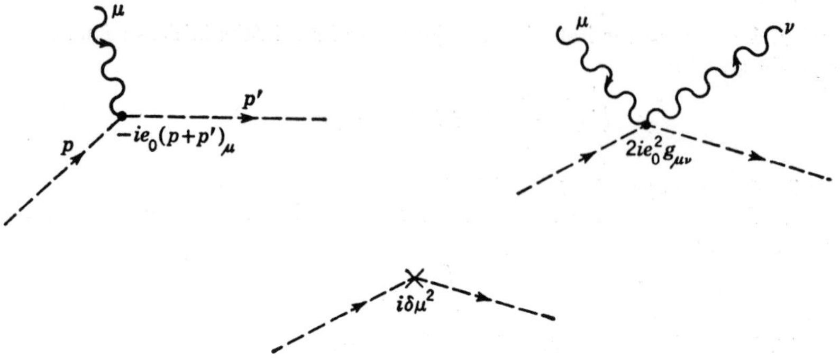

Fig. B.2

action lagrangian density

$$\mathcal{L}_I = -ie_0 {:} \varphi^\dagger \left(\frac{\overrightarrow{\partial}}{\partial x_\mu} - \frac{\overleftarrow{\partial}}{\partial x_\mu} \right) \varphi {:} A_\mu + e_0{}^2 {:} A^2 {:}{:} \varphi^\dagger \varphi {:} + \delta\mu^2 {:} \varphi^\dagger \varphi {:}$$

The rules for these vertices are:

1. A factor $-ie_0(p + p')_\mu$, where p and p' are the momenta in the charged line.
2. A factor $+2ie_0{}^2 g_{\mu\nu}$ for each "seagull" graph.
3. A factor $i\delta\mu^2$ for each mass counterterm.
4. A factor ½ for each closed loop containing only two photon lines, as shown in Fig. B.3.

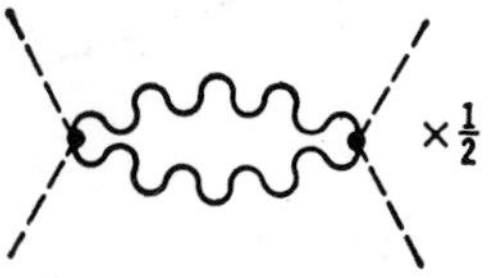

Fig. B.3

5. Renormalize the charge as in spinor electrodynamics.

γ_5 Meson-Nucleon Scattering

There are four interaction terms in the charge-independent theory:

$$\mathcal{H}_I = -\mathcal{L}_I = :ig_0\bar{\Psi}\gamma_5\boldsymbol{\tau}\cdot\boldsymbol{\phi}\Psi: - \delta m:\bar{\Psi}\Psi: - \tfrac{1}{2}\delta\mu^2:\boldsymbol{\phi}\cdot\boldsymbol{\phi}: + \tfrac{1}{4}\delta\lambda:(\boldsymbol{\phi}\cdot\boldsymbol{\phi})^2:$$

as illustrated in Figs. B.4 and B.5. The dotted line signifies that $I = 0$ only is

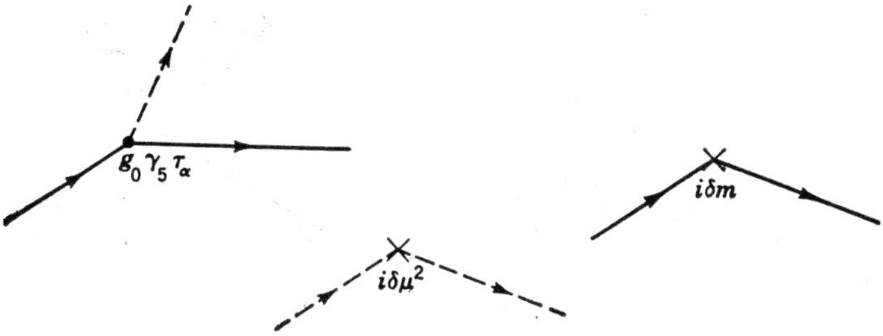

Fig. B.4

transmitted from the meson pair ij to the pair rs, as shown by rule 2 below. The mass counterterms are treated as before and there is:

Fig. B.5

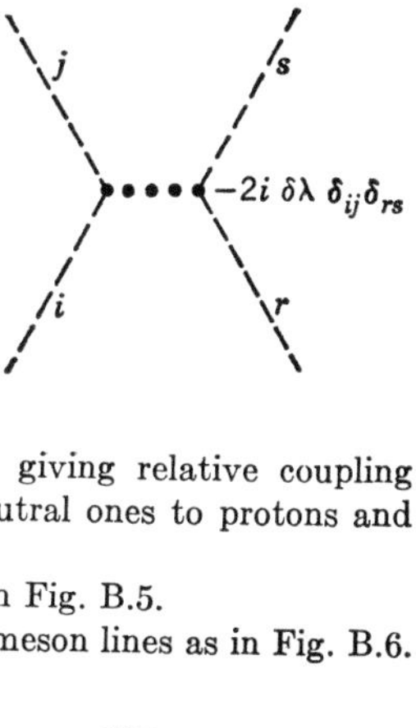

1. A factor $g_0\gamma_5\tau_\alpha$ at each meson-nucleon vertex giving relative coupling strengths of $\sqrt{2}\,g_0$ for charged mesons and ± 1 for neutral ones to protons and neutrons, respectively.
2. A factor $-2i\delta\lambda\delta_{ij}\delta_{rs}$ at each four-meson vertex in Fig. B.5.
3. A factor $\frac{1}{2}$ for each closed loop containing two meson lines as in Fig. B.6.

Fig. B.6

Electrodynamics of Spin-one Boson

A vector boson propagator is $[-g_{\mu\nu} + k_\mu k_\nu/m^2](k^2 - m^2)^{-1}$ in place of the $-g_{\mu\nu}/k^2$ for massless photons, and the external line has a polarization factor ϵ_μ as for photons.

There are electrodynamics vertices shown in Fig. B.7 corresponding to an

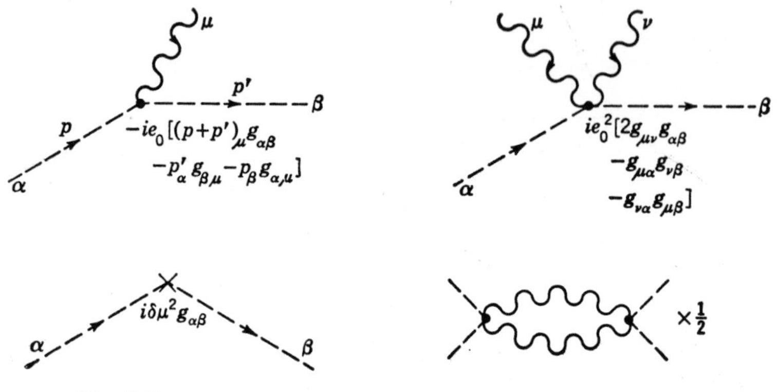

Fig. B.7

interaction lagrangian density

$$\mathcal{L}' = -ie_0 \colon \left[\left(\frac{\partial \varphi_\nu^*}{\partial x_\mu} \right) (A^\nu \varphi_\mu - A_\mu \varphi^\nu) - \left(\frac{\partial \varphi_\nu}{\partial x_\mu} \right) (A^\nu \varphi_\mu^* - A_\mu \varphi^{\nu *}) \right] \colon$$

$$+ e_0^2 \colon [A_\mu A^\mu \varphi_\nu^* \varphi^\nu - A_\mu \varphi^\mu A^\nu \varphi_\nu^*] \colon + \delta\mu^2 \colon \varphi_\nu^* \varphi^\nu \colon$$

The rules for these vertices as illustrated are:

1. A factor $-ie_0(p' + p)_\mu g_{\alpha\beta} + ie_0 g_{\beta\mu} p'_\alpha + ie_0 p_\beta g_{\alpha\mu}$.
2. A factor $+ ie_0^2[2g_{\mu\nu}g_{\alpha\beta} - g_{\mu\alpha}g_{\beta\nu} - g_{\mu\beta}g_{\alpha\nu}]$.
3. A factor $i\delta\mu^2 g_{\alpha\beta}$ for each mass counterterm.
4. A factor ½ for each closed loop containing only two photon lines.
5. For the derivation of these rules from canonical theory, effects of an anomalous magnetic moment term, and a regularization scheme see T. D. Lee and C. N. Yang, *Phys. Rev.*, **128,** 885 (1962).

In all above examples matrices are arranged in "natural order." For closed loops this means taking a trace. Isotopic indices are contracted with their mate at the other end of a boson line. In taking polarization sums for photons

$$\sum_\lambda \epsilon_\mu(k,\lambda)\epsilon_\nu(k,\lambda) \Rightarrow -g_{\mu\nu}$$

and for vector mesons

$$\sum_\lambda \epsilon_\mu(k,\lambda)\epsilon_\nu(k,\lambda) \Rightarrow -g_{\mu\nu} + \frac{k_\mu k_\nu}{m^2}$$

Appendix C

Commutator and Propagator Functions

$ds^2 = dt^2 - d\vec{x}^2$

$k \cdot x = \omega t - \vec{p} \cdot \vec{x}$

The invariant commutator and propagator functions encountered in the text are collected below.

The Feynman propagator for the free Dirac equation is

$$S_F(x' - x) = -i \int \frac{d^3p}{(2\pi)^3} \frac{m}{E} [\theta(t' - t)\Lambda_+(p)e^{-ip\cdot(x'-x)} + \theta(t - t')\Lambda_-(p)e^{ip\cdot(x'-x)}]$$
$$= \int \frac{d^4p}{(2\pi)^4} e^{-ip\cdot(x'-x)} \frac{\not{p} + m}{p^2 - m^2 + i\epsilon}$$

The $i\epsilon$ in the denominator is understood in the limit $\epsilon \to 0^+$ and signifies use of the contour in Fig. C.1 in doing the integral over p_0. $\theta(t' - t)$ denotes the unit step function and is given by

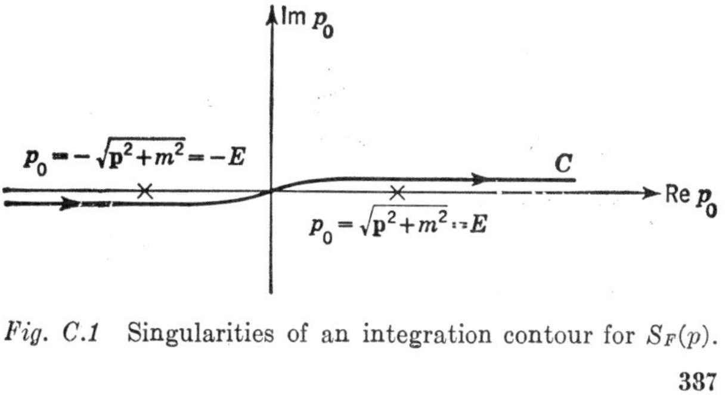

Fig. C.1 Singularities of an integration contour for $S_F(p)$.

$$\theta(t' - t) = \begin{Bmatrix} +1 & t' > t \\ 0 & t' < t \end{Bmatrix} = \frac{-1}{2\pi i}\int_{-\infty}^{\infty} \frac{d\omega}{\omega + i\epsilon} e^{-i\omega(t'-t)}$$

S_F satisfies the differential Green's function equation

$$(i\nabla_{x'} - m)S_F(x' - x) = \delta^4(x' - x)$$

In momentum space

$$S_F(p) \equiv \frac{\not{p} + m}{p^2 - m^2 + i\epsilon} \equiv \frac{1}{\not{p} - m + i\epsilon}$$

The Feynman propagator for the Klein-Gordon equation is

$$\begin{aligned}\Delta_F(x' - x) &= -i\int \frac{d^3k}{(2\pi)^3 2\omega_k}[\theta(t' - t)e^{-ik\cdot(x'-x)} + \theta(t - t')e^{ik\cdot(x'-x)}] \\ &= \int \frac{d^4k}{(2\pi)^4} e^{-ik\cdot(x'-x)} \frac{1}{k^2 - m^2 + i\epsilon}\end{aligned}$$

and satisfies

$$(\Box_{x'} + m^2)\Delta_F(x' - x) = -\delta^4(x' - x)$$

sign different from Hartle & me.

In momentum space $\Delta_F(k) = \dfrac{1}{k^2 - m^2 + i\epsilon}$. Δ_F and S_F are related by

$$S_F(x' - x) = +(i\nabla_{x'} + m)\Delta_F(x' - x) \tag{C.1}$$

For the d'Alembertian equation with $m \to 0$ we define (notice the change of sign relative to Δ_F)

$$\Box_{x'} D_F(x' - x) = +\delta^4(x' - x)$$

and have the Feynman propagator for photons

$$D_F(q)_{\mu\nu} = +g_{\mu\nu}D_F(q) = -\frac{g_{\mu\nu}}{q^2 + i\epsilon}$$

In terms of even and odd solutions of the homogeneous wave equation

$$\begin{aligned}\Delta(x' - x) &= -i\int \frac{d^3k}{(2\pi)^3 2\omega_k}(e^{-ik\cdot(x'-x)} - e^{ik\cdot(x'-x)}) \\ \Delta_1(x' - x) &= \int \frac{d^3k}{(2\pi)^3 2\omega_k}(e^{-ik\cdot(x'-x)} + e^{+ik\cdot(x'-x)})\end{aligned}$$

which satisfy

$$\begin{aligned}(\Box_{x'} + m^2)\Delta(x' - x) &= 0 \qquad & \Delta(x' - x) &= -\Delta(x - x') \\ (\Box_{x'} + m^2)\Delta_1(x' - x) &= 0 \qquad & \Delta_1(x' - x) &= +\Delta_1(x - x')\end{aligned}$$

we can write

$$2\Delta_F(x' - x) = -i\Delta_1(x' - x) + \epsilon(t' - t)\Delta(x' - x)$$

where

$$\epsilon(t' - t) \equiv 2\theta(t' - t) - 1 = \begin{cases} +1 & t' > t \\ -1 & t' < t \end{cases}$$

The odd function Δ vanishes outside the light cone

$$\Delta(x' - x) = 0 \qquad \text{for all } (x' - x)^2 < 0$$

is singular on the light cone

$$\left(\frac{\partial\Delta(x' - x)}{\partial x_0'}\right)_{x_0' - x_0 = 0} = -\delta^3(\mathbf{x}' - \mathbf{x})$$

and equals in general

$$\Delta(x' - x) = \frac{1}{4\pi r}\frac{\partial}{\partial r}\begin{cases} J_0(m\sqrt{t^2 - r^2}) & t > r \\ 0 & -r < t < r \\ -J_0(m\sqrt{t^2 - r^2}) & t < -r \end{cases}$$

where $\quad r \equiv |\mathbf{x}' - \mathbf{x}| \qquad t \equiv x_0' - x_0$

and J_0 is a regular Bessel function.

The even function Δ_1 does not vanish outside of the light cone but falls off exponentially:

$$\left(\frac{\partial\Delta_1(x' - x)}{\partial x_0'}\right)_{x_0' - x_0 = 0} = 0$$

$$\Delta_1(x' - x) = \frac{1}{4\pi r}\frac{\partial}{\partial r}\begin{cases} Y_0(m\sqrt{t^2 - r^2}) & |t| > r \\ -\dfrac{2}{\pi}K_0(m\sqrt{r^2 - t^2}) & r > |t| \end{cases}$$

where Y_0 and K_0 are the cylindrical functions as given in G. N. Watson's "Theory of Bessel Functions," 2d ed., Cambridge University Press, London, 1952.

For the electromagnetic field $m \to 0$ and

$$-D(x' - x) = \lim_{m\to 0}\Delta(x' - x) = -\frac{1}{4\pi r}[\delta(r - t) - \delta(r + t)]$$

$$-D_1(x' - x) = \lim_{m\to 0}\Delta_1(x' - x) = \frac{1}{4\pi^2 r}\left(P\frac{1}{r - t} + P\frac{1}{r + t}\right)$$

where $r \equiv |\mathbf{x}' - \mathbf{x}|$, $t \equiv x_0' - x_0$, and P denotes the principal value.

We may also write Δ and Δ_1 in terms of their separate positive- and negative-frequency parts corresponding to the two terms in the above integrals:

$$\Delta_1(x' - x) = \Delta_+(x' - x) + \Delta_-(x' - x) \qquad i\Delta(x' - x) = \Delta_+(x' - x) - \Delta_-(x' - x)$$

For the Dirac equation we define

$$\begin{aligned} S(x' - x) &= -(i\nabla_x + m)\Delta(x' - x) \\ S_1(x' - x) &= -(i\nabla_x + m)\Delta_1(x' - x) \end{aligned} \tag{C.2}$$

and

$$2S_F(x' - x) = +iS_1(x' - x) - \epsilon(t' - t)S(x' - x)$$

The difference in sign appearing between definition (C.1) and (C.2) is the result of the choice of sign convention in our rules for Feynman graphs which were designed to minimize minus signs in practical calculations. These functions are related to time-ordered products for free Klein-Gordon, Maxwell, and Dirac fields by

$$
\begin{aligned}
\langle 0|T(\varphi_i(x')\varphi_j(x))|0\rangle &= i\delta_{ij}\Delta_F(x'-x) \\
\langle 0|T(\varphi(x')\varphi^*(x))|0\rangle &= i\Delta_F(x'-x) \\
\langle 0|T(A_\nu(x')A_\mu(x))|0\rangle &= iD_F{}^{\text{tr}}(x'-x)_{\nu\mu} = i\int \frac{d^4k}{(2\pi)^4}\frac{e^{-ik\cdot(x'-x)}}{k^2+i\epsilon}\sum_{\lambda=1}^{2}\epsilon_\nu(k,\lambda)\epsilon_\mu(k,\lambda) \\
&= ig_{\nu\mu}D_F(x'-x) + \text{gauge terms} - \text{Coulomb term} \\
\langle 0|T(\psi_\beta(x')\bar\psi_\alpha(x))|0\rangle &= iS_{F_{\beta\alpha}}(x'-x)
\end{aligned}
$$

In addition the retarded and advanced Green's functions Δ_{ret} and Δ_{adv} are defined by

$$
\begin{aligned}
\Delta_{\text{ret}}(x'-x) &= -\Delta(x'-x)\theta(t'-t) \\
\Delta_{\text{adv}}(x'-x) &= +\Delta(x'-x)\theta(t-t')
\end{aligned}
$$

Δ_{ret} vanishes outside the forward light cone; whereas Δ_{adv} vanishes outside the backward light cone. They satisfy the inhomogeneous wave equation

$$
(\Box_{x'} + m^2)\Delta_{\substack{\text{ret}\\ \text{adv}}}(x'-x) = \delta^4(x'-x)
$$

Likewise

$$
S_{\substack{\text{ret}\\ \text{adv}}}(x'-x) = -(i\nabla\!\!\!\!/_{x'} + m)\Delta_{\substack{\text{ret}\\ \text{adv}}}(x'-x)
$$

and

$$
D_{\substack{\text{ret}\\ \text{adv}}}(x'-x) = +\lim_{m\to 0}\Delta_{\substack{\text{ret}\\ \text{adv}}}(x'-x)
$$

Finally, in terms of vacuum expectation values of a free hermitian spinless field

$$
\begin{aligned}
\langle 0|[\varphi(x'),\varphi(x)]|0\rangle &= i\Delta(x'-x) \\
\langle 0|\{\varphi(x'),\varphi(x)\}|0\rangle &= \Delta_1(x'-x) \\
\langle 0|\varphi(x')\varphi(x)|0\rangle &= \Delta_+(x'-x) \\
\langle 0|\varphi(x)\varphi(x')|0\rangle &= \Delta_-(x'-x) \\
\langle 0|T(\varphi(x')\varphi(x))|0\rangle &= i\Delta_F(x'-x) \\
\langle 0|\bar T(\varphi(x')\varphi(x))|0\rangle &= +i\bar\Delta_F(x'-x) \\
\langle 0|[\varphi(x'),\varphi(x)]|0\rangle\theta(t'-t) &= -i\Delta_{\text{ret}}(x'-x) \\
\langle 0|[\varphi(x'),\varphi(x)]|0\rangle\theta(t-t') &= +i\Delta_{\text{adv}}(x'-x)
\end{aligned}
$$

with similar results for Dirac and Maxwell fields.

Index